METHODS IN MOLECULAR BIOLOGY

Series Editor
John M. Walker
School of Life and Medical Sciences
University of Hertfordshire
Hatfield, Hertfordshire, UK

For further volumes:
http://www.springer.com/series/7651

For over 35 years, biological scientists have come to rely on the research protocols and methodologies in the critically acclaimed *Methods in Molecular Biology* series. The series was the first to introduce the step-by-step protocols approach that has become the standard in all biomedical protocol publishing. Each protocol is provided in readily-reproducible step-by-step fashion, opening with an introductory overview, a list of the materials and reagents needed to complete the experiment, and followed by a detailed procedure that is supported with a helpful notes section offering tips and tricks of the trade as well as troubleshooting advice. These hallmark features were introduced by series editor Dr. John Walker and constitute the key ingredient in each and every volume of the *Methods in Molecular Biology* series. Tested and trusted, comprehensive and reliable, all protocols from the series are indexed in PubMed.

Bacterial Vaccines

Methods and Protocols

Edited by

Fadil Bidmos, Janine Bossé, and Paul Langford

Department of Infectious Disease, Imperial College London, London, United Kingdom

Humana Press

Editors
Fadil Bidmos
Department of Infectious Disease
Imperial College London
London, United Kingdom

Janine Bossé
Department of Infectious Disease
Imperial College London
London, United Kingdom

Paul Langford
Department of Infectious Disease
Imperial College London
London, United Kingdom

ISSN 1064-3745 ISSN 1940-6029 (electronic)
Methods in Molecular Biology
ISBN 978-1-0716-1899-8 ISBN 978-1-0716-1900-1 (eBook)
https://doi.org/10.1007/978-1-0716-1900-1

This Humana imprint is published by the registered company Springer Science+Business Media, LLC, part of Springer Nature.
The registered company address is: 1 New York Plaza, New York, NY 10004, U.S.A.

Preface

Vaccination is considered the second most effective public health intervention available, only following the provision and access to clean water for disease prevention. Compared to the peak number of deaths prior to introduction of bacterial vaccines against diphtheria, tetanus, pertussis (whooping cough), and disease caused by *Haemophilus influenzae* type b, it is estimated that c. 1.3 million lives, predominantly of children, have been saved annually. The associated reduction in morbidity by prevention of disease through herd immunity, and reduction in secondary infections that complicate vaccine-preventable diseases, is estimated to save 386 million life years and 96 million disability-adjusted life years (DALYs) globally. In addition to reduction of mortality and morbidity, the use of vaccines has substantial social benefits. The latter include equity of healthcare, strengthening of health and social care infrastructure, increased social mobility, improved life expectancy, and empowerment of women. Economically, in lower- and middle-income countries, an investment of $34 billion in vaccines resulted in savings of $586 billion from the cost associated with direct illness. Cost-effectiveness or cost/benefit analyses do not typically consider prevention of long-term morbidity following acute infection, such as hearing loss or amputation of limbs as sequelae of meningococcal disease, and the calculated figures can be considered underestimates.

Given their benefits, it seems a paradox that historically pharmaceutical manufacturers have been wary of investing in vaccines because of concerns about legal liability and a comparative low return on investment compared to pharmaceuticals. That situation has recently started to change in the bacterial arena because of the threats posed by long-standing diseases where there is still a need for improved vaccines (e.g., for tuberculosis). In addition, with an aging worldwide population, vaccination is increasingly part of a life-course strategy to meet the needs of the elderly. However, the main driver is the increase in worldwide antimicrobial resistance (AMR). As Dame Sally Davies, the UK Chief Medical Officer, stated: *"The world is facing an antibiotic apocalypse. Unless action is taken to halt the practices that have allowed antimicrobial resistance to spread and ways are found to develop new types of antibiotics, we could return to the days when routine operations, simple wounds or straightforward infections could pose real threats to life."* In particular, there is currently much research focus on vaccine development for the so-called bacterial ESKAPE pathogens: *Enterococcus faecium*, *Staphylococcus aureus*, *Klebsiella pneumoniae*, *Acinetobacter baumannii*, *Pseudomonas aeruginosa*, and *Enterobacter* spp. Furthermore, new preventions and/or antibiotics are urgently needed for other World Health Organization-designated critical or high priority pathogens such as *Neisseria gonorrhoeae*, where strains have already been isolated that are resistant to current recommended treatment regimes. The solution to AMR will also involve improvement in and introduction of novel bacterial vaccines to reduce antimicrobial use in animals to prevent transfer of resistance genes through natural transformation, transconjugation, transduction, and vesiduction in zoonotic pathogens that cause disease in humans.

Thus, a book on bacterial vaccine methods is extremely timely. We have organized the chapters into three sections: (1) vaccine antigen discovery; (2) vaccine production and delivery; and (3) immunology of vaccine candidates. The first two chapters (Ong and He; Leow et al.) describe reverse vaccinology methods—the use of bioinformatic approaches to identify vaccine candidates, which have been successfully applied in development of the commercially available *Neisseria meningitidis* serogroup B Bexsero vaccine. With over 200,000 bacterial genomes publicly available, e.g., through NCBI and EMBL, the use of reverse vaccinology approaches can be a valuable starting point for immunogenic antigen and/or epitope discovery, and crucially is not hypothesis-driven. Four further chapters describe different proteomic approaches for vaccine antigen discovery. These include classic methods to identify potential immunogenic epitopes, such as immunoprecipitation (Reglinski), enzymatic surface shaving (Luu and Lan), two-dimensional electrophoresis combined with western blotting (Obradovic and Wilson), as well as pan-proteomic array technology (Campo and Oberai) which is also a bottom-up approach starting with genome data, like reverse vaccinology. Nine chapters are devoted to vaccine production and delivery systems, illustrating the range and ingenuity of platforms available. These include a relatively simple yet elegant bacterin inactivation mechanism involving low-energy electron irradiation (Fertey et al.), and bioengineering approaches such as those described for Gram-negative metal ion transporters (Chauduri et al.), Gram-positive extracellular membrane vesicles (Stentz et al.), and biological conjugation (Terra and Kay). Different methods for preparation of outer membrane vesicles/generalized modules for membrane antigens are presented, including those for secretory production of heterologous antigens (Kawamoto and Kurihara), for reduction of potential toxicity (Hirayama and Nakao), and for ensuring quality and stability (Micoli et al.). In addition, two chapters describe platforms that can be used to present peptide subunits (trimethyl chitosan-based polyelectrolyte complexes) (Zhao et al.) or multi-epitope fusion antigens (MEFA) (Li et al.), respectively. The final set of chapters provide methods for assessment of antigens at the discovery, delivery, and post-licensure phases, including monitoring of immune responses to inhaled antigens (Ashhurst et al.) and in infants following maternal immunization (Rice and Holder). The production of a universal human complement source (Alexander et al.) is described, which could be used to ensure reproducibility of the functional opsonophagocytic activity and serum bactericidal assays detailed by Semchenko et al. and Wagstaffe et al. Furthermore, informative infection models, including the ex vivo organ co-perfusion model (Hames et al.) and controlled human infection model (Dale et al.) described, as well as techniques such as multi-color flow cytometry and high-dimensional data analysis for analyses of vaccine responses (Cole et al.) will improve evaluation of prospective vaccines and enable prioritization of candidates for field testing, thus speeding up the time to market.

We hope that researchers will not only find specific chapters of interest but also take the time to review others detailed in the book, as overall they give a snapshot of the variety of methods available and the extremely high innovative nature of the bacterial vaccine field. We would like to express our thanks to all of the authors for their valuable contributions, meeting deadlines, and their understanding through the editing process, and to the series editor, John Walker, for his excellent guidance.

London, UK

Fadil Bidmos
Janine Bossé
Paul Langford

Contents

Contributors

ABU BAKAR ABDUL MAJEED • *Faculty of Pharmacy, Universiti Teknologi MARA, Kuala Selangor, Selangor, Malaysia*
FRANCES ALEXANDER • *Public Health England, Porton Down, UK*
RENZO ALFINI • *GSK Vaccines Institute for Global Health, Siena, Italy*
LAUREN ALLEN • *Public Health England, Porton Down, UK*
ANNELIESE S. ASHHURST • *School of Chemistry, The University of Sydney, Sydney, NSW, Australia; Tuberculosis Research Program Centenary Institute, The University of Sydney, Camperdown, NSW, Australia; School of Medical Sciences, Faculty of Medicine and Health, The University of Sydney, Camperdown, NSW, Australia*
SAHRA BASHIRI • *School of Chemistry & Molecular Biosciences, The University of Queensland, St Lucia, QLD, Australia*
JANA BECKMANN • *Fraunhofer Institute for Organic Electronics, Electron Beam and Plasma Technology FEP, Dresden, Germany*
WARWICK J. BRITTON • *Tuberculosis Research Program Centenary Institute, The University of Sydney, Camperdown, NSW, Australia; School of Medical Sciences, Faculty of Medicine and Health, The University of Sydney, Camperdown, NSW, Australia; Department of Clinical Immunology, Royal Prince Alfred Hospital, Camperdown, NSW, Australia*
EMILY BRUNT • *Public Health England, Porton Down, UK*
JOSEPH J. CAMPO • *Antigen Discovery, Inc., Irvine, CA, USA*
SIMON R. CARDING • *Quadram Institute, Norwich, UK*
DAVID CARRENO • *Department of Genetics and Genome Biology, University of Leicester, Leicester, UK*
BREEZE CAVELL • *Public Health England, Porton Down, UK*
SOMSHUKLA CHAUDHURI • *Department of Microbiology, Immunology, and Infectious Diseases, Cumming School of Medicine, University of Calgary, Calgary, AB, Canada*
HANNAH M. CHEESEMAN • *Department of Infectious Disease, Imperial College London, London, UK*
CANDY CHUAH • *Faculty of Medicine and Health Sciences, Universiti Tunku Abdul Rahman, Kajang, Selangor, Malaysia*
WEN Y. CHUNG • *Department of Hepatobiliary and Pancreatic Surgery, University Hospitals of Leicester, Leicester, UK*
MEGAN E. COLE • *The Pirbright Institute, Surrey, UK*
ADAM P. DALE • *Clinical and Experimental Sciences, University of Southampton, Southampton, UK*
ASHLEY R. DENNISON • *Department of Hepatobiliary and Pancreatic Surgery, University Hospitals of Leicester, Leicester, UK*
NIKOLAS F. EWASECHKO • *Department of Microbiology, Immunology, and Infectious Diseases, Cumming School of Medicine, University of Calgary, Calgary, AB, Canada*
JAMIE E. FEGAN • *Department of Molecular Genetics, University of Toronto, Toronto, ON, Canada*
JASMIN FERTEY • *Fraunhofer-Institute for Cell Therapy and Immunology IZI, Leipzig, Germany*

DIANE F. GBESEMETE • *Clinical and Experimental Sciences, University of Southampton, Southampton, UK; NIHR Clinical Research Facility, University Hospital Southampton NHS Foundation Trust, Southampton, UK*
CARLO GIANNELLI • *GSK Vaccines Institute for Global Health, Siena, Italy*
DAVID GOLDBLATT • *Great Ormond Street Institute of Child Health, University College London, London, UK*
ANDREW GORRINGE • *Public Health England, Porton Down, UK*
THOMAS GRUNWALD • *Fraunhofer-Institute for Cell Therapy and Immunology IZI, Leipzig, Germany*
YANPING GUO • *Cancer Research UK Flow Cytometry Translational Technology Platform, Cancer Institute, UCL, London, UK*
RACHEL HALKERSTON • *Public Health England, Porton Down, UK*
RYAN G. HAMES • *Department of Genetics and Genome Biology, University of Leicester, Leicester, UK*
CAMERON C. HANNA • *School of Chemistry, The University of Sydney, Sydney, NSW, Australia*
YONGQUN HE • *Center of Computational Medicine and Bioinformatics, Unit for Laboratory Animal Medicine, Department of Microbiology and Immunology, University of Michigan Medical School, Ann Arbor, MI, USA*
SATORU HIRAYAMA • *Department of Bacteriology I, National Institute of Infectious Diseases, Tokyo, Japan; Division of Microbiology and Infectious Diseases, Niigata University Graduate School of Medical and Dental Sciences, Niigata, Japan*
BETH HOLDER • *Institute of Reproductive and Developmental Biology (IRDB), Department of Metabolism, Development and Reproduction (MDR), Imperial College London, London, UK*
HOLLY HUMPHRIES • *Public Health England, Porton Down, UK*
ZYDRUNE JASIUNAITE • *Department of Genetics and Genome Biology, University of Leicester, Leicester, UK*
FREDA E.-C. JEN • *Institute for Glycomics, Griffith University, Gold Coast, QLD, Australia*
MICHAEL P. JENNINGS • *Institute for Glycomics, Griffith University, Gold Coast, QLD, Australia*
MARINA JOHNSON • *Great Ormond Street Institute of Child Health, University College London, London, UK*
SCOTT JONES • *Great Ormond Street Institute of Child Health, University College London, London, UK*
JUN KAWAMOTO • *Institute for Chemical Research, Kyoto University, Uji, Kyoto, Japan*
EMILY J. KAY • *Faculty of Infectious and Tropical Diseases, Department of Infection Biology, London School of Hygiene & Tropical Medicine, Keppel Street, London, UK*
TATSUO KURIHARA • *Institute for Chemical Research, Kyoto University, Uji, Kyoto, Japan*
RUITING LAN • *School of Biotechnology and Biomolecular Sciences, University of New South Wales, Sydney, NSW, Australia*
JAY R. LAVER • *Clinical and Experimental Sciences, University of Southampton, Southampton, UK*
KUO HAO LEE • *Computational Chemistry and Molecular Biophysics Unit, National Institute of Health/National Institute on Drug Abuse, Baltimore, MD, USA*
CHIUAN HERNG LEOW • *Institute for Research in Molecular Medicine, Universiti Sains Malaysia, Penang, Malaysia*
CHIUAN YEE LEOW • *School of Pharmaceutical Sciences, Universiti Sains Malaysia, Penang, Malaysia*

ELODIE LESNE • *Public Health England, Porton Down, UK*
STEPHANIE LEUNG • *Public Health England, Porton Down, UK*
SIQI LI • *Department of Pathobiology, University of Illinois at Urbana-Champaign, Urbana, IL, USA*
LAURENCE DON WAI LUU • *School of Biotechnology and Biomolecular Sciences, University of New South Wales, Sydney, NSW, Australia*
FRANCESCA MICOLI • *GSK Vaccines Institute for Global Health, Siena, Italy*
ARIADNA MIQUEL-CLOPÉS • *Quadram Institute, Norwich, UK*
NORAZMI MOHD NOR • *School of Health Sciences, Universiti Sains Malaysia, Kubang Kerian, Kelantan, Malaysia*
RYOMA NAKAO • *Department of Bacteriology I, National Institute of Infectious Diseases, Tokyo, Japan*
AMIT OBERAI • *Antigen Discovery, Inc., Irvine, CA, USA*
MILAN OBRADOVIC • *Faculté de médecine vétérinaire/Faculty of Veterinary Medicine, University of Montreal, St-Hyacinthe, QC, Canada*
MARCO R. OGGIONI • *Department of Genetics and Genome Biology, University of Leicester, Leicester, UK; Department of Pharmacy and Biotechnology, University of Bologna, Bologna, Italy*
EDISON ONG • *Department of Computational Medicine and Bioinformatics, University of Michigan Medical School, Ann Arbor, MI, USA; GlaxoSmithKline Vaccines, Rixensart, Belgium*
RICHARD J. PAYNE • *School of Chemistry, The University of Sydney, Sydney, NSW, Australia; Australian Research Council Centre of Excellence for Innovations in Peptide and Protein Science, The University of Sydney, Sydney, NSW, Australia*
ELIZABETH PENN • *Public Health England, Porton Down, UK*
KATRINA M. POLLOCK • *Department of Infectious Disease, Imperial College London, London, UK*
ROBERT C. READ • *Clinical and Experimental Sciences, University of Southampton, Southampton, UK; NIHR Clinical Research Facility, University Hospital Southampton NHS Foundation Trust, Southampton, UK; NIHR Southampton Biomedical Research Centre, Southampton, UK*
MARK REGLINSKI • *Division of Molecular Microbiology, School of Life Sciences, University of Dundee, Dundee, UK*
THOMAS RICE • *Centre for Endocrinology, William Harvey Research Institute, Barts and The London School of Medicine and Dentistry, Queen Mary University of London, London, UK*
LUISA SAMANIEGO-BARRON • *Department of Microbiology, Immunology, and Infectious Diseases, Cumming School of Medicine, University of Calgary, Calgary, AB, Canada*
ANTHONY B. SCHRYVERS • *Department of Microbiology, Immunology, and Infectious Diseases, Cumming School of Medicine, University of Calgary, Calgary, AB, Canada*
KATE L. SEIB • *Institute for Glycomics, Griffith University, Gold Coast, QLD, Australia*
EVGENY A. SEMCHENKO • *Institute for Glycomics, Griffith University, Gold Coast, QLD, Australia*
MARIUSZ SKWARCZYNSKI • *School of Chemistry & Molecular Biosciences, The University of Queensland, St Lucia, QLD, Australia*
BASTIAN STANDFEST • *Fraunhofer Institute for Manufacturing Engineering and Automation IPA, Stuttgart, Germany*
RÉGIS STENTZ • *Quadram Institute, Norwich, UK*
STEPHEN TAYLOR • *Public Health England, Porton Down, UK*

VANESSA S. TERRA • *Faculty of Infectious and Tropical Diseases, Department of Infection Biology, London School of Hygiene & Tropical Medicine, Keppel Street, London, UK*
MARTIN THOMA • *Fraunhofer Institute for Manufacturing Engineering and Automation IPA, Stuttgart, Germany*
STEPHEN THOMAS • *Public Health England, Porton Down, UK*
ISTVAN TOTH • *School of Chemistry & Molecular Biosciences, The University of Queensland, St Lucia, QLD, Australia; Institute for Molecular Bioscience, The University of Queensland, St. Lucia, QLD, Australia; School of Pharmacy, The University of Queensland, Woolloongabba, QLD, Australia*
SEBASTIAN ULBERT • *Fraunhofer-Institute for Cell Therapy and Immunology IZI, Leipzig, Germany*
HELEN WAGSTAFFE • *Great Ormond Street Institute of Child Health, University College London, London, UK; Department of Infectious Disease, Imperial College London, London, UK*
JOSEPH J. WANFORD • *Department of Genetics and Genome Biology, University of Leicester, Leicester, UK*
HEATHER L. WILSON • *Department of Veterinary Microbiology and the School of Public Health, Vaccine & Infectious Disease Organization (VIDO), University of Saskatchewan, Saskatoon, SK, Canada*
WEIPING ZHANG • *Department of Pathobiology, University of Illinois at Urbana-Champaign, Urbana, IL, USA*
LILI ZHAO • *School of Chemistry & Molecular Biosciences, The University of Queensland, St Lucia, QLD, Australia*

Chapter 1

Vaccine Design by Reverse Vaccinology and Machine Learning

Edison Ong and Yongqun He

Abstract

Reverse vaccinology (RV) is the state-of-the-art vaccine development strategy that starts with predicting vaccine antigens by bioinformatics analysis of the whole genome of a pathogen of interest. Vaxign is the first web-based RV vaccine prediction method based on calculating and filtering different criteria of proteins. Vaxign-ML is a new Vaxign machine learning (ML) method that predicts vaccine antigens based on extreme gradient boosting with the advance of new technologies and cumulation of protective antigen data. Using a benchmark dataset, Vaxign-ML showed superior performance in comparison to existing open-source RV tools. Vaxign-ML is also implemented within the web-based Vaxign platform to support easy and intuitive access. Vaxign-ML is also available as a command-based software package for more advanced and customizable vaccine antigen prediction. Both Vaxign and Vaxign-ML have been applied to predict SARS-CoV-2 (cause of COVID-19) and *Brucella* vaccine antigens to demonstrate the integrative approach to analyze and select vaccine candidates using the Vaxign platform.

Key words Vaccine, Antigen, Reverse vaccinology, Machine learning, Vaxign, Vaxign-ML, Vaxitop

1 Introduction

The conventional vaccine development process, which goes from one wet-lab research to another is time-consuming. The advances in high-throughput sequencing technologies have provided the foundation for Reverse Vaccinology (RV) to predict potential vaccine candidates from the whole pathogen genome using bioinformatics methods. The first RV study, pioneered by Rappuoli et al. successfully selected and verified 28 immunogenic proteins using bioinformatics analyses followed by experimental validation [1]. Eventually, 5 out of these 28 protein candidates were formulated in Bexsero®, licensed in the USA and Europe [2, 3].

In the first-generation RV technology development, a general workflow included the prediction of: (a) subcellular localization, (b) transmembrane helix, (c) adhesin probability, (d) signaling peptide, (e) protein function, (f) conserved domain, and (g) similarity

Fadil Bidmos et al. (eds.), *Bacterial Vaccines: Methods and Protocols*,
Methods in Molecular Biology, vol. 2414, https://doi.org/10.1007/978-1-0716-1900-1_1,

Table 1
Vaxign system components

Vaxign system component	Included analysis
Vaxign	Subcellular localization Transmembrane helix Adhesin probability Similarity to host (human, mouse, and pig)
Vaxign-ML	Extreme gradient boosting machine learning model to predict vaccine candidate
Vaxitop	Vaccine epitope prediction and analysis system based on the principle of reverse vaccinology
Postprediction analysis	IEDB epitope search IEDB population coverage EggNOG function EggNOG orthologs
Pangenome analysis	Orthologs' phylogeny Multiple sequence alignment

to host (Table 1). These properties were shown to correlate to vaccine protection significantly [4], and several programs (NERVE [5], Vaxign [6], and VacSol [7]) utilized these properties as the filtering criteria for vaccine candidate prediction. On the other hand, Doytchinova and Flower developed the first machine learning-based RV program, VaxiJen, which applied discriminant analysis by partial least square trained on the auto cross-covariance transformed physicochemical properties of the pathogen protein sequences [8]. By adopting and expanding the VaxiJen training samples, Bowman et al., and later on, Heinson et al. applied a support vector machine to predict vaccine candidates [9, 10]. Another major difference between the VaxiJen program and the Bowman-Heinson method was that the former used physicochemical features while the latter used biological parts derived from the pathogen protein sequences. A thorough benchmarking study was conducted by Dalsass et al. to systematically evaluate the RV prediction tools/methods [11], but the best performer's accuracy was still suboptimal and there is room for improvement.

Motivated by the advance of machine learning in other biomedical research fields, including vaccine candidate prediction, the Vaxign program was extended to include a machine learning component, Vaxign-ML, in 2020 [12]. Vaxign-ML is a supervised machine learning classification tool to predict protective antigens. To identify the best machine learning method with optimized conditions, five machine learning algorithms (logistic regression, support vector machine, k-nearest neighbors, random forest, and extreme gradient boosting) were tested with biological and physicochemical features extracted from the manually annotated Protegen protective vaccine antigen database [13]. Nested fivefold cross-validation and leave-one-pathogen-out validation were used to ensure unbiased performance assessment and the capability to predict vaccine candidates for a new emerging pathogen. The Vaxign-ML (extreme gradient boosting trained on all Protegen data) was the best performing model compared to three publicly available reverse vaccinology programs with a high-quality benchmark dataset (*see* **Note 1**) and showed superior performance in predicting protective antigens.

This article illustrates the procedures of Vaxign and Vaxign-ML implementation. Furthermore, we have applied Vaxign and Vaxign-ML to predict vaccine candidates for two pathogens, SARS-CoV-2 (cause of COVID-19) and *Brucella abortus* (cause of zoonotic brucellosis).

2 Vaxign-ML

The Vaxign-ML models (individual models for bacteria, virus, and parasite) were trained on the protective antigens available in Protegen [13]. All protective antigens stored in Protegen are manually collected and curated using the in-house semiautomatic annotation system [13]. For each specific protective antigen, the Protegen database contains the detailed reference citation information from PubMed and extracted general information of protective antigens from the NCBI databases. The current Protegen database contained 1432 protective antigens from 44 bacteria, 40 viruses, 19 parasites, and other noninfectious diseases (cancer and allergy) (Table 2), which is near double the number since the initial release in 2010. The Vaxign-ML tool was also applied to predict SARS-CoV-2 vaccine candidate [14].

2.1 Vaxign-ML Stand-Alone Web Tool

The most convenient and straightforward way to use Vaxign-ML for vaccine candidate prediction is to access the Vaxign-ML standalone web tool http://www.violinet.org/vaxign2/vaxign-ml. A typical Vaxign-ML web interface is provided in Fig. 1. The standard procedure for running Vaxign-ML is as follows:

1. Input the protein sequence(s). The following formats are supported: (a) FASTA format, (b) UniProtKB Protein ID, (c) NCBI Protein ID, (d) NCBI Protein RefSeq, (e) NCBI Gene ID, and (f) FASTA File Download Link. Users can also upload the protein sequence(s) in FASTA format instead of typing into the input text field.

Table 2
Protegen protective antigen statistics

Pathogen name	# Protective antigens
Gram-positive bacteria	195
Bacillus anthracis (14)	
Clostridium botulinum (10)	
Listeria monocytogenes (2)	
Mycobacterium tuberculosis (26)	
Staphylococcus aureus (29)	
Streptococcus agalactiae (22)	
Streptococcus equi (17)	
Streptococcus pneumoniae (26)	
Streptococcus pyogenes (44)	
Gram-negative bacteria	465
Actinobacillus pleuropneumoniae (20)	
Bordetella pertussis (11)	
Borrelia burgdorferi (11)	
Brucella spp. (26)	
Burkholderia pseudomallei (13)	
Campylobacter jejuni (19)	
Chlamydia muridarum (15)	
Chlamydia trachomatis (14)	
Chlamydophila abortus (12)	
Chlamydophila pneumoniae (14)	
Coxiella burnetii (12)	
Edwardsiella tarda (22)	
Escherichia coli (50)	
Francisella tularensis (12)	
Haemophilus influenzae (16)	
Haemophilus parasuis (28)	
Helicobacter pylori (23)	
Leptospira spp. (11)	
Neisseria meningitidis (20)	
Pseudomonas aeruginosa (15)	
Rickettsia spp (14)	
Shigella (11)	

(continued)

Table 2
(continued)

Pathogen name	# Protective antigens
Treponema pallidum (17)	
Yersinia pestis (26)	
Viruses	417
African Swine Fever Virus (7)	
Bovine Herpesvirus (7)	
Dengue Virus (11)	
Ebola Virus (19)	
Foot-and-Mouth Disease Virus (8)	
Hantavirus (10)	
Hepatitis B Virus (8)	
Hepatitis C Virus (3)	
Herpes Simplex Virus type 1 and 2 (12)	
Human Immunodeficiency Virus (41)	
Influenza Virus (50)	
Japanese Encephalitis Virus (10)	
Marburg Virus (9)	
Measles Virus (5)	
Mumps Virus (2)	
Porcine Respiratory and Reproductive Syndrome Virus (7)	
Pseudorabies Virus (8)	
Rotavirus (9)	
SARS-CoV (3)	
Vaccinia Virus (12)	
Western Equine Encephalomyelitis Virus (6)	
Yellow Fever Virus (5)	
Parasites	184
Leishmania donovani (15)	
Leishmania infantum (8)	
Leishmania major (13)	
Neospora caninum (10)	
Plasmodium spp. (33)	
Toxoplasma gondii (18)	
Trypanosoma cruzi (19)	

(continued)

Table 2
(continued)

Pathogen name	# Protective antigens
Fungi	9
Cancer	73
Other diseases (e.g., *allergy, arthritis, and diabetes)*	28
Total	1371

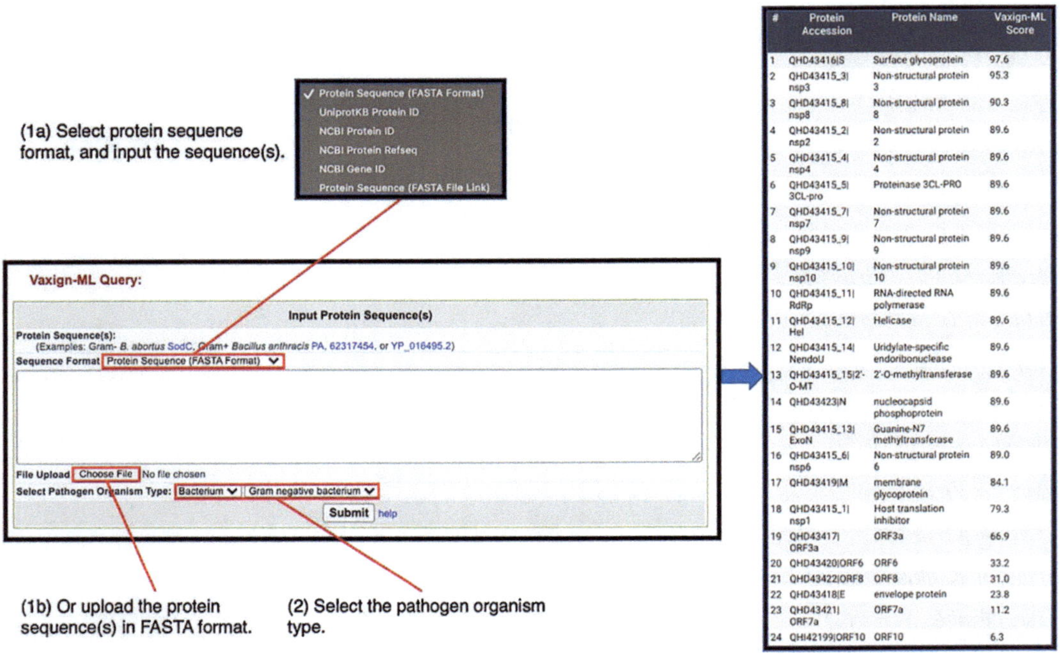

Fig. 1 Vaxign-ML web interface

2. Input pathogen organism type. The current Vaxign-ML supports "Gram positive bacterium," "Gram negative bacterium," and "Virus."
3. Click the submit button.
4. A Vaxign-ML prediction status page will be displayed, which will be automatically refreshed every 30 s.
5. Once the result is ready, the overall result summary page will appear, where the Vaxign-ML Protegenicity score(s) will be displayed, and the result is also available for download in Excel, CSV, and PDF formats. Users can perform additional analyses to aid the vaccine candidate selection (*see* Subheading 3.4). The Protegenicity score represents the level of protective antigenicity, that is, the extent of protective antigenicity a protein candidate can stimulate in vivo [12].

2.2 Vaxign-ML Stand-Alone Command-Line Tool

The stand-alone Vaxign-ML command-line tool is executable in Docker containers (tested on Docker version 1.13.1 and Docker API version 1.26). Docker can be downloaded from https://docs.docker.com/get-docker/, and installed with the appropriate platform. The source code of the command-line tool is also available in GitHub: https://github.com/VIOLINet/Vaxign-ML-docker. Here is how the stand-alone Vaxign-ML can be conducted.

1. Download the prediction script from https://raw.githubusercontent.com/VIOLINet/Vaxign-ML-docker/master/VaxignML.sh, and allow this script to be executed by the system. Then, execute the downloaded VaxignML.sh script with the following parameters: (a) input FASTA file, (b) output directory to store Vaxign-ML prediction result, and (c) pathogen organism type (currently support Gram-positive and -negative bacterium, and virus). There is also an optional fourth parameter to specify the custom Vaxign-ML model (see next step for the instruction to create custom Vaxign-ML model).
2. (Optional) Instead of using the pretrained Vaxign-ML model, users can also create a custom Vaxign-ML model using their own dataset. This customized model can be created by first downloading the Train.sh shell script from https://raw.githubusercontent.com/VIOLINet/Vaxign-ML-docker/master/Train.sh, and allow this script to be executed by the system. Second, prepare two protein sequences files in FASTA format for the positive (protective antigens) and negative (nonprotective antigens). Execute the downloaded Train.sh script with the following parameters: (a) positive data FASTA file; (b) negative data FASTA file; (c) output directory to store the custom Vaxign-ML model; and (d) pathogen organism type (currently support gram-positive and -negative bacterium, and virus).

2.3 SARS-CoV-2 Example

The Vaxign-ML was used to predict COVID-19 vaccine candidates from the SARS-CoV-2 proteome [14]. As expected, the top candidate indicated by Vaxign-ML was the SARS-CoV-2 spike protein (Table 3), which has been the primary target of many vaccines currently in clinical trials. However, Vaxign-ML also predicted several nonstructural proteins as a potential vaccine candidate. Among which, the nsp3 protein was extensively studied in the Vaxign-ML COVID-19 vaccine prediction study. A separate study reported that the PL-PRO, as part of the nsp3, was associated with innate immunity and virulence factor [15]. Overall, Vaxign and Vaxign-ML offer a valuable resource to select vaccine candidates and prioritize them for experimental verification.

Table 3
Predicted SARS-CoV-2 vaccine candidates using Vaxign-ML

SARS-CoV-2 Protein	Vaxign-ML Score
Surface glycoprotein	97.623*
Nonstructural protein 3 (including PL-pro domain)	95.283*
Nonstructural protein 8	90.349*
Nonstructural protein 2	89.647
Nonstructural protein 4	89.647
Proteinase 3CL-PRO	89.647
Nonstructural protein 7	89.647
Nonstructural protein 9	89.647
Nonstructural protein 10	89.647
RNA-directed RNA polymerase	89.647
Helicase	89.647
Uridylate-specific endoribonuclease	89.647
2′-O-methyltransferase	89.647
Nucleocapsid phosphoprotein	89.647
Guanine-N7 methyltransferase	89.629
Nonstructural protein 6	89.017
Membrane glycoprotein	84.102
Host translation inhibitor	79.312
ORF3a	66.925
ORF6	33.165
ORF8	31.023
Envelope protein	23.839
ORF7a	11.199
ORF10	6.266

*Vaxign-ML predicted vaccine antigen candidates

3 Vaxign with Vaxign-ML Integration

The Vaxign-ML has been fully integrated into the Vaxign system. The overall Vaxign system framework includes the primary Vaxign analyses, Vaxign-ML vaccine candidate prediction, Vaxitop epitope prediction, postprediction analysis, and pangenome analysis (Fig. 2).

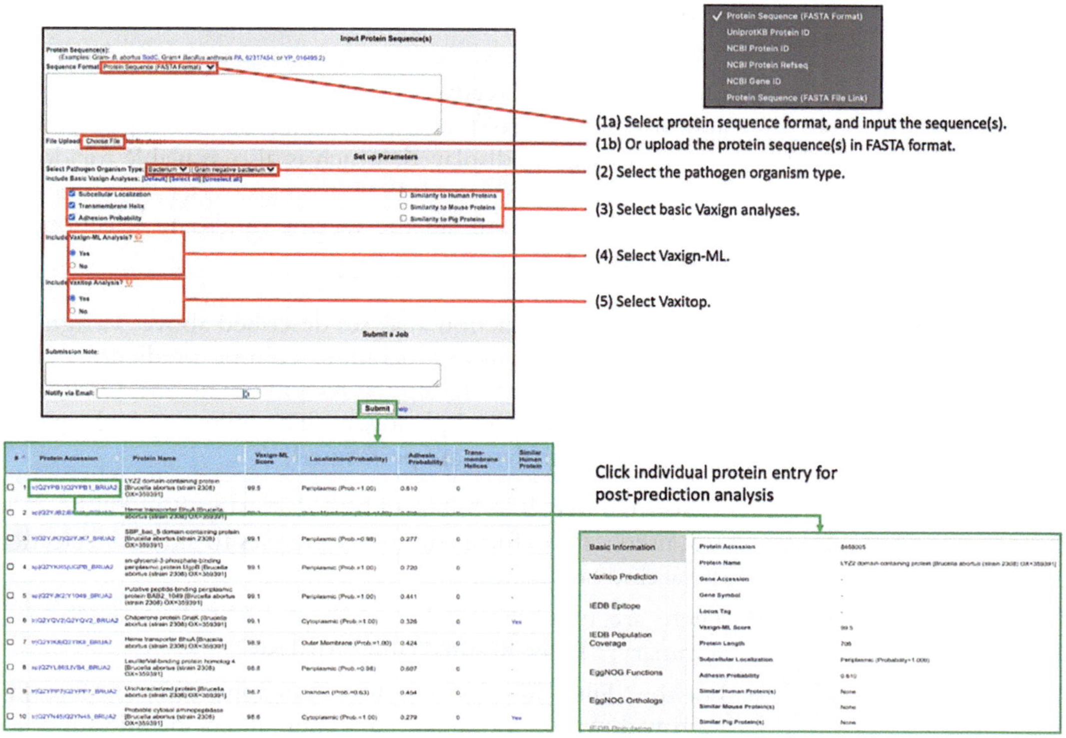

Fig. 2 Vaxign dynamic analysis web interface

3.1 Vaxign Main Interface

1. From the Vaxign main page (http://www.violinet.org/vaxign2), users can submit a protein sequence(s) for prediction. The following formats are supported: (a) FASTA format, (b) UniProtKB Protein ID, (c) NCBI Protein ID, (d) NCBI Protein RefSeq, (e) NCBI Gene ID, and (f) FASTA File Download Link. Users can also upload the protein sequence(s) in FASTA format instead of typing into the input text field.
2. Select the pathogen organism type. The current Vaxign-ML supports "Gram positive bacterium", "Gram negative bacterium", "Virus", and "Parasite".
3. Select whether the primary Vaxign analyses include (a) subcellular localization, (b) transmembrane helix, (c) adhesion probability, and (d) similar to host (human, mouse, and pig) proteins.
4. Users can choose whether the Vaxign-ML is to be included in the pipeline.
5. Besides the primary Vaxign and Vaxign-ML analyses, the system also supports epitope prediction via the in-house Vaxitop program.
6. Click the "submit" button. Users can optionally provide their email address and be notified to the email address once the result is ready.

7. A status page will be displayed, which will be automatically refreshed every 30 s.
8. Once the result is ready, the overall result summary page will appear. The selected analysis results of the input protein sequences will be displayed, which is also available for download in Excel, CSV, and PDF formats. Users can perform additional analyses to aid the vaccine candidate selection (*see* Subheading 3.4).

3.2 Vaxign Analyses After Vaccine Candidate Prediction

In addition to the prediction analyses described above, Vaxign also includes additional analyses, including epitope predictions using Vaxitop epitope prediction, EggNOG Functions and Orthologs, and pangenome orthologs' phylogeny. The following is the procedure of how such analyses are performed.

1. Following **step 5** in Subheading 3.1 or **step 8** in Subheading 3.3, users can perform additional analyses to aid their selection of vaccine candidates.
2. There are two tabs in the upper panel of the overall result summary: "Filter Results" and "Analysis."
3. For the "Filter Results" tab, users can apply different filters to narrow down the candidate list. These filters include subcellular localization, number of transmembrane helices, adhesin probability, and similarity to human/mouse/pig proteins. Besides, users can also include or exclude candidates with orthologs in specific pathogen strain(s). (For details of obtaining the ortholog analysis, *see* Subheading 3.3)
4. For the "Analysis" tab, users can browse the predicted epitopes and the corresponding population coverage of the restricted epitopes if the Vaxitop option is selected in Subheading 3.3 **step 5** (*see* **Note 2**). Users can also view the complete orthologs table of the pangenome analysis for specific pathogen strain(s). (For details of obtaining the pangenomic analysis, *see* Subheading 3.3)
5. Users can also click the individual input protein, and a web page will be opened with detailed information of the protein, including basic information, Vaxitop prediction (if not selected, a new Vaxitop analysis will be performed automatically), IEDB epitope (*see* **Note 3**), IEDB population coverage (*see* **Note 2**), EggNOG Functions and Orthologs (*see* **Note 4**), and pangenome orthologs' phylogeny and multiple sequence alignment. (For details of obtaining the pangenomic analysis, *see* Subheading 3.3)

3.3 Vaxign Pangenome Analysis

1. The Vaxign pangenome analysis can be accessed http://www.violinet.org/vaxign2/project.
2. For each pangenome analysis, create a new project.

3. Enter into the project summary page, and click "Start Vaxign Dynamic Analysis" button.
4. Users can submit protein sequence(s) for prediction, and the following formats are supported: (a) FASTA format, (b) UniProtKB Protein ID, (c) UniProt Proteome ID, (d) NCBI Protein ID, (e) NCBI Protein RefSeq, (f) NCBI Gene ID, (g) NCBI BioProject ID, (h) NCBI Nucleotide ID, and (i) FASTA File Download Link. In addition, users can also upload the protein sequence(s) in FASTA format instead of typing into the input text field.
5. Enter the genome group or pathogen name (e.g., *Mycobacterium tuberculosis*), and genome or strain name (e.g., *Mycobacterium tuberculosis* H37Rv).
6. Select the pathogen organism type, the current Vaxign-ML supports "Gram positive bacterium", "Gram negative bacterium", "Virus", and "Parasite".
7. Select whether the basic Vaxign analyses includes (a) subcellular localization, (b) transmembrane helix, (c) adhesin probability, and (d) similarity to host (human, mouse, and pig) proteins.
8. Users can choose whether Vaxign-ML is to be included in the pipeline.
9. Besides the basic Vaxign and Vaxign-ML analyses, the system also supports epitope prediction via the in-house Vaxitop program.
10. Click the submit button. Users will be directed to the project summary page with the Vaxign analysis status displayed.
11. Repeat **steps 3–10** for the pathogen strains to be included in the pangenome analysis.
12. Once the protein sequences of all the pathogen strains are submitted and the corresponding Vaxign analyses are finished, users can click the "Run Vaxign Ortholog Analysis" for the pangenome analysis.
13. Once the analysis is completed, users can click on the "result" button for each submitted query. Then follow the same procedures (*see* Subheading 3.4) to include or exclude candidates with orthologs in specific pathogen strain(s).
14. To facilitate easier Vaxign usage, a set of precomputed queries were created, which can be accessed from the home page (or http://www.violinet.org/vaxign2/precompute). The precomputed Vaxign results from either the protein level or genome level. Users are prompted to set up preferred query criteria; the output data are then provided. The query of precomputed Vaxign results is fast.

3.4 Prediction of Brucella Vaccine Antigens Using Vaxign and Vaxign-ML

Brucella is a facultative intracellular bacterium that causes brucellosis, one of the most common zoonotic diseases in humans and various domestic and wildlife animals [16]. *B. abortus*, *B. melitensis*, and *B. suis* can cause brucellosis in cattle, goat, and pigs, respectively. These three *Brucella* strains can also cause brucellosis in humans. Although several live attenuated *B. abortus* vaccines such as RB51 and S19 are available to protect cattle against brucellosis, there is still no safe and effective vaccine against human brucellosis. A major reason is related to the safety of live attenuated whole organism vaccines. These vaccines may still cause human brucellosis. We have previously applied Vaxign to predict *Brucella* vaccine targets (Ref. PMC1539029). This study used an updated Vaxign tool and the new Vaxign-ML tool to support *Brucella* vaccine antigen prediction.

Figure 3 illustrates the results of the *Brucella* vaccine prediction using Vaxign and Vaxign-ML. In this case, virulent *Brucella abortus* strain 2308 has 3023 proteins. From this proteome, Vaxign-ML predicted 482 proteins being protective vaccine antigens. Vaxign identified 44 out of the 3033 proteins as being outer membrane (OM) or extracellular proteins; similarly, 29 out of the 482 Vaxign-ML predicted vaccine antigens were also identified as OM or extracellular proteins. Since proteins with more than one transmembrane helices are often difficult to express recombinantly [6], we further selected 43 out of the 44 OM proteins from the Vaxign prediction with less than or equal to one transmembrane helix. None of the 29 Vaxign-ML OM vaccine antigens predicted as vaccine antigens had more than one transmembrane protein. Given the adhesin criterion, 31 out of the 43 Vaxign OM proteins,

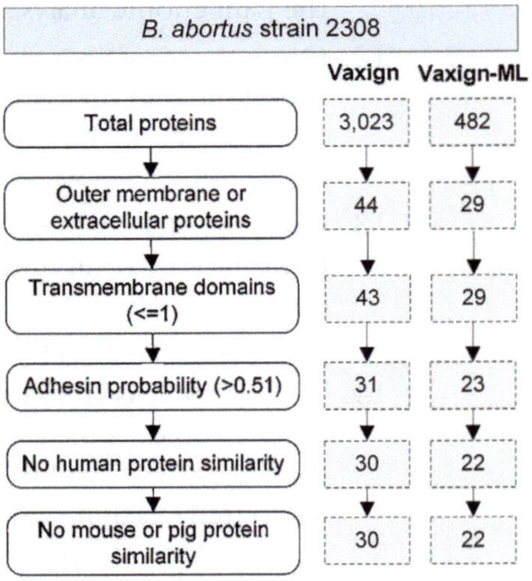

Fig. 3 *B. abortus* 2308 Vaxign and Vaxign-ML analyses. Both Vaxign and Vaxign-ML were applied to analyze the genome of *B. abortus* strain 2308. Five different filtering criteria were used for the filtering analysis

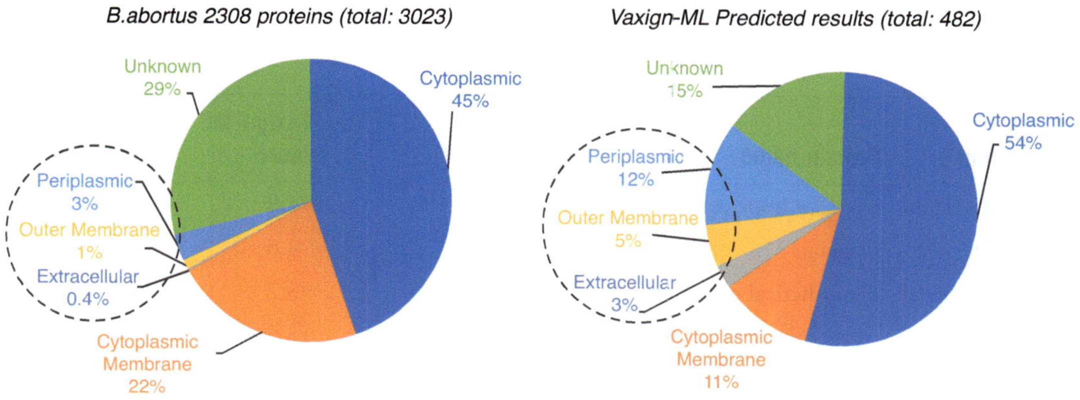

Fig. 4 Comparison of subcellular locations of all proteins of *B. abortus* strain 2308 and the 482 vaccine candidates predicted by Vaxign-ML

and 23 out of the 29 Vaxign-ML OM vaccine antigens were predicted to be adhesins. Eventually, after filtering with no host protein similarity, we found 22 proteins that meet all filtering criteria and ML cutoff (Fig. 3).

Figure 4 illustrates a comparison of subcellular locations of all the 3023 proteins in the proteome and the 482 vaccine candidates predicted by Vaxign-ML. Out of the 3023 proteins, only 0.4% are extracellular proteins, 1% are OM proteins, and 3% are periplasmic proteins. In comparison, out of the 482 Vaxign-ML vaccine antigens, 3% are extracellular proteins, 5% are OM proteins, and 12% are periplasmic proteins. It is clear that the OM proteins, periplasmic proteins, and extracellular proteins are significantly enriched in the Vaxign-ML predicted set compared to the whole proteome. It is also noted that there are considerably more percentages of cytoplasmic membrane proteins and proteins with unknown functions.

Table 4 lists all the 22 proteins predicted by Vaxign and Vaxign-ML. This list includes three flagella related proteins (FlgE, flagellin, and FlgK), two TonB-dependent receptors (BAB_RS22470 and BAB_RS31825), and eight porin family proteins. The porin family proteins form the primary group of proteins being protective antigens for *Brucella abortus* vaccine development. The flagella-related and TonB-dependent proteins are also favorable candidates based on our Vaxign and Vaxign-ML analyses.

4 Notes

1. The high-quality benchmark dataset was collected and curated [8–11]. Any duplicated data presented in the Protegen database (which is used as the training data of Vaxign-ML) was removed from this dataset to ensure a nonbiased performance evaluation of the Vaxign-ML. As a quality check of the negative

Table 4
Predicted *B. abortus* vaccine candidates using Vaxign and Vaxign-ML

Gene name	Protein name	Vaxign-ML score	Localization
BAB_RS31580	Flagellar hook protein FlgE	94.8	EC
BR_RS15470	Flagellin	98.6	EC
BAB_RS25510	Hypothetical protein	95.7	EC
BR_RS08885	AprI/Inh family metalloprotease inhibitor, omp19	93.6	OM
flgK	Flagellar hook-associated protein FlgK	91.7	OM
BAB_RS19300	LPS-assembly protein LptD	97.5	OM
BOV_RS07445	OmpW family protein	94.2	OM
BAB_RS26700	Outer membrane beta-barrel protein	91.9	OM
BR_RS06570	Outer membrane protein assembly factor BamD	92.5	OM
BAB1_0907	Peptidoglycan-binding LysM:Peptidase M23/M37	96.8	OM
BAB_RS22155	Porin	93.9	OM
BOV_RS00570	Porin family protein	91.1	OM
BR_RS00555	Porin family protein	92.3	OM
BAB_RS27875	Porin family protein	95.2	OM
BR_RS03235	Porin family protein, OMP_b-brl domain-containing protein	96.4	OM
BAB_RS23735	Porin family protein, OMP_b-brl domain-containing protein	94	OM
BAB_RS19090	Porin Omp2a	94.9	OM
BAB_RS19100	Porin Omp2b	96.9	OM
ybgF	Tol-pal system protein YbgF	96.8	OM
BAB_RS22470	TonB-dependent receptor, Heme transporter BhuA	97.9	OM
BAB_RS31825	TonB-dependent receptor, Heme transporter BhuA	99.3	OM
BAB_RS31975	YadA-like family protein	98.1	OM

Note: The genome of *B. abortus* strain 2308 was used for this analysis

samples, a manual search of all negative samples in the literature found no reported experimental evidence for the negative samples to induce immune responses.

2. The population coverage is calculated using the IEDB population coverage too l [17].
3. The full-length input protein will be searched against the IEDB epitope database for reported T cell and B cell epitopes.
4. The EggNOG functions and orthologs are predicted using EggNog mapper [18, 19].

Acknowledgments

This work was supported by NIH-NIAID grants 1R01AI081062 and 1UH2AI13293.

References

1. Pizza M, Scarlato V, Masignani V, Giuliani MM, Aricò B, Comanducci M, Jennings GT, Baldi L, Bartolini E, Capecchi B, Galeotti CL, Luzzi E, Manetti R, Marchetti E, Mora M, Nuti S, Ratti G, Santini L, Savino S, Scarselli M, Storni E, Zuo P, Broeker M, Hundt E, Knapp B, Blair E, Mason T, Tettelin H, Hood DW, Jeffries AC, Saunders NJ, Granoff DM, Venter JC, Moxon ER, Grandi G, Rappuoli R (2000) Identification of vaccine candidates against serogroup B meningococcus by whole-genome sequencing. Science 287:1816–1820. https://doi.org/10.1126/science.287.5459.1816
2. Vernikos G, Medini D (2014) Bexsero H chronicle. Pathog Glob Health 108:305–311. https://doi.org/10.1179/2047773214Y.0000000162
3. Folaranmi T, Rubin L, Martin SW, Patel M, MacNeil JR (2015) Use of serogroup B meningococcal vaccines in persons aged≥ 10 years at increased risk for serogroup B meningococcal disease: recommendations of the advisory committee on immunization practices, 2015. MMWR Morb Mortal Wkly Rep 64:608–612
4. Ong E, Wong MU, He Y (2017) Identification of new features from known bacterial protective vaccine antigens enhances rational vaccine design. Front Immunol 8:1–11. https://doi.org/10.3389/fimmu.2017.01382
5. Vivona S, Bernante F, Filippini F (2006) NERVE: new enhanced reverse vaccinology environment. BMC Biotechnol 6:35. https://doi.org/10.1186/1472-6750-6-35
6. He Y, Xiang Z, Mobley HLT (2010) Vaxign: the first web-based vaccine design program for reverse vaccinology and applications for vaccine development. J Biomed Biotechnol 2010:1–15. https://doi.org/10.1155/2010/297505
7. Rizwan M, Naz A, Ahmad J, Naz K, Obaid A, Parveen T, Ahsan M, Ali A (2017) VacSol: a high throughput in silico pipeline to predict potential therapeutic targets in prokaryotic pathogens using subtractive reverse vaccinology. BMC Bioinformatics 18:1–7. https://doi.org/10.1186/s12859-017-1540-0
8. Doytchinova IA, Flower DR (2007) VaxiJen: a server for prediction of protective antigens, tumour antigens and subunit vaccines. BMC Bioinformatics 8:4. https://doi.org/10.1186/1471-2105-8-4
9. Bowman BN, McAdam PR, Vivona S, Zhang JX, Luong T, Belew RK, Sahota H, Guiney D, Valafar F, Fierer J, Woelk CH (2011) Improving reverse vaccinology with a machine learning approach. Vaccine 29:8156–8164. https://doi.org/10.1016/j.vaccine.2011.07.142
10. Heinson AI, Gunawardana Y, Moesker B, Denman Hume CC, Vataga E, Hall Y, Stylianou E, McShane H, Williams A, Niranjan M, Woelk CH (2017) Enhancing the biological relevance of machine learning classifiers for reverse vaccinology. Int J Mol Sci 18:312. https://doi.org/10.3390/ijms18020312
11. Dalsass M, Brozzi A, Medini D, Rappuoli R (2019) Comparison of open-source reverse vaccinology programs for bacterial vaccine antigen discovery. Front Immunol 10:1–12. https://doi.org/10.3389/fimmu.2019.00113
12. Ong E, Wang H, Wong MU, Seetharaman M, Valdez N, He Y (2020) Vaxign-ML: supervised machine learning reverse vaccinology model for improved prediction of bacterial protective antigens. Bioinformatics 36:3185–3191. https://doi.org/10.1093/bioinformatics/btaa119
13. Yang B, Sayers S, Xiang Z, He Y (2011) Protegen: a web-based protective antigen database and analysis system. Nucleic Acids Res 39:1073–1078. https://doi.org/10.1093/nar/gkq944
14. Ong E, Wong MU, Huffman A, He Y (2020) COVID-19 coronavirus vaccine design using reverse vaccinology and machine learning. Front Immunol 11:1581
15. Shin D, Mukherjee R, Grewe D, Bojkova D, Baek K, Bhattacharya A, Schulz L, Widera M, Mehdipour AR, Tascher G, Geurink PP, Wilhelm A, van der Heden van Noort GJ, Ovaa H, Müller S, Knobeloch K-P, Rajalingam K, Schulman BA, Cinatl J, Hummer G, Ciesek S, Dikic I (2020) Papain-like protease regulates SARS-CoV-2 viral spread and innate immunity. Nature 587

(7835):657–662. https://doi.org/10.1038/s41586-020-2601-5

16. Xiang Z, Zheng W, He Y (2006) BBP: Brucella genome annotation with literature mining and curation. BMC Bioinformatics 7:347. https://doi.org/10.1186/1471-2105-7-347
17. Bui HH, Sidney J, Dinh K, Southwood S, Newman MJ, Sette A (2006) Predicting population coverage of T-cell epitope-based diagnostics and vaccines. BMC Bioinformatics 7:1–5. https://doi.org/10.1186/1471-2105-7-153
18. Huerta-Cepas J, Szklarczyk D, Heller D, Hernández-Plaza A, Forslund SK, Cook H, Mende DR, Letunic I, Rattei T, Jensen LJ, Von Mering C, Bork P (2019) EggNOG 5.0: a hierarchical, functionally and phylogenetically annotated orthology resource based on 5090 organisms and 2502 viruses. Nucleic Acids Res 47:D309–D314. https://doi.org/10.1093/nar/gky1085
19. Huerta-Cepas J, Forslund K, Coelho LP, Szklarczyk D, Jensen LJ, Von Mering C, Bork P (2017) Fast genome-wide functional annotation through orthology assignment by eggNOG-mapper. Mol Biol Evol 34:2115–2122. https://doi.org/10.1093/molbev/msx148

Chapter 2

Application of Reverse Vaccinology and Immunoinformatic Strategies for the Identification of Vaccine Candidates Against *Shigella flexneri*

Chiuan Yee Leow, Candy Chuah, Abu Bakar Abdul Majeed, Norazmi Mohd Nor, and Chiuan Herng Leow

Abstract

Reverse vaccinology (RV) was first introduced by Rappuoli for the development of an effective vaccine against serogroup B *Neisseria meningitidis* (MenB). With the advances in next generation sequencing technologies, the amount of genomic data has risen exponentially. Since then, the RV approach has widely been used to discover potential vaccine protein targets by screening whole genome sequences of pathogens using a combination of sophisticated computational algorithms and bioinformatic tools. In contrast to conventional vaccine development strategies, RV offers a novel method to facilitate rapid vaccine design and reduces reliance on the traditional, relatively tedious, and labor-intensive approach based on Pasteur"s principles of isolating, inactivating, and injecting the causative agent of an infectious disease. Advances in biocomputational techniques have remarkably increased the significance for the rapid identification of the proteins that are secreted or expressed on the surface of pathogens. Immunogenic proteins which are able to induce the immune response in the hosts can be predicted based on the immune epitopes present within the protein sequence. To date, RV has successfully been applied to develop vaccines against a variety of infectious pathogens. In this chapter, we apply a pipeline of bioinformatic programs for identification of *Shigella flexneri* potential vaccine candidates as an illustration immunoinformatic tools available for RV.

Key words Antigens, Bacteria, Epitope, Immunoinformatics, Outer membrane protein (OMP), *Shigella flexneri*, Reverse vaccinology (RV), Vaccine

1 Introduction

Shigellosis is an acute inflammatory bowel disease caused by the gram-negative intracellular enterobacterial genus, *Shigella*. Food and water sources that are contaminated with human or animal waste contribute primarily toward the transmission of *Shigella* [1]. *Shigella* spp. are increasingly exhibiting resistance against currently available antibiotics, and it was recently reported that tetracycline and trimethoprim–sulfamethoxazole are no longer used for

Fadil Bidmos et al. (eds.), *Bacterial Vaccines: Methods and Protocols*,
Methods in Molecular Biology, vol. 2414, https://doi.org/10.1007/978-1-0716-1900-1_2,

treatment of severe diarrhea and dysentery in some areas [2]. Thus, development of a protective vaccine for the prevention of Shigella infection, especially in epidemic areas, is a high priority. As yet, no practical vaccine for shigellosis has been licensed. Vaccine candidates currently in clinical trial are either not adequately attenuated or have proven to be less immunogenic in the host [3]. In order to develop an effective vaccine for long term protection, it is essential to accurately identify the immunogenic protective antigens. Development of vaccines through conventional techniques can take decades. With the increasing availability of whole genome sequences, in silico analysis of genomic data can be used to identify potential vaccine targets in a process termed "reverse vaccinology" (RV) [4].

In the RV strategy, candidate antigens likely to elicit protective immune responses are identified at the gene level. Following genome-wide analysis, a high-throughput cloning and expression platform is used to generate and purify recombinant proteins predicted as outer membrane, invasion, and other virulence-related antigens. The purified recombinant antigens are then evaluated in vivo and in vitro to shortlist protective antigens for vaccine research. The RV approach was firstly introduced by Rappuoli and his colleagues to develop a safe and protective vaccine against serogroup B *Neisseria meningitidis* (MenB) [5]. With an expanding wealth of genomic data, the RV approach has been used for the discovery of potential vaccine candidates in *Streptococcus pneumoniae* [6], *Campylobacter jejuni* [7], *Shigella flexneri* [8], and COVID-19 [9], among others.

In 2008, a genome-based computational vaccine discovery tool known as Vaxign was introduced by He and colleagues [10]. Vaxign is a vaccine target prediction and analysis system developed based on the principles of RV and is composed of a sophisticated computational pipeline that utilizes bioinformatic technology to find potential antigenic proteins from genome data for vaccine development [10, 11]. The major predicted features include identification of subcellular location of proteins, transmembrane domains, adhesion probability, sequence similarity to host proteome, and MHC class I and II epitope binding. Among these features, subcellular localization is considered as one of the main criteria for target prediction [10, 11]. Vaxign is a part of web-based system called Vaccine Investigation and Online Information Network (VIOLIN, http://www.violinet.org) and has been widely used for the identification of vaccine targets against various pathogens including *Brucella* spp. [12, 13], *Acinetobacter baumannii* [14], and *Mycobacterium tuberculosis* [15].

The immunogenicity of an antigen is associated with its ability to interface with the humoral (B-cell) and cellular (T-cell) immune systems. A vaccine consisting of both B- and T-cell epitopes is of importance to *effectively elicit strong immune responses for long term protection* [16, 17]. The integration of genomic, proteomic, and bioinformatic strategies has paved way for the development of immunoinformatics. To date, a number of B- and T-cell epitope prediction tools have been developed using various computational algorithms. This has led us toward the evolution of designing modern vaccines based on epitope-focused recombinant antigens [18, 19].

In this chapter, we describe our pipeline for identification of the most conserved and immunogenic outer membrane proteins (OMPs) from the *S. flexneri* genome, which involves identification of vaccine targets using Vaxign (*see* Subheading 3.1), analysis of conserved identity with other strains of the same species (*see* Subheading 3.2) and human homologs of the predicted proteins (*see* Subheading 3.3), analysis of antigenicity using VaxiJen v2.0 (*see* Subheading 3.4), prediction of linear B-cell epitopes using BCPREDS (*see* Subheading 3.5), and prediction of HLA Class I and Class II T-cell epitopes using HLApred (*see* Subheading 3.6). An overview of the reverse vaccinology steps adopted in this chapter is summarized in Fig. 1.

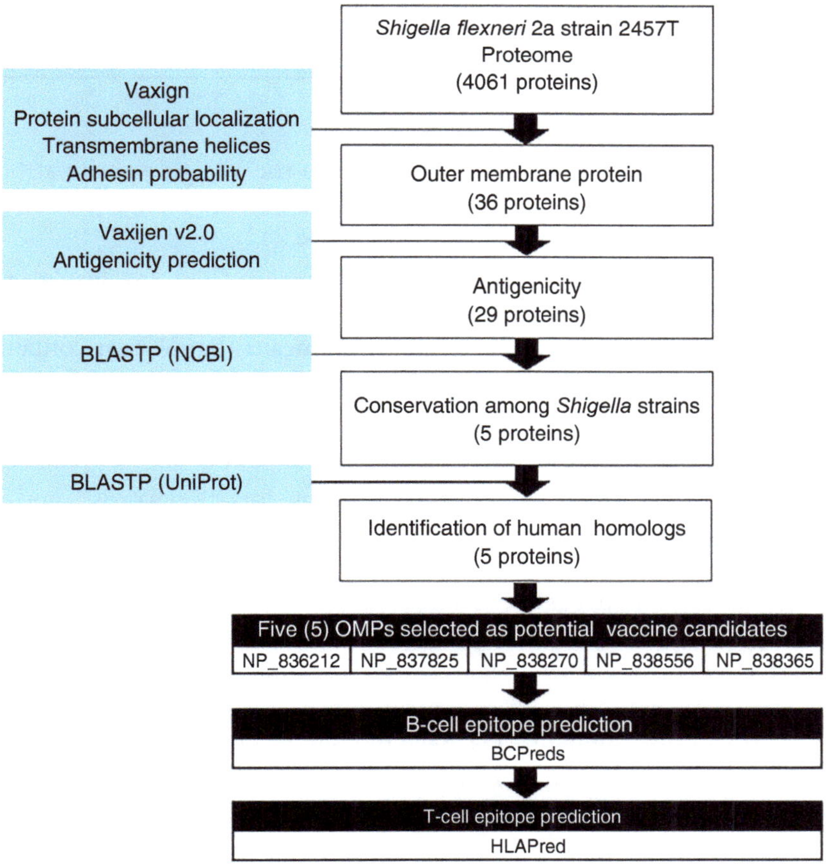

Fig. 1 Flowchart of the use of reverse vaccinology approach for the prediction of vaccine candidates against *Shigella flexneri*

2 Materials

1. Windows or Mac computer.
2. Google Chrome or other web browser (*see* **Note 1**).
3. Vaxign: Vaccine Design Pipeline [10, 11].http://www.violinet.org/vaxign/
4. BLASTP: Basic Local Alignment Search Tool for Protein Sequences (NCBI) [20, 21].https://blast.ncbi.nlm.nih.gov/Blast.cgi?PAGE=Proteins
   https://www.uniprot.org/blast/
5. VaxiJen v2.0: Prediction of Protective Antigens and Subunit Vaccines [22].http://www.ddg-pharmfac.net/vaxijen/VaxiJen/VaxiJen.html
6. BCPREDS: B-cell Epitope Prediction Server [23, 24].http://ailab-projects1.ist.psu.edu:8080/bcpred/
7. HLAPred: Identification and Prediction of HLA Class I and Class II T-cell Epitope [25, 26].http://crdd.osdd.net/raghava/hlapred/

3 Methods

3.1 Identification of Vaccine Targets Using Vaxign

3.1.1 Select a Genome

1. Point the web browser to the Vaxign website at http://www.violinet.org/vaxign/. The Vaccine Design section of the Vaxign page appears (*see* **Note 2**).
2. In the "Vaxign Query" section (Fig. 2a), first go to "Select a Genome Group." Select <Shigella (5) > in the drop-down menu. The number (5) means that Vaxign contains five genomes corresponding to *Shigella* species (*see* **Note 3**).
3. Next, go to "Select a Genome." Select <*Shigella flexneri* 2a strain 2457T genome> in the drop-down menu (*see* **Note 4**).
4. Keep the default settings for "Keywords" and "Sort by" options. Proceed to "Filter Options."

3.1.2 Select Filter Options

1. Set subcellular localization by selecting <Outer Membrane> in the scrollable selection menu (*see* **Notes 5** and **6**).
2. Restrict the number of transmembrane helices. The default setting for protein transmembrane number is less than or equal to 1. Check the box to include proteins having less than one transmembrane helix (*see* **Note 7**).
3. For adhesin probability analysis, the default cutoff is 0.51. Check the box to include proteins having an adhesin probability score equal or more than 0.51 (*see* **Note 8**).

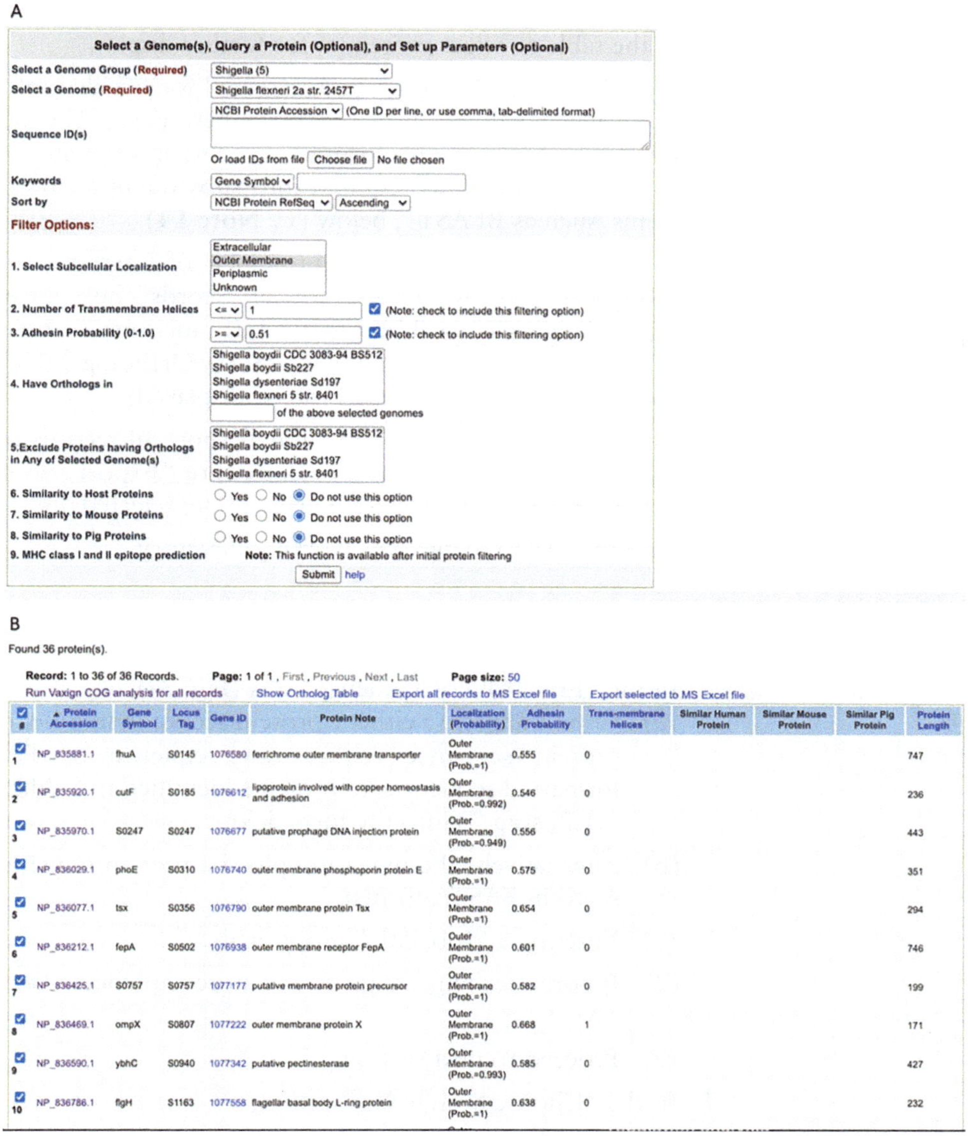

Fig. 2 Vaxign query submission for genome-based vaccine candidate prediction. (**a**) Vaxign query web interface. (**b**) An example of potential vaccine candidates predicted using Vaxign pipeline

4. The remaining Filter Options (i.e., options 4–9 shown in Fig. 2a) for analysis of protein orthologue comparisons, sequence similarity to host proteome, and epitope prediction were not included in this case and they were substituted using other online software described in the following section (*see* **Notes 9–13**).
5. Click the "Submit" button and wait. The result of analysis is shown in Fig. 2b.

6. In the result page, check one or more box at the left-most side of the table to select protein(s) for further display.
7. Click link(s) in the "Protein Accession" for detailed information, which will be displayed on a new page. The desired FASTA protein sequences can be saved to appropriate files, or copied and pasted, as required into subsequent analysis platforms, such as BLASTP, below (*see* **Note 14**).
8. If biological function of the protein is of interest, click the "Run Vaxign COG analysis for all records" link above the Table. A new page will be displayed. If functional orthologues of the protein are of interest, click "Show Ortholog Table" link above the Table. A new page will be displayed.
9. Click the link above the table for "Export all records to MS Excel file" (for all proteins are selected) or "Export selected to MS Excel file" (if only certain proteins are selected).
10. Save the MS Excel file on a working directory.

3.2 Identification of Conserved Identity with Other Shigella Strains

1. Point the internet browser to the NCBI BLASTP website at https://blast.ncbi.nlm.nih.gov/Blast.cgi?PAGE=Proteins
2. In the "Enter Query Sequence" box (Fig. 3a),
   (a) Copy and paste a either a protein accession number or the protein sequence (either raw sequence or FASTA-formatted sequences of proteins identified in Subheading 3.1.2, **step** 7) into the form field.
   (b) Alternatively, the user can upload a protein sequence file saved in FASTA-format.
   (c) Entering a "Job Title" is optional.
   (d) Ignore checking "Align two or more sequences" (*see* **Note 15**).
   (e) Proceed to next section.
3. In the "Choose Search Set" section,
   (a) Select the <non-redundant protein sequences (nr) > database in the drop-down menu.
   (b) Enter organism(s) to search against. In this case, <*Shigella sonnei* (taxid:624)>; <*Shigella dysenteriae* (taxid:622)>; and < *Shigella boydii* (taxid:621) > were selected (*see* **Note 16**).
   (c) Do not check the "Exclude" boxes.
   (d) Proceed to next section.
4. For "Program Selection," use the default blastp (protein–protein BLAST) program.
5. Click the "BLAST" button at the bottom of the page. The results will appear as in Fig. 3b (*see* **Note 17**).

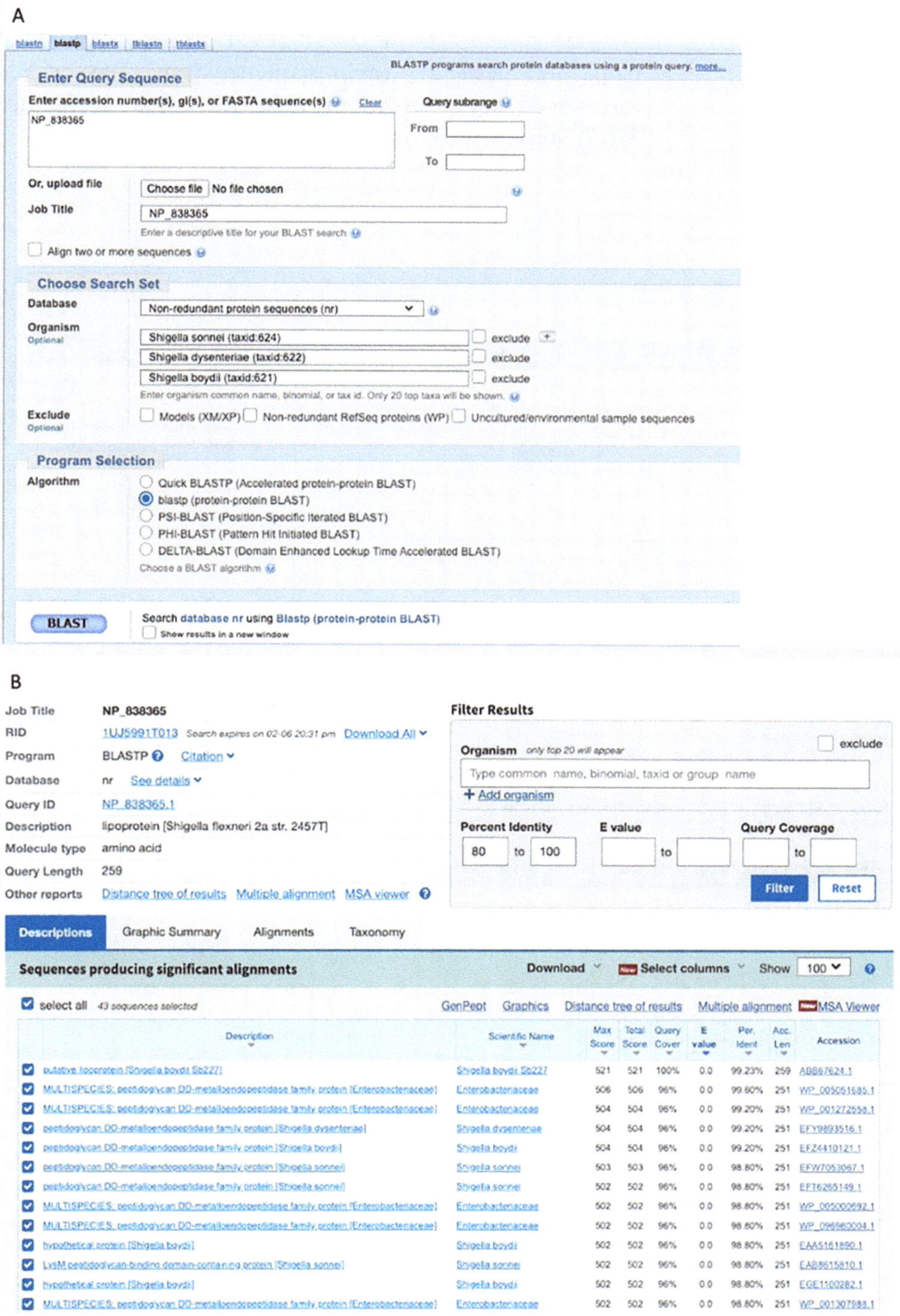

Fig. 3 An example of BLASTP protein sequence similarity analysis among *Shigella* spp. proteomes. (**a**) Protein sequence submission interface. Outcome of BLASTP analysis. (**b**) The result showing NP_838365 of *S. flexneri* is orthologous in *S. dysenteriae*

3.3 Identification of Human Homologs

1. Point the internet browser to the BLAST function within the UniProt website at https://www.uniprot.org/blast [27].
2. Copy and paste a protein sequence (raw data or FASTA-formatted sequences of proteins identified in Subheading 3.1.2, **step 7**) into the form field (Fig. 4a).

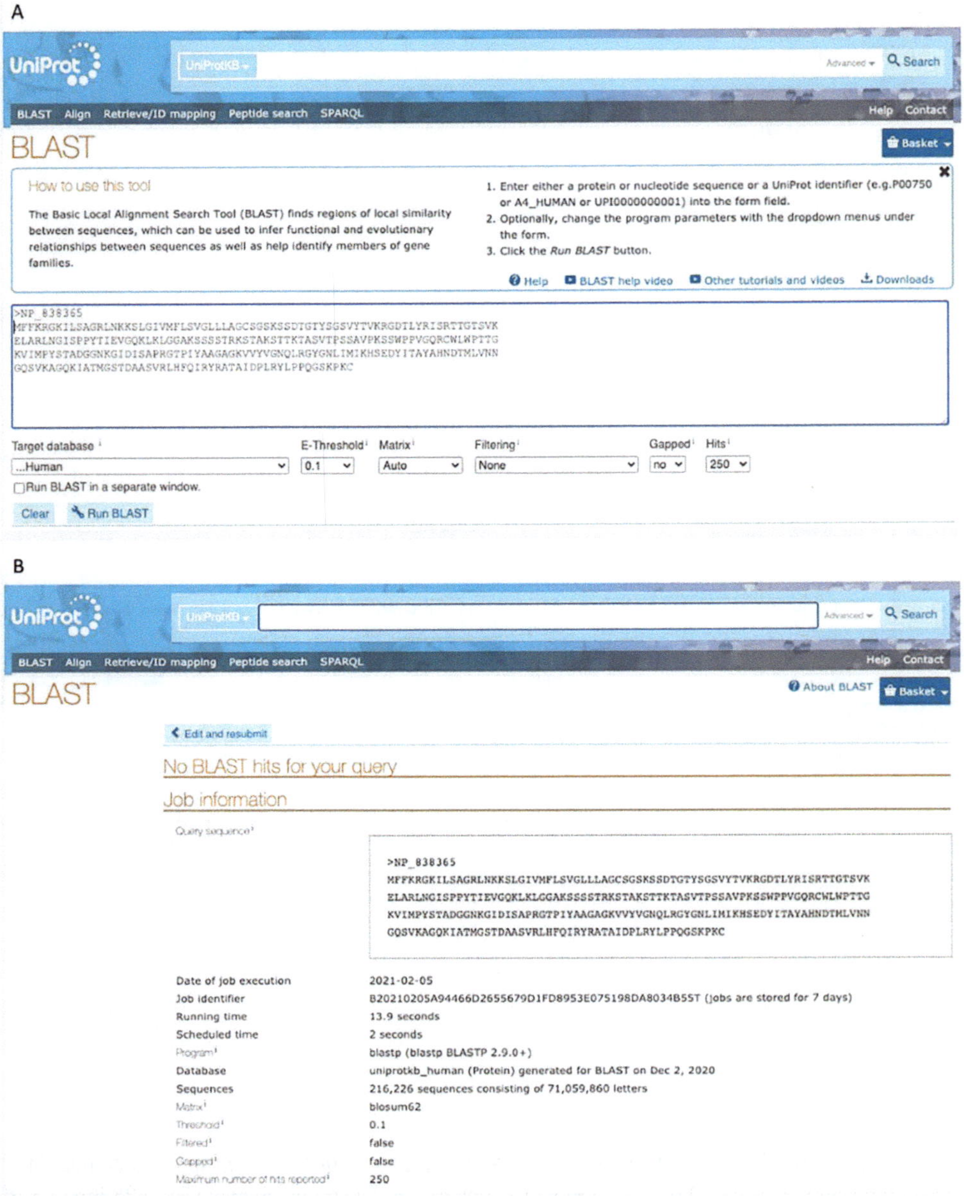

Fig. 4 An example of protein sequence similarity analysis against human whole proteome using BLAST tool in UniProt. (**a**) Protein sequence submission interface. (**b**) Outcome of BLAST analysis showing NP_838365 has no similar protein sequence found in human proteome

3. To identify the most related human proteins, change the program parameters in the drop-down menus under the form as follows.
   (a) Target database: Human.
   (b) E-Threshold: 0.1.
   (c) Matrix: Auto.
   (d) Filtering: None.
   (e) Gapped: No.
   (f) Hits: 250.
4. Click the "Run BLAST" button at the bottom of the page. The results appear as in Fig. 4b (*see* **Note 18**).

3.4 Antigenicity Analysis

1. Point the internet browser to the VaxiJen v2.0 website at http://www.ddg-pharmfac.net/vaxijen/VaxiJen/VaxiJen.html
2. At "Enter a PROTEIN sequence here," copy and paste a protein sequence (plain format) in the form field. Alternatively, a file containing multiple protein sequences saved in FASTA-format can be submitted using the "Choose file" button (Fig. 5a).
3. At "Select a TARGET ORGANISM," select <Bacteria> in the scrollable selection menu.
4. Keep default setting for the "THRESHOLD" (cutoff at 0.4) (*see* **Note 19**).
5. Check the "Sequence Output" box. Check the other provided boxes if user would like to learn more results for the submitted protein sequence (*see* **Note 20**).
6. Click the "Submit" button at the bottom of the page. The results appear as in Fig. 5b.
7. Five outer membrane proteins predicted to be potential vaccine candidates against *S. flexneri* were selected for further analysis, and the relevant results for these are shown in Table 1. Each protein was subsequently analyzed to identify potential B- and T-cell epitopes.

3.5 Prediction of Linear B-Cell Epitopes

1. Point the internet browser to the predictions page of the BCPREDS website at http://ailab-projects1.ist.psu.edu:8080/bcpred/predict.html
2. Copy and paste a protein sequence (plain format) in the form field (Fig. 6a).
3. At "Methods," select the "Fixed length epitope prediction" option <BCPred> with a window (epitope length) of 20 amino acids, as set by the default parameters (*see* **Note 21**).
4. At "Specificity," select <75% > in the drop-down menu (*see* **Note 22**).

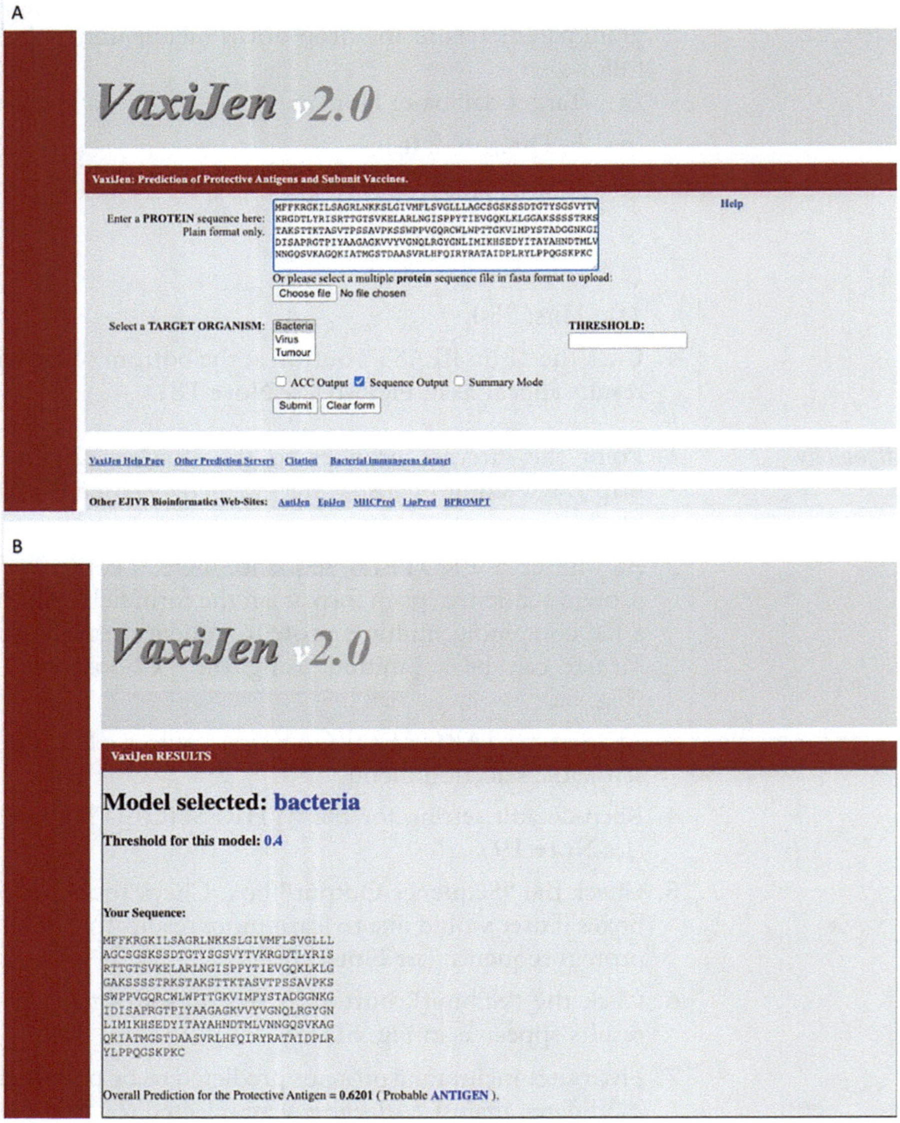

Fig. 5 VaxiJen v2.0 query submission for protein antigenicity prediction. (**a**) VaxiJen v2.0 query web interface. (**b**) An example of result page. The protein (NP_838365) was predicted to be antigenic. Threshold for this model is 0.4; overall antigen prediction is 0.6201 (predicted to be probable ANTIGEN)

5. Check the box for "report only non-overlapping epitopes."
6. Click the "Submit query" button at the bottom of the page. The result of analysis is shown in Fig. 6b.

3.6 Prediction of HLA Class I and Class II T-Cell Epitope

1. Point the internet browser to the HLApred website at http://crdd.osdd.net/raghava/hlapred/ (*see* **Note 23**).
2. Copy and paste a protein sequence (plain format) into the "Paste Your Protein Sequence" form field, or use the "Upload Sequence File" option to choose a saved sequence file (Fig. 7).

Table 1
List of potential vaccine candidates against *S. flexneri* predicted using RV approach

No.	Protein name	Amino acid sequence homology percentage in *Shigella* strains			Antigenicity score (Vaxijen) Threshold > 0.4	Human proteome homology
		Shigella boydii	*Shigella dysentery*	*Shigella sonnei*		
1	NP_836212 Outer membrane receptor	99%	99%	99%	0.7454	No significant similarity found
2	NP_837825 Outer membrane porin protein C	93%	93%	95%	0.6986	No significant similarity found
3	NP_838270 Lipoprotein NlpD	99%	100%	99%	0.7117	No significant similarity found
4	NP_838556 Outer membrane channel protein	99%	99%	99%	0.5494	No significant similarity found
5	NP_838365 Lipoprotein	99%	99%	99%	0.6168	No significant similarity found

3. At "Choose Format of input sequence," select <Amino acids in single letter code>. Alternatively, standard formats (e.g., EMBL, PIR, FASTA, GCG) are optionally chosen if the protein sequence submitted is based on FASTA-format.
4. At "Select HLA allele," select multiple alleles by pressing the "Alt" (Window OS) or "cmd" (MACS OS) key followed by "mouse click." Alternatively, select "ALL" if the entire set of alleles are to be analyzed.
5. To include HLA-I and HLA-II binders in the analysis, click the Class I and Class II alleles through "Both" option in selection box.
6. At "Method," select the <Predict Binder> as set by default.
7. At "Prediction Parameter," select <3% > in the "Threshold" drop-down menu; select <HTML Mapping> in the "Display Format"; and "Display Top <4 > Peptides" (*see* **Note 24**).
8. Click the "Submit sequence" button at the bottom of the page. The result of analysis for HLA class I and HLA class II binders are shown in Figs. 8 and 9, respectively.

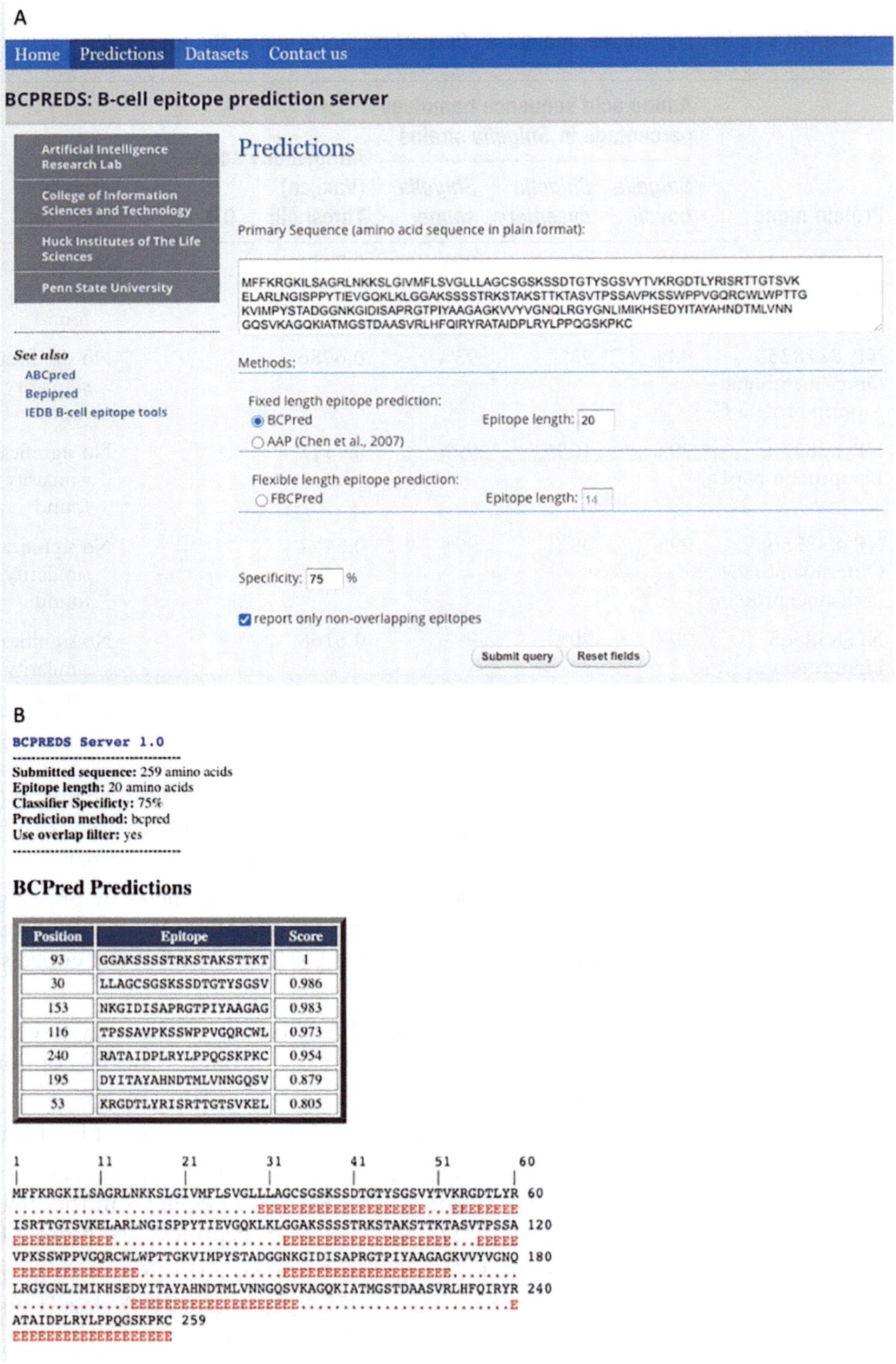

Fig. 6 BCPreds server for B-cell linear epitope prediction. (**a**) Cover page of BCPreds. (**b**) An example of result page for the prediction of B-cell epitope performed using BCPreds. Table showing seven 20-mers B-cell epitopes were predicted for NP_838365. Peptide (20 mers) with score >0.75 were predicted to be potential B-cell linear epitope

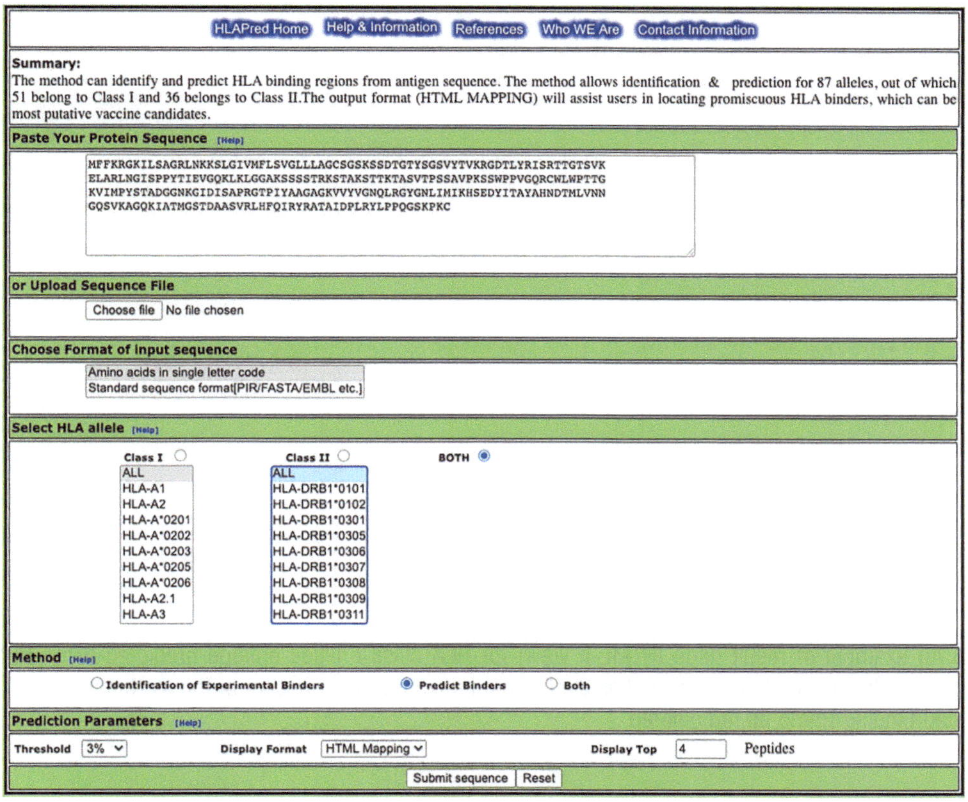

Fig. 7 A web interface of HLAPred. A server for prediction and identification of HLA class I and class II T-cell epitope using quantitative matrix-based prediction method

4 Notes

1. We have tested and confirmed all the programs are compatible with Google Chrome; Internet Explorer or Mozilla Firefox. We have chosen to use Google Chrome in our protocol. Furthermore, the type of computer and operating system do not affect the online programs used in the protocol.
2. Two methods are available for running Vaxign: (a) Vaxign query is for precomputed results (b) Dynamic analysis is for use of own sequence input (sequence format can be chosen in the drop-down menu).
3. The current genomes available in the Vaxign are limited to 350. If a genome of interest is not currently available in the pre-computed Vaxign database, the user may request addition of the genome by contacting the Vaxign server developer on the provided page (http://www.violinet.org/contact_us.php).

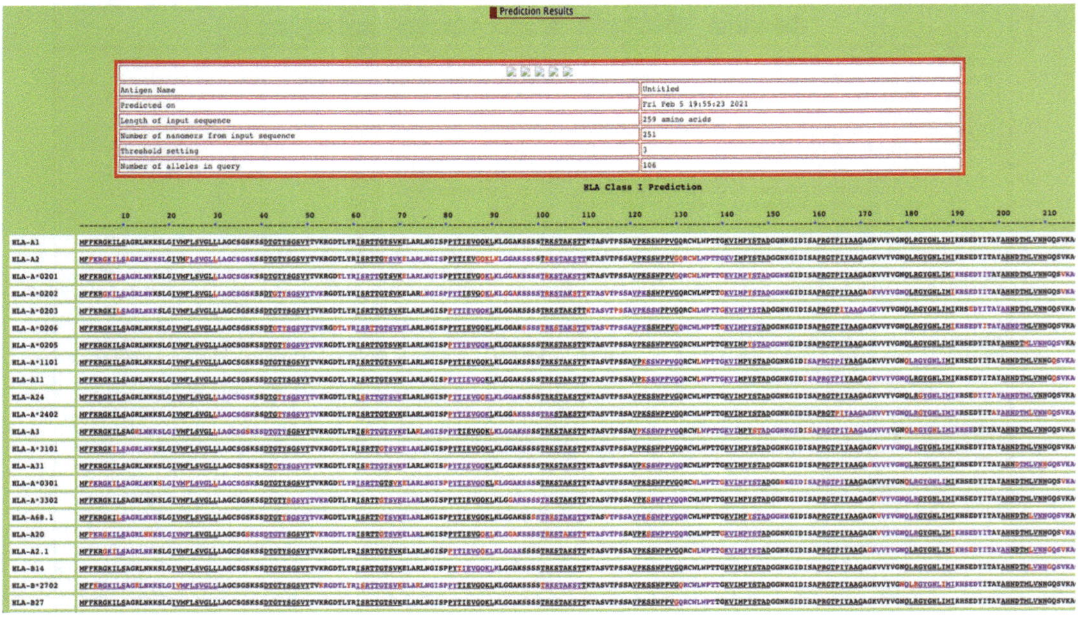

Fig. 8 An example of result page (HTML-II view) for HLA class I T-cell epitope predicted using HLAPred. Predicted binders (9-mer) are displayed as blue colored region, with P1 anchor or the starting residue of each predicted binding frame as red colored

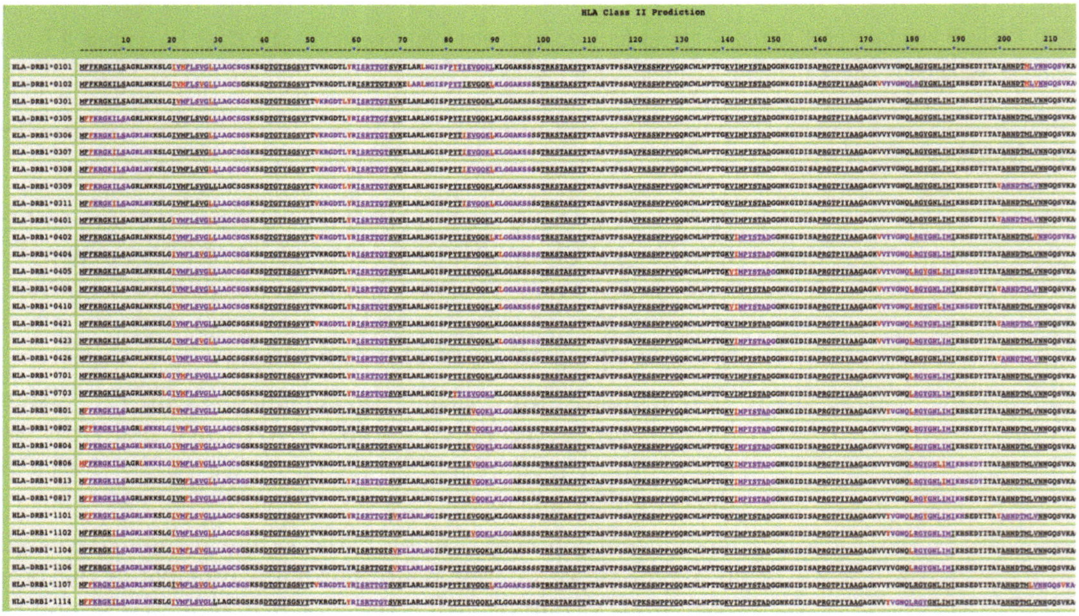

Fig. 9 An example of result page (HTML-II view) for HLA class II T-cell epitope predicted using HLAPred. Predicted binders (9-mer) are displayed as blue colored region, with P1 anchor or the starting residue of each predicted binding frame as red colored

4. Sequenced genomes used in Vaxign were retrieved from NCBI RefSeq database and used in Vaxign vaccine prediction pipeline.
5. Subcellular localization provides important information for diagnostics, drug, and vaccine design. PSORTb (v.3.0) [28], a subcellular localization prediction tool, is included in Vaxign vaccine prediction pipeline in order to identify the protein functional annotation based on the protein localization in the bacteria.
6. If the user is interested to explore the total proteomes in the chosen genome, the user can choose "Any Localization" in the scrollable selection menu. Alternatively, the user can select more than two options in the drop-down menu by pressing the "Ctrl" key + mouse left click.
7. The prediction of transmembrane helices and topology of proteins in Vaxign pipeline is performed using HMMTOP [29]. A protein comprising more than one transmembrane helix is considered to have difficulty in protein production and purification when it is expressed as a recombinant protein. It is wise to exclude those proteins with multiple transmembrane spanning regions from the selection process.
8. The prediction of adhesion characteristic of the protein is performed using SPAAN software [30]. Proteins predicted to have adhesin capacities are known to be involved in host-bacterial interaction. These proteins play vital role in bacterial invasion. The adhesin-like proteins appear to be potential vaccine candidates since they can be recognized by immune cells in the host [31].
9. Analysis of protein orthologue comparisons and sequence similarity to host proteome were not included in the protocol described in this chapter because the Vaxign server only provides qualitative results if these options are chosen. In order to analyze the antigen similarity with higher stringency settings for Expectation (E) values and percent identity, it is recommended to replace this step using alternative online servers such as BLASTP on NCBI or UniProt, as described in this chapter. If stringency or quantitation of the sequence similarity comparison is not required, user can proceed with the built-in options in Vaxign.
10. Epitope prediction option at Vaxign is only limited to MHC class I and MHC class II analysis. This option was not used in our current protocol, where the shortlisted antigens are subjected to B-cell epitope prediction followed by T-cell epitope prediction using alternative online software. However, the user can use the built-in option in Vaxign if only T-cell epitopes are to be analyzed.

11. "Have Orthologs In" is used for identifying protein conserved regions among a selected list of strains corresponding to the specific genome, for example, *Shigella* as in this case.
12. "No Similarity to Host Proteins" is used to exclude those protein targets that present either in humans, mouse and pigs.
13. "MHC class I and II epitope prediction" is performed using Vaxitop, a built-in vaccine epitope prediction and analysis system based on the principle of reverse vaccinology. This feature is available after initial protein filtering.
14. Even though the Vaxign pipeline saves time for users by shortlisting the proteins of interest, the user still needs to analyze the shortlisted proteins manually for best results. Compared to the total proteomes which are likely made up of >3000 proteins, dealing with selected proteins in the result page is comparatively time saving and manageable. We highly recommend the user to click into the details of each protein before proceeding to the next step.
15. Check this box if alignment of query with a known subject is required. This option is also known as BLAST 2 Sequences.
16. Initially, a single box is available to enter choice of organism. If more organisms are to be selected, the user can add in additional rows by clicking the "+" button on the right.
17. Conserved immunogenic antigens are promising vaccine candidates provided they are able to induce broad-spectrum protection against bacterial pathogens [32]. BLAST is a promising tool for the comparative analysis of the sequence similarity qualitatively and quantitatively. In general, the similarity between the sequences corresponds to the E value. The lower the E value, the more likely the match is to be significant. E values between 0.1 and 10 are generally more stringent (less false positive), whereas over 10 are unlikely to have biological significance. In this case, E value <0.1 was chosen. This option can be found in the parameter setting at the bottom of the page. In the result page, identity percentages equal to 80% or higher are considered to be orthologues. Proteins sharing identity percentages <80% are considered not conserved and are eliminated from the dataset.
18. The proteins having no significant identity (<35%) are considered to be sufficiently distant to the human proteome and will not interfere with normal host immune mechanism when used as a vaccine candidate [8].
19. For the antigenicity prediction by VaxiJen, the default threshold set for bacterial antigen analysis is 0.4. The percentage threshold parameter allows the user to select for different

stringency levels. As the threshold window increases (i.e., > 0.4), the specificity increases and the sensitivity decreases.

20. VaxiJen is a server designed to analyze protective antigens using alignment-independent prediction. The server contains models derived by auto cross covariance (ACC) preprocessing of amino acids properties. It has been trained to identify potential antigenic epitopes for bacterial, viral, tumor, parasite, and fungal origin. By default, the serve displays only the Sequence Output. In this mode, the threshold value, protein sequence and prediction result are displayed. If the user would like to learn more about the antigen ACC value, ACC Output can be selected. Alternatively, user can choose the Summary Output. In this mode, only the prediction result is provided.
21. BCPREDS server allows the user to select among three prediction methods: (a) BCPred, (b) AAP method, and (c) FBCPred. It is recommended to use at least two of these methods, as identification of overlapping epitopic regions identified using multiple B-cell epitope prediction algorithms enhances the accuracy of prediction.
22. Increasing specificity percentage corresponds to a high stringency prediction.
23. HLApred is a quantitative matrix-based method that allows the prediction of HLA binding sites in an antigenic sequence for 51 HLA class-I alleles and 36 HLA class-II alleles. Instead of choosing all alleles, user can also focus on the superalleles which are reported to cover more than 99% of the human population. For HLA Class I, supertype HLA alleles are "HLA-A*01:01," "HLA-A*02:01," "HLA-A*03:01," "HLA-A*24:02," "HLA-B*07:02," and "HLA-B*44:03" whereas for HLA class II, supertype HLA alleles are "HLA-DRB1*01:01," "HLA-DRB1*03:01," "HLA-DRB1*04:01," "HLA-DRB1*07:01," "HLA-DRB1*08:01," "HLA-DRB1*11:01," "HLA-DRB1*13:01," and "HLA-DRB1*15:01" [33]. The promiscuous epitopes are those which bind with many HLA alleles. For instance, in the example protein (NP_838365) shown in Fig. 7, region ^{59}YRISRT TGT^{67} has binding affinity for most of the HLA class II alleles used in the query (see results shown in Fig. 9), suggesting it can be a potential T-cell epitope candidate. However, due to discrepancies between different algorithms, users are recommended to employ >2 alternative epitope prediction algorithms to increase the likelihood of the predicted epitope(s).
24. The percentage threshold parameter allows the user to select for different stringency levels; a lower threshold corresponds to a high stringency prediction.

Acknowledgments

The authors acknowledge the funding support provided for this work by USM Research University (Individual) Grant (no.: 1001. CIPPM.8011078) and The Malaysian Ministry of Higher Education of the Higher Institutions Centre of Excellence Program under Grant (no: 311/CIPPM/4401005). The authors would like to thank Associate Professor Dr. Oliver He Yongqun at University of Michigan Medical School for providing his excellent guidance on the Vaxign Vaccine Design platform.

References

1. Kotloff KL, Riddle MS, Platts-Mills JA et al (2018) Shigellosis. Lancet 391 (10122):801–812. https://doi.org/10.1016/S0140-6736(17)33296-8
2. Banga Singh KK, Ojha SC, Deris ZZ, Rahman RA (2011) A 9-year study of shigellosis in Northeast Malaysia: antimicrobial susceptibility and shifting species dominance. J Public Health 19(3):231–236. https://doi.org/10.1007/s10389-010-0384-0
3. Ashkenazi S, Cohen D (2013) An update on vaccines against Shigella. Ther Adv Vaccines 1 (3):113–123. https://doi.org/10.1177/2051013613500428
4. Tettelin H (2009) The bacterial pan-genome and reverse vaccinology. In: de Reuse H, Bereswill S (eds) Microbial pathogenomics, Genome Dyn, vol 6. Karger, Basel, pp 35–47. https://doi.org/10.1159/000235761
5. Pizza M, Scarlato V, Masignani V et al (2000) Identification of vaccine candidates against serogroup B meningococcus by whole-genome sequencing. Science 287(5459):1816–1820. https://doi.org/10.1126/science.287.5459.1816
6. Mamede LD, de Paula KG, de Oliveira B et al (2020) Reverse and structural vaccinology approach to design a highly immunogenic multi-epitope subunit vaccine against *Streptococcus pneumoniae* infection. Infect Genet Evol 85:104473. https://doi.org/10.1016/j.meegid.2020.104473
7. Gupta N, Kumar A (2020) Designing an efficient multi-epitope vaccine against *Campylobacter jejuni* using immunoinformatics and reverse vaccinology approach. Microb Pathog 147:104398. https://doi.org/10.1016/j.micpath.2020.104398
8. Leow CY, Kazi A, Hisyam Ismail CMK et al (2020) Reverse vaccinology approach for the identification and characterization of outer membrane proteins of *Shigella flexneri* as potential cellular- and antibody-dependent vaccine candidates. Clin Exp Vaccine Res 9 (1):15–25. https://doi.org/10.7774/cevr.2020.9.1.15
9. Ong E, Wong MU, Huffman A, He Y (2020) COVID-19 coronavirus vaccine design using reverse vaccinology and machine learning. Front Immunol 11:1581. https://doi.org/10.3389/fimmu.2020.01581
10. He Y, Xiang Z, Mobley HL (2010) Vaxign: the first web-based vaccine design program for reverse vaccinology and applications for vaccine development. J Biomed Biotechnol 2010:297505. https://doi.org/10.1155/2010/297505
11. Xiang Z, He Y (2013) Genome-wide prediction of vaccine targets for human herpes simplex viruses using Vaxign reverse vaccinology. BMC Bioinformatics 14 Suppl 4:S2. https://doi.org/10.1186/1471-2105-14-S4-S2
12. He Y (2012) Analyses of Brucella pathogenesis, host immunity, and vaccine targets using systems biology and bioinformatics. Front Cell Infect Microbiol 2:2. https://doi.org/10.3389/fcimb.2012.00002
13. Gomez G, Pei J, Mwangi W, Adams LG, Rice-Ficht A, Ficht TA (2013) Immunogenic and invasive properties of *Brucella melitensis* 16M outer membrane protein vaccine candidates identified via a reverse vaccinology approach. PLoS One 8(3):e59751. https://doi.org/10.1371/journal.pone.0059751
14. Garg N, Singh R, Shukla G, Capalash N, Sharma P (2016) Immunoprotective potential of in silico predicted *Acinetobacter baumannii* outer membrane nuclease, NucAb. Int J Med Microbiol 306(1):1–9. https://doi.org/10.1016/j.ijmm.2015.10.005
15. Hossain MS, Azad AK, Chowdhury PA, Wakayama M (2017) Computational identification and characterization of a promiscuous T-cell epitope on the extracellular protein 85B

of *Mycobacterium spp.* for peptide-based subunit vaccine design. BioMed Res Int 2017:4826030. https://doi.org/10.1155/2017/4826030

16. He Y, Rappuoli R, De Groot AS, Chen RT (2010) Emerging vaccine informatics. J Biomed Biotechnol 2010:218590. https://doi.org/10.1155/2010/218590
17. Moise L, Cousens L, Fueyo J, De Groot AS (2011) Harnessing the power of genomics and immunoinformatics to produce improved vaccines. Expert Opin Drug Discov 6(1):9–15. https://doi.org/10.1517/17460441.2011.534454
18. Kazi A, Chuah C, Majeed ABA et al (2018) Current progress of immunoinformatics approach harnessed for cellular- and antibody-dependent vaccine design. Pathog Glob Health 112(3):123–131. https://doi.org/10.1080/20477724.2018.1446773
19. Kazi A, Hisyam Ismail CMK, Anthony AA et al (2020) Designing and evaluation of an antibody-targeted chimeric recombinant vaccine encoding *Shigella flexneri* outer membrane antigens. Infect Genet Evol 80:104176. https://doi.org/10.1016/j.meegid.2020.104176
20. Mount DW (2007) Using the Basic Local Alignment Search Tool (BLAST). CSH Protocols 2007:pdb top17. https://doi.org/10.1101/pdb.top17
21. Altschul SF, Gish W, Miller W, Myers EW, Lipman DJ (1990) Basic local alignment search tool. J Mol Biol 215(3):403–410. https://doi.org/10.1016/S0022-2836(05)80360-2
22. Doytchinova IA, Flower DR (2007) VaxiJen: a server for prediction of protective antigens, tumour antigens and subunit vaccines. BMC Bioinformatics 8:4. https://doi.org/10.1186/1471-2105-8-4
23. El-Manzalawy Y, Dobbs D, Honavar V (2008) Predicting linear B-cell epitopes using string kernels. J Mol Recognit 21(4):243–255. https://doi.org/10.1002/jmr.893
24. El-Manzalawy Y, Dobbs D, Honavar V (2008) Predicting flexible length linear B-cell epitopes. Comput Syst Bioinformatics Conf 7:121–132
25. Singh H, Raghava GP (2003) ProPred1: prediction of promiscuous MHC class-I binding sites. Bioinformatics 19(8):1009–1014. https://doi.org/10.1093/bioinformatics/btg108
26. Singh H, Raghava GP (2001) ProPred: prediction of HLA-DR binding sites. Bioinformatics 17(12):1236–1237. https://doi.org/10.1093/bioinformatics/17.12.1236
27. UniProt C (2019) UniProt: a worldwide hub of protein knowledge. Nucleic Acids Res 47 (D1):D506–D515. https://doi.org/10.1093/nar/gky1049
28. Yu NY, Wagner JR, Laird MR et al (2010) PSORTb 3.0: improved protein subcellular localization prediction with refined localization subcategories and predictive capabilities for all prokaryotes. Bioinformatics 26 (13):1608–1615. https://doi.org/10.1093/bioinformatics/btq249
29. Tusnady GE, Simon I (2001) The HMMTOP transmembrane topology prediction server. Bioinformatics 17(9):849–850. https://doi.org/10.1093/bioinformatics/17.9.849
30. Sachdeva G, Kumar K, Jain P, Ramachandran S (2005) SPAAN: a software program for prediction of adhesins and adhesin-like proteins using neural networks. Bioinformatics 21 (4):483–491. https://doi.org/10.1093/bioinformatics/bti028
31. Raynes JM, Young PG, Proft T et al (2018) Protein adhesins as vaccine antigens for Group A Streptococcus. Pathog Dis 76(2). https://doi.org/10.1093/femspd/fty016
32. Nagy G, Emody L, Pal T (2008) Strategies for the development of vaccines conferring broad-spectrum protection. Int J Med Microbiol 298 (5–6):379–395. https://doi.org/10.1016/j.ijmm.2008.01.012
33. De Groot AS, Ardito M, McClaine EM et al (2009) Immunoinformatic comparison of T-cell epitopes contained in novel swine-origin influenza A (H1N1) virus with epitopes in 2008-2009 conventional influenza vaccine. Vaccine 27(42):5740–5747. https://doi.org/10.1016/j.vaccine.2009.07.040

Chapter 3

Purification of Prospective Vaccine Antigens from Gram-Positive Pathogens by Immunoprecipitation

Mark Reglinski

Abstract

Immunoprecipitation is an affinity purification technique that exploits the highly specific interactions formed between antibodies and their cognate antigens to purify molecules of interest from complex biological solutions. The generation of an effective humoral response provides protection against a wide range of gram-positive pathogens, and thus immunoprecipitation using antibodies purified from immune humans or animals provides a simple but effective means of isolating prospective vaccine antigens from fractionated bacterial cells for downstream identification. The commercial availability of antibody preparations from donated human plasma, containing antibodies against many common gram-positive pathogens, allows the protocol to be performed in the absence of bespoke vaccination experiments. Thus, immunoprecipitation has the potential to reduce the number of animals used in vaccine studies by allowing an initial screen for promising antigens to be conducted in vitro.

Key words Immunoprecipitation, Vaccine antigen, Gram-positive, Affinity purification, IVIG

1 Introduction

Immunoprecipitation (or immunoaffinity purification, *see* **Note 1**) is a technique that uses immobilized antibodies to purify specific proteins from complex biological solutions [1]. Today immunoprecipitation is commonly employed to isolate proteins that have been genetically engineered to include specific peptide tags for which commercial immunoprecipitation resins are available [2]. However, the technique was initially developed to isolate naturally occurring antibodies or antigens implicated in the pathophysiology of infection, allergy, or autoimmunity for downstream characterisation [3–5].

The protocol below provides a simple framework for applying immunoprecipitation to the identification of prospective vaccine antigens from gram-positive pathogens and can be performed using standard laboratory equipment and consumables. The

Fadil Bidmos et al. (eds.), *Bacterial Vaccines: Methods and Protocols*,
Methods in Molecular Biology, vol. 2414, https://doi.org/10.1007/978-1-0716-1900-1_3,

protocol as described uses pooled human immunoglobulin (Intravenous Immunoglobulin G, IVIG), a clinical antibody infusion prepared from the plasma of more than 3000 blood donors [6]; however, it could equally be performed using antiserum from experimentally vaccinated animals. As both IVIG and animal antiserum will contain antibodies raised in response to many pathogens, antibodies targeted toward the pathogen selected for study are first purified to produce a pathogen-reactive antibody pool (PRAP) which is then coupled to a commercially available resin. The resulting immunoprecipitation resin is then used to purify prospective vaccine antigens from bacterial preparations (e.g., cell wall extracts or cell lysates) for downstream characterisation.

2 Materials

2.1 Preparation of Gram-Positive Cell Wall Extracts

1. 30% w/v raffinose (*see* **Note 2**): Dissolve 30 g of raffinose in 80 ml of deionized water using a magnetic hotplate and a stir bar. Add deionized water to 100 ml and autoclave. Store at room temperature (RT).
2. 1 M Tris–HCl (pH 8): Dissolve 12.1 g of tris base ($NH_2C(CH_2OH)_3$) in 80 ml of deionized water and adjust to pH 8 with concentrated HCl. Add deionized water to 100 ml and store at RT.
3. 10 kU/ml mutanolysin: Add 1 ml of molecular grade water to a vial containing 10,000 U of mutanolysin. Vortex vigorously to ensure that the powder has completely dissolved and store at −20 °C in 100 μl aliquots.
4. 100 mg/ml lysozyme: Dissolve 100 mg of lysozyme in 1 ml of molecular grade water. Vortex vigorously to ensure that the powder has completely dissolved and store at −20 °C in 100 μl aliquots.
5. Protease Inhibitor Cocktail Set III (Millipore).
6. Cell wall extraction buffer: 30% raffinose, 10 mM Tris–HCl, 100 U/ml mutanolysin, 1 mg/ml lysozyme, 1% protease inhibitor cocktail. Combine 960 μl of 30% raffinose with 10 μl of Tris–HCl, mutanolysin, lysozyme, and protease inhibitor cocktail solutions to give 1 ml of cell wall extraction buffer (*see* **Note 3**).
7. 20 ml polypropylene syringes with Luer slip.
8. 18-gauge, 1-in. beveled needles.
9. Slide-A-Lyzer Dialysis Cassettes, 20 K MWCO, 12 ml (Thermo-Fisher).
10. Vivaspin 20 centrifugal concentrators, MWCO 10 kDa (Sartorius).

11. Millex-GP Syringe Filter Units, 0.22 μm (Millipore).
12. Colorimetric protein quantification assay reagents. For example, Pierce BCA Protein Assay Kit or Pierce Coomassie (Bradford) Protein Assay Kit.

2.2 Coupling of Proteins to Cyanogen Bromide Activated Sepharose

1. Coupling buffer (*see* **Note 4**): 0.1 M sodium bicarbonate, 0.5 M sodium chloride (pH 8.3). Dissolve 8.4 g of sodium bicarbonate ($NaHCO_3$) and 29 g of sodium chloride (NaCl) in 800 ml of deionized water. Adjust to pH 8.3 with 5 M sodium hydroxide (NaOH, *see* **Note 5**) and add deionized water to 1 l. Store at RT.
2. 1 mM hydrochloric acid (HCl, *see* **Note 6**): Dilute 83 μl of concentrated HCl in 1 L of deionized water. Store at 4 °C.
3. 0.2 M glycine: Dissolve 7.5 g of glycine (NH_2CH_2COOH) in 400 ml of deionized water. Add deionized water to 500 ml and store at RT.
4. Wash buffer: 0.1 M acetate, 0.5 M sodium chloride (pH 4). Dissolve 4.2 g of anhydrous sodium acetate ($C_2H_3O_2Na$) and 14.5 g of NaCl in 400 ml of deionized water. Add 2.2 ml of glacial acetic acid and adjust to pH 4 using concentrated HCl (*see* **Note 5**). Add deionized water to 500 ml and store at RT.
5. Cyanogen bromide-activated Sepharose (CBr-Sepharose), lyophilized powder (Sigma-Aldrich).
6. Econo-Column® Chromatography Column, 1.5 × 5 cm (Bio-Rad).
7. Horizontal tube rotator capable of supporting 15 ml centrifuge tubes.
8. Sodium azide: 1% in aqueous solution, purchased premade from commercial vendors.

2.3 Affinity Purification of Pathogen-Reactive Antibodies and Immunoprecipitation of Prospective Vaccine Antigens

1. Loading buffer: 0.1 M sodium phosphate monobasic, 0.15 M sodium chloride (pH 7). Dissolve 12 g of anhydrous sodium phosphate monobasic (NaH_2PO_4) and 8.7 g of NaCl in 800 ml of deionized water. Adjust to pH 7 with 5 M NaOH. Add deionized water to 1 L and store at RT.
2. 1 M acetic acid: Dilute 5.74 ml of glacial acetic acid in 94.26 ml of deionized water. Store at RT.
3. Neutralization buffer: 3 M Tris–HCl (pH 8.8). Dissolve 36.3 g of Tris base in 80 ml of deionized water and adjust to pH 8.8 with concentrated HCl. Add deionized water to 100 ml and store at RT.
4. Pooled human immunoglobulin (e.g., Privigen 100 mg/ml solution for infusion from CSL Behring UK, *see* **Note 7**). 5 ml at 5 mg/ml in loading buffer.

3 Methods

3.1 Preparation of Gram-Positive Cell Wall Extracts

1. Culture the target bacterium to the desired growth phase under appropriate conditions in 500 ml aliquots (*see* **Note 8**).
2. Centrifuge the culture at 4000 × *g* for 10 min and resuspend the cell pellet in 10 ml of 30% raffinose.
3. Centrifuge the cell suspension at 4000 × *g* for 10 min and resuspend the cell pellet in 10 ml of cell wall extraction buffer.
4. Incubate at 37 °C for 3 h with occasional agitation.
5. Pellet the protoplasts at 14,000 × *g* for 10 min and aspirate the supernatant (cell wall extract).
6. Pass the cell wall extract through a 0.22 μm syringe filter and transfer to a rehydrated 20 kDa MWCO cassette using a 20 ml syringe and 18-gauge needle (*see* **Note 9**).
7. Dialyze the sample against 2 l of coupling buffer for at least 4 h at 4 °C.
8. Discard the spent dialysate and replace with 2 l of fresh coupling buffer. Dialyze the sample overnight at 4 °C.
9. Transfer the cell wall extract to a centrifugal filter unit and concentrate to approximately 1 ml through multiple rounds of centrifugation at 3000 × *g* in a refrigerated centrifuge (*see* **Note 10**).
10. Measure protein concentration of cell wall extract using a colorimetric protein assay (*see* **Note 11**) and store at −20 °C.

3.2 Coupling the Cell Wall Proteins to CBr-Sepharose

1. Gradually add 250 mg of CBr-Sepharose to 50 ml of ice cold 1 mM HCl and incubate at RT for 30 min with occasional agitation (*see* **Note 12**).
2. Transfer the resin suspension to a 1.5 × 5 cm chromatography column and allow HCl to flow through the column.
3. Wash the resin bed with 5 ml of deionized water then with 5 ml of coupling buffer, and immediately resuspend the resin in 2–4 ml of coupling buffer containing approximately 1 mg of cell wall extract (*see* **Note 13**).
4. Incubate the resin slurry overnight at 4 °C using a horizontal tube rotator to maintain the liquid–solid suspension (*see* **Note 12**).
5. Allow the resin to settle and collect the flow through for downstream analysis (*see* **Note 14**).
6. Block the unreacted groups with 5 ml of 0.2 M glycine for 2 h at RT using a horizontal tube rotator to maintain the liquid–solid suspension.

7. Allow the blocking solution to flow through the column and wash the resin with 10 ml of coupling buffer then with 10 ml of wash buffer. Complete this wash cycle of high and low pH buffer solutions four more times to ensure removal of all unreacted ligand.
8. Wash the column with 10 ml of loading buffer and resuspend the affinity purification resin in 2 ml of loading buffer supplemented with 100 μl of 1% sodium azide solution (final concentration 0.05%). Store at 4 °C.

3.3 Affinity Purification of Pathogen-Reactive Antibodies

1. Wash the affinity purification resin extensively with loading buffer to remove the sodium azide.
2. Apply 5 ml of pooled human immunoglobulin (25 mg of IgG) in loading buffer and incubate at RT for 2 h using a horizontal tube rotator to maintain the liquid–solid suspension.
3. Allow resin to settle and collect flow-through for downstream PRAP purifications (*see* **Note 15**).
4. Wash the resin with 20 ml of loading buffer and elute the bound antibody (PRAP) directly into 8 ml of neutralization buffer using 8 ml of 1 M acetic acid (*see* **Note 16**).
5. Wash the column with 10 ml of loading buffer and strip the remaining antibody using 10 ml of 0.5 M NaOH (*see* **Note 17**).
6. Wash the column with loading buffer until the pH of the flow-through is neutral.
7. Resuspend the resin in 2 ml of loading buffer supplemented with 100 μl of 1% sodium azide solution (final concentration 0.05%). Store at 4 °C.
8. Transfer the eluted PRAP to a rehydrated 20 kDa MWCO cassette using a 20 ml syringe and 18-gauge needle.
9. Dialyze the sample against 2 l of coupling buffer for at least 4 h at 4 °C.
10. Discard the spent dialysate and replace with 2 l of fresh coupling buffer. Dialyze the sample overnight at 4 °C.
11. Transfer the PRAP to a centrifugal filter unit and concentrate to 1 mg/ml through multiple rounds of centrifugation at 3000 × *g* in a refrigerated centrifuge (*see* **Note 10**).
12. Measure protein concentration of the PRAP and store at 4 °C.

3.4 Coupling of Pathogen-Reactive Antibodies to CBr-Sepharose and Immunoprecipitation of Prospective Vaccine Antigens

1. Couple 1 mg of the PRAP to 1 ml of swollen CBr-Sepharose in a 1.5 × 5 cm chromatography column according to the protocol above to produce the immunoprecipitation resin.
2. Resuspend the immunoprecipitation resin in 5 ml of concentrated cell wall extract supplemented with 0.1% protease inhibitor cocktail and incubate for 2 h at RT using a horizontal tube rotator to maintain the liquid–solid suspension.
3. Allow the immunoprecipitation resin to settle and collect the flow-through (containing unbound surface proteins) for downstream analysis.
4. Wash the immunoprecipitation resin with 20 ml of loading buffer and elute the bound protein fraction (containing prospective vaccine antigens) directly into 8 ml of neutralization buffer using 8 ml of 1 M acetic acid (*see* **Note 16**).
5. Wash the immunoprecipitation resin with 10 ml of loading buffer and strip the remaining protein using 10 ml of 0.5 M NaOH (*see* **Note 17**).
6. Wash the immunoprecipitation resin with loading buffer until the pH of the flow through is neutral.
7. Resuspend the immunoprecipitation resin in 2 ml of loading buffer supplemented with 100 μl of 1% sodium azide solution (final concentration 0.05%). Store at 4 °C.
8. Transfer the immunoprecipitated protein fraction to a rehydrated 20 kDa MWCO cassette using a 20 ml syringe and 18-gauge needle.
9. Dialyze the sample against 2 l of loading buffer for at least 4 h at 4 °C.
10. Discard the spent dialysate and replace with 2 l of fresh loading buffer. Dialyze the sample overnight at 4 °C.
11. Transfer the immunoprecipitated protein fraction to a centrifugal filter unit and concentrate to approximately 50 μl through multiple rounds of centrifugation at 3000 × *g* in a refrigerated centrifuge (*see* **Note 10**).
12. Measure the protein concentration of the immunoprecipitated protein fraction using a colorimetric protein assay (*see* **Note 11**) and store at −20 °C for downstream characterisation.

4 Notes

1. Immunoprecipitation is a technique that arose from a series of classical immunological protocols such as the capillary precipitin and double diffusion tests. Such assays rely on antibody-mediated cross-linking of high molecular weight branching

antigens such as capsular polysaccharides or crude cellular preparations to generate insoluble "precipitates" that could be visualized in solution or in specialized agarose gels [7]. However, the purification of antigens using immobilized antibodies (or vice versa) does not cause antigen precipitation and thus while the term "immunoprecipitation" has been widely adopted, it does not accurately describe the procedure.

2. Less expensive sugars such as sucrose can be used for osmotic stabilization of the bacterial protoplasts generated following digestion of the cell wall. However, high molecular weight polysaccharides such as raffinose have demonstrated superior protoplast stabilizing capacities in comparative studies [8].
3. The lysozyme/mutanolysin based cell wall extraction buffer can be modified for species specific applications. For example, affinity purification resins have been prepared from *S. aureus* using 1 mg/ml of lysostaphin in place of lysozyme [9]. Affinity purification resins have also been prepared using whole *S. pneumoniae* lysates in place of cell wall extracts [10]. While no direct comparison has been made, it is likely that affinity purification resins produced using whole cell lysates are less efficient than resins produced using cell wall extracts for identification of prospective vaccine antigens. Where whole cell lysates are used, much of the binding capacity of the resin will be lost to prominent intracellular proteins that may have less contact with the immune response. This may reduce both the level of circulating antibodies against such proteins in the IVIG and the likelihood that their recognition by antibodies will promote opsonophagocytic killing of the bacterium in downstream vaccination studies.
4. While other buffers can be used, amine containing reagents such as Tris and glycine must be avoided as these will react with the active groups on the resin, reducing its protein binding capacity.
5. As the coupling and acetate buffers are supplemented 500 mM of NaCl, adjusting the pH with HCl and NaOH will not affect the ionic strength of the solution appreciably.
6. 1 mM HCl is used to swell the resin in place of deionized water to prevent hydrolysis of the reactive groups which occurs at an alkaline pH. The HCl dilution outlined is based on the use of 37% HCl which has a molarity of approximately 12. The concentration of commercial, concentrated HCl can vary from 36 to 38% (approximately 11.8 M to 12.5 M); however, achieving an exact HCl concentration of 1 mM is not critical to the success of the coupling protocol.
7. Intravenous immunoglobulin G, a clinical preparation of pooled human immunoglobulin, is a prescription only

medication that is not available for purchase through normal laboratory suppliers. In previous publications IVIG has been obtained from clinical suppliers (e.g., Privigen 100 mg/ml solution for infusion from CSL Behring UK) via a centralized hospital pharmacy. IgG from human serum can be purchased from Sigma-Aldrich (catalog number I4506); however, purification of PRAPs from this reagent has not been tested.

8. Previously cell wall extracts from cells grown to logarithmic phase (defined as an A_{600} of 0.4–0.8) in standard laboratory broth have been successfully used to purify opsonic PRAPs from several species [9, 11].
9. The dialysis cassettes will swell substantially due to the osmotic pressure generated by the 30% raffinose solution. The cassettes have previously been filled with 20 ml of cell wall extract with no adverse effects. Dialysis cassettes with a molecular weight cut off of 20 kDa or higher are recommended to facilitate removal of residual lysozyme and mutanolysin.
10. The amount of time and number of rounds of centrifugation required to concentrate the samples will vary. An initial centrifugation time of 20 min is recommended. The filtrate reservoirs of Vivaspin centrifugal filter units can be removed and emptied allowing volumes greater than 20 ml to be concentrated using a single unit over multiple spins. For large cell wall extract volumes, tangential flow filtration may be useful for both raffinose removal and protein concentration.
11. The protein concentration of the resulting cell wall extract will vary from species to species and the protocol may need to be scaled up considerably to reach the 1 mg of material recommended for affinity resin preparation. In previous studies, cell wall extracts from multiple strains of the same species have been pooled to increase the repertoire of reactive antigens available for IgG recognition [9, 11]. Prior to CBr-Sepharose coupling nonspecific IgG binding proteins (e.g., Sbi and protein A from *S. aureus*) that predominantly bind the Fc region of IgG can be removed using a column prepared using a cleaved IgG preparation [9]. Suitable antibody fragments can also be purchased directly from several commercial suppliers including Sigma-Aldrich (e.g., catalog number AG714). Quantification of the total protein present in the cell wall extracts by ultraviolet absorption (e.g., using a Nanodrop) is not recommended as the complexity of the samples complicates the selection of an appropriate extinction coefficient.
12. 250 mg of lyophilized CBr-Sepharose will swell to give a resin bed of approximately 1 ml. Pouring the HCl onto the dehydrated CBr-Sepharose will result in irreversible clumping and should be avoided. End over end rotation or the use of a

magnetic stir bar to maintain the liquid–solid suspension should also be avoided as this may adversely affect the integrity of the beads.

13. A resin-to-buffer ratio of 1:2 to 1:4 is required to maintain the liquid–solid suspension during tube rotation.
14. The protein concentration of the cell wall extract solution can be measured before and after resin incubation to confirm that the immobilization has been successful.
15. The available data suggests that several PRAPs targeting surface antigens from different bacterial species can be sequentially purified from a single pool of human immunoglobulin [9]. However, it should be noted that antibodies recognizing conserved epitopes may be absent from, or drastically reduced within the secondary and tertiary preparations.
16. The bound antibodies/antigens are eluted directly into 3 M Tris–HCl (pH 8.8) to neutralize the acetic acid before irreversible denaturation occurs. The neutralization buffer should be agitated to ensure that the eluant is rapidly incorporated.
17. The column can be stripped and reused at least five times.

References

1. Wofsy L, Burr B (1969) The use of affinity chromatography for the specific purification of antibodies and antigens. J Immunol 103 (2):380–382
2. DeCaprio J, Kohl TO (2019) Tandem Immunoaffinity purification using anti-FLAG and anti-HA antibodies. Cold Spring Harb Protoc 2019(2). https://doi.org/10.1101/pdb.prot098657
3. Del Villano BC, Defendi V (1973) Characterization of the SV40 T antigen. Virology 51 (1):34–46. https://doi.org/10.1016/0042-6822(73)90363-2
4. Alspaugh MA, Talal N, Tan EM (1976) Differentiation and characterization of autoantibodies and their antigens in sjögren's syndrome. Arthritis Rheum 19(2):216–222. https://doi.org/10.1002/art.1780190214
5. Ohman JL, Lowell FC, Bloch KJ (1974) Allergens of mammalian origin. J Immunol 113 (6):1668
6. Seite JF, Shoenfeld Y, Youinou P, Hillion S (2008) What is the contents of the magic draft IVIg? Autoimmun Rev 7(6):435–439. https://doi.org/10.1016/j.autrev.2008.04.012
7. Mehta PD (1972) Immunological techniques. Indian J Ophthalmol 20(2):49–54
8. Marquis RE (1965) Osmotic stability of bacterial protoplasts related to molecular size of stabilizing solutes. Biochem Biophys Res Commun 20(5):580–585. https://doi.org/10.1016/0006-291x(65)90438-9
9. Reglinski M, Sriskandan S (2019) Treatment potential of pathogen-reactive antibodies sequentially purified from pooled human immunoglobulin. BMC Res Notes 12(1):228. https://doi.org/10.1186/s13104-019-4262-8
10. Wilson R, Cohen JM, Reglinski M, Jose RJ, Chan WY, Marshall H, de Vogel C, Gordon S, Goldblatt D, Petersen FC, Baxendale H, Brown JS (2017) Naturally acquired human immunity to pneumococcus is dependent on antibody to protein antigens. PLoS Pathog 13(1):e1006137. https://doi.org/10.1371/journal.ppat.1006137
11. Reglinski M, Gierula M, Lynskey NN, Edwards RJ, Sriskandan S (2015) Identification of the *Streptococcus pyogenes* surface antigens recognised by pooled human immunoglobulin. Sci Rep 5:15825. https://doi.org/10.1038/srep15825

Chapter 4

Rapid Surface Shaving for Proteomic Identification of Novel Surface Antigens for Vaccine Development

Laurence Don Wai Luu and Ruiting Lan

Abstract

The bacterial cell surface (surfaceome) is the first site encountered by immune cells and is thus an important site for immune recognition. As such, the characterization of bacterial surface proteins can lead to the discovery of novel antigens for potential vaccine development. In this chapter, we describe a rapid 5-min surface shaving proteomics protocol where live bacterial cells are incubated with trypsin and surface peptides are "shaved" off. The shaved peptides are subsequently identified with liquid chromatography--tandem mass spectrometry (LC-MS/MS). Several checkpoints, including colony forming unit (CFU) counts, flow cytometry, and a false positive unshaved control, are introduced to ensure cell viability/membrane integrity are maintained and that proteins identified are true surface proteins. The protein topology of shaved peptides can be bioinformatically confirmed for surface location. Surface shaving facilitates identification of surface proteins expressed under different conditions, by different strains as well as highly abundant essential and immunogenic bacterial surface antigens for potential vaccine development.

Key words Surface shaving, Surfaceome, Surface proteins, Antigens, Vaccines, Proteomics, Cell surface, Mass spectrometry, LC-MS/MS, Epitopes

1 Introduction

The bacterial cell surface or "surfaceome" is the site where many important biological processes and host-pathogen interactions occurs including nutrient transport, cellular adhesion, delivery of cytotoxic effectors and immune recognition. As a result, this site is rich with many potential vaccine antigens. The advancement of proteomics and liquid chromatography–tandem mass spectrometry (LC-MS/MS), along with different surface extraction and enrichment techniques, has facilitated the characterization of these surface proteins.

Initially, surface protein identification was performed using subcellular fractionation methods which involve either differential lysis of different cellular compartments and/or differential

Fadil Bidmos et al. (eds.), *Bacterial Vaccines: Methods and Protocols*,
Methods in Molecular Biology, vol. 2414, https://doi.org/10.1007/978-1-0716-1900-1_4,

centrifugation in various buffers to separate surface proteins from other compartments [1]. However, these methods often had cytoplasmic contamination and were long and tedious to perform. As a result, an alternate method for bacterial surface protein identification was developed, initially using *Streptococcus pyogenes*, called "surface shaving" [2]. This method involves the incubation of live whole bacterial cells with trypsin (or other proteases) to "shave" peptides off the cell surface without lysing the cell. This method is comparatively easier to perform and provides important information on protein topology and surface exposed epitopes. Knowledge of surface exposed epitopes can be used to identify peptide regions accessible to immune recognition and inform vaccine design. However, one concern from surface shaving is the potential for inadvertent cell lysis, during incubation with trypsin, leading to the release of contaminating cytoplasmic proteins [3]. As a result, surface shaving has been more commonly used in gram-positive bacteria, which have a thicker peptidoglycan cell wall to withstand shaving, compared to gram-negatives [4].

To address the question of how to determine whether proteins identified by surface shaving are truly surface localized, or released as a result of cell lysis during incubation, Solis et al. [5] developed an "unshaved control" strategy to detect peptides released in the absence of trypsin digestion and eliminate false-positive prediction of these as surface proteins. To improve surface shaving for use in gram-negative bacteria and surface antigen discovery, we further refined the technique by reducing the amount of trypsin needed and the shaving incubation time, as well as incorporating colony forming unit (CFU) counts and flow cytometry checkpoints to ensure cell viability/membrane integrity. We have demonstrated the feasibility of using rapid surface shaving in the gram-negative bacterium *Bordetella pertussis*, which causes whooping cough, to compare surface protein expression between two representative strains and to identify novel surface antigens for vaccine development [6].

In this chapter, we detail the steps involved for our rapid surface shaving protocol of live bacterial cells for proteomic identification and quantification of surface antigens (Fig. 1). Although *B. pertussis* is used as an example, rapid surface shaving can be similarly applied to both gram-negative and gram-positive bacteria, as well as eukaryotes [7]. The protocol involves preparation of cells (*see* Subheading 3.1), rapid surface shaving and establishment of a false positive unshaved control (*see* Subheading 3.2), determining cell viability/membrane integrity after surface shaving (*see* Subheading 3.3) and preparing peptides for proteomic identification with LC-MS/MS and subsequent analysis to identify highly expressed, essential, and immunogenic surface antigens for further vaccine development (*see* Subheadings 3.4 and 3.5).

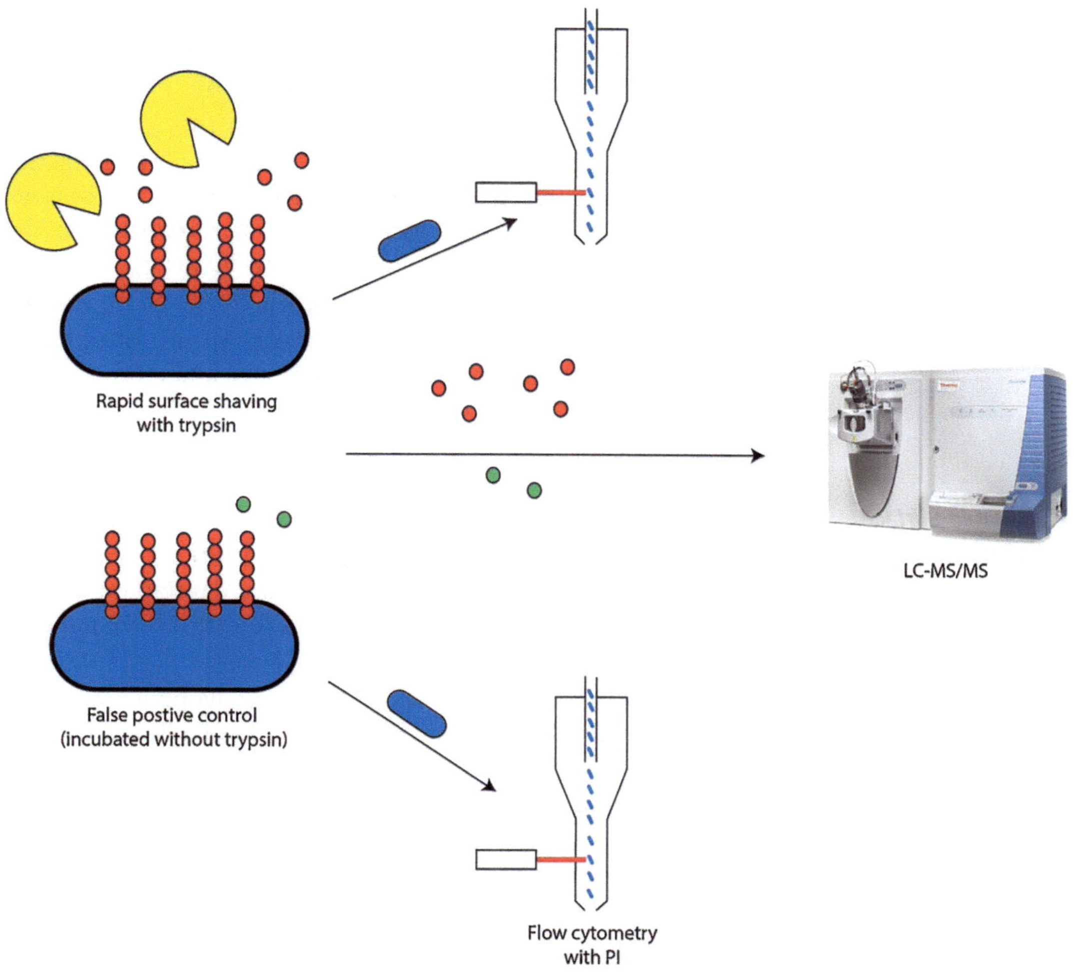

Fig. 1 Overview of the rapid surface shaving protocol for proteomic identification of surface antigens from live cells. Live whole cells are incubated with trypsin for 5 min to shave off surface peptides. An unshaved control is incubated under identical conditions without trypsin to identify peptides released during incubation which could lead to false positive prediction of surface exposure. Cell viability/membrane integrity is assessed using flow cytometry with propidium iodide (PI) and CFU counts. Shaved peptides are desalted and concentrated with C18 StageTips and analysed with LC-MS/MS.

2 Materials

2.1 Preparation of B. pertussis *Cells*

1. *B. pertussis* culture (*see* **Notes 1** and **2**).
2. Bordet–Gengou (BG) agar: 14 g of Bordet–Gengou agar base, 340 mL of ultrapure water, 6 mL of 50% glycerol (v/v). Autoclave-sterilize and cool to 50 °C. Add 25 mL of sterile defibrinated horse blood and pour into sterile Petri dishes.
3. Thalen–IJssel (THIJS) medium: 3.319 g NaCl, 0.107 g NH_4Cl, 0.5 g KH_2PO_4, 0.5 g KCl, 0.1 g $MgCl_2.6H_2O$, 1.525 g Tris, and 1.87 g Na glutamate.H_2O. Add

994.169 mL of ultrapure water, 3.76 mL of L-lactate (40% w/v), and 2.071 mL of 5 M NaOH. Autoclave-sterilize and store at 4 °C for up to 6 months (*see* **Note 3**).

4. 100x THIJS supplement: 0.2 g L-cystine, 0.13 g $CaCl_2.2H_2O$, 0.5 g L-glutathione reduced, 0.05 g $FeSO_4.7H_2O$, 0.02 g nicotinic acid, and 0.1 g L-ascorbic acid. Add 44 mL of ultrapure water and 6 mL of 1 M HCl. Filter-sterilize with a 0.22-μ m filter and store as 1 mL aliquots at -20 °C for up to 1 year.
5. 100× heptakis (also known as (2,6-di-O-methyl)-β-cyclodextrin): 1 g of heptakis, 10 mL of ultrapure water. Filter-sterilize with a 0.22-μm filter and store as 1 mL aliquots at -20 °C for up to 1 year.
6. TPP TubeSpin bioreactor tube (50 mL) (*see* **Note 4**).

2.2 Rapid Surface Shaving

1. 1× Phosphate buffered saline (PBS), pH 7.4.
2. 10 mM HEPES, pH 7.4: 0.238 g HEPES, 80 mL of ultrapure water. Adjust pH to 7.4 with NaOH and add ultrapure water to a final volume of 100 mL. Filter-sterilize with a 0.22-μm filter and store at 4 °C.
3. 1μg/μL trypsin: Reconstitute 20 μg of lyophilized trypsin (Sequencing Grade Modified) in 20 μL of ultrapure water. Store at −80 °C.
4. Charcoal blood agar (CBA): Dissolve 25 g of charcoal agar base in 450 mL of ultrapure water. Autoclave-sterilise and cool to 50 °C. Add 50 mL of sterile defibrinated horse blood and pour into sterile Petri dishes.
5. Formic acid, LC-MS grade.

2.3 Flow Cytometry

1. FACS buffer (1% BSA and 0.1% sodium azide in 1× PBS): Add 0.5 g BSA and 0.05 g sodium azide to 50 mL of 1× PBS. Filter-sterilise with a 0.22-μm filter and store at 4 °C (*see* **Note 5**).
2. 0.5 mg/mL propidium iodide (PI) (*see* **Note 6**).

2.4 C18 Clean Up and Liquid Chromatography–Tandem Mass Spectrometry (LC-MS/MS)

1. Pierce 200 μL C18 StageTips from Thermo Fisher to desalt and concentrate peptides for LC-MS/MS.
2. Centrifuge adapter from GL Sciences to hold the StageTips in place in 1.5 mL microcentrifuge tubes (Fig. 2).
3. 0.2% heptafluorobutyric acid (HFBA), HPLC grade.
4. 80% acetonitrile, 5% formic acid: 800 μL of acetonitrile (HPLC grade), 50 μL of formic acid (LC-MS grade), and 150 μL of ultrapure water.

3. 0.1% formic acid.

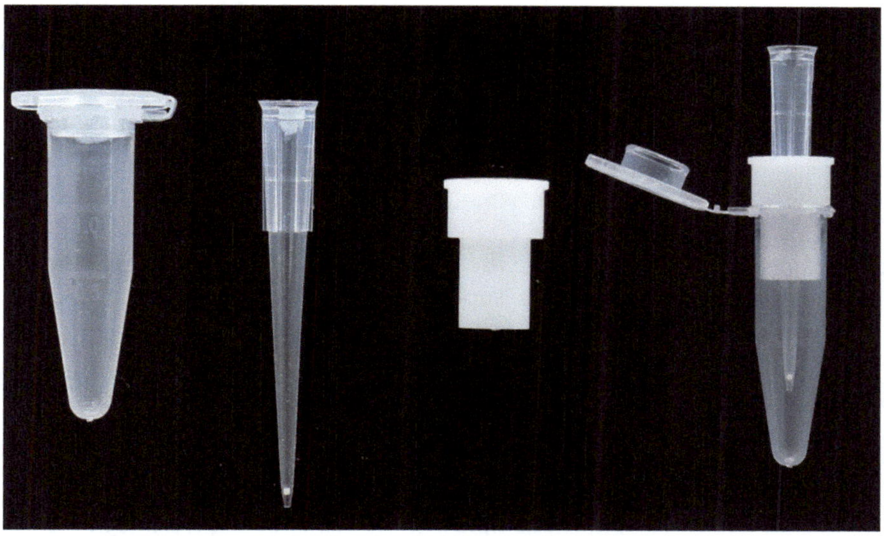

Fig. 2 Set up of 200 μL C18 StageTips in the centrifuge adapter and 1.5 mL microcentrifuge tube

2.5 Bioinformatic Analysis and Identification of Surface Proteins

1. Mascot Daemon (v.2.5.1)
2. Scaffold (v4.4.5)
3. R package (qvalue v2.24.0)
4. PSORTb v3.0 [8]
5. SignalP v5.0 [9]
6. TMHMM v2.0 [10]
7. Phobius [11]
8. Protter v1.0 [12]
9. KofamKOALA [13]
10. eggNOG-mapper v2 [14]
11. DAVID v6.8 [15]
12. IEDB [16]
13. DEG [17]

3 Methods

3.1 Preparation of* B. pertussis *Cells

1. Inoculate one loopful of Bvg+ *B. pertussis* grown on BG agar into a TPP TubeSpin bioreactor tube with 20 mL of THIJS medium supplemented with 1× THIJS supplement and 1x heptakis (*see* **Note 7**).
2. Incubate the culture for 24 h at 37 °C with shaking at 180 rpm.
3. Measure the OD_{600} with 1 mL of the 24 h culture.

4. Adjust the starting OD_{600} to 0.05 in a total of 20 mL of fresh THIJS medium (supplemented with 1× THIJS supplement and 1× heptakis) in a new TPP TubeSpin bioreactor tube.
5. Incubate the culture for 12 h at 37 °C with shaking at 180 rpm.
6. Optional: record the OD_{600} after 12 h incubation (*see* **Notes 8** and **9**).

3.2 Rapid Surface Shaving with Trypsin

1. Centrifuge the 12 h culture at 3300 × *g* for 15 min at 4 °C and discard the supernatant (*see* **Note 10**).
2. Gently wash cells with 1 mL of ice-cold 1× PBS and centrifuge at 3300 × *g* for 15 min at 4 °C. Discard supernatant and repeat wash step for a total of 3 washes (*see* **Note 11**).
3. Gently resuspend cells in 2 mL of 10 mM HEPES (*see* **Note 12**).
4. Split the sample into two by aliquoting 1 mL of the sample into a TPP TubeSpin bioreactor tube labeled "shaved" and another labeled as "unshaved control" (*see* **Notes 13** and **14**).
5. Add 1 μL of 1 μg/μL trypsin to the "shaved" sample (*see* **Note 15**).
6. Incubate both "shaved" and "unshaved control" samples for 5 min at 37 °C with gentle agitation (*see* **Notes 16** and **17**).
7. Transfer both samples into two new, appropriately labeled, 1.5 mL microcentrifuge tubes.
8. Optional: aliquot 10 μL of each sample for colony forming unit (CFU) counts (*see* **Note 18**).
9. Pellet cells in both samples by centrifugation at maximum speed (20,238 × *g*) for 1 min at 4 °C (*see* **Note 19**).
10. For both the "shaved" and "unshaved control" samples, transfer the respective supernatants into new, appropriately labeled, 1.5 mL microcentrifuge tubes. The cell pellets from each sample are kept on ice until processed for flow cytometry (*see* Subheading 3.3).
11. Add 1 μL of 1 μg/μL trypsin to the supernatant of the "unshaved control" sample and incubate for 5 min at 37 °C. Keep the "shaved" sample supernatant on ice to prevent further trypsin digestion while the "unshaved control" is undergoing trypsin digestion.
12. Add 1% of formic acid (10 μL) to both samples to inactivate trypsin digestion (*see* **Note 20**).

3.3 Determining Cell Viability After Rapid Surface Shaving Using Propidium Iodide (PI).

1. Prepare a positive control for cell death by resuspending 1 loopful of Bvg+ *B. pertussis* grown on BG agar in 1 mL of 1x PBS and heat at 70 °C for 30 min. Centrifuge at maximum speed (20,238 × *g*) for 1 min at 4 °C. Discard supernatant and resuspend cells in 900 μL of FACS buffer (*see* **Note 21**).

2. Prepare an unstained control by resuspending 1 loopful of Bvg+ *B. pertussis* grown on BG agar in 1 mL of 1× PBS.
3. Gently resuspend the "shaved" and "unshaved control" cells (from Subheading 3.2, **step 9**) in 1 mL ice-cold 1x PBS to wash. Centrifuge at maximum speed (20,238 × *g*) for 1 min at 4 °C. Discard supernatant. Repeat wash step once for a total of 2 washes.
4. Gently resuspend cells in 900 μL of FACS buffer.
5. Add 100 μL of PI (0.5 mg/mL) to the "shaved" and "unshaved control" cells, as well as to the "positive control for cell death" sample. Incubate on ice for 5 min in the dark.
6. Analyse cell viability by flow cytometry (*see* **Note 22**).

3.4 C18 Peptide Clean-Up and LC-MS/MS

1. Prepare the "shaved" and "unshaved control" supernatant samples (from Subheading 3.2, **step 12**) for C18 clean-up by mixing 50 μL of sample with 20 μL of 0.2% HFBA in appropriately labeled 1.5 mL microcentrifuge tubes (*see* **Note 23**).
2. For each sample, place a C18 StageTip into a centrifuge adapter and insert into an appropriately labeled 1.5 mL microcentrifuge tube (Fig. 2).
3. Initialize the C18 StageTips by adding 20 μL of 80% acetonitrile, 5% formic acid. Centrifuge at 200 × *g* for 2 min (*see* **Note 24**).
4. Reequilibrate the C18 StageTips with 20 μL of 0.2% HFBA and centrifuge at 200 × *g* for 2 min. Discard flowthrough.
5. Load each 70 μL sample (from Subheading 3.4, **step 1**) into a prepared C18 StageTip and centrifuge at 200 × *g* for 2 min (*see* **Notes 25** and **26**).
6. Wash the samples on the C18 StageTips with 20 μL of 0.2% HFBA and centrifuge at 200 × *g* for 2 min.
7. Transfer the C18 StageTips with bound sample, along with their centrifuge adaptors, to new appropriately labeled 1.5 mL microcentrifuge tubes and elute samples with 20 μL of 80% acetonitrile, 5% formic acid and centrifuge at 200 × *g* for 2 min.
8. Vacuum-dry samples (*see* **Note 27**).
9. Resuspend dried peptides in 10 μL of 0.1% formic acid and perform LC-MS/MS (*see* **Note 28**).

3.5 Bioinformatics Analysis and Identification of Surface Proteins

1. Load the resulting LC-MS/MS raw datafiles into Mascot Daemon (v.2.5.1) for protein identification with the following search parameters: instrument type = ESI-type; peptide tolerance = 4 ppm; MS/MS tolerance = 0.4 Da; variable modifications = Carbamidomethyl (C) and Oxidation (M); enzyme

specificity = trypsin; max number of missed cleavages = 1; and search database = custom *B. pertussis* database (*see* **Note 29**).

2. Load the Mascot output into Scaffold (v4.4.5), which implements the ProteinProphet algorithm to validate protein identifications. Choose "spectrum counting" as the quantitative technique. Biological replicates (and shaved or unshaved control samples) can be grouped together by providing the same sample category name (e.g., Sample category 1: *B. pertussis* shaved, Sample category 2: *B. pertussis* unshaved control). Set the threshold for identification in Scaffold as: protein threshold = 99%, peptide threshold = 95% and minimum number of peptides per protein = 1 (Fig. 3a) (*see* **Notes 30** and **31**).

3. Remove the peptides identified in the "unshaved control" dataset from matched false-positive predicted surface peptides in the "shaved" dataset by unticking the valid box of the matched peptide sequence in the Proteins tab (Fig. 3b) (*see* **Note 32**).

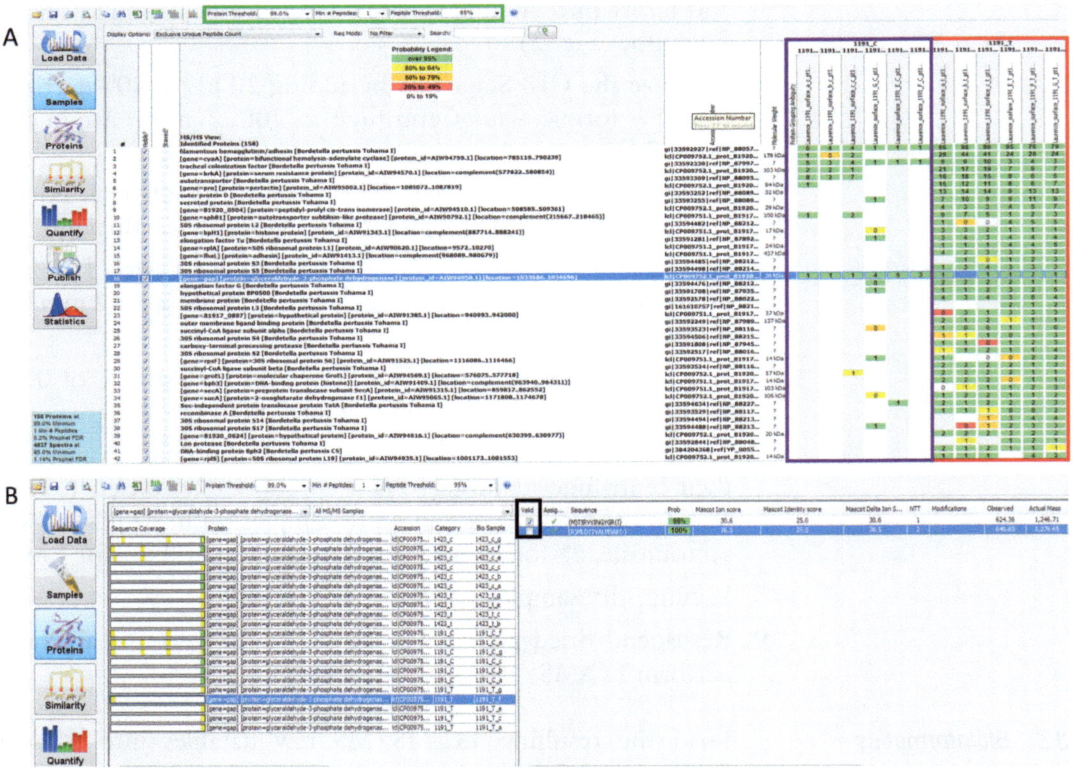

Fig. 3 (**a**) Example screenshot of the Samples tab in Scaffold illustrating the number of proteins and peptides identified in the unshaved control (1191_C in the purple box outline) and shaved samples (1191_T in the red box outline). The settings for protein threshold, peptide threshold, and minimum number of peptides per protein are shown in the green box outline. (**b**) The highlighted peptide (MLDTTVALMSAK) from the cytoplasmic protein glyceraldehyde-3-phosphate dehydrogenase was identified in the unshaved control sample and is removed from the shaved samples by unticking the "valid" box (black box outline) in the Proteins tab

4. Perform relative quantitation analysis using normalized total spectral counts in Scaffold's quantitative analysis setup page. Normalized total spectral counts can be used to quantify and identify significant surface protein expression changes between two strains or conditions. Select the "shaved" samples as input with the following settings: statistical test = *T*-test; multiple test correction = no correction; significance level = $p < 0.05$; fold change = fold change by category; use normalization = yes; minimum value = 0.0; and quantitative method = Total spectra. In the Export menu bar, click "Samples Report" to export the samples report to excel for further analysis.
5. Perform multiple test correction using the Storey–Tibshirani method in R by importing Scaffold p values and calculating the q value using the qvalue R package (*see* **Notes 33** and **34**).
6. Calculate the normalized spectral abundance factor (NSAF) for each surface protein to identify the most abundant surface proteins (*see* **Note 35**).

$$\text{Normalised spectral abundance factor (NSAF)} = \frac{\text{Normalised total spectral counts}}{\text{Amino acid length of protein}}.$$

7. Other web-based bioinformatic analyses that can be performed include: PSORTb v3.0 to predict the protein cellular location; SignalP v5.0 to identify proteins with signal peptides for translocation across membranes; TMHMM v2.0 to identify transmembrane helices; Phobius to predict transmembrane topology; and Protter v1.0 to visualize the topology of shaved peptides in surface proteins (Fig. 4). Functional category annotation of surface proteins can be performed with tools such as KofamKOALA, eggNOG-mapper v2 and/or DAVID v6.8. IEDB can be used to identify known and predicted immune epitopes of surface proteins/epitopes identified through rapid surface shaving. DEG can be used to identify which surface antigens are essential for "survival".

4 Notes

1. *B. pertussis* is a human respiratory pathogen. To limit the risk of infection, all work involving live cells should be performed in a class II biological safety cabinet and all personnel working with *B. pertussis* should be immunized against whooping cough.
2. *B. pertussis* is grown on BG agar at 37 °C for 4 days. Ensure that all *B. pertussis* colonies on the plate are in the Bvg+ phase when virulence factors are expressed. Bvg+ colonies are small (1 mm) and hemolytic, while Bvg- colonies are larger and nonhemolytic.

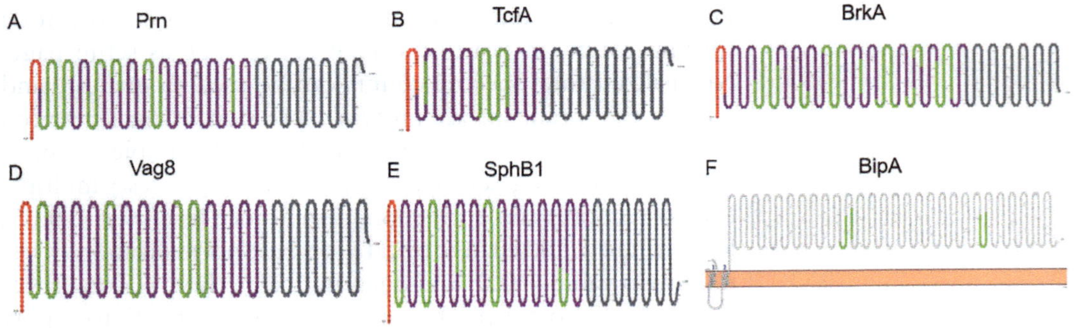

Fig. 4 Topology of peptides identified by rapid surface shaving (in green) were mapped to their proteins using Protter v1.0. (**a-e**) display 5 well-characterized *B. pertussis* autotransporter proteins that are either current antigens found in the pertussis acellular vaccine, or proposed to be good candidates as future vaccine antigens (A. Pertactin (Prn); B. Tracheal colonization factor A (TcfA); C. *Bordetella* resistance to killing A (BrkA); D. Virulence associated gene 8 (Vag8); and E. Subtilisin (SphB1). For A-E, N-signal peptides are shown in red, autotransporter passenger domains, which are known to be surface exposed, are colored purple while channel domains, which are embedded into the membrane, are shown in grey. In (**f**), the *Bordetella* Bvg-intermediate phase protein (BipA) is displayed. BipA is an outer membrane ligand binding protein and another proposed potential vaccine antigen. The shaved peptides (green) from BipA are mapped to the predicted surface exposed area from Phobius. The bacterial outer membrane is depicted as an orange bar and the numbers 1 and 2 are the predicted BipA transmembrane domains that anchor the protein to the outer membrane. (Reprinted from Vaccine, 38 (3), Luu LDW, Octavia S, Aitken C, Zhong L, Raftery MJ, Sintchenko V and Lan R, Surfaceome analysis of Australian epidemic *Bordetella pertussis* reveals potential vaccine antigens, 539–548, 2020, with permission from Elsevier)

3. THIJS is a chemically defined medium that was modified from the original Stainer-Scholte (SS) medium [18]. THIJS is designed for the optimal growth of *B. pertussis* and is prepared according to Thalen et al. [19]. Our study found that THIJS promotes the production of virulence factors in *B. pertussis* and therefore is the growth medium of choice for the liquid culture of *B. pertussis* [20]. However, SS medium may also be used.
4. *B. pertussis* is a strict aerobe and a respiratory pathogen, therefore 50 mL TPP TubeSpin bioreactor tubes, which have lids fitted with 0.22-μm filters, are used for liquid cultures. Other vessels can be used, provided that there is enough oxygen exchange and the required safety standards are met.
5. BSA is difficult to dissolve. To dissolve BSA in 1× PBS, lightly dust BSA over the surface of the liquid and allow it to sink and dissolve over time. Do not stir or vortex.
6. PI is a dye which binds DNA and cannot translocate across intact membranes. It can be used to determine cell viability and membrane integrity after rapid surface shaving. PI is resuspended in 1× PBS and stored in the dark at 4 °C.
7. 200 μL of THIJS supplement and 200 μL of heptakis are added to 20 mL of THIJS medium immediately prior to each use.

8. To obtain enough power, a minimum of 6 biological replicates for each strain/condition is needed if relative quantitation of surface proteins is performed for *B. pertussis.*
9. From our previous studies [21], the *B. pertussis* strains used are known to be in the early exponential growth phase after 12 h incubation in THIJS. The OD_{600} for *B. pertussis* clinical strains L1423 and L1191 should be 0.4–0.6 [21].
10. Rapid surface shaving **steps 1–4** should be completed as fast as possible (within 90 min) to minimize cell lysis during the sample preparation stage. After incubation, the cultures should always be kept on ice throughout the experiment (unless otherwise indicated) to minimize cell lysis.
11. Ensure cells are gently washed and resuspended throughout the experiment to minimize cell lysis from vigorous pipetting. It is also recommended to avoid pipetting liquid directly onto the cells.
12. 10 mM HEPES was found to minimize cell lysis during surface shaving by Walters et al. [3].
13. An "unshaved control" sample is established to identify false positive predictions of surface peptides, as described in Solis et al. [5]. The control is incubated under identical conditions as the "shaved" sample, except in the absence of trypsin. The purpose of the "unshaved control" is to identify proteins/peptides falsely predicted to be surface exposed (likely released due to secretion, cell lysis and/or membrane leakage during rapid surface shaving), and to remove them from the proteins/peptides identified in the shaved sample.
14. *B. pertussis* is a strict aerobe, therefore, to maintain oxygen exchange and minimize the chance of cell death, surface shaving is performed in the 50 mL TPP TubeSpin bioreactor tubes; however, other vessels may be used for other organisms (e.g., 1.5 mL microcentrifuge tubes).
15. We previously tested different trypsin concentrations (1 μg, 2.5 μg, and 5 μg) and digestion times (5, 10, and 15 min) to optimize surface shaving parameters for *B. pertussis* and obtain enough shaved peptides for identification, without compromising membrane integrity. We assessed cell/membrane integrity using CFU counts and flow cytometry with PI, as well as by determining the number of known outer membrane peptides/proteins identified in shaved samples and the number of cytoplasmic peptides/proteins identified in the unshaved control samples. Rapid surface shaving using 1 μg of trypsin for 5 min identified the highest proportion of known outer membrane proteins with the least amount of cell lysis in *B. pertussis* [6]. For other organisms or strains, digestion conditions may need to be optimized.

16. Samples can be incubated in a 37 °C water bath with gentle agitation at 60 rpm. Due to the short incubation period, a water bath is recommended over an incubator as the water ensures efficient heat transfer.
17. Longer incubation times may be used for other organisms, however they must be validated with CFU counts and flow cytometry (*see* Subheading 3.3) to ensure cell viability and membrane integrity are not compromised.
18. Ensure tubes are placed immediately on ice to prevent further trypsin digestion. CFU counts are performed in triplicate on charcoal blood agar (CBA) using the drop plate method [22], with incubation at 37 °C for 5–7 days, or until colonies appear.
19. Quick centrifugation at maximum speed for 1 min is used to pellet cells quickly and prevent further trypsin digestion.
20. Flow cytometry to assess cell viability should be performed first. Peptides in the "shaved" and "unshaved control" sample supernatants can be stored in -80 °C (following addition of 1% of formic acid) before proceeding to C18 clean-up and mass spectrometry. Ideally, C18 clean-up should be performed on the day of shaving or on the next day, as storage of peptides in solution should not be longer than a few days at −80 °C.
21. The positive control for cell death can be prepared up to several days in advance and stored at 4 °C.
22. Cell viability was analysed on the BD FACSCanto II using a blue laser 488 nm excitation and a 670 LP bandpass filter. Ten thousand events were acquired for each sample and data analysed using FlowJo software. Ensure that there is minimal cell lysis and disruption to the membrane as a result of rapid surface shaving; cell viability should be at least 98% (Fig. 5).
23. For C18 peptide clean-up, three 1.5 mL tubes will be required per "shaved" and "unshaved control" sample: one for the supernatant samples to be loaded into the C18 StageTip (*see* Subheading 3.4, **step 1**); one to initialize the C18 StageTips and for flowthrough collection after supernatant samples are loaded (*see* Subheading 3.4, **steps 2–6**); and one for the elution of samples (*see* Subheading 3.4, **step** 7). These can be organized and labeled ahead of time with simple labels such as "SL" and "UL" for load, "SFT" and "UFT" for flowthrough collection and "SE" and "UE" for elution, for the respective shaved (S) and unshaved (U) samples. If there are multiple strains and/or biological replicates being processed at one time, the strain name and biological replicate (indicated by a hyphen and number) can also be added to the labels, for example, "L1423–1 SL" and "L1191–1 SL".

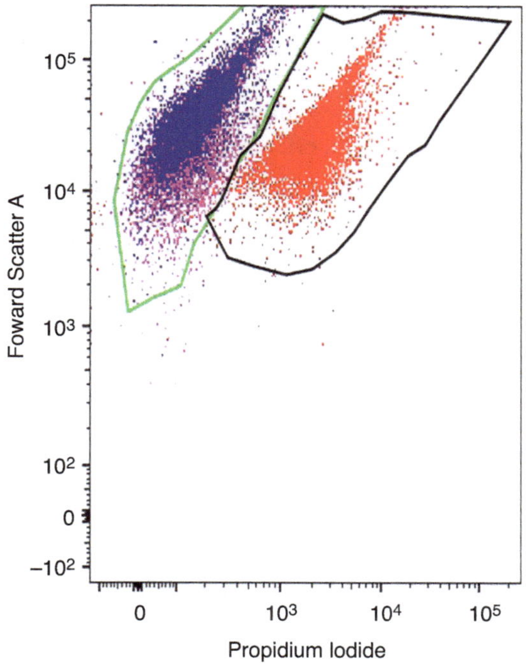

Fig. 5 Flow cytometry dot plots of propidium iodide stained cells indicating intact membrane integrity and minimal cell lysis from cells subjected to rapid surface shaving protocol (purple dots). Unshaved control cells (incubated in the absence of trypsin) are illustrated as pink dots, while the heated treated cells (positive control for cell death) are shown as red dots. (Reprinted from Vaccine, 38 (3), Luu LDW, Octavia S, Aitken C, Zhong L, Raftery MJ, Sintchenko V and Lan R, Surfaceome analysis of Australian epidemic *Bordetella pertussis* reveals potential vaccine antigens, 539–548, 2020, with permission from Elsevier)

24. For each centrifugation step, check that all liquid has flowed through the StageTip. If there is remaining liquid, increase centrifugal force by $100 \times g$ and centrifuge again until all liquid has passed.
25. Centrifuge samples at maximum speed for 1 min prior to loading onto the StageTips to remove precipitates which may block liquid from passing through.
26. Keep the flowthrough from the load and wash steps in case peptides have not bound and store it at −80 °C. If after LC-MS/MS of the C18-eluted samples (*see* Subheading 3.4, **step 9**), no/minimal peptides are detected especially in the "shaved" samples, this may indicate that the peptides did not bind to the C18 StageTips. If this is the case, the flowthrough can be reloaded into a new C18 StageTip that has been initialized and reequilibrated (*see* Subheading 3.4, **steps 3–5**) and cleaned up again.

27. Dried peptides can be stored at −80 °C before proceeding with mass spectrometry and should remain stable long-term (>6 months) if proper storage and handling conditions are followed [23]. However, we have not directly assessed the effect of long-term storage on results.

28. LC-MS/MS was performed using the LTQ-Orbitrap Velos with 0.1 μL of samples separated over a 30 min gradient, as detailed in Luu et al. [20]. Due to the low quantity of peptides released by rapid surface shaving, it is difficult to determine the concentration of proteins/peptides (e.g., with Qubit) loaded for LC-MS/MS.

29. Generic databases, such as NCBI or SWISS-PROT, can be searched for sequence matches. However, to reduce search time, a specific custom database can be created. In this case, we used a custom *B. pertussis* database containing amino acid FASTA files for Tohama I (NC_002929.2), CS (NC_017223.1), B1917 (CP009751.1) and B1920 (CP009752.1) strains.

30. The protein identification threshold for number of peptides identified per protein is set to 1, as rapid digestion may not result in a large number of peptides [5].

31. We typically obtain an average of ~250 unique peptide identifications in the shaved samples and an average ~20 peptides identified in the false-positive unshaved control for *B. pertussis* (Fig. 3a). If there is a large amount of cytoplasmic peptides/proteins identified in the unshaved control, it may be a sign of cell lysis/compromised membrane integrity.

32. Peptides corresponding to those identified in the "unshaved control," representing false-positive predictions of surface exposure, can be removed from the matched "shaved" sample, or from the entire dataset of "shaved" samples. We were more conservative and removed the predicted false-positive peptides for the whole dataset.

33. Multiple test correction can also be performed in Scaffold by setting multiple test correction as Benjamini–Hochberg (or other methods).

34. Upregulated surface proteins can be defined as those showing >1.2 fold change and downregulated surface proteins as those showing <0.8 fold change, with significance set at p value <0.05 and q value <0.05.

35. Normalized spectral abundance factor (NSAF) [24] normalizes spectral count to amino acid length to quantify relative protein abundance within a sample. Identification of highly abundant surface antigens as targets of potential vaccines is important since they may have a greater potential for immune recognition and antibody binding [25].

References

1. Molloy MP, Herbert BR, Slade MB, Rabilloud T, Nouwens AS, Williams KL, Gooley AA (2000) Proteomic analysis of the Escherichia coli outer membrane. Eur J Biochem 267(10):2871–2881
2. Rodríguez-Ortega MJ, Norais N, Bensi G, Liberatori S, Capo S, Mora M, Scarselli M, Doro F, Ferrari G, Garaguso I (2006) Characterization and identification of vaccine candidate proteins through analysis of the group A *Streptococcus* surface proteome. Nat Biotechnol 24(2):191–197
3. Walters MS, Mobley HL (2009) Identification of uropathogenic *Escherichia coli* surface proteins by shotgun proteomics. J Microbiol Methods 78(2):131–135
4. Olaya-Abril A, Jiménez-Munguía I, Gómez-Gascón L, Rodríguez-Ortega MJ (2014) Surfomics: shaving live organisms for a fast proteomic identification of surface proteins. J Proteome 97:164–176
5. Solis N, Larsen MR, Cordwell SJ (2010) Improved accuracy of cell surface shaving proteomics in *Staphylococcus aureus* using a false-positive control. Proteomics 10 (10):2037–2049
6. Luu LDW, Octavia S, Aitken C, Zhong L, Raftery MJ, Sintchenko V, Lan R (2019) Surfaceome analysis of Australian epidemic *Bordetella pertussis* reveals potential vaccine antigens. Vaccine 38(3):539–548
7. Bark SKN, Ahmad R, Dantzler K, Lukens AK, De Niz M, Szucs MJ, Jin X, Cotton J, Hoffmann D, Bric-Furlong E (2018) Quantitative proteomic profiling reveals novel *plasmodium falciparum* surface antigens and possible vaccine candidates. Mol Cell Proteomics 17 (1):43–60
8. Nancy YY, Wagner JR, Laird MR, Melli G, Rey S, Lo R, Dao P, Sahinalp SC, Ester M, Foster LJ (2010) PSORTb 3.0: improved protein subcellular localization prediction with refined localization subcategories and predictive capabilities for all prokaryotes. Bioinformatics 26(13):1608–1615
9. Armenteros JJA, Tsirigos KD, Sønderby CK, Petersen TN, Winther O, Brunak S, von Heijne G, Nielsen H (2019) SignalP 5.0 improves signal peptide predictions using deep neural networks. Nat Biotechnol 37 (4):420–423
10. Krogh A, Larsson B, Von Heijne G, Sonnhammer EL (2001) Predicting transmembrane protein topology with a hidden Markov model: application to complete genomes. J Mol Biol 305(3):567–580
11. Käll L, Krogh A, Sonnhammer EL (2004) A combined transmembrane topology and signal peptide prediction method. J Mol Biol 338 (5):1027–1036
12. Omasits U, Ahrens CH, Müller S, Wollscheid B (2013) Protter: interactive protein feature visualization and integration with experimental proteomic data. Bioinformatics 30 (6):884–886
13. Aramaki T, Blanc-Mathieu R, Endo H, Ohkubo K, Kanehisa M, Goto S, Ogata H (2020) KofamKOALA: KEGG ortholog assignment based on profile HMM and adaptive score threshold. Bioinformatics 36 (7):2251–2252
14. Huerta-Cepas J, Forslund K, Coelho LP, Szklarczyk D, Jensen LJ, Von Mering C, Bork P (2017) Fast genome-wide functional annotation through orthology assignment by eggNOG-mapper. Mol Biol Evol 34 (8):2115–2122
15. Sherman BT, Lempicki RA (2009) Systematic and integrative analysis of large gene lists using DAVID bioinformatics resources. Nat Protoc 4 (1):44
16. Vita R, Mahajan S, Overton JA, Dhanda SK, Martini S, Cantrell JR, Wheeler DK, Sette A, Peters B (2019) The immune epitope database (IEDB): 2018 update. Nucleic Acids Res 47 (D1):D339–D343
17. Luo H, Lin Y, Gao F, Zhang C-T, Zhang R (2014) DEG 10, an update of the database of essential genes that includes both protein-coding genes and noncoding genomic elements. Nucleic Acids Res 42(D1):D574–D580
18. Thalen M, van den IJssel J, Jiskoot W, Zomer B, Roholl P, de Gooijer C, Beuvery C, Tramper J (1999) Rational medium design for *Bordetella pertussis*: basic metabolism. J Biotechnol 75(2):147–159
19. Thalen M, van der Ark A, van den Ijssel J, van Straaten I, Jansen D, Beuvery C, Martens D, Tramper J (2008) Improving the cellular pertussis vaccine: increased potency and consistency. Vaccine 26(5):653–663. https://doi.org/10.1016/j.vaccine.2007.11.047
20. Luu LDW, Octavia S, Zhong L, Raftery M, Sintchenko V, Lan R (2017) Characterisation of the *Bordetella pertussis* secretome under different media. J Proteome 158:43–51
21. Luu LDW, Octavia S, Zhong L, Raftery MJ, Sintchenko V, Lan R (2018) Proteomic

adaptation of Australian epidemic *Bordetella pertussis*. Proteomics 18(8):1700237

22. Herigstad B, Hamilton M, Heersink J (2001) How to optimize the drop plate method for enumerating bacteria. J Microbiol Methods 44 (2):121–129
23. Hoofnagle AN, Whiteaker JR, Carr SA, Kuhn E, Liu T, Massoni SA, Thomas SN, Townsend RR, Zimmerman LJ, Boja E (2016) Recommendations for the generation, quantification, storage, and handling of peptides used for mass spectrometry–based assays. Clin Chem 62(1):48–69
24. Paoletti AC, Parmely TJ, Tomomori-Sato C, Sato S, Zhu D, Conaway RC, Conaway JW, Florens L, Washburn MP (2006) Quantitative proteomic analysis of distinct mammalian mediator complexes using normalized spectral abundance factors. Proc Natl Acad Sci U S A 103(50):18928–18933
25. Keasey S, Pugh C, Tikhonov A, Chen G, Schweitzer B, Nalca A, Ulrich RG (2010) Proteomic basis of the antibody response to monkeypox virus infection examined in cynomolgus macaques and a comparison to human smallpox vaccination. PLoS One 5 (12):e15547

Chapter 5

Two-Dimensional Electrophoresis Coupled with Western Blot as a Method to Detect Potential Neutralizing Antibody Targets from Gram-Negative Intracellular Bacteria

Milan Obradovic and Heather L. Wilson

Abstract

Antigen selection is a critical step in subunit vaccine design, especially if the goal is to identify antigens that can be bound by neutralizing antibodies to prevent invasion of cells by intracellular bacteria. Here, we describe a method involving two dimensional gel electrophoresis (2-DE) coupled with western blotting (WB) and mass spectrometry (MS) to identify bacterial proteins that: (1) interact with the host target cell proteins, and (2) are targeted by antibodies from sera from infected animals. Subsequent steps would be performed to validate that the bacteria are targeted by neutralizing antibodies to prevent invasion of the eukaryotic cells.

Key words *Lawsonia intracellularis*, Two-dimensional gel electrophoresis, Neutralizing antibodies, Antigen, Western blotting, Gram-negative bacteria

1 Introduction

2-DE coupled with an immunoblotting technique and MS is an efficient, robust, and proven methodology to detect antigens from bacteria, cancer cells, and fungi that are recognized by the human immune system [1–5]. Multiple bacterial species have been analyzed using immunoproteomics which led to important discoveries for *Helicobacter pylori*, uropathogenic *Escherichia coli*, *Staphylococcus aureus*, *Streptococcus thermophilus*, and other bacteria research [6–9]. Bacterial cells have lower complexity compared to mammalian cells which readily allow for successful protein separation and analysis using 2-DE and immunoproteomics [1]. Although 2-DE coupled with western blotting and MS has been used for decades, there are still limitations in the precise reproduction of gels and difficulties in separating proteins that are hydrophobic, very acidic or basic [3]. One of the solutions to mitigate these limitations is to lower the complexity of the sample by analyzing specific regions of

Fadil Bidmos et al. (eds.), *Bacterial Vaccines: Methods and Protocols*,
Methods in Molecular Biology, vol. 2414, https://doi.org/10.1007/978-1-0716-1900-1_5,

the bacterial or secreted bacterial products. We have developed a protein selection procedure that, coupled with 2-DE, WB, and MS, was able to identify proteins from intracellular bacteria recognized by antibodies from hyperimmune serum and to determine which of these immunogenic proteins bind to host cells [10].

Lawsonia intracellularis is an important pig pathogen that only causes disease when it invades intestinal epithelial cells. The identification of *L. intracellularis* proteins that bind to and/or interact with pig intestinal epithelial cell proteins will help identify neutralizing antibody targets that, once bound by hyperimmune sera antibodies, will prevent invasion. The proteins of *L. intracellularis* detected with this method were further validated experimentally by cloning the genes from proteins of interest to produce and purify recombinant proteins. These recombinant proteins were then used to generate protein-specific hyperimmune sera antibodies. We then assessed whether *L. intracellularis* coated with each antigen-specific hyperimmune sera failed to invade the intestinal pig epithelial cells (IPECs). These data helped us identify which antigens were targeted by neutralizing antibodies. These antigens were then used for subunit vaccine development that was validated using in vitro and in vivo studies [10, 11]. This immunoproteomic methodology was efficient and practical in the selection of immunogenic proteins from intracellular bacteria and their utilization as subunit vaccine antigens.

2 Materials

All reagents are analytical grade unless otherwise specified and solutions were made using deionized water. All reagents, solutions and gels were used at room temperature, unless specified otherwise. All steps are performed in 1.5 mL Eppendorf tubes unless specified otherwise.

2.1 Selection of L. intracellularis Proteins that Interact with Intestinal Pig Epithelial Cells (IPEC)

1. Cy5 dye.
2. Sodium bicarbonate buffer ($NaHCO_3$): 100 mM $NaHCO_3$ buffer consisting of 8.401 g in 1 L ddH_2O adjusted to pH 8.2.
3. 3 kilodalton (kDa) molecular weight cutoff (3 K MWCO) Amicon filters, 15 mL volume.
4. Bicinchoninic acid protein assay kit.
5. IPEC medium: Dulbecco's Modified Eagle Medium (DMEM)/F-12 with 5% Fetal bovine serum (FBS), insulin (10 μg/mL), transferrin (5.5 μg/mL), selenium (5 ng/mL) and 5 ng/mL of epidermal growth factor (EGF).
6. Radioimmunoprecipitation assay (RIPA) buffer (50 mM Tris pH 8, 150 mM sodium chloride, 0.1% SDS, 1% deoxycholic acid, 1% Nonidet P-40 substitute, distilled water), complete with 0.1 mM phenylmethylsulfonyl fluoride (PMSF) in isopropanol.

2.2 Rehydration of Immobilized pH Gradient (IPG) Strips

1. Immobilized pH gradient (IPG strip; Immobiline™ DryStrip, pH 4–7, 13 cm, GE Healthcare) (*see* **Note 1**).
2. Rehydration buffer, 10 mL volume: 9 M urea, 2% CHAPS (Fisher BioReagents), 1% Dithiothreitol (DTT), 2% Pharmalyte pH 5–8 (GE Healthcare) (*see* **Notes 2** and **3**), bromophenol blue.
3. Dry Strip Cover fluid (GE Healthcare).

2.3 Isoelectric Focusing (IEF) of Rehydrated Strips Using IPGphor Device

1. Dry Strip Cover fluid (GE Healthcare).
2. IPGphor device (GE Healthcare-Amersham Biosciences).
3. Ettan IPGphor rehydration strip apparatus (Pharmacia Biotech).

2.4 10% SDS PAGE Gel

1. Sodium dodecyl sulfate (SDS) equilibration buffer 2× 210 mL: 6 M urea, 75 mM Tris–HCl pH 8.8, 29.3% glycerol, 2% SDS, 0.002% bromophenol blue (can be stored at −20 °C in 20 mL aliquots).
2. DTT, 100 mg per 210 mL equilibration buffer (*see* **Note 4**).
3. 2.5% iodoacetamide, 250 mg per 210 mL equilibration buffer (*see* **Note 4**).
4. Resolving gel buffer (total 25.32 mL for two gels; *see* **Note 5**): 25.32 mL double-distilled water (ddH_2O), 21.36 mL 30% acrylamide mix, 16.04 mL 1.5 M Tris pH 8.8, 0.64 mL 10% SDS, 0.64 mL 10% ammonium persulphate (APS), and 25.6 μL tetramethylethylenediamine (TEMED).
5. Silver stain kit.

2.5 Semidry Transfer

1. Nitrocellulose membranes.
2. Immerse Blot Absorbent Filter paper.
3. Towbin transfer buffer: 25 mM Tris, 192 mM glycine, pH 8.3 with 20% methanol (v/v).
4. Plastic container.
5. Bio-Rad Trans-Blot SD semidry transfer cell.

2.6 Immunoblotting

1. Nitrocellulose membranes.
2. Plastic containers.
3. 10× Tris-buffered saline (TBS): 1.5 M NaCl, 100 mM Tris–HCl, pH 7.4.
4. TBS containing 0.05% Tween 20 (TBST).
5. Blocking solution: 5% milk in TBS.
6. Diluent solution: 5% milk in TBST.

7. Rabbit hyperimmune serum against *L. intracellularis* 1:500 in 5% milk in TBST (acquired from naïve rabbits before immunization [negative sera] and from rabbits immunized with whole inactivated bacteria) as primary antibodies.
8. Anti-rabbit IR 800 antibody (1 μg/mL).
9. Odyssey scanner (LI-COR®).

3 Methods

3.1 Selection of L. intracellularis Proteins that Interact with the IPEC Cells and Their Preparation for 2-DE

1. *L. intracellularis* propagation in McCoy cells, lysis, and protein extraction is out of the scope of this chapter and further information is available in previously published protocols [10, 12].
2. Add Cy5 dye to *L. intracellularis* proteins resuspended in $NaHCO_3$ buffer in dye/proteins molar ratio of 8:1. Incubate at room temperature for 4 h in the dark with gentle shaking on the nutator.
3. After the 4-h incubation, add Cy5-stained bacterial proteins into the 3 K MWCO 15 mL volume filters and centrifuge at 4700 × *g* for 15 min. Wash with ddH_2O three times using the same settings (*see* **Note 6**).
4. After the final wash, carefully resuspend the Cy5-stained proteins with ddH_2O, pipetting up/down from the walls of the filter chamber. Transfer resuspended Cy5-*L. intracellularis* proteins to a new Eppendorf tube.
5. Determine the concentration of Cy5-labeled proteins using a BCA protein assay kit.
6. Add 700 μg of Cy5-*L. intracellularis* proteins into the antibiotic- and FBS-free IPEC medium with 10^6 IPEC cells.
7. Incubate for 3 h with gentle shaking on a nutator at 4 °C in the dark.
8. After the 4-h incubation, centrifuge the mixture at 500 × *g* for 10 min to pellet the cells. Discard the supernatant (*see* **Note 7**).
9. Add RIPA buffer with PMSF to the cell pellet to lyse the IPEC cells. Freeze/thaw the cells using liquid nitrogen and a water bath at 37 °C.
10. Centrifuge 10,000 × *g* for 10 min and preserve the supernatant.
11. Add 4× volumes of ice-cold acetone to the supernatant, vortex, and store at −20 °C for 1 h.
12. Centrifuge at 14,000 × *g* for 10 min. Resuspend the pellet in $NaHCO_3$ buffer. Perform the BCA protein assay and store at −20 °C (*see* **Note 8**).

13. For analytical gels, add 250 μg IPEC + Cy-5 labeled *L. intracellularis* proteins to rehydration buffer up to the 250 μL volume. For MS preparatory gels, add 600 μg of *L. intracellularis* proteins in rehydration buffer up to the 250 μL volume.

3.2 Rehydration of Two 13 Cm pH 4–7 IPG Strips (See Note 9)

1. Pipet 250 μL of protein sample in rehydration buffer into one lane of the Ettan IPGphor rehydration strip apparatus. Slowly discharge the sample into the rack lane about the same length of the strip. Avoid creating bubbles (*see* **Note 10**).
2. Remove strip from a −20 °C freezer.
3. Carefully peel off the protective backing from the strip.
4. Gently place the strip (gel side down) onto the sample, being careful not to create bubbles and ensuring the entire strip is in contact with the sample.
5. Cover strip with 3–4 mL of Dry Strip Cover fluid.
6. Place the lid on a rack and leave on a level bench for at least 10 h or overnight, in the dark.

3.3 IEF of Rehydrated Strips Using IPGphor

1. Place strip in a strip holder, gel side up—positive end of gel at the pointed end of the holder and negative end all the way to the flat side.
2. Place 1–1.5 cm long paper strips (presoaked with water and blotted almost completely dry) on each end of the strip ensuring some overlap with the gel.
3. Place electrodes on the ends of a filter paper not in contact with the gel.
4. Completely cover with Dry Strip Cover fluid (around 4.5 mL).
5. Ensure that pointed ends of holders (with the positive end of the strip) are in the positive area of the apparatus.
6. Run focusing: 150 V step and hold for 3 h.
   300 V step and hold for 1200 Volt hours (Vh).
   1000 V gradient for 3900 Vh.
   8000 V gradient for 13,500 Vh.
   8000 V step and hold for 25,000 Vh.
7. After focusing, strips are placed in a tube, gel side up, and stored at −80 °C until ready for the next step (*see* **Note 11**).

3.4 10% SDS-PAGE Gel, Using Bio-Rad Protean II Xi Cell Apparatus

1. Run the second dimension on a medium size 10% SDS-PAGE gel, using Bio-Rad PROTEAN II xi cell apparatus connected to a circulating water bath to maintain temperature at around 10 °C.
2. Thaw the required number of strips at room temperature (RT) for 15–30 min.

3. For each strip, prepare 2× 10 mL portions of the SDS equilibration buffer in 15 mL tubes.
4. Label one tube as "DTT in equilibration buffer" and add in 10 mg DTT to 10 mL of equilibration buffer.
5. Label second tube as "Iodoacetamide in equilibration buffer" and add in 25 mg iodoacetamide to 10 mL equilibration buffer.
6. Add each strip to the tube labeled "DTT in equilibration buffer" and gently rock for 15 min at RT.
7. Remove strip and add to the tube labeled "Iodoacetamide in equilibration buffer" and gently rock for 15 min at RT.
8. Prepare the resolving buffer in a 100 mL glass flask. Add a magnetic stir bar and de-gas with a vacuum while stirring the buffer for 10 min, being careful not to aspirate the buffer. Add APS and TEMED, as described in Subheading 2.4 above, and mix while avoiding the creation of bubbles. Using a 50 mL glass pipet, slowly pour the buffer within a 20 cm × 21 cm × 1 mm gel glass cassette to cast the gel. Leave space for a stacking gel and gently overlay with water or 0.1% SDS solution to prevent excessive drying of the upper part of the gel.
9. Dip a strip briefly into SDS PAGE running buffer and apply to the well with the positive end farthest from the protein ladder well and with the gel facing the shorter glass plate. Gently push the strip to the bottom of the well with a comb, pipette tip or forceps.
10. Overlay with 1–1.5 mL of 0.5% agarose in running buffer for each strip using the minimum volume to submerge the strip. Allow the agarose 1 min to cool and solidify.
11. Add protein ladder to marker lane and run the gel at 110 V for 16 h. Alternatively, run the gel at 90 V overnight.
12. After electrophoresis, carefully separate the gel plates with the use of a plastic spatula. The gel should remain intact on one of the glass plates. Use gloves and avoid touching the center of the gel in order to prevent contamination. Wash the analytical gel with ddH_2O and transfer it carefully to a container with a western blot transfer buffer.
13. Transfer the preparative gel to another container for silver staining (*see* **Note 12**).
14. Perform silver staining of the preparative gel using a Silver stain kit (Sigma-Aldrich) and following the manufacturer's protocol. After the silver staining, the preparative gel should be covered with 1% acetic acid and stored at 4 °C, overnight.

3.5 Semidry Transfer

1. Cut a nitrocellulose membrane and absorbent filter paper to the size of the analytical gel and immerse in Towbin transfer buffer for 30 min at RT (*see* **Note 13**).
2. Place the presoaked filter paper at the middle of the plate surface of a semidry transfer cell.
3. Place the precut nitrocellulose membrane, from **step 1** above, on top of the filter paper and ensure that two surfaces match in size.
4. Carefully place the analytical gel above the nitrocellulose membrane to avoid creating bubbles or wrinkles.
5. Place the filter paper over the analytical gel. Use a glass tube or roller to gently roll over the upper filter paper and remove bubbles or excessive buffer. Close the lid of the machine.
6. Semidry transfer for this size of the gel is achieved using the Bio-Rad Trans-Blot SD semidry transfer cell at 15 V for 60 min (*see* **Note 14**).

3.6 Immunoblotting

1. Block the nitrocellulose membrane with 50 mL of 5% milk in TBS for 1 h.
2. Remove 5% milk in TBS and wash 3 times for 10 min with TBST.
3. Add 50 mL of a 1:500 dilution of hyperimmune rabbit serum in 5% milk with TBST.
4. Incubate with gentle shaking on a nutator overnight at 4 °C.
5. Decant milk (*see* **Note 15**) and wash 3 times for 10 min with TBST.
6. Add secondary antibody IR800 (1:10,000 dilution) to 50 mL of 5% milk in TBST and incubate with gentle shaking, for 30 min to 1 h, covered to protect from the light.
7. Following incubation, remove the secondary antibody and wash 3 times for 10 min with TBST.
8. Scan with LI-COR® Odyssey Scanner in IR700 and IR800 channels (*see* Fig. 1a).
9. Save the digital copy of the image for further analysis or print it on A4 paper (*see* **Note 16**).

3.7 Excising Gel Plugs for Mass Spectrometry

1. Excise gel with silver-stained dots which correspond to IR-800 labeled proteins detected by WB analysis using a sterile biopsy punch (3 mm diameter) in a biological cabinet to avoid contamination of gel samples with environmental proteins (Fig. 1b).
2. Collect gel plugs and store them in Eppendorf tube with 100 μL ultrapure water at −20 °C. Send gel plug samples for MS analysis (*see* **Notes 17** and **18**).

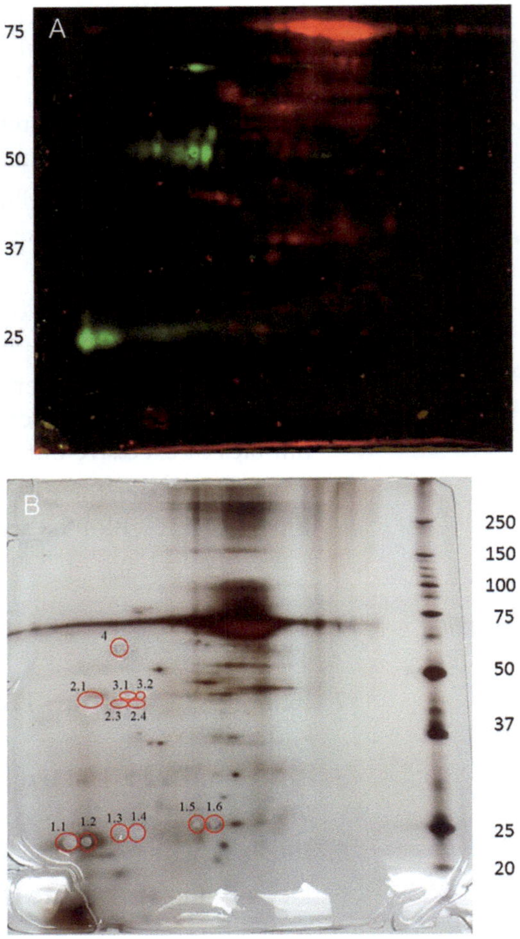

Fig. 1 Protein separation by 2-dimensional electrophoresis and selection of spots for mass spectrometry, cropped image. Cy5-labeled *Lawsonia intracellularis* proteins that bound to IPEC1 cells were subjected to isoelectric focusing using IPG strip 4–7 (horizontal plane) followed by SDS-PAGE using 10% SDS-PAGE gel (vertical plane). Molecular weight markers are indicated (kDa). (**a**) Proteins were transferred to a nitrocellulose membrane and incubated with hyperimmune rabbit serum as primary antibody and anti-rabbit IR800 secondary antibody. Proteins visible in IR700 channel are red and indicate all Cy5-labeled bacterial proteins. The proteins visible in IR800 channel are green and indicate proteins bound by rabbit antibodies from rabbits immunized with whole-cell *L. intracellularis.* Note, all green proteins are also red and therefore should appear yellow in colour but they are overwhelmed by the green fluorescence. (**b**) A replicate gel was stained with PROTSil-1 silver stain kit. Position and numbering of gel spots are indicated by red circles. Gel plug samples 1.4, 2.3, 3.1, 3.2, and 4 were submitted for mass spectrometry. (Reproduced from [10] with permission from Elsevier)

4 Notes

1. The adequate pH and type of the IPG strip should be evaluated experimentally before the main experiment.
2. If the protocol requires IPG strips pH 3–10, then Pharmalyte pH 3–7 should be used.
3. DTT and Pharmalyte should be added last, just before mixing with the protein sample. Store DTT at −20 °C and Pharmalyte at 4 °C.
4. Equilibration buffer with DTT or iodoacetamide is aliquoted in 20 mL volume, in 50 mL conical tubes, and stored at −20 °C. Use of aliquots are suggested to avoid frequent freeze-thaw cycles, which could damage the properties of the buffer.
5. We ran two 10% SDS PAGE gels (analytical and preparatory) at the same time to have consistent results.
6. Washing of Cy5-labeled proteins is important to remove the unbound Cy5 dye from the buffer. The remaining Cy5-labeled proteins are visible as a blue pellet at the bottom of the filter chamber. The size of the filter is based on the size of the proteins that you want to extract and analyze downstream.
7. The speed and the time of centrifugation depend on the type of cells and should be experimentally standardized. Targeted Cy5-labeled *L. intracellularis* proteins bind to the IPEC cells and the unbound proteins are left in the supernatant.
8. To further concentrate the protein sample use 3 kDa centrifugal filters (14,000 × *g* for 15 min).
9. Strip 1 is for the analytical gel and strip 2 is for the preparatory gel.
10. For gels, use 250 μg proteins for analytical gel and 600 μg proteins for preparatory gel in rehydration solution, total 250 μL volume, for each per IPG strip.
11. The strip can be imaged after the IEF to confirm that IEF worked well as evidenced by sharp bands. We used the Odyssey scanner (Li-COR) with IR700 channel since Cy5 is fluorescent at 700 nm.
12. The preparatory gel was left on one of the glass plates during silver staining and storage to ensure that the gel did not break and to allow easier cutting of gel plugs with a biopsy puncher.
13. Readers can perform the semidry transfer using their own lab kits.

14. Shorter time could be used but with higher voltage such as 25 V for 30 min. The time and voltage should be established based on the size of the gel and semidry apparatus protocol.
15. From our experience, hyperimmune rabbit serum in 5% milk in TBST could be used again up to 4 times if stored at −20 °C.
16. There are multiple image software and projectors available for processing of immunoblot image and correlation of protein spots to the preparative gel. This is useful if larger and complex proteome is analyzed. Although we did not have this equipment, we performed selection and Cy5 staining of proteins (*see* Subheading 3.1), thus reducing the number of proteins on the membrane and increasing their detection. We were able to compare the printed image of the immunoblot to the silver stained gel and select gel plugs for MS with great accuracy.
17. Gel plug samples (annotated as 1.4, 2.3, 3.1, 3.2, and 4 (Fig. 1b)) were sent to Plateforme Protéomique Centre de Recherche du CHU de Québec CHUL, Québec, Canada for MS analysis.

Acknowledgments

This chapter is published with permission of the director of VIDO as journal series #916.

References

1. Rabilloud T, Strub JM et al (2001) A comparison between Sypro ruby and ruthenium II tris (bathophenanthroline disulfonate) as fluorescent stains for protein detection in gels. Proteomics 1(5):699–704
2. Lahner E, Bernardini G et al (2011) Immunoproteomics of *Helicobacter pylori* infection in patients with atrophic body gastritis, a predisposing condition for gastric cancer. Int J Med Microbiol 301(2):125–132
3. Havlasova J, Hernychova L et al (2005) Proteomic analysis of anti-*Francisella tularensis* LVS antibody response in murine model of tularemia. Proteomics 5(8):2090–2103
4. Kellner R, Lichtenfels R et al (2002) Targeting of tumor associated antigens in renal cell carcinoma using proteome-based analysis and their clinical significance. Proteomics 2 (12):1743–1751
5. Pitarch A, Pardo M et al (1999) Two-dimensional gel electrophoresis as analytical tool for identifying *Candida albicans* immunogenic proteins. Electrophoresis 20 (4–5):1001–1010
6. Haas G, Karaali G et al (2002) Immunoproteomics of *Helicobacter pylori* infection and relation to gastric disease. Proteomics 2 (3):313–324
7. Hagan EC, Mobley HLT (2007) Uropathogenic *Escherichia coli* outer membrane antigens expressed during urinary tract infection. Infect Immun 75(8):3941–3949
8. Enany S, Yoshida Y et al (2013) Two dimensional electrophoresis of the exo-proteome produced from community acquired methicillin resistant *Staphylococcus aureus* belonging to clonal complex 80. Microbiol Res 168 (8):504–511
9. Salzano AM, Novi G et al (2013) Monodimensional blue native-PAGE and bi-dimensional blue native/urea-PAGE or/SDS-PAGE combined with nLC-ESI-LIT-MS/MS unveil membrane protein heteromeric

and homomeric complexes in *Streptococcus thermophilus*. J Proteome 94:240–261

10. Obradovic M, Pasternak JA et al (2019) Immunoproteomic analysis of *Lawsonia intracellularis* identifies candidate neutralizing antibody targets for use in subunit vaccine development. Vet Microbiol 235:270–279

11. Obradovic MR, Wilson HL (2020) Immune response and protection against *Lawsonia intracellularis* infections in pigs. Vet Immunol Immunopathol 219:109959

12. Lawson GH, McOrist S et al (1993) Intracellular bacteria of porcine proliferative enteropathy: cultivation and maintenance in vitro. J Clin Microbiol 31(5):1136–1142

Chapter 6

Panproteome Analysis of the Human Antibody Response to Bacterial Vaccines and Challenge

Joseph J. Campo and Amit Oberai

Abstract

High-density protein microarray is an established technology for characterizing host antibody profiles against entire pathogen proteomes. As one of the highest throughput technologies for antigen discovery, proteome microarrays are a translational research tool for identification of vaccine candidates and biomarkers of susceptibility or protection from microbial challenge. The application has been expanded in recent years due to increased availability of bacterial genomic sequences for a broader range of species and strain diversity. Panproteome microarrays now allow for fine characterization of antibody specificity and cross-reactivity that may be relevant to vaccine design and biomarker discovery, as well as a fuller understanding of factors underlying themes of bacterial evolution and host–pathogen interactions. In this chapter, we present a workflow for design of panproteome microarrays and demonstrate statistical analysis of panproteomic human antibody responses to bacterial vaccination and challenge. Focus is particularly drawn to the bioinformatics and statistical tools and providing nontrivial, real examples that may help foster hypotheses and rational design of panproteomic studies.

Key words Protein microarray, Panproteome, Antigenic diversity, Antibody profiling, Differential analysis

1 Introduction

Advances in proteomics tools have been drivers of vaccine antigen discovery. High throughput gene cloning and cell-free in vitro transcription and translations ("IVTT") of hundreds to thousands of proteins at a time allows proteome-wide analysis of specific antibody binding in a single assay. Expressed proteins are immobilized as spots in a 2D grid format on a planar surface, such as nitrocellulose-coated glass microscope slides. These protein arrays are typically fabricated using robotic microarray printers utilizing contact print pins or contactless droplet jet printers. Antibody-containing biological specimens can be probed over the printed arrays, and antibody binding to specific, known proteins can be revealed using fluorescent or colorimetric secondary antibody

Fadil Bidmos et al. (eds.), *Bacterial Vaccines: Methods and Protocols*,
Methods in Molecular Biology, vol. 2414, https://doi.org/10.1007/978-1-0716-1900-1_6,

detection systems. Proteome microarray is now an established technology. For specific details about the laboratory protocols, we refer the readers to a comprehensive chapter in a previous volume of this series by Driguez et al. for schistosome protein microarrays [1]. The same principles described by Driguez et al. apply for bacterial proteomes, as well, and we follow a standard development pipeline for protein microarray development and antibody profiling (*see* Fig. 1).

Genomic sequence information is readily accessible for numerous pathogens, often for multiple laboratory strains and clinical isolates encompassing broad species and strain coverage of bacterial populations. This advance in microbial genomics is driven by improvements in whole genome sequencing and reduction in costs of sequencing a bacterial genome [2]. Characterization of bacterial genomic diversity allows for the definition of a pangenome and, similarly, the protein-coded features that constitute a panproteome. However, the definition of a panproteome with regards to microarray development varies depending on the scientific questions addressed and availability of genomic sequence data. At the species level, a panproteome may encompass all genes that are common between species of interest, as well as the genes that are unique to each species. Within species, at the strain level, a panproteome may encompass genes encoding the common "core proteome," those displaying allelic diversity and unique genes in the accessory genome. And at the individual protein level, focus can be drawn to the repertoire of protein fragments or epitopes representing genetic polymorphisms. Broadly defined, the panproteome used for microarray development comprises protein sequences providing enough coverage of genomic diversity to address the scientific questions.

Using whole genome sequencing data on *Streptococcus pneumoniae* clinical isolates from pediatric nasopharyngeal swabs [3], we developed our first panproteome microarray for profiling pneumococcal antibodies in participants of a trial of a pneumococcal whole cell vaccine [4, 5]. The array contents included proteins present in nearly all isolates sequenced, constituting the core pneumococcal proteome, as well as genes present in at least 20% of isolates. Also included in the array was an expanded repertoire of variants of zinc metalloproteases ZmpA and ZmpB and pneumococcal surface proteins A (PspA) and C (PspC), constituting the "diverse core loci" (Fig. 2). This approach has been applied for development of panproteome microarrays for other pathogens, such as *Brucella* spp. and *Treponema pallidum* subsp. *pallidum* strains. It has also highlighted the complexities of targeting highly diverse bacteria such as *Clostridium* spp. commonly found as

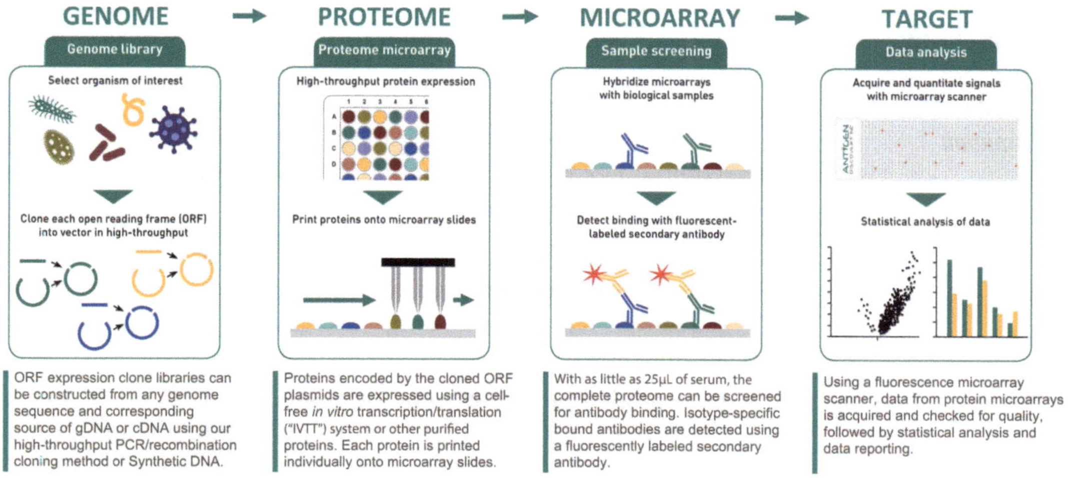

Fig. 1 Proteome microarray development and antibody profiling workflow (reproduced from https:/antigendiscovery.com/adi-proteome-microarray-technology/ with permission from Antigen Discovery, Inc.)

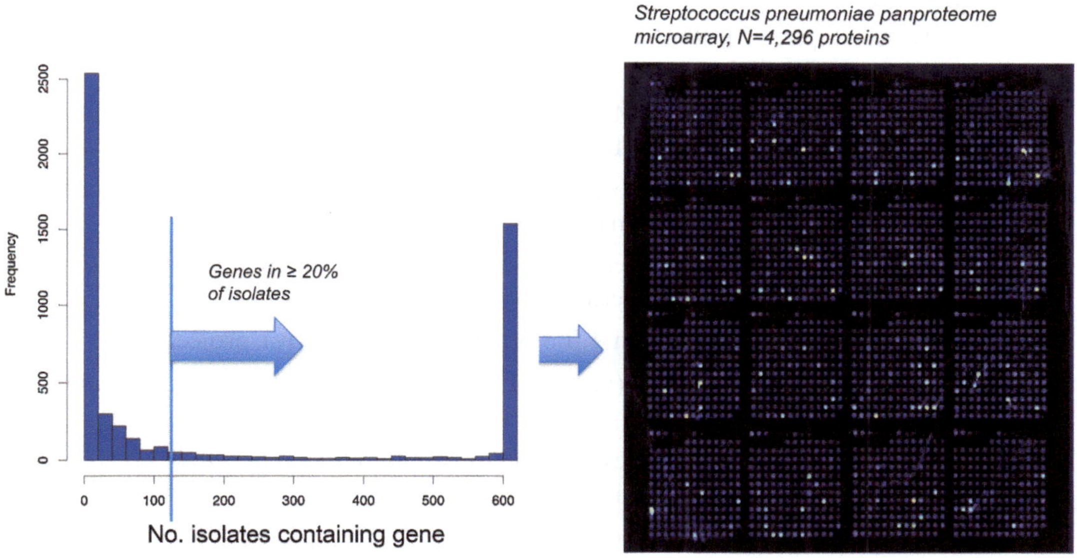

Fig. 2 Selection of *S. pneumoniae* genes for inclusion in panproteome microarray. Among 616 sequenced pediatric clinical isolates, genes present in at least 20% of the bacterial population were included in an array containing 4296 proteins

commensal bacteria in the human gut microbiome. This chapter describes the process of developing panproteome microarrays from concept to protein selection and the specific bioinformatics analyses. Wet lab protocols have been published previously [1] and will not be detailed herein, but statistical methods pertinent to panproteome-wide analysis are presented.

2 Materials

2.1 Genomic Information

A prerequisite for panproteome microarray development is access to whole genome sequencing information; at least, two genomes are required (*see* **Note 1**). The source of data can be public databases or privately curated datasets.

1. Genbank (GCA) is an annotated collection of all publicly available nucleotide and protein sequences derived from public sequence repositories and, thereby containing potential redundancies.
2. Refseq (GCF) provides a complete set of nonredundant, extensively cross-referenced and annotated nucleotide and protein sequences for each species and strain, with assembly records maintained by the NCBI.

2.2 Data Analysis Software (See Note 2 for Descriptions)

1. BLAST+ (version 2.11.0 or later) https://ftp.ncbi.nlm.nih.gov/blast/executables/blast+/LATEST/
2. PSORTb (version 3.0.2 or later) [6] https://www.psort.org/psortb/
3. TMHMM (version 2.0 or later) [7] http://www.cbs.dtu.dk/services/TMHMM/
4. SignalP (version 5.0 or later) http://www.cbs.dtu.dk/services/SignalP/
5. NCBI web CD-Search tool [8] https://www.ncbi.nlm.nih.gov/Structure/bwrpsb/bwrpsb.cgi
6. R (version 4.0.3 or later) https://www.r-project.org/
7. R studio (version 1.3 or later) https://www.rstudio.com/ (*see* **Note 3**).
8. R packages for analysis of differential reactivity.
   (a) limma (*see* **Note 4**) [9].
   (b) Lme4 (*see* **Note 5**).
   (c) ROCR.
   (d) Rtsne (*see* **Note 6**) [10].
9. R packages for data visualization.
   (a) ggplot2 (https://ggplot2.tidyverse.org/),
   (b) gplots,
   (c) ComplexHeatmap https://jokergoo.github.io/ComplexHeatmap-reference/book/

3 Methods

3.1 Select Pathogen

Identify an organism or set of organisms from which to perform genome-wide homology analysis prior to designing the contents of a panproteome microarray. Base the selection of organisms on the research questions, but balance factors such as limitation of array space (i.e., number of proteins), and quality and source of genomic information. *See* **Note 7** for considerations in selecting organisms.

3.2 Identify Conserved and Diverse Proteins

The goal is to capture species or strain diversity by identifying the panproteome, or the highly diverse proteins, for example those under immune pressure. The steps involved in capturing the panproteome are as follows, using an example of 3 selected species:

1. Download whole proteome sequences for the group of species or strains of the pathogens of interest that will constitute the panproteome microarray. The fasta format files "cds_from_genomic.fna" and "translated_cds.faa" (nucleotide and protein files, respectively) are downloaded from NCBI Genbank or Refseq, or they may be obtained from private curated whole genome sequencing data (*see* **Note 8**).
2. Run BLASTp for all vs. all sequence homology comparison for the selected species or strains. The stand-alone BLAST is run in the windows command prompt (cmd) application. The first step in running BLASTp is to format each of the proteome fasta files for BLAST using the "makeblastdb" command. For comparing 3 species A, B, and C, with proteome sequence files A_translated_cds.faa, B_translated_cds.faa, and C_translated_cds.faa, the makeblast db command would be as follows:

```
makeblastdb -in A_translated_cds.faa -dbtype prot
makeblastdb -in B_translated_cds.faa -dbtype prot
makeblastdb -in C_translated_cds.faa -dbtype prot
```

3. In BLASTp, align all sequences in each species to all other species in both directions, that is,
   - A is compared to itself, B and C;
   - B is compared to itself, A and C; and.
   - C is compared to itself, A and B.

   There will therefore be nine output alignment files generated for a three species comparison:
   - with respect to A: AA, AB, AC,
   - with respect to B: BB, BA, BC,
   - with respect to C: CC, CA, CB,

The BLASTp command used for one of these comparisons, for example A vs B (the entire command should be typed in one continuous line):

```
"C:\Program Files\NCBI\blast-2.11.0+\bin\blastp" -query
A_translated_cds.faa -subject B_translated_cds.faa -outfmt "6
qseqid sseqid pident qstart qend qlen sstart send slen evalue"
-max_hsps 1 -max_target_seqs 1 -evalue 1e-10 > AB.out
```

In the BLASTp command line, `-outfmt 6` refers to tabular alignment output format. The parameters chosen here for output are:

- `qseqid` (query sequence ID),
- `sseqid` (subject sequence ID),
- `pident` (percent identical matches in the aligned region),
- `qstart` (query protein start position in the alignment),
- `qend` (query protein end position in the alignment),
- `qlen` (length of the query protein sequence),
- `sstart` (subject protein start position in the alignment),
- `send` (subject protein end position in the alignment),
- `slen` (length of the subject protein sequence),
- `max_hsps 1` (show only the best high scoring pair or alignment for a query-subject pair),
- `max_target_seqs 1` (show only one aligned query-subject pair),
- `e-value` (a statistical score for the alignment equal to the number of expected hits of similar quality (score) that could be found just by chance—when *e*-value<0.01 it is almost identical to *p*-value), and,
- `AB.out` (is the name of the output alignment file in tab delimited format).

4. Open the R program and find the working drive by typing the `getwd()` command into the command line, or set the working drive using the `setwd()` command.
5. Move the homology search output `A_vs_B.out` into the R working drive and import the data into R using the `read.table()` command as follows:

```
> blastAB.df <- read.table("AB.out", sep = "\t", stringsAs-
Factors = FALSE, check.names = FALSE)
```

6. Calculate an overall percent identity for each alignment as:

```
> overall_pid_query <- blastAB.df$pident * (blastAB.df$qend - blastAB.df$qstart + 1) / blastAB.df$qlen
> overall_pid_subject <- blastAB.df$pident * (blastAB.df$send - blastAB.df$sstart + 1) / blastAB.df$slen
```

7. Group the sequences that are common to all species as follows. A threshold percent identity value is set (typically 80% or 90%, *see* **Note 9**) using the overall percent identity of aligned proteins returned by BLASTp, and the aligned protein pair is required to have identity above that threshold value for consideration to be added to the group.

   The hits are first sorted into categories or groups as:

   – For Species A: **A**-B-C, else **A**-B, else **A**-C else **A**-A.
   – For Species B: A-**B**-C, else A-**B**, else **B**-C, else **B**-B.
   – For Species C: A-B-**C**, else A-**C**, else B-**C**, else **C**-C.

   Note that the red bold text (above) is intended to highlight the reference proteome with respect to which comparison is being done. For example, for species A, **A**-B-C is a protein that has a blast hit between species A and B, and between species A and C, above the required threshold percent identity. If not, then the protein is sorted at the next level **A**-B which means the protein has a blast hit between A and B, and if not then at the next level **A**-C for a blast hit between A and C, and finally **A**-A. **A**-A is a singleton which did not have any hits above the required threshold with species B or C. The same is done for species B and for species C. In this way, all hits for the 3 species are sorted into one of these shown categories. Redundancy of repeating a protein between different category comparisons is avoided by removing the protein from subsequent consideration for grouping once it has been counted into one of the groups or categories. Note that the complexity of groupings increases factorially with number of species selected. An example of setting the threshold identity for grouping in R is coded as follows:

```
> perc <- 80 #threshold for homology
> blastAB_grp.df <- blastAB.df[overall_pid_query >= perc & overall_pid_subject >= perc, ] #80%ID threshold over full length of query sequence and subject sequence
> blastABq.vec <- blastAB_grp.df[, 1] #AB is the query IDs for hits with species 2
> blastABs.vec <- blastAB_grp.df[, 2] #sseqid ABs is the subject ID's
```

3.3 Classify Proteins in the Core and Panproteome

Classify proteins by intersections between groups, here using the 3 species example homology analysis in Subheading 3.2.

1. The proteins that intersect in groups **A**-B-C, A-**B**-C and A-B-**C** are the protein sequences that are common to all species. This group of proteins forms the core proteome. Define vectors `group_abc.vec` as the vector containing proteins sequence IDs from proteins in group **A**-B-C, `group_bac.vec` as the vector containing protein sequence IDs from proteins in group A-**B**-C, and `group_cab.vec` as vector containing the protein sequence IDs from proteins in group A-B-**C**, as follows:

```
> group_abc.vec <- c()
> for(idx in 1:length(blastAA.vec)) {
 if ((blastAA.vec[idx] %in% blastAB.vec) & (blastAB.vec[idx] %
in% blastAC.vec)) { ##all 3 overlap 123
 idx2 <- grep(blastAA.vec[idx], blastABq.vec)
 idx3 <- grep(blastAA.vec[idx], blastACq.vec)
 x3a <- paste(blastAA.vec[idx], blastABs.vec[idx2], blastACs.
vec[idx3], sep = "~") ##all 3 sequence ID's in the hit with
species 1
 group_abc.vec <- c(group_abc.vec, x3a) }}
```

2. Calculate the intersections of all group vectors, as follows:

```
> intersect_all.vec <- intersect(intersect(group_abc.vec,
group_bac.vec), group_cab.vec)
```

3. For proteins not common to all species, the largest group of species to which they are common is identified. This process of identifying the largest group of species to which a protein is found in common continues as proteins common to groups of 2 species are defined. If we define vectors `group_ab.vec` as the vector containing protein sequence IDs from proteins in group **A**-B, `group_ba.vec` as the as the vector containing protein sequence IDs from proteins in group A-**B**, `group_ac.vec` as the vector containing protein sequence IDs from proteins in group **A**-C, `group_ca.vec` as the vector containing protein sequence IDs from proteins in group A-**C**, `group_bc.vec` as the vector containing protein sequence IDs from proteins in group **B**-C, and `group_cb.vec` as the vector containing protein sequence IDs from proteins in group B-**C**, then:

```
> intersect_ab.vec <- intersect(group_ab.vec, group_ba.vec)
> intersect_ac.vec <- intersect(group_ac.vec, group_ca.vec)
> intersect_bc.vec <- intersect(group_bc.vec, group_cb.vec)
```

4. The remaining proteins are singletons, or those that are not common to any other species. These groups of proteins that are outside the core proteome form the panproteome. Tabulate the hits for each intersection. For example, during our development of a panproteome microarray for *Brucella* spp., five organisms were selected and assessed for core and panproteomes, with the results tabulated in Table 1 (*see* **Note 7**).
5. For larger selections of organisms, for example capturing strain diversity among numerous clinical isolates, the sequences of proteins encoded by all unique clusters of orthologous genes (COGs) in the sequenced bacterial population are queried against each isolate genome, and the number of genome hits above the percent identity threshold for each protein is tabulated. COGs that are to be included in the panproteome microarray are filtered by prevalence in the bacterial population. For example, a threshold of 20% was used to gate for genes included in our *S. pneumoniae* panproteome microarray (Fig. 2), that is, all COGs present in at least 20% of pediatric clinical isolates were included in the array design (*see* **Note 10**) [4]. Additionally, highly diverse proteins ZmpA, ZmpB, PspA, and PspC present in all isolates were assessed for divergence in order to select an expanded repertoire of these "diverse core loci" (Fig. 3).

While the intended goal is to capture as much species and strain diversity as possible on the panproteome microarray, a major consideration is the number of available spots on the array. In the case of close homology between the included species, the core proteome may define the bulk of the proteins involved, as in the case of the Nichols and SS14 strains of syphilis (*Treponema pallidum* subsp. *pallidum*, Table 2). In this case, the diversity (or the pan-proteome) will be small, and capturing the diversity on the array will be easier. However, in the case where there is low homology between species leading to failure to identify a core proteome, a large fraction of the proteomes will constitute the panproteome. Consequently, it may not be possible to be as inclusive of the diversity due to the limitation of the number of spots on the array. An example of such a failure is shown in our analysis of the pathogen and gut commensal *Clostridium* spp. (Table 3), where the pairwise distribution of shared genes greater than a permissive 70% sequence identity threshold was low between species (Fig. 4).

Table 1

***Brucella* species homology analysis for classification of the core proteome and panproteome**

No. of Strains Shared (Refseq)	*B. melitensis* (M16) N=3,090 genes	*B. canis* (RM6/66) + 21 strains (97% non-redundant) N=3,337	*B. abortus* (RB51) N=3,138	*B. suis* (1330) N=3,144	*B. ovis* (*ATCC25840*) N=2,971	Total no. of genes shared between the 5 species
5	■	■	■	■	■	**2347**
4	■	■	■	■		**269**
4	■	■	■		■	**26**
4	■	■		■	■	**105**
4	■		■	■	■	**29**
3	■	■	■			**14**
3	■	■		■		**40**
3	■	■			■	**3**
3	■		■	■		**7**
3	■		■		■	**58**
3	■			■	■	**5**
2	■	■				**9**
2	■		■			**49**
2	■			■		**5**
2	■				■	**14**
1	■					**110**
4		■	■	■	■	**128**
3		■	■	■		**35**
3		■	■		■	**4**
3		■		■	■	**50**
2		■	■			**6**
2		■		■		**90**
2		■			■	**11**
1		■				**200**
3			■	■	■	**8**
2			■	■		**3**
2			■		■	**13**
1			■			**142**
2				■	■	**3**
1				■		**20**
1					■	**167**
	90% Homology				**No. of Proteins**	**3970**

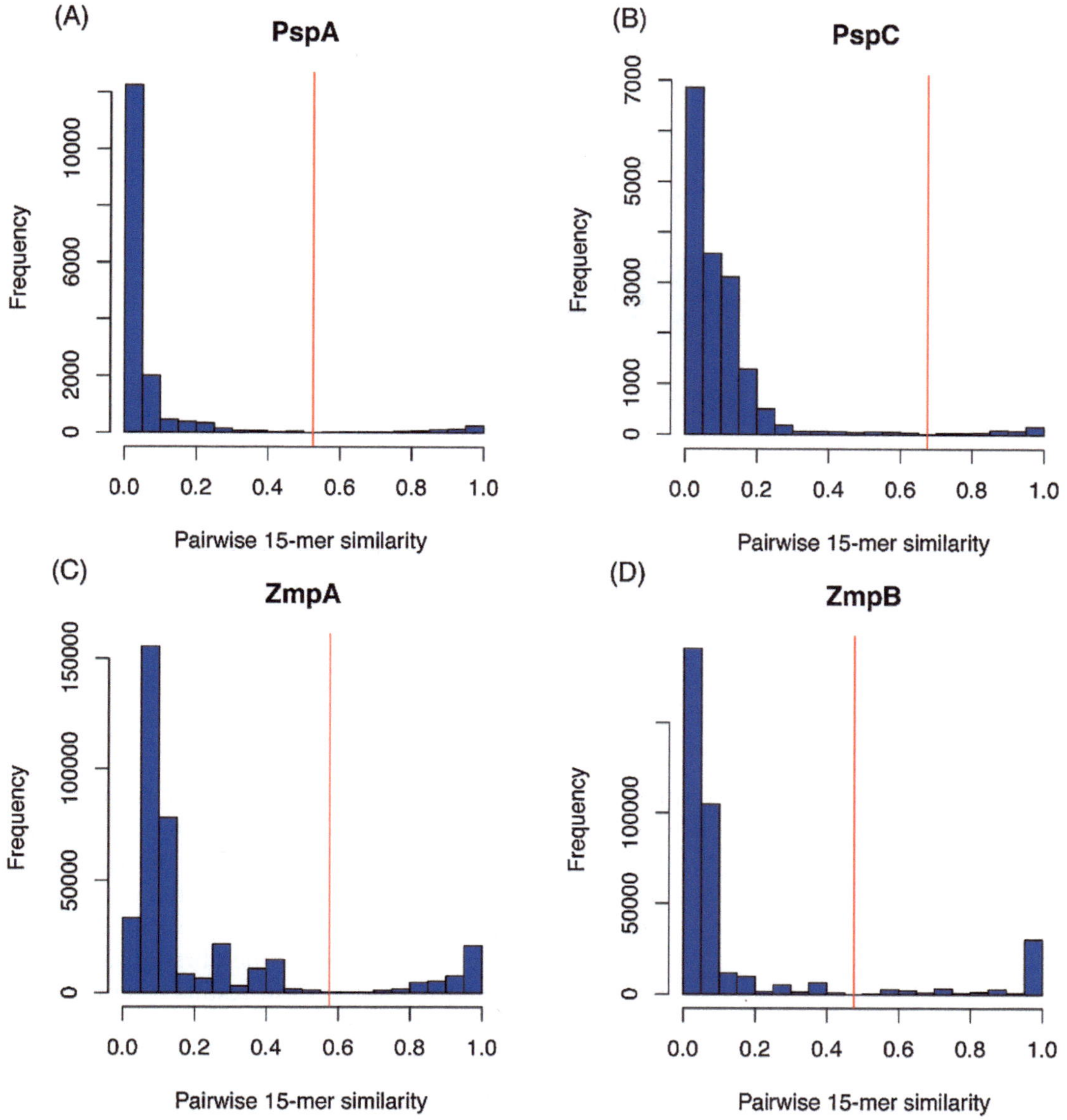

Fig. 3 Defining variants for the four core variable *Streptococcus pneumoniae* antigens. Each histogram shows the distribution of pairwise similarities between representatives of (**a**) PspA, (**b**) PspC, (**c**) ZmpA, and (**d**) ZmpB. The vertical red lines on each plot represent the thresholds that were used as a cut-off to define distinct variants: 0.525 for PspA; 0.675 for PspC; 0.575 for ZmpA, and 0.475 for ZmpB (reproduced from [4] with permission from PNAS)

3.4 Down-Select Proteins from the Core and Panproteome if Necessary

Given limitations in the number of proteins on the microarray, we can downselect the proteins in the panproteome using a set of criteria to allow as much inclusivity as possible to capture maximum diversity, while limiting the proteins to the number of available spots. Firstly, the defined core proteome will constitute a part of the array. Secondly, the remaining spots that are available are assigned to proteins using a scoring scheme (*see* **Note 11**) for top

Table 2
***T. pallidum* subsp. *pallidum* (*T.p.p.*) Nichols and SS14 strain homology analysis for classification of the core proteome and panproteome**

2-way homology (Refseq)	*T.p.p.* Nichols Strain	*T.p.p.* SS14 Strain	No. of genes shared between species
2			987
1			16
1			15
	95% Homology	No. of proteins	1018

ranked proteins that is made from tabulating results of protein feature prediction software, including PSORTb (*see* **Note 12**), TMHMM, SignalP, and the NCBI web CD-Search tool. Additionally, a dictionary of keywords (*see* **Note 13**) is supplemented to these criteria for scoring the top antigens and picking out antigens missed by the above scoring method. The process is executed as follows:

1. Annotate proteins using protein feature prediction software.
   (a) PSORTb: Open a web browser and navigate to the PSORTb webserver (*see* Subheading 2.2, **item 2**). Select options from scroll down menus: (1) bacteria or archaea; (2) Gram stain negative or Gram stain positive with sub-option of with or without outer membrane; (3) output format long or short; and (4) show results by email or on the web. Upload the sequence file in fasta format or paste fasta format sequences in the box and click "Submit." Use the "Localization" header in the output file for scoring proteins by priority, depending on the design of the study (e.g., "Cell Membrane" > "Cytoplasmic"). Set the weighting of scores depending on the objectives of the study.
   - Use Deeploc 1.0 for Eukaryote subcellular localization prediction http://www.cbs.dtu.dk/services/DeepLoc/

   (b) TMHMM: Open a web browser and navigate to the TMHMM webserver (*see* Subheading 2.2, **item 3**). Upload the sequence file in fasta format or paste fasta format sequences in the box and click "Submit." Use the "PredHel" field in the output file, which indicates the number of predicted transmembrane domains, for scoring

Table 3
***Clostridium* species homology analysis for classification of the core proteome (yellow highlighted for 3 or more species) and panproteome**

No. of strains shared	*Paraclostridium scindens*, str. ATCC 35704 N=3647 proteins	*P. bifermentans*, str. cbm N=3546 proteins	*P. leptum*, str. dsm 753 N=2930 proteins	*P. ramosum*, str. dsm 1402 N=3037 proteins	Total no. genes shared between species
4					6
3					2
3					2
3					4
3					0
2					11
2					55
2					26
2					5
2					3
2					7
1					3541
1					3517
1					2851
1					2989
	70% Homology			**No. of proteins**	13019

proteins by priority, depending on the design of the study (e.g., PredHel >0 yields a score of 1 for a study prioritizing any transmembrane domains).

(c) SignalP: Open a web browser and navigate to the SignalP webserver (*see* Subheading 2.2, **item 4**). Upload the sequence file in fasta format or paste fasta format sequences in the box. Select the organism group from options: Eukarya, Gram-positive, Gram-negative, or Archaea. Select output format from options: Long output and Short output (preferred). Click "Submit." Click the "Download" button to download the "Prediction summary" as a text file. Use the "Prediction" field in the

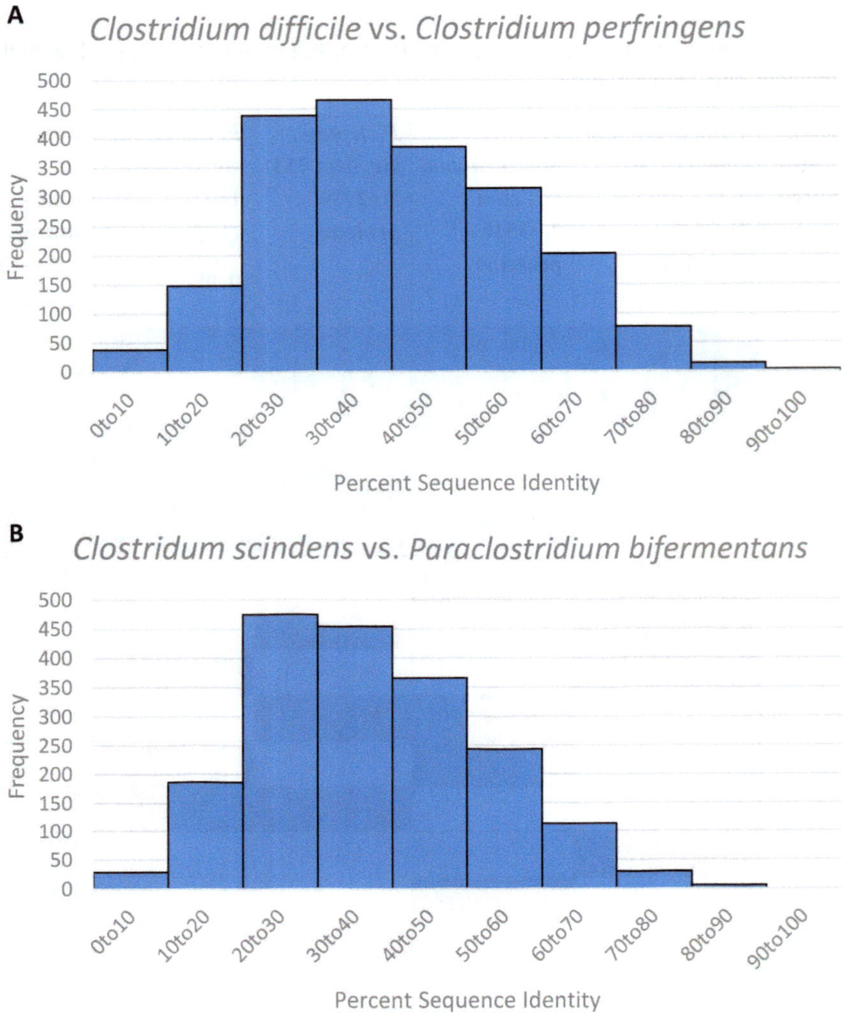

Fig. 4 Distribution of percent identity between *Clostridium* species. The histograms show for pairwise BLASTp results for (**a**) *C. difficile* vs. *C. perfringens* and (**b**) *C. scindens* vs. *P. bifermentans* showing that most genes are poorly conserved between species

output file to score proteins by presence or absence of any of the signal peptides: Sec signal peptide (Sec/SPI), Lipoprotein signal peptide (Sec/SPII) or Tat signal peptide (Tat/SPI). If "Other" has the highest likelihood, then there are no predicted signal peptides.

(d) CDSearchTool: Open a web browser and navigate to the CD Search Tool webserver (*see* Subheading 2.2, **item 5**). Upload the sequence file in fasta format or paste fasta format sequences in the box and click submit. You may change the "Adjust search options" if needed or use the default settings. The retrieved ontologies are given *e*-value scores. Select an *e*-value threshold acceptable for the study (e.g., $<1 \times 10^{-10}$) and score proteins based on significant ontologies that are included in a keyword list, defined by the study objectives (*see* **Note 13**).

2. After running all protein feature predictions, tabulate results by adding the weighted scores for each protein and sorting proteins by highest score.
3. Use the ranked list of proteins to fabricate panproteome microarrays using the desired methods for expressing and printing proteins, or following the published protocols [1].

3.5 Statistical Analysis of the Panproteome Antibody Response to Vaccines and Challenge

Typical methods for univariate differential reactivity analysis have been reported extensively for proteome microarrays (see https://antigendiscovery.com/publications/), which also apply for panproteome analyses. Examples are given from our studies of panproteome antibody responses to a *S. pneumoniae* whole cell vaccine, showing responses to both the core proteome and diverse core loci [5].

- To investigate homology between whole cell vaccine strains and proteins present on the panproteome microarray, follow the methods described in Subheading 3.2, **steps 1–6** to perform pairwise BLASTp using amino acid sequences of each protein on the array as queries against the vaccine strain genome as the subject. This provides a distribution of homology (% sequence identity) or, conversely, divergence (100 – (% sequence identity)) from the vaccine strain (Fig. 5).
- To investigate panproteomic immunogenicity profiles, differential reactivity analysis can be performed using standard parametric and nonparametric methods such as *t*-tests, Wilcoxon's rank-sum tests, empirical Bayes models or multivariable linear models, as appropriate for the study design. Example code for performing empirical Bayes moderated *t*-tests on array data and example output in a study of immunogenicity of a *S. pneumoniae* whole cell vaccine (Fig. 6):

```
> library(limma)
> # sample.classes is a vector of classes for group comparison
> # int.data.df is a data frame containing the matrix of data
with rows as proteins and columns as samples
> trt <- as.factor(sample.classes)
> design <- model.matrix(~trt) # Create a design matrix to fit
limma model
> model <- lmFit(int.data.df, design = design) # Fit ANOVA
model and build contrasts
> result <- eBayes(model)
> ebayes <- topTable(result, coef = 2, number = nrow(int.data.
df),
```

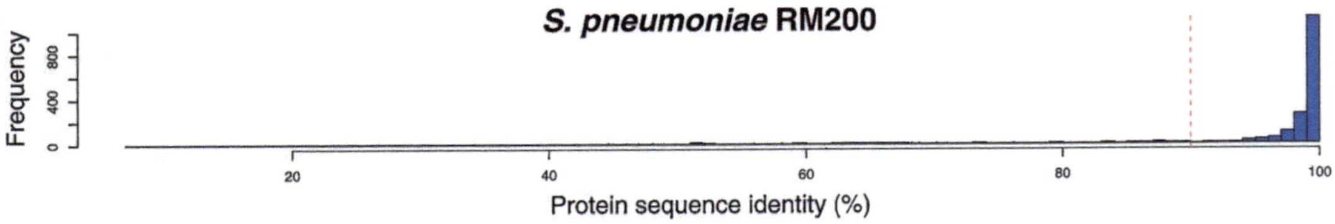

Fig. 5 Distribution of percent identity between *S. pneumoniae* panproteome microarray proteins and the *S. pneumoniae* whole cell vaccine strain RM200 showing high level sequence identity for most proteins constituting the core proteome and greater diversity for a subset of proteins on the microarray (reproduced from [5] under a CC-BY license)

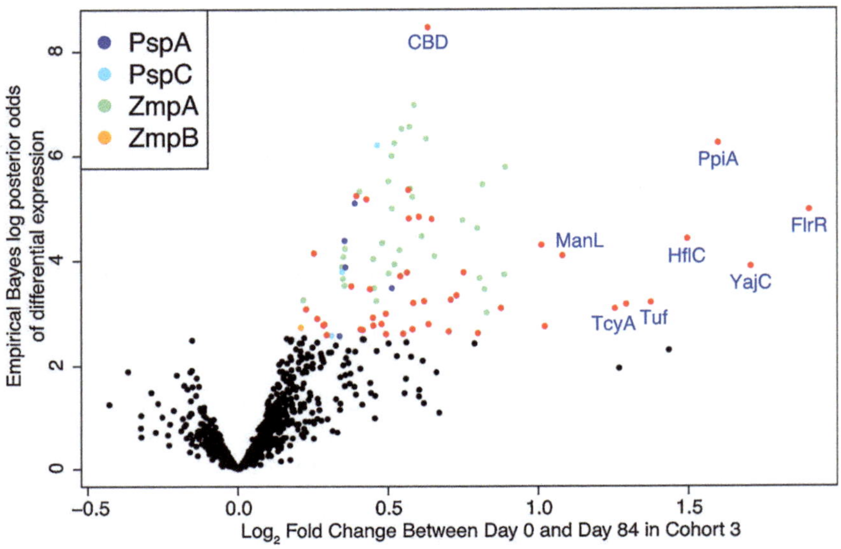

Fig. 6 Volcano plot showing the statistical and biological significance of changes in IgG binding following *S. pneumoniae* whole cell vaccination, showing increase in antibodies targeting diverse proteins in the panproteome. Points corresponding to variants of diverse core loci PspA, PspC, ZmpA, or ZmpB are colored in purple, blue, green and orange, respectively (reproduced from [5] under a CC-BY license)

```
genelist = row.names(int.data.df),
sort.by = "none", p.value = 1.1)
```

– Proteome array data is commonly visualized with heat maps. Multiple R packages facilitate the rendering of heat maps using the data organized in matrix format with signal intensity values arranged in columns by sample and rows by protein (or transposed to invert rows and columns). Order of rows and columns can be ordered manually or by clustering algorithms. Heat maps can be annotated with additional data to highlight associations. For example, to show the association of amino acid sequence similarity of the diverse core loci of a panproteome microarray with a strain used for vaccination or challenge,

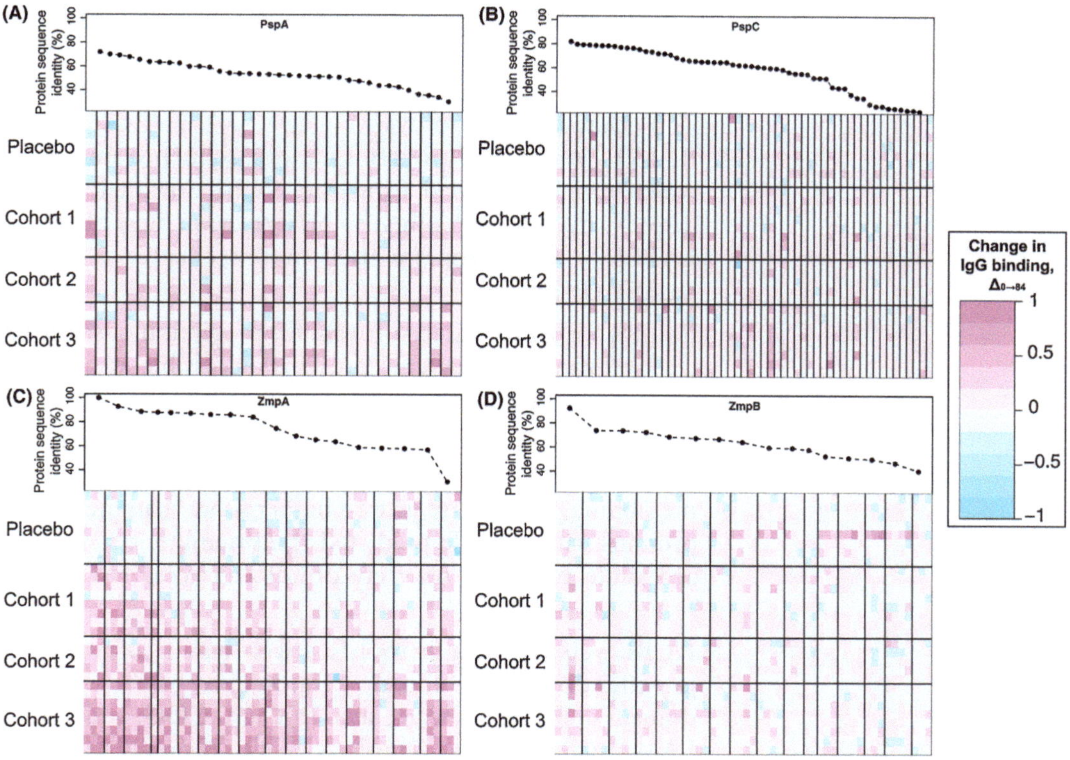

Fig. 7 Changes in IgG binding to *S. pneumoniae* diverse core loci following vaccination with a whole cell vaccine. The variants with higher similarity to the RM200 vaccine strain tended to have greater immunogenicity for ZmpA, showing a strain-transcendent immune response (reproduced from [5] under a CC-BY license)

proteins can be ordered by pairwise BLASTp results, sorting by percent sequence identity (Fig. 7).

- All *p*-values should be reported with appropriate adjustment, such as the false discovery rate described by Benjamini and Hochberg (*see* **Note 14**) [11].

4 Notes

1. There is no fast rule for the genomes selected for panproteome microarray development and will depend on the unique characteristics of the taxa and study questions for the organism of interest. For example, the genomic structure of *T. pallidum* subsp. *pallidum*, the causal bacteria of syphilis, can be divided into two genetic subclusters that can be represented by the Nichols and SS14 strains (Table 2) [12, 13]. Thus, our panproteome approach for syphilis selected only two genomes. In contrast, a total of 91 clinical isolates of *S. pneumoniae* were

selected for development of a panproteome microarray out of a total of 616 sequenced isolates [4].

2. Definitions/descriptions of the software tools from Subheading 2.2 (Data Analysis Software):
   - BLAST+ is a suite of programs for local sequence alignment, which can be installed from the NCBI download site at https://ftp.ncbi.nlm.nih.gov/blast/executables/blast+/LATEST/. The program used in this chapter is protein BLAST (BLASTp).
   - PSORTb is a web based software for prediction of subcellular localization of bacterial proteins, accessible at https://www.psort.org/psortb/.
   - TMHMM is a web based and stand-alone software for prediction of transmembrane helices in proteins. Multipass membrane proteins are given a higher score than membrane proteins with a single predicted transmembrane helix. It is accessible at http://www.cbs.dtu.dk/services/TMHMM/ .
   - NCBI web CD-Search tool is a web based software that provides domain and function annotations based on several domain defining software, such as Pfam, COG, and TIGRFAM (includes Gene Ontology "GO" terminology). The CD-Search conserved domain database (CDD) search returns the domain defined region of the query protein, along with their domain annotations, as well as an associated bit score and *e*-value. It is accessible at https://www.ncbi.nlm.nih.gov/Structure/bwrpsb/bwrpsb.cgi.
   - R is a free software environment for statistical computing and graphics, which can be downloaded from the official website of the R Project for Statistical Computing (https://www.r-project.org/).
   - R studio is an integrated development environment (IDE) for R (https://www.rstudio.com/). This is an optional software that provides a user-friendly interface that for some users of other statistical software will provide greater familiarity than the standard R interface (*also see* **Note 2**).
   - R packages are provided by the R environment in addition to base R functions. Several R packages provide convenient functions for groupwise comparison of antibody measurements on proteome microarrays.
   - Limma is a library built for the analysis of gene expression data, limma contains useful functions for performing empirical Bayesian methods for differential analysis, providing variance shrinkage and robustness for small sample sizes on large arrays (*also see* **Note 3**).

- Lme4 is a package for fitting linear and generalized linear mixed-effects models, useful for analysis of longitudinal datasets (e.g., vaccine/challenge clinical trials) where subjects have repeated measurements (*also see* **Note 4**).
- ROCR is a package for analyzing and visualizing performance of classifiers, particularly useful in calculating the area under the receiver operating characteristic ("ROC") curve.
- Rtsne is a package for performing t-distributed stochastic neighbor embedding (t-SNE) [10], which has been employed in proteome microarray studies for dimension reduction and visualization (*also see* **Note 5**).
- R packages for data visualization are in addition to base R plotting functions. Several R packages provide comprehensive, modular graphics functions that can be used for visualization of panproteome microarray data.
- ggplot2 is a comprehensive system of creating graphics with a learning curve, but well-documented help and examples of graphical parameterization (https://ggplot2.tidyverse.org/).
- gplots contains the function "heatmap.2" which is useful for producing heat maps of array data.
- ComplexHeatmap is a package for more complex datasets. This package includes capabilities for overlaying protein annotation and sample phenotype metadata. Taking full advantage of heat map customization requires a steep learning curve, but detailed instructions and examples are available on the website (https://jokergoo.github.io/ComplexHeatmap-reference/book/).

3. Other programs that facilitate script-writing include Sublime Text (https://www.sublimetext.com/) and Notepad++ (https://notepad-plus-plus.org/).
4. For most purposes, a Student's *t*-test with *p*-value adjustment for the false discovery rate will perform as well as empirical Bayes moderated *t*-tests, given that the goal in most protein array studies for vaccine antigen discovery is to identify the top N antigens.
5. There is ambiguity about the calculation of *p*-values for linear mixed effects regression, and the lme4 package in R does not compute *p*-values purposefully (discussion by the package author shown here: https://stat.ethz.ch/pipermail/r-help/2006-May/094765.html). To approximate the likelihood of an effect estimate being nonzero, null models and full models are fit, and likelihood ratios are tested with an ANOVA test.

6. Other dimension reduction methods such as principal component analysis (PCA) can be substituted and is a generally more familiar method. t-SNE is a nonlinear embedding technique that has shown good clustering of longitudinal data from multiple time points for individuals' proteome-wide antibody profiles [5, 14].
7. Selection of organisms for the panproteome is not trivial in that much depends on the research question, desired breadth of coverage and practical matters such as space for proteins on the array and costs. For example, one of our research questions for characterizing the antibody response to a *S. pneumoniae* whole cell vaccine was whether vaccination induced strain-transcendent antibodies [5]. Thus, we used genomic information from 616 *S. pneumoniae* pediatric clinical isolates for development of a panproteome microarray targeted to within-species diversity. On the other hand, our work with *B. melitensis* asked whether cross-reactive and specific antibody profiles exist between multiple species of *Brucella*, thus we used reference strains of 5 *Brucella* species for homology analysis.
8. The preference for using Genbank or Refseq is subjective to the task. In our experience, the latter provides greater annotation and removes redundancy, while the former is more often referenced in literature and more readily cross-referenced with gene IDs from other studies.
9. There is no fast rule for a percent identity threshold. A lower threshold will permit more divergent proteins to be classified as "core." The decision is subjective to the characteristics of the bacterial population. In the examples presented in Tables 1–3, three different threshold were used (90%, 95%, and 70%, respectively).
10. There is no fast rule for the threshold of prevalence of a protein in the bacterial population (present in % of isolates) for selection. In the case of a U-shaped distribution, as seen in Fig. 2, a cutoff can be empirically defined, assuming that the lower prevalence genes include rare genes in the bacterial population and also sequencing misreads. However, distributions as seen in *Clostridium* spp. (Fig. 4) may require cutoffs based on practical considerations such as availability of spots on the array.
11. The score system used in our down-selection of antigens uses an arbitrary point system, whereby points are tallied based on results of predictive models such as PSORTb, TMHMM, and SignalP, as well as keyword searches in the protein annotations. The overarching goal in our studies is to enrich selection for surface-exposed proteins.
12. PSORTb scoring hierarchy (with SignalP prediction) used in order of highest to lowest score is (1) Extracellular + predicted

signal peptide, (2) Extracellular, (3) Outer membrane/periplasmic + predicted signal peptide (or cell wall in case of Gram positive), (4) Outer membrane/periplasmic (or cell wall in case of Gram positive), (5) Inner membrane or cytoplasmic/membrane + predicted signal peptide, (6) Inner membrane or cytoplasmic/membrane, (7) Cytoplasmic + predicted signal peptide, and (8) Cytoplasmic.

13. Keywords are selected with respect to the organism; for example, "flagella" is a keyword for bacteria with flagellar proteins. Example keywords include flagellin, flagella, flagellar, Fli, fimbrial, flag, membrane, secreted, adhesin, cell wall, surface, hemagglutinin, transport, cilia, pilus, pili, Pil, porin, holin, Ton, Omp, receptor, transfer protein, toxin, antigen, trigger factor, protease, lysin, and adhesin.

14. Consideration should be given to the independence assumption of measurements. If an outsized proportion of the panproteome microarray content is variants of the same protein, it follows that the measurements are not independent and that the degree of false discovery may be overestimated by the Benjamini–Hochberg method. *P*-value adjustment methods should follow best practices for the type of dataset analyzed.

References

1. Driguez P, Doolan DL, Molina DM et al (2015) Protein microarrays for parasite antigen discovery. Methods Mol Biol 1201:221–233. https://doi.org/10.1007/978-1-4939-1438-8_13
2. Köser CU, Ellington MJ, Cartwright EJP et al (2012) Routine use of microbial whole genome sequencing in diagnostic and public health microbiology. PLoS Pathog 8: e1002824. https://doi.org/10.1371/journal.ppat.1002824
3. Croucher NJ, Finkelstein JA, Pelton SI et al (2013) Population genomics of post-vaccine changes in pneumococcal epidemiology. Nat Genet 45:656–663. https://doi.org/10.1038/ng.2625
4. Croucher NJ, Campo JJ, Le TQ et al (2017) Diverse evolutionary patterns of pneumococcal antigens identified by pangenome-wide immunological screening. Proc Natl Acad Sci 114: E357–E366. https://doi.org/10.1073/pnas.1613937114
5. Campo JJ, Le TQ, Pablo JV et al (2018) Panproteome-wide analysis of antibody responses to whole cell pneumococcal vaccination. eLife 7:e37015. https://doi.org/10.7554/eLife.37015
6. Yu NY, Wagner JR, Laird MR et al (2010) PSORTb 3.0: improved protein subcellular localization prediction with refined localization subcategories and predictive capabilities for all prokaryotes. Bioinformatics 26:1608–1615. https://doi.org/10.1093/bioinformatics/btq249
7. Krogh A, Larsson B, von Heijne G, Sonnhammer EL (2001) Predicting transmembrane protein topology with a hidden Markov model: application to complete genomes. J Mol Biol 305:567–580. https://doi.org/10.1006/jmbi.2000.4315
8. Marchler-Bauer A, Derbyshire MK, Gonzales NR et al (2015) CDD: NCBI's conserved domain database. Nucleic Acids Res 43: D222–D226. https://doi.org/10.1093/nar/gku1221
9. Smyth GK (2004) Linear models and empirical bayes methods for assessing differential expression in microarray experiments. Stat Appl Genet Mol Biol 3:1–69
10. van der Maaten L, Hinton G (2008) Visualizing high-dimensional data using t-SNE. J Mach Learn Res 9:2579–2605
11. Benjamini Y, Hochberg Y (1995) Controlling the false discovery rate: a practical and powerful

approach to multiple testing. J Royal Stat Soc Ser B 57:289–300

12. Pětrošová H, Pospíšilová P, Strouhal M et al (2013) Resequencing of *Treponema pallidum* ssp. *pallidum* strains Nichols and SS14: correction of sequencing errors resulted in increased separation of syphilis Treponeme subclusters. PLoS One 8:e74319. https://doi.org/10.1371/journal.pone.0074319

13. Arora N, Schuenemann VJ, Jäger G et al (2016) Origin of modern syphilis and emergence of a pandemic *Treponema pallidum* cluster. Nat Microbiol 2:16245. https://doi.org/10.1038/nmicrobiol.2016.245

14. Camponovo F, Campo JJ, Le TQ et al (2020) Proteome-wide analysis of a malaria vaccine study reveals personalized humoral immune profiles in Tanzanian adults. Elife 9:e53080. https://doi.org/10.7554/eLife.53080

Chapter 7

Low-Energy Electron Irradiation (LEEI) for the Generation of Inactivated Bacterial Vaccines

Jasmin Fertey, Bastian Standfest, Jana Beckmann, Martin Thoma, Thomas Grunwald, and Sebastian Ulbert

Abstract

Vaccines consisting of whole inactivated bacteria (bacterins) are generated by incubation of the pathogen with chemicals. This is a time-consuming procedure which may lead to less immunogenic material, as critical antigenic structures can be altered by chemical modification. A promising alternative approach is low-energy electron irradiation (LEEI). Like other types of ionizing radiation, it mainly acts by destroying nucleic acids but causes less damage to structural components like proteins. As the electrons have a limited penetration depth, LEEI is currently used for sterilization of surfaces. The inactivation of pathogens in liquids requires irradiation of the culture in a thin film to ensure complete penetration. Here, we describe two approaches for the irradiation of bacterial suspensions in a research scale. After confirmation of inactivation, the material can be directly used for vaccination, without any purification steps.

Key words Bacterial inactivation, Low-energy electron irradiation, LEEI, Bacterial vaccine, Bacterin, Electron beam

1 Introduction

Prophylactic vaccines provide an alternative to the use of antibiotics by preventing and controlling infectious diseases in humans and animals. Some bacterial vaccines contain only parts of the pathogen (e.g., polysaccharide structures or other recombinantly expressed proteins), or inactivated bacterial toxins. Currently used veterinary vaccines mainly include attenuated live vaccines and inactivated vaccines [1]. Live attenuated vaccines provide protection through a limited infection of a living organism which elicits an immune response, similar to that of a natural infection. However, they have the very rare potential to revert to a pathogenic form leading to disease. Inactivated or killed vaccines are safe and show similar protection against systemic infections and disease. They consist of killed whole bacteria, which are usually produced by chemical

Fadil Bidmos et al. (eds.), *Bacterial Vaccines: Methods and Protocols*,
Methods in Molecular Biology, vol. 2414, https://doi.org/10.1007/978-1-0716-1900-1_7,

treatment by incubating the pathogens with toxic substances, for example binary ethyleneimine or formaldehyde. This is a time-consuming step and requires days or even weeks of incubation, depending on the organism. In many cases, the chemicals need to be inactivated or removed before the material can be administered as a vaccine. Additionally, the inactivation process often results in chemical modifications of critical antigenic structures and therefore leads to lower vaccine efficacy.

As an alternative strategy to chemical inactivation, ionizing radiation has been used to generate bacterial vaccines, but these approaches are still in the experimental stage, despite promising results [2–4]. Compared to chemical inactivation, that mostly acts by crosslinking nucleic acids and proteins, the major inactivation mechanism of ionizing irradiation is the damage to nucleic acids, while other structural components such as proteins remain largely intact [5–7]. As a rule of thumb, the dose which is required for inactivation is dependent on the genome size, with smaller genomes being more resistant to the effects. Larger genomes are more sensitive toward irradiation because the number of induced DNA double-strand breaks per cell at a given dose is proportional to the size of the genome. Therefore, more irreparable strand breaks occur upon irradiation in larger genomes, hence for the inactivation of bacteria (except for extremely resistant species, e.g., *Deinococcus radiodurans*, or dormant bodies like spores) generally lower doses are required than for the inactivation of viruses [8–10]. The inactivation doses are in the kilogray (kGy) range and currently used irradiation technologies (i.e., gamma-, X-rays, or high-energy electron beam) generate considerable amounts of radiation, either directly or as a side product. Therefore complex concrete and lead shielding constructions for its absorption are required to protect personnel and environment [11]. This has so far prevented the application of ionizing radiation for vaccine manufacturing processes. In contrast, low-energy electron irradiation (LEEI) uses a lower voltage, up to 300 kilo electron volts (keV), instead of high-energy electron irradiation that operates in the range of 1–10 mega electron volts (MeV). LEEI generates lower amounts of secondary radiation (Bremsstrahlung) than high-energy electrons and therefore requires only a few centimeters of protecting lead. This enables the use of LEEI in normal laboratory settings, including potential integration into GMP processes.

Due to high dose rates, LEEI acts very fast and the applied doses can be exactly adjusted via controlling the acceleration voltage. This leads to more reproducible results than, for example, with gamma irradiation, where the activity of the radioactive source decreases because of the constant decay [12]. Electron-beam treatment is already an FDA-approved sterilization procedure for food and pharmaceutical products. LEEI sources require low maintenance and devices are already commercially available (i.e., https://www.ebeamtechnologies.com/en/products). However, they are currently not suited for irradiating liquids and the technique is mainly used for surface sterilization. The major challenge for LEEI in inactivating liquid solutions is the limited penetration depth of the electrons. To overcome this limitation, the liquid has to be irradiated in a thin film, and manufacturing of larger volumes in a multiliter scale with this technique is challenging. With the currently available commercial devices specified for surface sterilization, this can be achieved by using Petri dishes, followed by covering the liquid with a piece of polypropylene foil [13, 14]. However, amounts that are required for larger study groups or larger animals are difficult to produce with this small-scale approach. Further development of irradiation technologies for larger volumes is therefore required.

We have recently described the development of the first research-scale prototype for LEEI of liquids in higher throughput [8], and this development will form the basis for commercially available devices. We have also shown that LEEI can be used to generate efficient vaccines against viruses, parasites, and gram-negative bacteria [13, 15, 16]. In this chapter, we demonstrate both small-scale (*see* Subheading 3.2) and large-scale (*see* Subheading 3.3) methods to generate LEEI-inactivated bacterial vaccines and to evaluate their antigenic components (*see* Subheading 3.4) before administration into animals. As an example, we chose the gram-negative bacterium *Rodentibacter pneumotropicus* to describe the procedures for small-scale and larger-scale irradiation experiments. *R. pneumotropicus* belongs to the family *Pasteurellaceae* and is related to other human and veterinary pathogens such as *Haemophilus influenzae*, *Actinobacillus pleuropneumoniae*, and *Pasteurella multocida*. It is an opportunistic pathogen commonly found in the microbiome of the respiratory tract of mice and rats. Primary or secondary infections can result in severe diseases, and make *R. pneumotropicus* one of the most prevalent infectious agents in laboratory rodents [17].

2 Materials

Prepare and store all reagents at room temperature (unless indicated otherwise).

2.1 Buffers, Reagents and Media

1. Brain Heart Infusion broth: 37.0 g per 1 L distilled water. Autoclave and store at 4 °C (*see* **Note 1**).
2. Phosphate buffered saline (PBS), 10× stock: 1.37 M NaCl, 270 mM KCl, 100 mM Na_2PO_4, 20 mM KH_2PO_4 pH 7.4 (*see* **Note 2**).

3. ELISA coating buffer: 15 mM Na_2CO_3, 35 mM $NaHCO_3$, pH 9.6. Filter through 0.22 μm membrane filter.
4. PBS–Tween: PBS containing 0.05% Tween 20.
5. Antibodies: polyclonal mouse serum against *R. pneumotropicus* and HRP-conjugated anti-mouse antibody (store at 4 °C or at −20 °C for long-term storage) (*see* **Note 3**).
6. Blocking solution: 5% skim milk in PBS–Tween (store at 4 °C).
7. Detection substrate: 3,3′,5,5′-tetramethylbenzidine (TMB) (*see* **Note 4**).
8. H_2SO_4: 1 M.
9. 70% ethanol.

2.2 Equipment

1. 37 °C orbital shaker
2. 37 °C static incubator
3. Tabletop centrifuge with rotor for 15 mL reaction tubes.
4. Microcentrifuge with rotor for 1.5 mL reaction tubes.
5. Biological safety cabinet (BSC).
6. Spectrophotometer for checking optical density at 600 nm (OD_{600}).
7. Cuvettes for absorbance measurement.
8. 0.22 μm membrane filters
9. Polyethylene terephthalate/polyethylene (PET/PE) material with 12 cm width (*see* **Note 5**).
10. Small round pieces (6 cm in diameter) of oriented polypropylene (OPP)-foil (*see* **Notes 5** and **6**).
11. Impulse sealer with a length of 80–100 cm.
12. Impulse sealer with a length of 30 cm (optional; *see* **Note 7**).
13. ELISA reader.
14. 1.5 mL tubes
15. Pipets.
16. Pipetting aid.
17. 15 mL and 50 mL tubes
18. Sterile scissors.
19. Petri dishes (10 cm diameter; *see* **Note 8**).
20. ELISA plates (96-well) (*see* **Note 9**).

2.3 Specialized Equipment

1. Electron irradiation device (Fig. 1a) with module for automated bag transport (Fig. 2), for bag irradiation experiments.

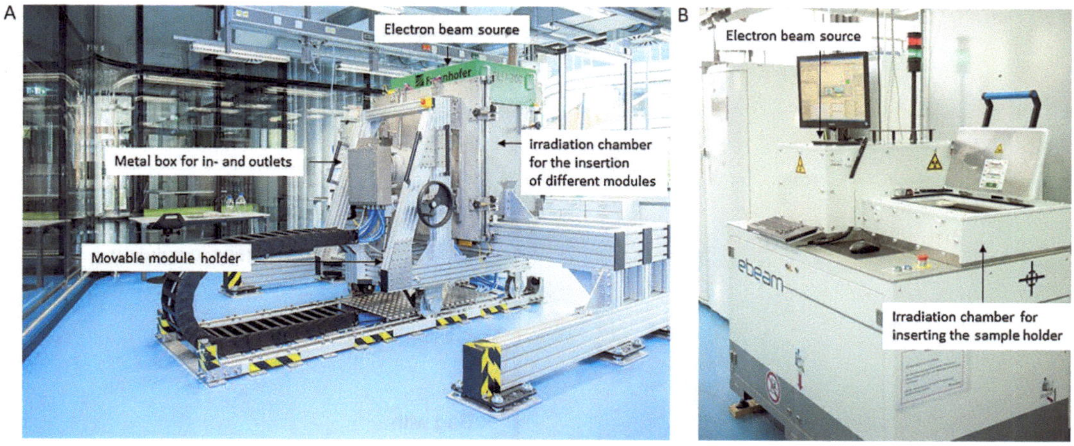

Fig. 1 (**a**) Photograph of the custom-built LEEI device. The picture shows the closed device with the electron beam source and the leaden irradiation chamber in the middle. Feed lines with flexible hoses for the in- and outlet of liquids (e.g., for cooling inside the irradiation chamber) are guided through the metal box followed by baffles to shield the environment from occurring Bremsstrahlung. The movable module holder can be rolled out of the irradiation chamber and carries the modules (shown in Fig. 3) for liquid handling. The irradiation chamber was constructed as a research-scale prototype and different experimental modules designed for liquid handling can be inserted completely [8]. This increases the overall footprint of the electron beam device. (**b**) Photograph of the commercially available EB-Lab 200 (https:/www.ebeamtechnologies.com/en/eblab). The picture shows the opened chamber, where the sample holder for Petri plates is placed, and the cover is closed before irradiation. The sample is transported to the left for irradiation where the electron beam source is located

2. EB Lab 200 or EB Lab 300 (https://www.ebeamtechnologies.com/en/eblab) or comparable electron irradiation device (Fig. 1b), for small-scale experiments.

3 Methods

This section explains the preparation of the irradiated bacterial sample for use in animal vaccination experiments. Two irradiation procedures, either in a 230 μL volume (small-scale in Petri dishes, covered with plastic foil; Fig. 3) or in a 20 mL volume (larger scale in disposable bags; Figs. 4 and 5) are described, followed by testing the antigenicity by ELISA.

All experimental steps involving living *R. pneumotropicus* must be performed in a laboratory authorized for the work with BSL-2 organisms.

3.1 Preparing the Bacterial Solution for Irradiation

1. Inoculate bacteria in 50 mL growth medium and incubate over night at 37 °C, 180 rpm (*see* **Note 10**).
2. Harvest bacteria by centrifugation (3452 × *g*, 4 °C, 10 min) (*see* **Note 11**).

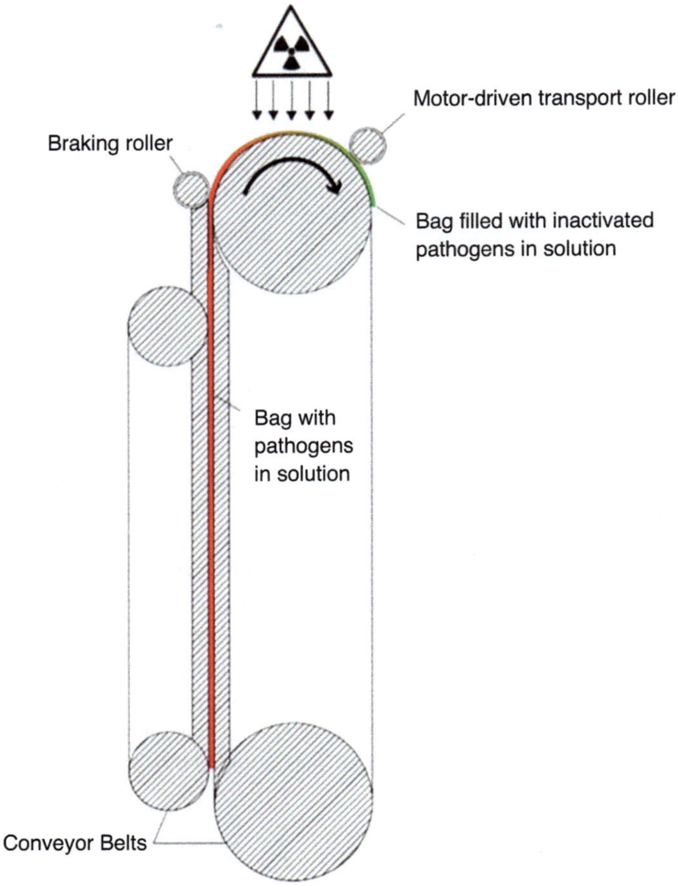

Fig. 2 Schematic drawing of the module. The disposable bag containing the pathogen solution is placed on the left side between the two conveyor belts (marked by the red line). The roller transports the bag through the irradiation zone (marked by radiation warning sign). The bag is stretched by the braking roller to limit the liquid height and to enable the low-energy electrons to penetrate the bag completely. The bag with the irradiated solution (marked in green) is transported to the right and can be taken out of the module after the irradiation is completed

3. Discard supernatant and resuspend bacteria in 50 mL PBS. Centrifuge again at 3452 × *g*, 4 °C, 10 min (*see* **Note 12**).
4. Repeat step 3 twice and resuspend in a final volume of 50 mL PBS (*see* **Note 13**).
5. Check the OD_{600} in a spectrophotometer and adjust with PBS to a maximum value between 1 and 2 (*see* **Note 14**).
6. Store washed bacteria on ice, or at 4 °C, until irradiation (*see* **Note 15**).

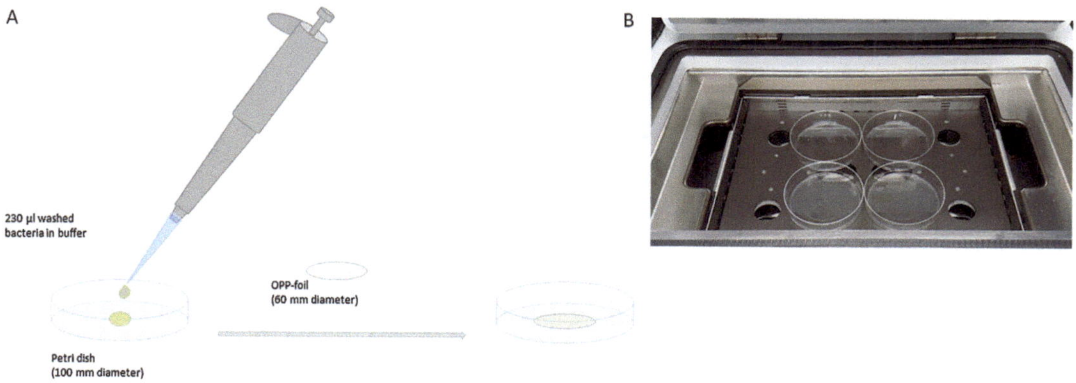

Fig. 3 (**a**) Sample preparation for LEEI in a small-scale experimental setup. 230 μL of bacteria are pipetted to the center of a Petri dish and covered with a piece of round, sterilized oriented polypropylene (OPP)-foil forming the thin liquid film. (**b**) Photograph of sample holder with prepared Petri dishes, placed in the center of the sample holder. The picture shows the sample holder already inserted into the irradiation device

3.2 Irradiation in a Small-Scale Experimental Setup

1. Take the sample holder out of the irradiation device and put into a biological safety cabinet (BSC). Discard lids from the Petri dishes, as they are not needed anymore. A maximum of six dishes can be placed on the sample holder and processed at the same time.
2. For each sample, pipet 230 μL of washed bacteria to the center of a Petri dish and cover the drop with a piece of sterilized OPP-foil (Fig. 3a). Be careful to avoid bubbles during pipetting and spreading the liquid, since they might affect the height of the liquid film (*see* **Note 16**).
3. Put the Petri dishes in the center of the sample holder and insert the sample holder in the device (Fig. 3b) (*see* **Note 17**).
4. Close the cover of the irradiation chamber, set the desired parameters and start the irradiation process (*see* **Note 18**).
5. Take the sample holder out of the irradiation device and put it back into the BSC.
6. Recover the liquid from Petri dishes by carefully tapping the edge of the dish on the surface of the BSC and collecting the liquid that runs down to the edge using a pipet. Transfer the sample to a sterile 1.5 mL tube.
7. For inactivation testing, inoculate 10% of the recovered sample into fresh growth medium and incubate at 37 °C overnight. Include a nonirradiated sample as positive control for growth (*see* **Note 19**).
8. Save 10–20% of the recovered sample for checking the antigenicity (*see* Subheading 3.4). Ensure to save the same amount of the nonirradiated sample, since this will be the reference for the antigenicity testing.
9. Store the residual sample at −80 °C until use.

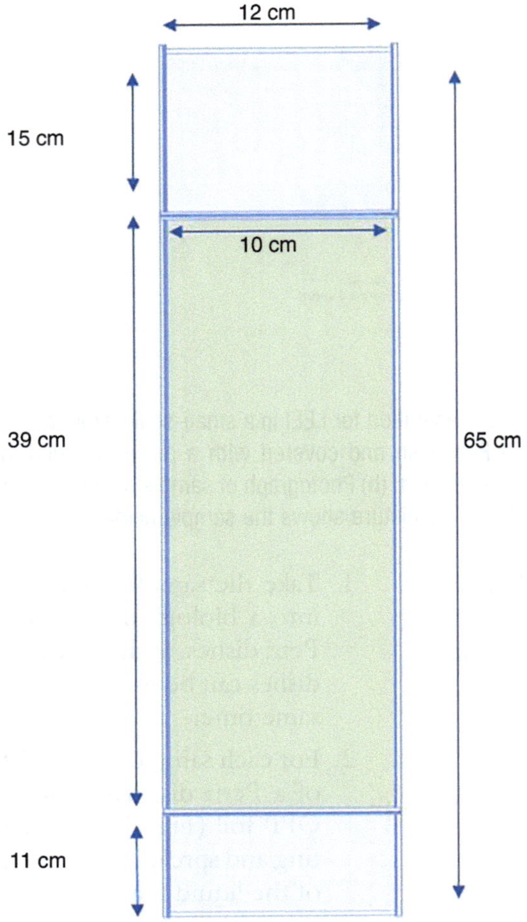

Fig. 4 Dimensions of PET/PE-bags for irradiation of 20 mL bacterial solution in bags. The area containing the liquid (pale green) is 39 cm × 10 cm. Although the bag could contain more than 20 mL, this volume ensures the optimal liquid height for complete penetration by LEEI

3.3 Irradiation in Bags

1. Prepare irradiation bags by placing two layers of PET/PE film carefully over one another. Seal the long (65 cm) sides with the 80–100 cm impulse sealer, taking care that the two layers are properly aligned on top of each other. Then seal the bottom of the bag, creating the seam about 10–11 cm from the bottom edge (Fig. 4), using the 30 cm impulse sealer (*see* **Note 7**). Make sure that all three sides are completely sealed (Figs. 4 and 5) and that the sealing seams are intact to ensure that no infectious content spills out (*see* **Note 20**). Leave the top of the bag open for filling (Fig. 5).
2. In a biological safety cabinet (BSC), fill 20 mL of washed bacteria in PBS or other buffer into a prepared irradiation bag using a 10 mL or 20 mL pipet and a pipetting aid (Fig. 5). Use one bag per irradiation dose to be tested and include an extra

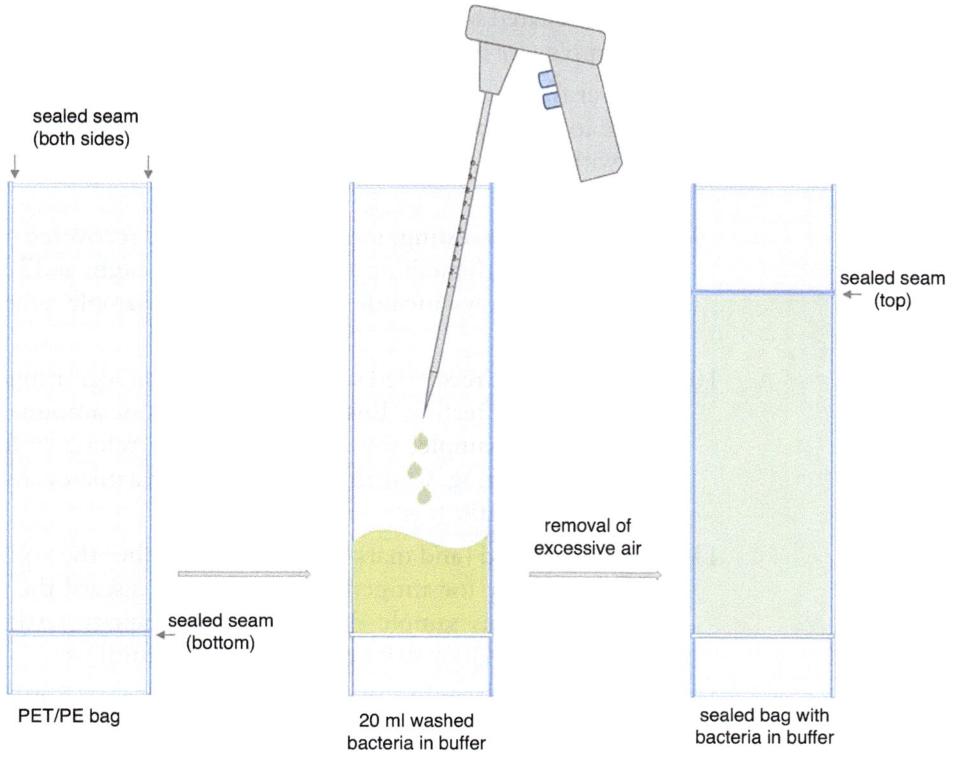

Fig. 5 sample preparation for LEEI in disposable bags. 20 mL of washed bacteria are transferred to PET/PE-bags (*see* Fig. 1 for dimensions). Air is removed from the bag and the top is sealed using an impulse sealing machine

bag/sample for the nonirradiated control (i.e., sample that receives no irradiation, but is processed). If processing effects are observed in downstream experiments, include an "untreated control" without processing as well for optimization of the irradiation process.

3. Remove excessive air from the bag before sealing by pulling the bag top down over the rim of the BSC to squeeze out residual air, then seal the upper part of the bag to close it (*see* **Note 21**). Disinfect the closed bag by wiping or spraying with disinfectant before transporting to the irradiation module, in case infectious content spilled out of the bag.
4. Insert the bag in the irradiation module between the conveyor belts and fix the bag between the two rollers (Fig. 2).
5. Close the irradiation chamber, set the desired parameters and start the irradiation process (*see* **Note 22**).
6. After irradiation, remove the processed bag from the device by loosening the rollers that hold the bag in place and pull the bag out of the module.

7. Store irradiated bags on ice or cool packs until ready to recover the sample (*see* **Note 23**).
8. Recover the bacterial solution in a BSC by squeezing the liquid to the lower part of the bag and cutting the upper part of the bag with sterile scissors, then pipet the content into a fresh, sterile vessel.
9. For inactivation testing, inoculate 10% of the recovered sample into fresh growth medium and incubate over night at 37 °C (*see* **Note 24**). Always include a nonirradiated sample as growth control.
10. Save 10% of the recovered sample into a 1.5 mL centrifuge tube for antigenicity testing. Ensure to save the same amount of the nonirradiated sample, since this will be the reference for the antigenicity testing. Centrifuge the sample in a microcentrifuge for 30 s at 24,100 × *g*.
11. Carefully record (and mark directly on the tube) the volume of the 10% sample for antigenicity testing and discard the supernatant. Use this sample directly for antigenicity testing (*see* Subheading 3.4), or store pellet at −80 °C until use.
12. Transfer the remaining recovered sample into a 15 mL tube. The bacteria can be stored at −80 °C, either as a pellet or in buffer. This depends on the intended administration or formulation (i.e., which buffer, concentration, etc.). If you wish to pellet the bacteria, use a tabletop centrifuge (3452 × *g*, 4 °C, 10 min), record the initial volume as described in **step 10**, and store pellet at −80 °C until use.

3.4 Determining the Antigenicity of the Irradiated Sample by ELISA

1. In a BSC, resuspend pelleted bacteria in the initially noted volume of buffer (PBS or coating buffer).
2. Always include the nonirradiated sample as positive control for the ELISA experiment. Make sure that the nonirradiated sample is diluted in the same way as the irradiated sample. Include the buffer (e.g., PBS) used for resuspending the bacteria before irradiation in the experiment as background negative control (*see* **Note 25**).
3. Dilute samples in coating buffer and transfer 100 μL of each dilution per well in duplicates or triplicates to the ELISA plate. Usually, 1–5 μL of sample (prepared from washed bacteria with an OD_{600} of 1–2) in 100 μL coating buffer per well are sufficient for obtaining optimal signals (*see* **Note 26**). Incubate the plate on a shaker over night at 4 °C.
4. On the next day, wash the plate 3 times with 200 μL per well PBS-T (*see* **Note 27**).

5. Block the plate by adding 100 μL blocking solution (5% skim milk in PBS-T) per well and incubate for 1–2 h at room temperature.
6. Remove blocking solution (*see* **Note 28**).
7. Dilute primary antibody in blocking solution and add 100 μL per well to the plate. Incubate for 2 h at room temperature (*see* **Note 29**).
8. Wash plate 3 times with 200 μL per well PBS-T.
9. Dilute secondary antibody in blocking solution and add 100 μL per well to the plate. Incubate for 1 h at room temperature.
10. Prepare fresh TMB solution (*see* **Note 4**) and add 100 μL per well. Incubate in the dark for 15 to max. 30 min (*see* **Note 30**).
11. Stop the reaction by adding 50 μL H_2SO_4 per well. You should observe a color change from blue to yellow.
12. Read absorbance at 450 nm in an ELISA reader.
13. Set the signal of the nonirradiated control as 100% and calculate the percentage of signal of the irradiated samples (*see* **Notes 31** and **32**).

4 Notes

1. BHI is the recommended growth medium for *R. pneumotropicus,* for other bacteria appropriate culture media should be used for growth and inactivation testing.
2. The buffer used for irradiation can be replaced by other buffering solutions, for example, Tris-based or Hepes-based buffers, if nonirradiated or untreated bacteria show loss of viability after storage in PBS. If concentrated stock solutions of buffers are prepared (e.g., 10x PBS), ensure to dilute them to 1× working concentration before use.
3. For our ELISA-experiments we used serum from a lab mouse experimentally infected with the homologous bacterial strain [17].
4. We use a commercially available TMB Substrate Set (e.g., Cat. No. 421101 from BioLegend) which contains TMB Substrate A and TMB Substrate B. Equal volumes of each substrate are mixed immediately before use. For example, for one 96-well plate, mix 5.5 mL TMB Substrate A with 5.5 mL of TMB Substrate B in a clean container. After mixing the reagents together, TMB substrate working solution should be colorless or very faint blue. The mixed substrate is not stable for a long

time, only the amount needed for each assay run should be prepared.

5. We recommend using polyethylene terephthalate/polyethylene (PET/PE) material for the bags and oriented polypropylene (OPP) for the round plastic-foils for irradiation [18]. In general, PET/PE and round pieces of OPP foil are available at specialized packaging companies that provide packaging solutions for medical or food products (e.g., sicht-pack HAGNER GmbH or tbs-pack GmbH, both in Germany). We recommend ordering PET/PE film rolls with the intended bag width (i.e., 12 cm), and enough length to generate the desired number of bags, each 65 cm long (Fig. 4).
6. For the small-scale irradiation experiments, it is crucial to avoid shadowing effects during irradiation, otherwise there is the chance that the sample is not evenly irradiated. Therefore, the OPP foil must have a smaller diameter than the Petri plates. It is recommended to sterilize the round OPP-foils before use by putting them into 70% ethanol for 15 min, then allow to dry on paper towels in a BSC. This treatment also removes any nonorganic contaminations.
7. It is possible to use the 80–100 cm impulse sealer for all sides. However, we prefer to have a smaller (30 cm) impulse sealer in the BSC to directly seal the bags after filling. This makes the disinfection (by spraying or wiping off the bag) much easier because no infectious content spills out.
8. We tried different Petri dishes and identified Corning® Primaria™ to be the best choice. The very hydrophilic surface treatment ensures that the liquid is evenly distributed under the OPP-foil, providing a homogenous liquid layer. In untreated Petri dishes the liquid stays more or less as a drop on the surface, making it difficult to achieve a liquid layer with an even height. We therefore recommend using specially treated Petri dishes (comparable to Corning® Primaria™).
9. In most cases, Nunc PolySorp® surface coated ELISA plates worked well for coating the irradiated samples; however, there have been some exceptions where other surfaces performed better. If weak ELISA-signals are observed due to insufficient coating, checking other plates (such as Nunc MaxiSorp® or MultiSorp®) might improve the results.
10. We have also performed experiments with other bacteria, for example, *Escherichia coli* and *Bacillus cereus.* We usually store the bacteria at −80 °C as cryostocks that can be used directly for inoculation. We observe optimal growth if the bacteria are precultured in a lower volume, such as 5 mL, at 37 °C, 180 rpm on the day before the inoculation for the experiment and use 0.5 mL of this culture for inoculation of the final culture.

11. Depending on the bacterial strain used, the centrifugation time might vary, depending on the consistency of the resulting pellet. Resuspending the bacteria should not be too difficult.
12. If decrease in bacterial viability is observed in nonirradiated control or after storage of the bacteria in PBS, the buffer used for washing and during irradiation should be exchanged. Furthermore, glycerol (up to a final concentration of 10%) can be added to the chosen buffer if handling stress is observed during antigenicity testing (marked by a strong decrease in ELISA signal intensity in the untreated vs. the nonirradiated sample; *see* **Notes 25** and **26**). The untreated sample is a sample which is just washed with buffer and stored, the nonirradiated sample is a sample that underwent all steps, except the irradiation (pipetting, filling, etc.). This optimization step should be done before the irradiation experiment. So far, the best results we have obtained were using PBS only. Irradiation in culture media is not recommended, as we have observed that the inactivation is less reproducible. It is possible that components in the media absorb the irradiation and bacterial aggregates may form as well.
13. It is crucial to have a homogenous resuspension to ensure complete inactivation. If clumps are visible, wash the bacterial pellet until no clumps are visible anymore. Note that many clinical isolates (especially *Pasteurellaceae* species) form strong biofilms. These cultures might require more washing and resuspension steps. Also note that some species, such as *A. pleuropneumoniae*, may lose this phenotype following multiple passage in broth culture.
14. Irradiation is possible with higher concentrations of bacteria, but it is difficult to ensure homogeneity in very dense solutions. We tested solutions with an OD_{600} of up to 3 with good inactivation results after irradiation. However, the readable range for some spectrophotometers may have a high uncertainty above 1.5. In this case it is better to dilute the suspended bacteria 1:2 or 1:5 prior to measuring the OD_{600}, and then back calculate the value for the undiluted stock. Adjusting the solution to a defined OD_{600} helps when comparing the input material between different irradiation experiments. Other assays for quantification like colony forming units per mL (CFU/mL) can be used as well, but require an additional overnight incubation step.
15. It is recommended to store the bacteria for no longer than 2 or 3 h at this stage, longer storage might lead to reduced growth in the positive controls afterward, which increases the risk of discrepant results when checking for CFU/mL and antigenicity in the ELISA.

16. Ensure that the liquid sample is in the center of the dish. Otherwise, shadowing effects from the side of the Petri dish might lead to incomplete inactivation.
17. Regardless of the number of Petri dishes that are used for each irradiation run, always ensure that they are in the center of the sample holder (Fig. 4b). If they are placed at the side, shadowing effects may occur, and the sample is not completely irradiated.
18. The small-scale setup is optimal for determining the required inactivation dose and for measuring the antigenicity of the inactivated sample. It can be used as a small-scale test before the actual experiment. In this case, several doses should be tested, and CFU/mL should be determined afterward in addition to liquid culture for checking the inactivation kinetics. In this way, a killing curve and the minimal dose required for inactivation can efficiently be established.
19. The nonirradiated control consists of an extra Petri dish with sample that is treated in the same way (covering with foil) like the samples, except without the irradiation process. The samples can stay in the BSC, while the other samples are irradiated.
20. Make sure that the sealer is not too hot and the seams are tightly sealed.
21. Removing air is crucial for processing of the bag and efficient inactivation. There should be as little air as possible in the closed bag. During irradiation, degassing of the buffer solution might be observed, in this case air bubbles will be visible in the irradiated bag. It is crucial that the thin liquid film that is generated by stretching the bag between two rollers (Fig. 2) is less than 150 μm thick to ensure complete penetration of the electrons to the bottom part of the bag. It is recommended that extra sealed seams (i.e., more than one) are added on the top of the bag to ensure that infectious content is not released when the bag is squeezed.
22. Depending on the velocity with which the bag is transported, the irradiation process takes approx. 2–5 min per bag. Since the applied dose is a product of applied energy and velocity of the sample, these parameters have to be determined empirically and controlled using commercially available, adjusted dosimeter films (e.g., Risø B3, Risø High Dose Reference Laboratory, Denmark). For commercially available irradiation devices, as used for the small-scale experiments, the dose application is precalibrated and usually no dosimetry is needed.
23. Cooling the irradiated solution might help to slow down residual hydrolysis reactions in the liquid after irradiation and improve antigenicity due to less formation of free radicals.

24. Bacterial growth should be drastically reduced after irradiation; however, there might be residual viability in the sample that is not visible after overnight incubation. Therefore, it is recommended to incubate the culture for longer periods (up to 3 days, or—depending on the doubling time of the organism—even longer).
25. We recommend testing each sample (including the nonirradiated sample, the resuspension buffer and the coating buffer) in triplicate. To check for handling effects during the irradiation procedure, it might be useful to include an untreated sample as well (*see* **Note 26**). It is not necessary to distribute replicates of samples and blanks in different parts of the plate. When testing only a few samples, it is possible to use only parts of the plate for the measurement. This is recommended to save reagents (such as primary or secondary antibodies) which may be expensive or in limited supply. The coating of the replicates (row by row or in columns) can be determined by the user. If only a part of the plate is coated and an ELISA-washer is used, it makes sense to coat the plate in a way that the washer has to wash only this part and not the whole plate. Also, if washing by hand is performed, it is useful to use a multichannel pipet (8-channel or 12-channel) and coat the plate in a way, that pipetting steps are reduced.
26. The amount of material to be coated per well has to be predetermined in a separate ELISA experiment and is dependent on the antibody that is used. It is recommended to initially use untreated, fresh bacteria in different amounts to find the linear range for the signal intensity. Signals should be in an OD_{450} range between 0.8 and 1.5. Signals above $OD_{450} = 1.8$ are saturated, signals below $OD_{450} = 0.5$ are often too close to the background, especially when working with animal sera where the antibodies have not been purified.
27. Washing can be performed by hand or by using an ELISA washer. It is recommended to tap plates on a paper towel to remove residual washing solution from the wells before adding antibodies to prevent dilution effects.
28. After blocking, plates can be washed 3 times with PBS-T, dried and sealed in a bag for storage. Blocked plates can be stored for several weeks in a cool and dry place until use.
29. It is recommended to use a polyclonal serum or specific antibodies for the detection of surface proteins to check the integrity of surface antigens. For testing disruption of the bacteria, antibodies raised against internally located antigens can be used to estimate the ratio of disrupted vs. intact bacteria. Working dilutions must be optimized for each antibody before the actual experiment.

30. Development of a blue color should be visible after 30 min. If the color development occurs directly after adding the substrate, check plate every 5 min and stop the reaction after 15 min.

31. In our hands, values of at least 70% were obtained for the irradiated, compared to nonirradiated, samples. This material resulted in a robust immune response after vaccination of animals.

32. The irradiated material can be directly used for vaccination or formulation with adjuvant. We found that, for gram-negative bacteria, it is not necessary to use adjuvants since the LPS structure is better conserved than in chemically treated bacteria and possibly acts as immune stimulant [15].

References

1. Hoelzer K, Bielke L, Blake DP et al (2018) Vaccines as alternatives to antibiotics for food producing animals. Part 1: challenges and needs. Vet Res 49:1–10. https://doi.org/10.1186/s13567-018-0560-8
2. Babb R, Chen A, Hirst TR et al (2016) Intranasal vaccination with γ-irradiated *Streptococcus pneumoniae* whole-cell vaccine provides serotype-independent protection mediated by B-cells and innate IL-17 responses. Clin Sci 130:697–710. https://doi.org/10.1042/CS20150699
3. Bordin AI, Pillai SD, Brake C et al (2014) Immunogenicity of an electron beam inactivated *Rhodococcus equi* vaccine in neonatal foals. PLoS One 9:e105367. https://doi.org/10.1371/journal.pone.0105367
4. Jwa MY, Jeong S, Ko EB et al (2018) Gamma-irradiation of *Streptococcus pneumoniae* for the use as an immunogenic whole cell vaccine. J Microbiol 56:579–585. https://doi.org/10.1007/s12275-018-8347-1
5. Alizadeh E, Orlando TM, Sanche L (2015) Biomolecular damage induced by ionizing radiation: the direct and indirect effects of low-energy electrons on DNA. Annu Rev Phys Chem 66:379–398. https://doi.org/10.1146/annurev-physchem-040513-103605
6. Brown F (1995) Formaldehyde as an inactivant. Vaccine 13:231. https://doi.org/10.1016/0264-410x(95)93142-v
7. Delrue I, Verzele D, Madder A et al (2012) Inactivated virus vaccines from chemistry to prophylaxis: merits, risks and challenges. Expert Rev Vaccines 11:695–719. https://doi.org/10.1586/erv.12.38
8. Fertey J, Thoma M, Beckmann J et al (2020) Automated application of low energy electron irradiation enables inactivation of pathogen- and cell-containing liquids in biomedical research and production facilities. Sci Rep 10:12786. https://doi.org/10.1038/s41598-020-69347-7
9. Espinosa AC, Jesudhasan P, Arredondo R et al (2012) Quantifying the reduction in potential health risks by determining the sensitivity of poliovirus type 1 chat strain and rotavirus SA-11 to electron beam irradiation of iceberg lettuce and spinach. Appl Environ Microbiol 78:988–993. https://doi.org/10.1128/AEM.06927-11
10. Skowron K, Paluszak Z, Olszewska H et al (2014) Effectiveness of high energy electron beam against spore forming bacteria and viruses in slurry. Radiat Phys Chem 101:36–40. https://doi.org/10.1016/j.radphyschem.2014.04.001
11. IAEA Safety standards (2010) Radiation safety of gamma, electron and x-ray irradiation facilities. IAEA Safety standards series. https://www-pub.iaea.org/MTCD/publications/PDF/Pub1454_web.pdf
12. Silindir M, Özer AY (2009) Sterilization methods and the comparison of E-beam sterilization with gamma radiation sterilization. FABAD J Pharm Sci 34:43–53
13. Fertey J, Bayer L, Grunwald T et al (2016) Pathogens inactivated by low-energy-electron irradiation maintain antigenic properties and induce protective immune responses. Viruses 8:319. https://doi.org/10.3390/v8110319
14. Bayer L, Fertey J, Ulbert S et al (2018) Immunization with an adjuvanted low-energy

electron irradiation inactivated respiratory syncytial virus vaccine shows immunoprotective activity in mice. Vaccine 36:1561–1569. https://doi.org/10.1016/j.vaccine.2018.02.014

15. Fertey J, Bayer L, Kähl S et al (2020) Low-energy electron irradiation efficiently inactivates the gram-negative pathogen *Rodentibacter pneumotropicus*-A new method for the generation of bacterial vaccines with increased efficacy. Vaccines 8:113. https://doi.org/10.3390/vaccines8010113
16. Thabet A, Schmäschke R, Fertey J et al (2019) *Eimeria tenella* oocysts attenuated by low energy electron irradiation (LEEI) induce protection against challenge infection in chickens. Vet Parasitol 266:18–26. https://doi.org/10.1016/j.vetpar.2019.01.001
17. Fornefett J, Krause J, Klose K et al (2018) Comparative analysis of humoral immune responses and pathologies of BALB/c and C57BL/6 wildtype mice experimentally infected with a highly virulent *Rodentibacter pneumotropicus* (*Pasteurella pneumotropica*) strain. BMC Microbiol 18:45. https://doi.org/10.1186/s12866-018-1186-8
18. Gotzmann G, Portillo J, Wronski S et al (2018) Low-energy electron-beam treatment as alternative for on-site sterilization of highly functionalized medical products – a feasibility study. Radiat Phys Chem 150:9–19. https://doi.org/10.1016/j.radphyschem.2018.04.008

Chapter 8

Design and Production of Hybrid Antigens for Targeting Integral Outer Membrane Proteins in Gram-Negative Bacteria

Somshukla Chaudhuri, Nikolas F. Ewasechko, Luisa Samaniego-Barron, Jamie E. Fegan, and Anthony B. Schryvers

Abstract

Metal ion transporters in the outer membrane of gram-negative bacteria that are responsible for acquiring iron and zinc are attractive vaccine targets due to their essential function. The core function is mediated by an integral outer membrane TonB-dependent transporter (TBDT) that mediates the transport of the metal ion across the outer membrane. Some TBDTs also have a surface lipoprotein (SLP) that assists in the efficient capture of the metal ion-containing host protein from which the metal ion is extracted. The challenges in producing the integral outer membrane protein for a commercial subunit vaccine prompted us to develop a hybrid antigen strategy in which surface loops of the TBDT are displayed on the lipoprotein, which can readily be produced as a soluble protein. The focus of this chapter will be on the methods for production of hybrid antigens and evaluating the immune response they elicit.

Key words Hybrid antigens, TonB-dependent transporter, Surface lipoprotein, Outer membrane proteins, ELISA, Transferrin binding proteins

1 Introduction

Our hybrid antigen approach was originally developed to target the surface epitopes of the integral outer membrane protein (OMP), transferrin binding protein A (TbpA) [1], and has since been used to target surface regions of other TonB-dependent transporters (TBDTs) [2]. TBDTs are a family of integral membrane proteins in the outer membrane of gram-negative bacteria and are primarily involved in the acquisition of metal ions or metal ion complexes [3]. TBDTs consist of a C-terminal 22-strand beta-barrel and a N-terminal plug region that interacts with TonB from an inner membrane complex that provides energy derived from ATP hydrolysis to drive the transport process [4]. The principles underlying the hybrid antigen approach could also be applied more generally

Fadil Bidmos et al. (eds.), *Bacterial Vaccines: Methods and Protocols*,
Methods in Molecular Biology, vol. 2414, https://doi.org/10.1007/978-1-0716-1900-1_8,

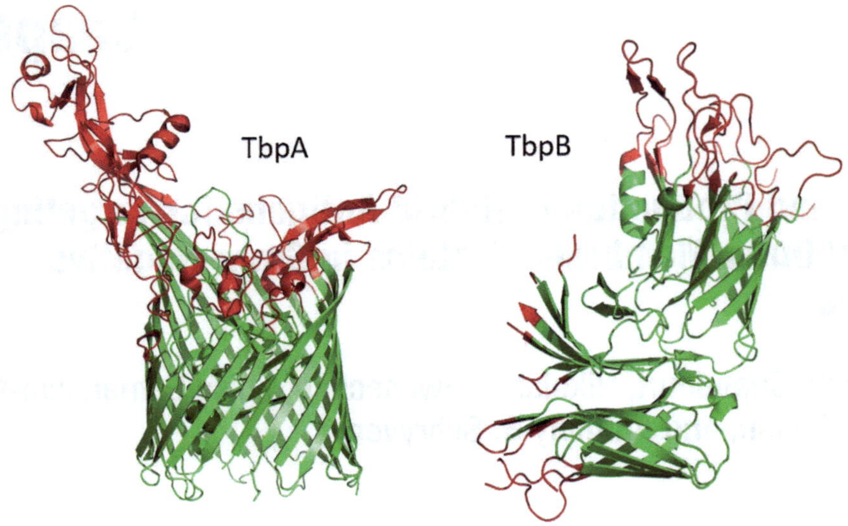

Fig. 1 TonB-dependent transporters (TBDTs) such as transferrin binding protein A (TbpA) and surface lipoproteins (SLP) such as TbpB both have loops (red) that are anchored by antiparallel beta-strands (green). Loops on the soluble SLP can be replaced with loops from TBDTs to generate novel hybrid antigens

to other integral OMPs in gram-negative bacteria. Integral OMPs consist of beta-barrels that have connecting extracellular surface "loop" regions that are potential targets for the immune response. These integral OMPs are typically not suitable for commercial vaccine production due to poor solubility. Thus, displaying the extracellular surface loops on soluble forms of surface lipoproteins (SLPs) that also possess beta-barrel type structures provide a potentially commercially viable means of antigen production (Fig. 1).

The first step in the design of hybrid antigens is the selection of an appropriate integral OMP to target for vaccine development. Ideally, the OMP is present in all strains of the bacterial pathogen that will be targeted with the vaccine and is required for survival in their mammalian or other vertebrate host. TBDTs are ideal targets due to the essential role they play in the survival of the bacteria and their expression under most conditions in their host [3, 5]. Several groups of gram-negative bacteria that reside exclusively in the upper respiratory or genitourinary tracts of their host possess TBDTs that are utilized for acquiring iron from host transferrin or lactoferrin [6]. These TBDTs have been shown to be essential for survival and disease pathogenesis [7–9] and thus fit the criteria for ideal vaccine targets.

The other requirement for the hybrid antigen approach is an SLP with the appropriate structural features for loop display—a characteristic of proteins transported to the surface by the surface lipoprotein assembly module (SLAM) system [10], which is present in most gram-negative species of interest. Receptors involved in acquisition of iron from host transferrin or lactoferrin have a bilobed SLP component, transferrin or lactoferrin binding protein B (TbpB or LbpB), that can readily serve as the loop display scaffold. In the first hybrid antigen study, a loopless C-lobe (LCL), derived from the TbpB of the human pathogen *Neisseria meningitidis*, and the intact TbpB from the porcine pathogen *Glaesserella parasuis* were used as display scaffolds [1]. Using the detailed structural information for the scaffold proteins [11, 12] and the TBDT, TbpA from *Neisseria* [13], single, modest-sized loop regions were selected for display. The successful production, using conventional systems, of soluble hybrid antigens in this study was likely a function of displaying single, modest-sized TbpA loop regions as opposed to using multiple, larger loops. One consequence of using the LCL in this study was the need to use a *Neisseria* strain deficient in TbpB to assess the contribution of the displayed TbpA loops to the functional properties of the immune response induced by the hybrid antigen. It is important to appreciate that most hybrid antigens that are designed for commercial use would employ a scaffold from the targeted bacterial pathogen and thus would resemble the *Neisseria* TbpA/B hybrid antigens described in this study.

The *N. meningitidis* LCL [1] was also used as the scaffold in a follow-up study for display of the surface loops of the zinc transport protein, ZnuD, from *Acinetobacter baumannii* [2]. Since a series of hybrid antigens displaying individual loops (or combinations of loops) were planned, an N-terminal maltose binding protein (MBP) was included to facilitate proper folding of the hybrid antigen. However, the expression from a T7 promoter resulted in the production of insoluble protein for most of the hybrid antigens, requiring solubilization with 8 M urea and refolding with a gradient of decreasing urea concentrations to obtain a soluble hybrid antigen [2]. In this study, the hybrid antigen was not purified from the N-terminal MBP after cleavage with tobacco etch virus (TEV) protease, but rather the entire recombinant protein complex was used in the immunization and challenge experiments. Fortunately, the control LCL scaffold with N-terminal MBP protein did not induce a protective immune response against *A. baumannii* in a mouse sepsis model, indicating that the protection induced by hybrid antigens could be attributed to the individual loop regions present on the scaffold.

One of the challenges with the hybrid antigen approach is the ability to determine the proportion of the antibody response that is induced against the specific loop being displayed, and particularly the titer of antibody directed against conformations of the loop present in the native TBDT/OMP. Although whole-cell enzyme-linked immunosorbent assays (ELISAs) using the target bacterium can provide some assessment of the titer of the relevant antibodies, results are influenced by numerous factors including the level of expression of the targeted TBDT/OMP, antigenicity of the targets (which could be affected by the method of inactivating or killing the bacterium for coating plates), and the presence of preexisting antibodies in tested sera that could react with a variety of antigens present on the bacterium.

In this chapter, we describe a method for preparing ELISA plates with the native TBDT that is designed to overcome these limitations and provide the titer of antibody directed against native conformation of the loops present in the hybrid antigen. The use of the N-terminal streptavidin binding peptide (SBP) and streptavidin-coated plates eliminates the need to isolate the TBDT from a crude extract and can be exploited to readily evaluate the cross-reactivity of antisera against heterologous TBDT variants. We have developed this protocol by modifying our high-throughput, nonbiased ELISA method which uses streptavidin-coated plates to capture and immobilize soluble antigens from crude *Escherichia coli* lysates in which the recombinant protein of interest is biotinylated in vivo during expression [14]. The extension of this approach with TbpA protein was feasible due to the ability to assess functionally folded protein by binding of labeled transferrin. Although this assay may not be available for other TBDTs, such as siderophore receptors, it could be performed in parallel or by coexpression of TbpA and the TBDT. For hybrid antigens displaying multiple loops, the sera would have to be pre-treated with hybrid antigens displaying individual loops, or the single-loop hybrids would have to be included in the incubation mixtures to determine the titer against individual loops.

In this chapter, we provide detailed methods for designing hybrid antigens and producing soluble and insoluble proteins with a conventional T7 expression system that clearly has many limitations. Alternative expression systems are being explored and the reader should not hesitate to use alternate systems for production and purification of the hybrid antigens. The method for preparation of ELISA plates for evaluating the antibody response against the loops displayed in the hybrid antigen can be implemented with a representative set of variant TBDTs to evaluate the cross-reactivity of the antibody response that would likely correlate with the cross-protective properties of the immune response. We hope that the additional information in the Notes regarding the design and production of hybrid antigens and assessment of the immune response will provide sufficient insight for successful implementation of this approach for other OMPs being considered as candidate vaccine antigens.

2 Materials

Prepare all solutions using double-distilled water. Prepare and store all solutions at room temperature, unless indicated otherwise. Any sterilization procedures, if applicable, will be listed in the following descriptions of each reagent.

2.1 Autoinduction Medium Components

1. ZY media: 1% tryptone, 0.5% yeast extract. Weigh out 15 g tryptone and 7.5 g yeast extract and transfer to a 4 L Erlenmeyer flask. Add 1.5 L of distilled water and place a piece of aluminum foil over the mouth of the flask. Autoclave, then store at room temperature.
2. 20× NPS: 0.5 M $(NH_4)_2SO_4$, 1 M KH_2PO_4, 1 M Na_2HPO_4. Weigh out 132 g $(NH_4)_2SO_4$, 272 g KH_2PO_4, and 284 g Na_2HPO_4 and transfer to a large beaker. Add 1400 mL of distilled water, dissolve solids with agitation, then transfer solution to a graduated cylinder and make up volume to 2 L with water. Transfer to a 2-L glass bottle, autoclave, then store at room temperature.
3. 50× 5052: 2.5% glucose, 10% lactose, 25% glycerol. Weigh out 50 g glucose, 200 g lactose, and 500 g (396.42 mL) glycerol and transfer to a large beaker. Add 1400 mL distilled water, dissolve solids with agitation, then transfer solution to a graduated cylinder and make up volume to 2 L with water. Transfer to a 2 L glass bottle, autoclave, then store at room temperature.
4. 1 M $MgSO_4$: Weigh out 61.62 g of $MgSO_4$ heptahydrate and transfer to a beaker. Add 200 mL of distilled water, dissolve solids with agitation, then transfer to a graduated cylinder and make up volume to 250 mL with water. Transfer to a 250 mL glass bottle, autoclave, then store at room temperature.

2.2 Other Media Used for Cultivating Bacteria

1. LB liquid media: Dissolve 25 g of commercial premixed solids in 1 L of distilled water by shaking manually or with a magnetic stir bar. Autoclave, then store at room temperature.
2. LB agar plates: Dissolve 32 g of commercial premixed solids in 1 L of distilled water by shaking manually or with a magnetic stir bar. Autoclave, let cool, and then add the appropriate antibiotic (if desired). Mix using a magnetic stir bar, then pour into Petri dish plates (~20–25 mL in each).

2.3 Stock Solutions of Components of Protein Purification Buffers

1. 1 M Tris pH 8.0: Weigh out 242.28 g of Tris and transfer to a large beaker. Add 1400 mL of distilled water, dissolve solids with agitation, then transfer solution to a graduated cylinder and make up volume to ~1950 mL to leave room for the

addition of acid to adjust to the desired pH. Filter-sterilize, adjust pH to 8.0 using concentrated HCl, then store at room temperature in a 2 L glass bottle.

2. 5 M NaCl: Weigh out 584.4 g of NaCl and transfer to a large beaker. Add 1400 mL of distilled water. Partially dissolve solids using heat and agitation, then gradually add more water until all solids have dissolved. If necessary, transfer solution to a graduated cylinder and make up volume to 2 L with water (*see* **Note 1**). Filter-sterilize, then store at room temperature in a 2-L glass bottle.
3. 1 M imidazole pH 7.4: Weigh out 68.08 g of imidazole and transfer to a large beaker. Add 700 mL of distilled water, dissolve solids with agitation, then transfer solution to a graduated cylinder and make up volume to ~950 mL to leave room for the addition of acid to adjust to the desired pH. Filter-sterilize, adjust pH to 7.4 using concentrated HCl, then store at room temperature in a 1 L glass bottle wrapped in aluminum foil to shield solution from exposure to light.
4. 10× phosphate-buffered saline (PBS) pH 7.4: Weigh out 80 g NaCl, 2 g KCl, 14.4 g Na_2HPO_4, 2.4 g KH_2PO_4 and transfer to a large beaker. Add 700 mL of distilled water, dissolve solids with agitation, then transfer solution to a graduated cylinder and make up volume to ~950 mL to leave room for the addition of acid to adjust to the desired pH. Filter-sterilize, adjust pH to 7.4 using concentrated HCl, then store at room temperature in a 1 L glass bottle.

2.4 Protein Purification Buffers

1. Resuspension Buffer: 50 mM Tris pH 8.0, 300 mM NaCl, 10 mM imidazole pH 7.4. Mix 50 mL of 1 M Tris pH 8.0 stock solution, 60 mL of 5 M NaCl stock solution, and 10 mL of 1 M imidazole pH 7.4 stock solution in a graduated cylinder. Make up volume to 1 L using distilled water. Filter-sterilize, then store at room temperature.
2. Wash Buffer: 50 mM Tris pH 8.0, 1 M NaCl, 20 mM imidazole pH 7.4. Mix 50 mL of 1 M Tris pH 8.0 stock solution, 200 mL of 5 M NaCl stock solution, and 20 mL of 1 M imidazole pH 7.4 stock solution in a graduated cylinder. Make up volume to 1 L using distilled water. Filter-sterilize, then store at room temperature.
3. Elution Buffer: 50 mM Tris pH 8.0, 300 mM NaCl, 300 mM imidazole pH 7.4. Mix 50 mL of 1 M Tris pH 8.0 stock solution, 60 mL of 5 M NaCl stock solution, and 300 mL of 1 M imidazole pH 7.4 stock solution in a graduated cylinder. Make up volume to 1 L using distilled water. Filter-sterilize, then store at room temperature.

4. Exchange Buffer: 50 mM Tris pH 8.0, 600 mM NaCl. Mix 100 mL of 1 M Tris pH 8.0 stock solution with 120 mL of 5 M NaCl stock solution in a graduated cylinder. Make up volume to 1 L using distilled water. Filter-sterilize, then store at room temperature.
5. Denaturing Lysis Buffer: 50 mM NaH_2PO_4, 300 mM NaCl, 6 M urea, pH 8.0. Weigh out 6.9 g NaH_2PO_4 monohydrate and 360.36 g urea and transfer to a large beaker. Add 60 mL of 5 M NaCl stock solution and 640 mL distilled water. Dissolve solids with agitation, then transfer to a graduated cylinder and make up volume to ~950 mL, leaving room for the addition of acid to adjust the pH. Adjust pH to 8.0 using concentrated HCl.
6. Denaturing Wash Buffer: 50 mM NaH_2PO_4, 300 mM NaCl, 6 M urea, 5 mM imidazole, pH 8.0. Weigh out 6.9 g NaH_2PO_4 monohydrate and 360.36 g urea and transfer to a large beaker. Add 60 mL of 5 M NaCl stock solution, 5 mL of 1 M imidazole pH 7.4 stock solution, and 635 mL of distilled water. Dissolve solids with heat and agitation, then transfer to a graduated cylinder and make up volume to ~950 mL, leaving room for the addition of acid to adjust the pH. Adjust pH to 8.0 using concentrated HCl.
7. Refolding Wash Buffer: 50 mM NaH_2PO_4, 300 mM NaCl, 5 mM imidazole, pH 8.0. Weigh out 6.9 g NaH_2PO_4 monohydrate and transfer to a large beaker. Add 60 mL of 5 M NaCl stock solution, 5 mL of 1 M imidazole pH 7.4 stock solution, and 635 mL of distilled water. Dissolve solids with agitation, then transfer to a graduated cylinder and make up volume to ~950 mL, leaving room for the addition of acid to adjust to the pH. Adjust pH to 8.0 using concentrated HCl.
8. Refolding Elution Buffer: 50 mM NaH_2PO_4, 300 mM NaCl, 400 mM imidazole, pH 8.0. Weigh out 6.9 g NaH_2PO_4 monohydrate and transfer to a large beaker. Add 60 mL of 5 M NaCl stock solution, 400 mL of 1 M imidazole pH 7.4 stock solution, and 240 mL of distilled water. Dissolve solids with agitation, then transfer to a graduated cylinder and make up volume to ~950 mL, leaving room for the addition of acid to adjust the pH. Adjust pH to 8.0 using concentrated HCl.

2.5 Other Materials for Protein Production and Purification

1. Protease inhibitor tablets: store at 4 °C or according to the manufacturer's instructions until ready to use.
2. Lysozyme: prepare 10 mg/mL solution by dissolving solids in ddH_2O and storing in 6-mL aliquots at −20 °C until ready to use.
3. Deoxyribonuclease (DNase) I: prepare 5 mg/mL solutions by dissolving solids in ddH_2O and storing in 200-μL aliquots at −20 °C until ready to use.

4. Nickel-nitrilotriacetic acid (Ni-NTA) resin: prepare according to the manufacturer's instructions and store at 4 °C until ready to use (*see* **Note 2**).
5. Dialysis tubing, 6–8 kDa molecular weight cutoff (MWCO): Store at room temperature until ready to use. Moisten the membrane with ddH_2O immediately prior to use.
6. Centrifugation concentrator, 50 kDa MWCO (*see* **Note 3**): Prime with 20 mL ddH_2O followed by 20 mL of the appropriate buffer immediately prior to use.

2.6 Materials for Hybrid Antigen Design

1. DNA and protein sequences of the SLP scaffold and the TBDT.
2. Access to programs/software for protein modeling, protein visualization, and gene cloning and visualization (*see* **Note 4**).

2.7 Materials for TBDT Extraction

1. 50 mM Tris pH 8.0: Add 50 mL of 1 M Tris pH 8.0 stock solution to a 1-L graduated cylinder. Make up volume to 1 L using distilled water. Filter-sterilize, adjust pH to 8.0 with HCl, transfer to a 1-L glass bottle, then store at room temperature.
2. Elugent detergent (*see* **Note 5**).

2.8 ELISA Reagents

1. PBST: Add 100 mL of 10× PBS to a 1-L graduated cylinder. Make up volume to 1 L using distilled water. Add 0.5 mL of Tween-20 (final concentration: 0.05%) using a wide-bore 1 mL pipette tip. Pipette up and down numerous times to ensure that all the viscous detergent is mixed into the PBS. Shake the bottle to mix well. Store at room temperature. This can be prepared several days in advance.
2. Blocking solution: Add 2.5 g of skim milk powder to 25 mL of PBST in a 50-mL conical tube and vortex to mix. Make up volume to 50 mL with more PBST. This can be prepared a day in advance and stored at 4 °C. Bring to room temperature prior to use. Alternatively, 5% w/v bovine serum albumin (BSA) can also be used as a blocking reagent: Add 50 g of BSA to 1 L of PBS and dissolve with stirring. Once dissolved, filter-sterilize and store at 4 °C.
3. Diluent solution: Add 1.25 g of skim milk powder to 25 mL of PBST in a 50-mL conical tube and vortex to mix. Make up volume to 50 mL with more PBST. This can be prepared a day in advance and stored at 4 °C. Bring to room temperature prior to use.
4. Transferrin solution: Using a stock solution of 0.5 mg/mL transferrin conjugated to horseradish peroxidase (Tf-HRP), prepare a 1:1000 dilution for a working solution in 2.5% skim milk solution in PBST just prior to use. Vortex to mix well.

5. 3,3′,5,5′-Tetramethylbenzidine (TMB) substrate solution: Prepare according to manufacturer's instructions and store at 4 °C until use.
6. 4 N HCl quenching solution: Make this solution in a fume hood. Measure out 200 mL of distilled water in a graduated cylinder, then transfer to a clean 500 mL glass bottle. Measure out 100 mL of 12 N HCl in a clean glass graduated cylinder. Gently and slowly pour the HCl into the water. Stir with a clean glass rod. Let sit for 2 min. Store at room temperature.
7. Serum samples from animals immunized with TbpA, TbpA-based hybrid antigens, or other TBDT-based antigens. Store at −20 °C until use. Thaw on ice for 20–30 min and vortex to mix prior to use.
8. Secondary antibody solution: HRP-conjugated antibody specific for detection of IgG of species immunized. Prepare according to the manufacturer's instructions, divide into 10-μL aliquots, and store at −20 °C until use. For use, thaw an aliquot on ice for 20 min and prepare working solution to 1:10,000 (or according to manufacturer's recommendations) in diluent solution just prior to use.
9. Other miscellaneous materials: streptavidin- or neutravidin-coated 96- or 384-well ELISA plates, multichannel pipettors, adhesive cover slips for ELISA plates, reagent boats/basins for containing ELISA solutions.

3 Methods

Carry out all procedures at room temperature unless stated otherwise. All techniques with live cells are to be performed in an aseptic manner either next to a Bunsen burner or in a biosafety cabinet.

3.1 Designing and Preparing Hybrid Antigen Genes

1. Acquire published structures of the SLP scaffold and the TBDT. If structures are not available, generate structural models of the scaffold and the TBDT in silico. Programs such as I-TASSER, RaptorX, or Phyre2 can all be used.
2. Visualize the models using PyMOL.
3. Identify the anchoring residue sites of loops on the SLP scaffold and the TBDT (*see* **Note 6**). Locate the antiparallel beta-strands in each model and label the residues in alternating color leading up to the "loop," as shown in Fig. 2 for the example with TBDT (*A. baumannii* ZnuD) and SLP scaffold (derivative of TbpB from *N. meningitidis*). The splicing site of the SLP scaffold is determined by selecting anchoring residues that are next to each other (parallel to each other) and have side chains that face the same direction on the scaffold ("out" and

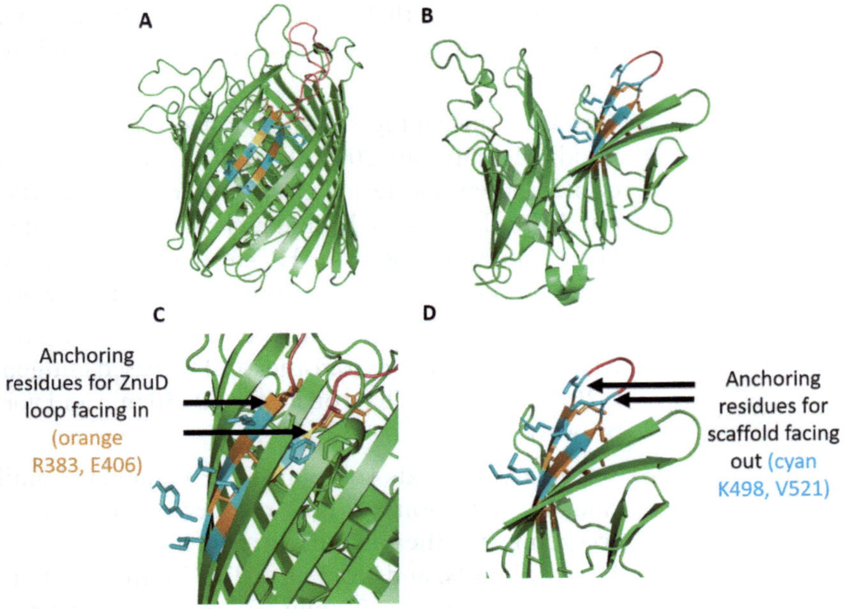

Fig. 2 Example of hybrid antigen design. (**a**) Computer generated model of the TBDT, ZnuD, from *Acinetobacter baumannii* with a loop highlighted (red) where the residues of the anchoring beta-barrel are colored differently depending on whether their side chains are oriented toward the inside of the barrel (orange) or the outside/ away from the barrel (cyan). (**b**) The scaffold, a derivative of the SLP TbpB C-lobe from *Neisseria meningitidis* strain M982 (PDB 5KKX) [1], is labeled similarly to **a**, with alternating residues colored. (**c**) The loop from the ZnuD is determined by selecting parallel anchoring residues (side chains facing the same direction) on the beta-barrel. If the anchoring residues of the loop are facing in (black arrow in **c**), then the anchoring residues of the scaffold face out (black arrow in **d**) to maintain the antiparallel beta-strand formation

labeled with arrow in Fig. 2d). These anchoring residues are included in the SLP scaffold (*see* **Note 6**).

4. Determine the splicing site of the loop by selecting parallel anchoring residues that face the opposite direction to that of the SLP scaffold on the TBDT ("in" and labeled with arrow Fig. 2c). These anchoring residues are included in the loop.
5. Using an in silico program for DNA cloning, identify the gene segments corresponding to the structural attributes of the SLP scaffold and TBDT identified in **steps 3** and **4**.
6. Generate the DNA sequence of the novel hybrid antigen by combining the different gene segments from both the SLP scaffold and the TBDT. Translate the gene sequence in silico to ensure the resulting protein is in frame.
7. The novel hybrid gene can either be synthesized by a commercial vendor or can be generated in the lab via splicing-by-overlap/extension polymerase chain reaction (SOE-PCR) using primers that anneal to the appropriate regions of the templates.

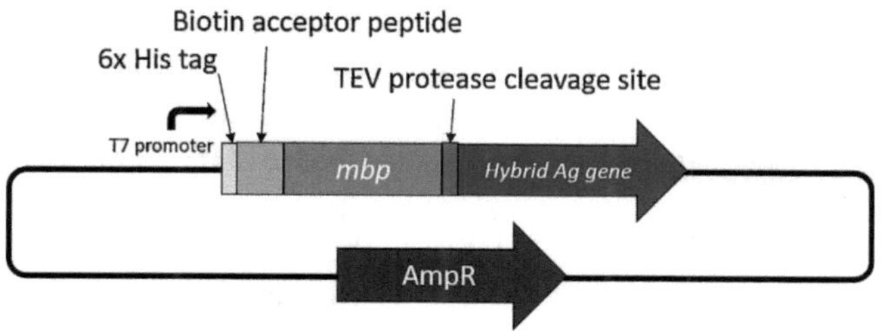

Fig. 3 A schematic of the His-Bio-MBP T7 expression vector (pE5770) with a gene of interest (hybrid antigen) cloned into the expression locus. The components of the vector (*see* **Note 7**) that are relevant to the protein production methods discussed in this manuscript are illustrated in the figure

8. Clone the hybrid gene fragment into plasmid pE5770 or any equivalent plasmid with similar elements (*see* **Note** 7 and Fig. 3).
9. In preparation for transformation of the newly constructed plasmid into *E. coli* TOP10 cells for long-term storage, thaw an aliquot of chemically competent TOP10 cells on ice.
10. Pipet 50–100 ng of the plasmid encoding the hybrid antigen gene into the thawed aliquot and incubate on ice for 30 min (*see* **Notes 8** and **9**).
11. Heat-shock the cells at 42 °C for 30–60 s. Promptly place the tube back on ice for 2 min.
12. Add 400 μL of cold sterile LB medium and then transfer the tube to a 37 °C shaking incubator for 1 h.
13. Plate 100 μL of transformation mixture onto an LB agar plate supplemented with 100 μg/mL ampicillin and incubate at 37 °C overnight. Colonies generated with the pE5770 plasmid can be streaked onto plates supplemented with 30 μg/mL kanamycin to screen for loss of the original insert.
14. Confirm by colony PCR or PCR of plasmid DNA prepared from positive clones and sequencing of insert or by restriction digestion of plasmid.
15. Isolate and store plasmid at −20 °C for subsequent use.
16. Inoculate 5-mL of LB medium supplemented with 100 μg/mL ampicillin with a colony confirmed to contain the desired plasmid. Incubate overnight (~16 h) at 37 °C in a shaking incubator.
17. The next day, create a 16% glycerol stock by mixing 800 μL of the broth culture with 200 μL of sterile 80% glycerol. Store at −80 °C.

3.2 Expressing Hybrid Protein Antigens in E. coli Using a T7 Expression Vector

1. Thaw out the plasmid encoding the desired hybrid protein designed in Subheading 3.1, as well as the competent cells derived from the strain of *E. coli* to be used for protein expression (*see* **Note 10**). The plasmid concentration should be ~100 ng/μL, and competent cells should be stored in 50- or 100-μL aliquots. After removing the competent cells from the −80 °C freezer, let the cells sit on ice for 20 min prior to transformation.
2. Add 1–2 μL of plasmid to the competent cells and let the cells sit on ice for 30 min. Next, transfer to a prechilled 14-mL round-bottom culture tube.
3. Prepare a 42 °C water bath or heat block, then "heat-shock" the cells at this temperature for 30–60 s, followed by a 2-min incubation on ice. Promptly add 700 μL of LB medium, then incubate the cells with shaking at 37 °C for 1 h.
4. Add 6 mL of LB containing the appropriate antibiotic (i.e., add 6 μL of 100 mg/mL ampicillin if you are using ampicillin resistance to select for the presence of your plasmid) to the culture and incubate with shaking at 37 °C for another 4 h.
5. Approximately 1–1.5 h before the end of the 4-h incubation period, prepare the autoinduction medium by aseptically adding 75 mL NPS, 30 mL 5052, and 1.5 mL $MgSO_4$ to 1.5 L of ZY media (*see* **Note 11**). Four 4-L flasks containing 1.5 L of media in each are used in a typical protein production run, but the culture volume can be adjusted depending on the expected yield of the protein being produced. Add the appropriate antibiotic at half the normal concentration (i.e., use 50 μg/mL instead of 100 μg/mL if using ampicillin). Prewarm the media in 37 °C shaking incubator at a shaking speed of ~50 rpm (*see* **Note 12**).
6. At the end of the 4-h incubation mentioned in **step 4**, inoculate 1.5 L of autoinduction media with 1.5 mL of the starter culture. Repeat as needed, depending on the number of flasks being used.
7. Incubate the flasks at 37 °C with vigorous shaking (175 rpm) for 18 h, then adjust the temperature to 20 °C and incubate for another 24 h (*see* **Note 13**).
8. Harvest cells by centrifugation at 5000 × *g* for 25 min at 4 °C.
9. Prepare Resuspension Buffer (25 mL per L of culture) and add 1 protease inhibitor tablet, 6 mL of 10 mg/mL lysozyme, and 200 μL of 5 mg/mL DNase I (increase the amounts of these reagents as necessary if the culture volume is greater than 6 L). Mix in an appropriate-sized beaker, keeping in mind that the cell pellets will subsequently be transferred to this beaker.

10. After centrifugation, decant the supernatant and resuspend the cell pellet(s) in the Resuspension Buffer mixture prepared in **step 9** (*see* **Note 14**). For optimal lysis, add a magnetic stir bar to the mixture and let stir at slow speed for 30 min at 4 °C to get rid of cell clumps.
11. Lyse the cells by passing the sample through a cell homogenizer four times (*see* **Note 15**).
12. After cell lysis, centrifuge at 35,000 × *g* for 90 min at 4 °C to separate out the cell debris.
13. Collect the supernatant and discard the pelleted cell debris. Filter the supernatant through a 0.2-μm filter (*see* **Note 16**).
14. Determine whether the fusion protein is present in the supernatant by performing sodium dodecyl sulfate-polyacrylamide gel electrophoresis (SDS-PAGE) using a 10-μL aliquot of the supernatant. If a band corresponding to the molecular weight of the fusion protein of interest is detected in the SDS-PAGE gel, proceed to Subheading 3.3. If the protein of interest is not detected, proceed to Subheading 3.4 (*see* **Note 17**).

3.3 Purification of Soluble Protein Antigens

1. Prepare 1 L each of Resuspension Buffer, Wash Buffer, and Elution Buffer.
2. Wash and equilibrate a 5-mL Ni-NTA column with Resuspension Buffer prior to use (*see* **Note 18**).
3. Using either a peristaltic pump or a fast-purification liquid chromatography (FPLC) system, continuously circulate the lysate prepared in Subheading 3.2 through the Ni-NTA column overnight (16–18 h) at 4 °C. If using a peristaltic pump, circulate lysate at ~0.5 mL/min (*see* **Note 19**).
4. Wash the column with at least 75 mL Wash Buffer, or until the protein concentration of the wash fractions is <0.1 mg/mL. Take one or more samples of the wash fractions for SDS-PAGE analysis.
5. Elute the protein bound to the column using high-imidazole Elution Buffer in 2-mL aliquots until the protein concentration reaches <0.1 mg/mL.
6. Take samples of the eluted fractions for SDS-PAGE to verify the presence of the desired protein antigen and determine which fractions to collect and pool for further processing (Fig. 4a). If desired, include the wash fractions as well to confirm that the protein is present in the eluted fraction and not the wash fractions and therefore is binding to the Ni-NTA resin.
7. Determine the amount of protein present in the pooled sample-containing fraction using a spectrophotometer or NanoDrop device (or equivalent method of determining protein concentration).

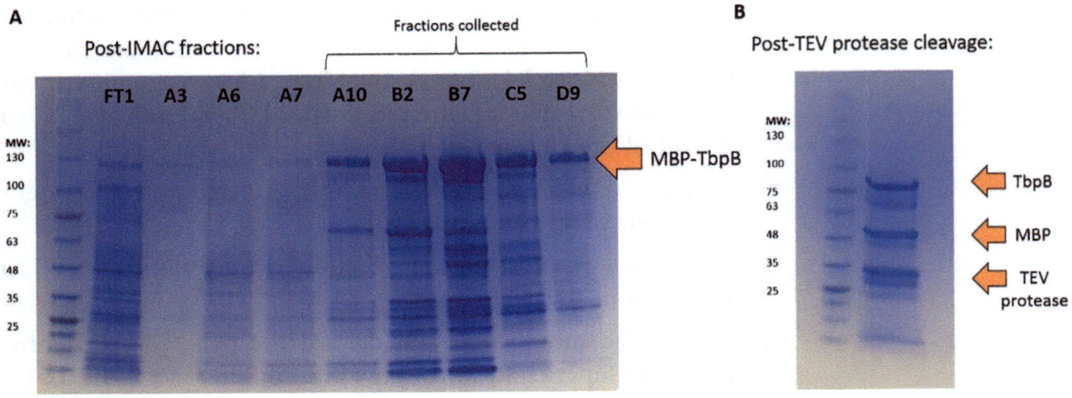

Fig. 4 Representative SDS-PAGE results for purification of a soluble hybrid antigen using the TbpB derived from *Haemophilus influenzae* as an example. (**a**) SDS-PAGE results for 10-μL aliquots of fractions from a Ni-NTA (IMAC—immobilized metal affinity chromatography) column. FT1 indicates a sample from the pooled flow through buffer containing proteins that did not bind to the column. Fractions A3-D9 represent the specific wells of the 96-well collection block that collected the fractions eluted with the imidazole gradient. Fractions A3–D9 were pooled based on the presence of the 130-kDa MBP-TbpB fusion protein for subsequent TEV cleavage. (**b**) SDS-PAGE results for a 10-μL aliquot of the MBP-TbpB fusion protein following cleavage with TEV protease. The gel confirms that TEV cleavage was complete as there is no MBP-TbpB band present with TbpB, MBP and TEV protease as the main protein bands. Applying this sample to a Ni-NTA column removes the MbpB and TEV, leaving TbpB as the main protein (similar to the purified LbpB example shown in Fig. 5)

8. In preparation for cleavage of the fusion protein using tobacco etch virus (TEV) protease—for which high-salt, low-imidazole conditions are optimal—transfer the eluted protein sample to dialysis tubing, ensure the tube is fastened at both ends, and place the tube in a beaker containing 2 L of Exchange Buffer. Place a magnetic stir bar in the beaker and let stir at a slow speed overnight (16–18 h) at 4 °C.
9. Remove the dialysis tubing from the 2-L beaker, open at one end of the tube, and add TEV protease (*see* **Note 20**). Then, place the tubing back in the Exchange Buffer (this can be carried over from **step 8**; there is no need to prepare fresh Exchange Buffer). Let stir overnight at 4 °C.
10. The next day, run a sample of the cleaved protein on an SDS-PAGE to determine the cleavage efficiency (Fig. 4b). If cleavage is incomplete—typically ascertained by the presence of the intact fusion protein in the SDS-PAGE gel—add more TEV protease.
11. In preparation for a second round of Ni-NTA chromatography to remove the cleaved MBP fusion partner and TEV protease from the protein sample, dialyze the sample overnight at 4 °C against 2 L of Resuspension Buffer. This step serves to replenish the imidazole that was removed from the sample during dialysis against Exchange Buffer in **step 8**.

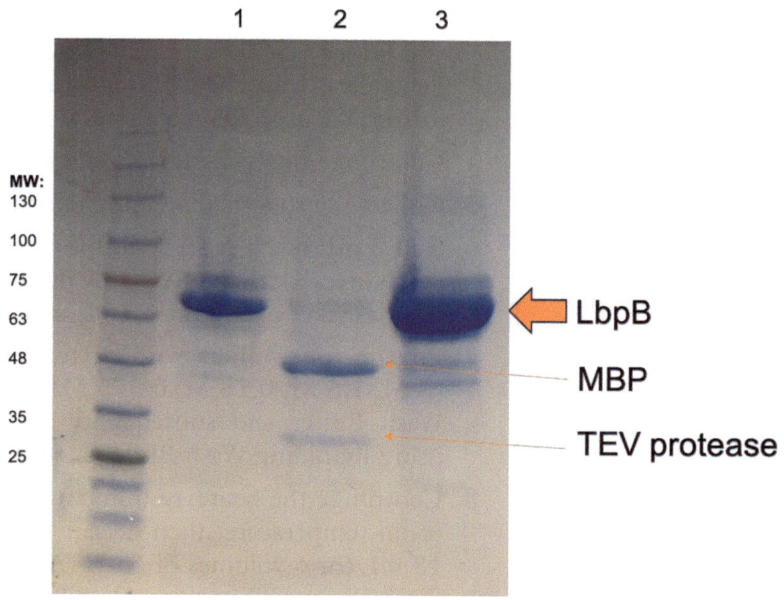

Fig. 5 Representative SDS-PAGE results following the separation of MBP and TEV protease from the recombinant protein of interest using a Ni-NTA column. A modified *Neisseria gonorrhoeae* lactoferrin binding protein B (LbpB) is provided as an example. Lane 1: flow through containing the LbpB; lane 2: pooled eluted fractions (MBP and TEV protease are both present here); lane 3: concentrated sample of flow through (lane 1). Each lane was loaded with a 10-μL aliquot

12. Circulate the cleaved protein through the Ni-NTA column as described in **step 3** (*see* **Note 21**). The polyhistidine-tagged MBP and TEV protease will bind to the resin and will thus separate these contaminants from the desired protein antigen.
13. After circulation through the Ni-NTA column, take a sample of the protein (which should not have bound to the column) and run an SDS-PAGE gel to confirm that the MBP and TEV protease have been separated from the desired protein antigen (Fig. 5).
14. Exchange the buffer containing the purified protein by dialysis against PBS overnight (16–18 h) at 4 °C, then concentrate the protein to a final concentration of 1–10 mg/mL by centrifugation using a concentrator with the appropriate MWCO.
15. Promptly store the protein in 100-μL aliquots at −80 °C until ready to use for animal immunizations (*see* **Note 22**).

3.4 Purification of Insoluble Protein Antigens

1. Perform **steps 1–8** described under Subheading 3.2.
2. Prepare Denaturing Lysis Buffer (60 mL per L of culture). After centrifugation (Subheading 3.2, **step 8**), decant the supernatant and resuspend the cell pellet using Denaturing Lysis Buffer.

3. Lyse the cells by sonicating for two 5 min periods on ice using conditions that do not result in overheating the samples. These generally involve short pulses with cooling periods of 2–4 times longer. We use a QSonica sonicator (pulse 03, amplitude 65%).
4. Centrifuge the lysate for 35 min at 20,440 × *g* at 4 °C.
5. Collect the supernatant and filter through a 0.2-μm filter.
6. Add 3 mL of Ni-NTA resin for every 50 mL of filtered lysate (*see* **Note 23**), then mix gently in 50-mL conical tubes using a rotary shaker overnight (~16 h) at room temperature.
7. Prepare four buffers with decreasing concentrations of urea (3 M, 1.5 M, 0.75 M, 0.375 M) by starting with Denaturing Wash Buffer and subsequently performing serial dilutions, using Refolding Wash Buffer as a diluent.
8. Centrifuge the lysate–resin mixture for 10 min at 3220 × *g* at room temperature, then decant the supernatant. Next, using 18 mL (or 6 volumes of Ni-NTA resin) of the initial buffer in the aforementioned serial dilution (50 mM NaH_2PO_4, 300 mM NaCl, 3 M urea, 5 mM imidazole, pH 8.0), resuspend the Ni-NTA resin and incubate the resulting mixture on a rotary shaker for 30 min at room temperature.
9. Repeat **step 8** for each of the subsequent buffers in the serial dilution series prepared in **step** 7.
10. Resuspend the Ni-NTA resin in 18 mL Refolding Wash Buffer, then transfer the mixture to a gravity column.
11. Elute any proteins bound to the Ni-NTA using 18 mL of Refolding Elution Buffer (*see* **Note 24**). The eluted volume can be collected in 1-mL fractions.
12. Verify the presence of the desired hybrid antigen using SDS-PAGE, then perform **steps 8–12** described under Section.

3.5 Cloning of Tbdt in Fusion with Streptavidin Binding Protein (SBP)

1. Identify mature sequence of *TBDT* using the online software SignalP (*see* **Note 25**).
2. Design primers to amplify the selected gene or synthesize the sequence.
3. Clone the *TBDT* gene into plasmid pE5771 or an alternate expression plasmid containing SBP (*see* **Note 26**).
4. Transform the newly constructed plasmid into *E. coli* TOP10 cells for long-term storage, as described in Subheading 3.1, **steps 9–15**.

3.6 Preparation of Detergent-Extracted SBP-TBDT Fusion Protein

1. Thaw plasmid DNA and an aliquot of chemically competent *E. coli* C43 cells on ice for 20 min (*see* **Note 27** and Fig. 6).
2. Pipette 50–100 ng of SBP-TBDT plasmid into the thawed aliquot and incubate on ice for 30 min.

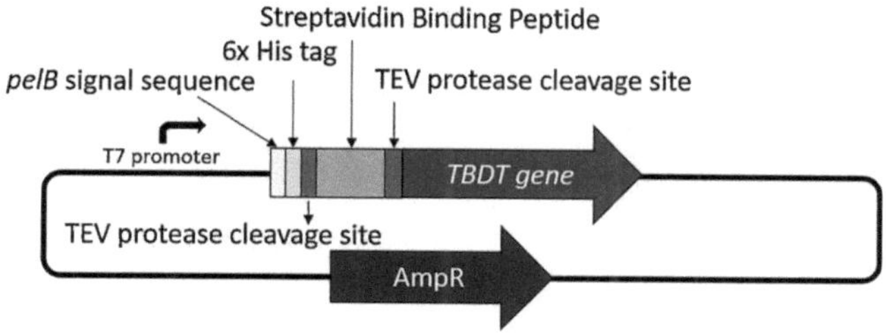

Fig. 6 A schematic of the His-SBP-TBDT T7 expression vector (pE5771) with a gene of interest (a TBDT to be used in ELISA assays) cloned into the expression locus. The components of the vector (*see* **Note 26**) that are relevant to the protein production methods discussed in this manuscript are illustrated in the figure

3. Heat-shock the cells at 42 °C for 30–60 s. Promptly place the tube back on ice for 2 min.
4. Add 400 μL of cold sterile LB medium.
5. Gently pipette out the entire mixture from the 1.5 mL tube and into a 50-mL conical tube. Incubate in a shaking incubator at 37 °C for 1 h.
6. Add 14.5 mL of autoinduction medium supplemented with 100 μg/mL ampicillin and incubate in a shaking incubator at 37 °C overnight.
7. The following day, centrifuge the culture at 3220 × *g* at 4 °C for 10 min.
8. Decant the supernatant, add 15 mL of cold 50 mM Tris pH 8.0, and resuspend the cell pellet.
9. Centrifuge again as described in **step 7**, decant the supernatant, then add 2 mL of cold 50 mM Tris pH 8.0 and 40 μL of Elugent detergent (final concentration: 1%, commercial Elugent is a 50% solution). Shake overnight at 4 °C (*see* **Note 28**).
10. The following day, centrifuge the samples at 16,100 × *g* at 4 °C for 30 min.
11. Carefully remove the supernatant containing the detergent extracted SBP-TBDT and pipette slowly into a 15-mL conical tube containing 8 mL of cold filtered PBST with 0.125% Elugent (0.25% of commercial preparation, final protein dilution: 1 in 5).
12. Gently rotate the tube by hand several times to ensure that it mixes well. Keep on ice until ready to use (*see* **Notes 29** and **30**).

3.7 Coating Streptavidin ELISA Plates with Detergent-Extracted SBP-TBDT

1. Pour the SBP-TBDT–PBST mixture into a reagent boat, and, using a multichannel pipette, pipette out 100 μL into each well of a streptavidin-coated 96-well plate (*see* **Note 31**).
2. Incubate at room temperature for 1 h, then wash with 250 μL/well of PBST three times (*see* **Notes 32** and **33**).
3. Add 250 μL of the prepared 5% blocking solution to each well after the wash step, incubate at room temperature again for 1 h, then wash with 250 μL/well of PBST three times.

3.8 Assessing Conformation of TbpA by Its Ability to Bind to Transferrin

1. During the blocking step (Subheading 3.7, **step 3**), add 1 μL Tf-HRP to 1 mL of diluent solution (i.e., final dilution of 1/1000).
2. Add 100 μL of the Tf-HRP solution to each of four wells (i.e., two test and two control wells), incubate for 1 h, then wash with 250 μL of PBST three times.
3. Add 50 μL of TMB to the four wells and develop in the dark (a cupboard/drawer is adequate) for 20 min, then add 25 μL of 4 N HCl to stop the reaction. Measure the optical density at 450 nm (*see* **Note 34** and Fig. 7).

3.9 Assessing antibody Titers in Serum Samples from Immunized Animals

1. During the blocking step (Subheading 3.7, **step 3**), thaw serum samples on ice for at least 20 min and vortex to mix well.
2. Add 100 μL of diluent solution to each well. Add additional diluent solution in the first well of each row (A to H). The amount of additional diluent solution will depend on the desired starting dilution for the serum used in the assay, such that the final volume of diluent with added serum is 200 μL.

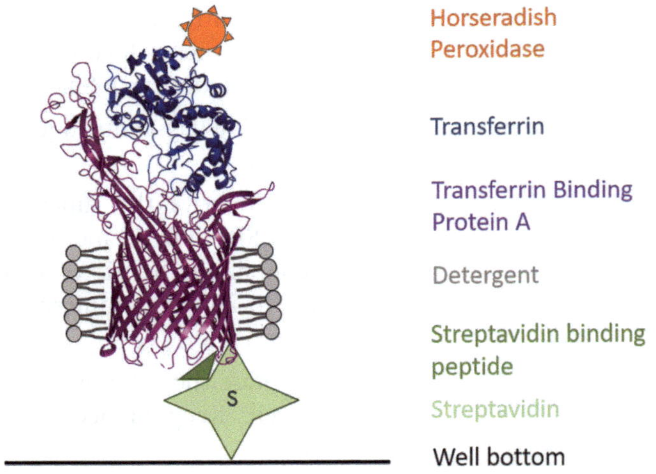

Fig. 7 Schematic of TbpA conformation assay. Detergent-extracted TbpA fused to streptavidin binding peptide (SBP) is immobilized on streptavidin-coated wells. Transferrin conjugated to horseradish peroxidase (Tf-HRP) is then added to the wells. If the immobilized TbpA is folded properly, this allows Tf binding to occur, which results in a signal from the HRP once the developing substrate is added

3. Add the desired amount of each serum sample to the appropriate wells containing diluent solution to achieve the desired initial serum dilution in a total volume of 200 μL in each well in the first column. Gently pipet up and down 10 times to mix.
4. Adjust a multichannel pipettor to 100 μL, then mix the serum solutions in the first column of the plate by pipetting up and down 5 times.
5. Remove 100 μL from the first column and transfer it the next column. Pipette up and down 5 times. Repeat with the next column.
6. Continue performing twofold serial dilutions until column 11 (*see* **Note 35**). Discard the leftover 100 μL.
7. Incubate at room temperature for 1 h, then wash with 250 μL of PBST three times.
8. Prepare the secondary antibody solution by adding 5 μL of anti-rabbit IgG or anti-mouse IgG (as appropriate for the animals in which the immune sera were generated) antibody conjugated to HRP to a 50-mL diluent solution (final antibody dilution: 1 in 10,000). Vortex to mix well.
9. Add 100 μL to each well, except for column 12.
10. In column 12, add 100 μL of diluent solution (no antibody) in wells 12A, 12B, and 12C. These are negative controls.
11. In wells 12D, 12E, and 12F, add 100 μL of the secondary antibody solution. These are the controls to check for background absorbance caused by the secondary antibody binding to TBDT.
12. If using TbpA, then in wells 12G and 12H, add 100 μL of the 1:1000 Tf-HRP solution described in Subheading 3.8, **steps 1** and **2**. These are the controls that ensure that the TbpA is properly folded and coating the wells at similar levels.
13. Incubate at room temperature for 1 h, then wash with 250 μL/well of PBST three times.

    Add 50 μL/well of TMB and develop in the dark (a cupboard/drawer is adequate) for 20 min, then add 25 μL of 4 N HCl to stop the reaction. Measure the optical density at 450 nm.

4 Notes

1. 5 M is close to the upper limit of solubility of NaCl in water; hence, when making the 5 M NaCl stock solution, in addition to the use of vigorous stirring and heat, it is necessary to add close to the full volume of water before all solids can be dissolved.

2. The preference of our group is to use either free Ni-NTA resin in a gravity column or a 5-mL affinity chromatography column containing Ni-NTA resin (i.e., the 5-mL HisTrap High Performance column from Cytiva) installed as part of an FPLC system or attached to a peristaltic pump.
3. Use a centrifugation concentrator with a lower MWCO if the protein of interest has a lower molecular weight than 50 kDa. Since our protein antigens are usually ~60–70 kDa, the protein concentrator most often used by our group is Cytiva's Vivaspin 20 with a MWCO of 50 kDa. The MWCO should ideally be just below the molecular weight of the desired protein antigen to facilitate the removal of any lower molecular weight contaminants that may be present in the sample.
4. Our group prefers to use the online platform I-TASSER for better models or more challenging projects (lower sequence identity) (https://zhanglab.ccmb.med.umich.edu/I-TASSER/) or use Phyre2 (http://www.sbg.bio.ic.ac.uk/~phyre2/html/page.cgi?id=index) for more rapid results in protein structure prediction/modelling. PyMOL is the preferred program for visualizing the models, and either Gene Construction Kit (http://www.textco.com/gene-construction-kit.php) or SnapGene (https://www.snapgene.com/) is used for in silico cloning.
5. We use a commercially available Elugent detergent (50% Solution) that is readily available from a variety of commercial suppliers. We have only used this method for TbpA and LbpA that enabled us to monitor isolation of functional protein by a solid-phase binding assay. However, this method should work effectively for other TBDTs.
6. We recommend selecting multiple loops of the TBDT to display on the scaffold as there is no empirical method for determining yet which loop will successfully fold with the scaffold and be sufficiently immunogenic and protective. For the scaffold, loops that are large can be selected to remove and then replace with a loop from a TBDT. We believe such positions are likely to accommodate TBDT loops of various sizes. We suggest designing and synthesizing various single and multiple loop hybrids using different loops in different positions initially. The number of hybrids can be narrowed down based on which constructs yield stable proteins for immunizations. Out of those hybrids, the one(s) that elicit a protective immune response can be used for future applications.
7. The custom expression vector pE5770 is available from Addgene (www.addgene.org) but any commercial or other available plasmids with similar components would be suitable.

The vector contains the T7 phage promoter for high level expression by the T7 RNA polymerase. It also contains an N-terminal polyhistidine (6 His residues) tag to allow for the purification of the fusion protein by Ni-NTA chromatography, a biotin acceptor peptide to facilitate the in vivo biotinylation by the *E. coli* biotin ligase BirA, MBP to promote the solubility and proper folding of the fusion protein and a cleavage site recognized by TEV protease to allow for the removal of the aforementioned components in the N-terminal fusion tag. The pE5770 plasmid has BamH1 and Xba1/HindIII restriction sites flanking a kanamycin resistance cassette that can be used to confirm successful cloning of the gene of interest (absence of kanamycin resistance). This expression system requires an *E. coli* strain (e.g., ER2566) that contains a chromosomal copy of the gene encoding T7 RNA polymerase under the control of the *lac* promoter and preferably is deficient in both *lon* and *ompT* proteases. In the presence of glucose, catabolite repression inhibits lactose transport so that expression of T7 RNA polymerase is inhibited by binding of the *lacI* repressor to the *lac* promoter. When the glucose in the autoinduction media is depleted, lactose transport is no longer inhibited, leading to activation of T7 RNA polymerase expression, and thereby triggers the expression of the protein of interest encoded by the T7 expression vector (Fig. 3). The rise in cAMP levels upon glucose depletion also increases expression by a cAMP-CAP complex binding to the CAP site upstream of the lac promoter.

8. After the addition of plasmid DNA into the tube containing chemically competent cells, we recommend flicking the tube gently to mix well.

9. Place the LB medium aliquot on ice and set the water bath/heat block at 42 °C immediately after starting the 30-min incubation on ice to ensure that the correct temperature is reached in time for the subsequent step.

10. Our preference is to use the *E. coli* strain ER2566; however, any strain that fits the criteria outlined in **Note 7** would suffice.

11. Due to problems encountered with strains containing the T7 polymerase gene, the transformed mixture is used directly for expression rather than initial plating and selection of colonies for expression experiments. An autoinduction medium without any antibiotics can be prepared beforehand in 4-L flasks and 1-L bottles and stored at room temperature for up to 1 month for use. For making 1-L of medium add 1 mL of $MgSO_4$, 50 mL of 20× NPS, and 20 mL of 50× 5052 to 929 mL of sterile ZY media in a sterile 1-L bottle.

12. This step is not always necessary; however, we find that pre-warming may enhance bacterial growth and protein yields compared to inoculation at room temperature.
13. The yield of protein after growth at 37 °C appears to predict whether an additional period of growth at low temperature will lead to an increased yield. The expression plasmids with lower yields after growth at 37 °C (perhaps unstable or prone to degradation) do not achieve enhanced yields with an additional incubation period at low temperature.
14. For efficient resuspension of the cell pellets, scoop out each cell pellet using a spatula and transfer it to the beaker containing the Resuspension Buffer mixture. Then, transfer ~20 mL of the Resuspension Buffer mixture into each centrifugation bottle using a serological pipettor to resuspend any remaining bacteria. Finally, transfer each of the resulting cell suspensions to the beaker containing the cell pellets and the remaining Resuspension Buffer mixture.
15. Perform four discrete passes through the homogenizer and avoid recirculating the sample as it results in uneven lysis—that is, collect the lysate in a beaker after each independent pass through the homogenizer instead of continuously directing the sample back through the homogenizer.
16. Use appropriate size filtration devices for filtering the supernatant. When using syringe filters, limit the volume to 25–50 mL of lysate, depending on viscosity. If the sample cannot be filtered easily using a 0.2-μm syringe filter, use a 0.45-μm filter first before passing the lysate through the 0.2-μm filter. If the lysate still cannot be easily filtered, an additional 200 μL of DNase I (5 mg/mL) can be added and mixed in with the lysate for 30 min using a magnetic stir bar prior to filtration.
17. If the protein antigen of interest is readily detected in the supernatant, this suggests that the recombinant protein is sufficiently soluble to allow for the purification of the protein using the protocol outlined in Subheading 3.3. If the SDS-PAGE gel indicates the absence of the protein in the supernatant, or "soluble fraction," this suggests that constitutive expression of the protein in the cytoplasm of *E. coli* has resulted in aggregation and misfolding of the protein. Thus, due to the apparent insolubility resulting from improper folding of the protein of interest, the protein is likely present alongside the cell debris in the pellet following centrifugation (Subheading 3.2, **step 12**). In this circumstance we recommend restarting the protein production run (beginning at the start of Subheading 3.2) and proceeding directly to Subheading 3.4 immediately following Subheading 3.2, **step 8**.

18. A 5-mL column is ideal for culture sizes of 6 L or less. If larger than 6 L, use two 5-mL columns in tandem. Our preference is to use the 5-mL HisTrap High Performance column from Cytiva. If a chromatography system is not available, the purification can be performed with Ni-NTA resin in a gravity fed column, by applying wash and elution buffers manually.
19. Circulation can be done at room temperature if 0.02% sodium azide is included in the lysate to prevent any microbial growth. For continuous circulation overnight, set up the column and any attached tubing such that the sample exiting the column is collected the same receptacle as the sample entering the column.
20. In lieu of commercial TEV protease we routinely use a TEV protease with an N-terminal polyhistidine tag produced in-house in *E. coli* using a T7 expression vector. When expressed with a polyhistidine tag, it can be purified using a Ni-NTA column as described in Subheading 3.2, concentrated to a final concentration of 1–10 mg/mL, mixed with sterile 100% glycerol to a final concentration of 50%, and stored at −20 °C. This glycerol mixture—typically containing ~5 mg of TEV protease for up to 100 mg of fusion protein—can then be added directly to the dialysis tubing containing the entirety of the protein antigen sample as part of the TEV cleavage step. Cleavage of the fusion protein by TEV protease is carried out efficiently during the overnight (16–18 h) incubation period at 4 °C—there is no need to extract the protein sample from the dialysis tubing for the cleavage step.
21. A single circulation would suffice and would reduce the risk of losing any uncleaved fusion protein in the process. If just circulating once, set up the column and any attached tubing such that the sample exiting the column is collected in a separate receptacle from the sample entering the column. Then, circulate the sample until the entire sample has passed through the column.
22. Our experience in preparation of antigens by these methods have not resulted in situations where lipopolysaccharide (LPS) toxicity was an issue or where levels measured by the limulus amoebocyte lysate coagulation (LAL) assay were of concern. Thus, aliquots were normally promptly stored at −80 °C; however, it may be prudent to test protein preparations with the LAL assay when using new strains or new proteins to ensure LPS removal is not required. If endotoxin removal is deemed necessary, the methods our group prefers include the MonoQ column from Cytiva, the CHT I or CHT II columns from Bio-Rad, or the Pierce High Capacity Endotoxin Removal Spin Columns.

23. Before adding 3 mL of Ni-NTA resin, wash the resin to eliminate the ethanol, and use 3 mL of Denaturing Lysis Buffer to equilibrate the resin.
24. Depending on the protein purification yield, a greater volume of Refolding Elution Buffer may be needed. Verify the protein concentration at the end of the elution step to confirm that all of the column-bound protein has been eluted.
25. As it is possible to assay SBP-TbpA for proper folding using the transferrin binding assay described in Subheading 3.8, we recommend cloning an SBP-TbpA fusion construct alongside cloning of the desired target SBP-TBDT construct (Subheading 3.5) as a control for steps in Subheadings 3.6 and 3.7.
26. The custom expression vector pE5771 is available from Addgene (www.addgene.org). Vector pE5771 was designed by our group for expression of TBDTs such as TbpA and LbpA. It has a kanamycin resistance cassette in the expression locus between BamH1 and XbaI sites so that colonies generated from cloning the gene of interest into the expression locus can be screened on plates containing kanamycin to confirm insertion (no growth on kanamycin-containing plates). This plasmid encodes the *pelB* signal sequence that results in substantial levels of periplasmic proteins being secreted into the periplasm of *E. coli* protein expression strains such as BL21. The signal sequence is followed by four amino acids preceding a polyhistidine (6× His) tag, a TEV protease cleavage site, SBP, and a second TEV cleavage site, upstream of the mature *tbdt* sequence (Fig. 6). This plasmid can also be used for large-scale TBDT protein preparations, where detergent-extracted TBDT membrane extracts can be purified using a Ni-NTA column and cleaved with TEV protease to generate purified TBDT protein.
27. We recommend using a cell line like *E. coli* C43, a derivative of the commonly used BL21 expression cell line. This cell line has the wild-type *lac* promoter in the chromosome instead of the mutated *lacUV5* promoter present in the parent. The result is reduced expression of the T7 RNA polymerase and subsequent reduction in expression of the target gene. Cell lines like C43 or the similar C41, also known as the Walker cell lines, have been shown to be more appropriate for expression of membrane proteins and toxins. Chemically competent cells can be premade and stored at −80 °C in 50- or 100-μL aliquots in 1.5-mL Eppendorf tubes until use. If the intention is to screen antisera for reactivity against different variants of the TBDT, it is advised to consider performing the procedure in Subheading 3.6 in parallel with plasmids encoding the different variants.

28. We typically remove the cell–Tris–Elugent mixture from the 50 mL conical tube and transfer it into a 2-mL Eppendorf tube and use a rotary shaker in the refrigerator overnight. A specific rpm is not required as long as it is gently shaking overnight.
29. We recommend using the TBDT–PBST mixture for coating ELISA plates the same day as we have not evaluated the impact of storage of the mixture with TbpA which would have enabled us to evaluate maintenance of native conformation by a binding assay.
30. The proportions described in this protocol amount to just enough TBDT for one 96-well ELISA plate coated with 100 μL/well of TBDT-containing solution. This can be scaled up or down based on how many plates or wells the user needs. We recommend making more than what is required in case some protein is lost in the process of coating the plates.
31. When using TbpA as the TBDT, either on its own or as a control for correct conformation of the detergent extracted TBDT, it is recommended to first perform a separate ELISA to test for correct conformation of TbpA prior to assessing antibody titers from immunized animals. For this conformation ELISA, only one column of a streptavidin-coated plate is needed, adding SBP-TbpA/PBST to only two wells (but blocking all wells in the column) following the instructions in Subheading 3.7, **steps 1–3**, and then proceeding to Subheading 3.8. If a weak signal is observed in this assay, anti-TBDT titres assessed in a subsequent ELISA should be interpreted with caution as the TBDT used to coat the plate may be misfolded. If a ligand known to bind to the TBDT under investigation is identified, this TBDT-ligand interaction can be exploited to give the user more confidence that the TBDT in question is properly folded (*see* **Note 34**).
32. When an entire plate is not needed, (e.g., when determining the ability of the TbpA to bind transferrin, as described in Subheading 3.8, *see* **Note 31**) only coat the wells to be used. Cover the unused portion of the plate, which can be used for a later experiment, with an adhesive slip. We typically use streptavidin-coated Greiner Bio-One plates in our experiments; however, other commercially available streptavidin/neutravidin-coated plates can also be used. Plates can also be coated with streptavidin or neutravidin in-house [14].
33. To wash, discard the PBST in the wells by inverting the 96-well plate over the sink and expelling the liquid with a sharp flick of the wrist. Following this, vigorously tap the plate face-down on a stack of paper towels to remove the remaining liquid. An automated plate washer may also be used for these steps, if available.

34. The TbpA/Tf-HRP ELISA is used to assess whether the TbpA is properly folded on the plate. For TBDTs that bind metal-containing proteins (hemoglobin, hemoglobin–haptoglobin, calprotectin), a similar approach could be used, with an additional step to add appropriate target protein, followed by HRP-conjugated antibody specific for the target protein; however, alternative approaches will be required for TBDTs that directly bind metal ions or metal ions bound by compounds (heme, siderophore–iron complexes).
35. We recommend performing twofold serial dilutions across the columns labeled 1–11, leaving the last column for any pertinent controls; however, if no controls are needed, up to 12 different dilutions can be used.

References

1. Fegan JE, Calmettes C, Islam EA, Ahn SK, Chaudhuri S, Yu RH, Gray-Owen SD, Moraes TF, Schryvers AB (2019) Utility of hybrid transferrin binding protein antigens for protection against pathogenic *Neisseria* species. Front Immunol 10:247. https://doi.org/10.3389/fimmu.2019.00247
2. Qamsari MM, Rasooli I, Chaudhuri S, Astaneh SDA, Schryvers AB (2020) Hybrid antigens expressing surface loops of ZnuD from *Acinetobacter baumannii* is capable of inducing protection against infection. Front Immunol 11:1–13. https://doi.org/10.3389/fimmu.2020.00158
3. Noinaj N, Guillier M, Barnard TJ, Buchanan SK (2010) TonB-dependent transporters: regulation, structure, and function. Annu Rev Microbiol 64:43–60. https://doi.org/10.1146/annurev.micro.112408.134247
4. Postle K, Kadner RJ (2003) Touch and go: tying TonB to transport. Mol Microbiol 49(4):869–882
5. Hood MI, Skaar EP (2012) Nutritional immunity: transition metals at the pathogen-host interface. Nat Rev Microbiol 10(8):525–537. https://doi.org/10.1038/nrmicro2836
6. Morgenthau A, Pogoutse A, Adamiak P, Moraes TF, Schryvers AB (2013) Bacterial receptors for host transferrin and lactoferrin: molecular mechanisms and role in host-microbe interactions. Future Microbiol 8(12):1575–1585
7. Cornelissen CN, Kelley M, Hobbs MM, Anderson JE, Cannon JG, Cohen MS, Sparling PF (1998) The transferrin receptor expressed by gonococcal strain FA1090 is required for the experimental infection of human male volunteers. Mol Microbiol 27(3):611–616
8. Baltes N, Hennig-Pauka I, Gerlach GF (2002) Both transferrin binding proteins are virulence factors in *Actinobacillus pleuropneumoniae* serotype 7 infection. FEMS Microbiol Lett 209(2):283–287
9. Anderson JE, Hobbs MM, Biswas GD, Sparling PF (2003) Opposing selective forces for expression of the gonococcal lactoferrin receptor. Mol Microbiol 48(5):1325–1337. https://doi.org/10.1046/j.1365-2958.2003.03496.x
10. Hooda Y, Lai CCL, Moraes TF (2017) Identification of a largefamily of Slam-dependent surface lipoproteins in gram-negative bacteria. Front Cell Infect Microbiol 7:207. https://doi.org/10.3389/fcimb.2017.00207
11. Frandoloso R, Martinez-Martinez S, Calmettes C, Fegan J, Costa E, Curran D, Yu R, Gutierrez-Martin CB, Rodriguez-Ferri EF, Moraes TF, Schryvers AB (2015) Nonbinding site-directed mutants of transferrin binding protein B enhances their immunogenicity and protective capabilities. Infect Immun 83(3):1030–1038
12. Calmettes C, Alcantara J, Yu R-H, Schryvers AB, Moraes TF (2012) The structural basis of transferrin sequestration by transferrin binding protein B. Nat Struct Mol Biol 19(3):358–360. https://doi.org/10.1038/nsmb.2251
13. Noinaj N, Buchanan SK, Cornelissen CN (2012) The transferrin-iron import system from pathogenic *Neisseria* species. Mol Microbiol 86(2):246–257. https://doi.org/10.1111/mmi.12002
14. Fegan JE, Yu R-H, Islam EA, Schryvers A (2021) Development of a non-biased, high-throughput ELISA for the rapid evaluation of immunogenicity and cross-reactivity. J Immunol Methods 493:113027

Chapter 9

Preparation of Trimethyl Chitosan-Based Polyelectrolyte Complexes for Peptide Subunit Vaccine Delivery

Lili Zhao, Sahra Bashiri, Istvan Toth, and Mariusz Skwarczynski

Abstract

A variety of delivery vehicles have been explored as adjuvanting/delivery platforms for peptide-based subunit vaccines. Polysaccharide-based systems have been found to be especially attractive due to their immune stimulating properties, biodegradability, biocompatibility, and low toxicity. Among them, chitosan and its derivatives are the most common cationic nanocarriers used for the delivery of antigens. Trimethyl chitosan (TMC) is a partially quaternized, water-soluble, and mucoadhesive derivative of chitosan. This chapter describes the preparation of a TMC-based polyelectrolyte complex as a delivery system for peptide subunit vaccines.

Key words Trimethyl chitosan, Alginate, Polyelectrolyte complexes, Peptide subunit vaccine, Nanoparticles

1 Introduction

Peptide subunit vaccines, which contain only the minimum necessary components required to stimulate immune responses, have been extensively investigated over the past few decades. Since peptides are mostly nonimmunogenic on their own, employing an adjuvant and/or a delivery system is vital to the efficacy of peptide vaccines [1, 2]. Delivery systems, such as lipid-based formulations [3], polymers [4], dendrimers [5], and inorganic nanoparticles [6], have been widely explored for peptide subunit vaccine delivery. Efficient delivery systems enable long-lasting release of peptide antigen through a depot effect or enhanced trafficking to the lymph nodes through the lymphatic system [7, 8]; targeting antigen presenting cells [9, 10], enhanced mucosa adhesion and mucus penetration of peptide antigen at mucosa sites [11, 12]; antigen protection from degradation [13]; and ultimately, stimulation of the desired immune responses.

Fadil Bidmos et al. (eds.), *Bacterial Vaccines: Methods and Protocols*,
Methods in Molecular Biology, vol. 2414, https://doi.org/10.1007/978-1-0716-1900-1_9,

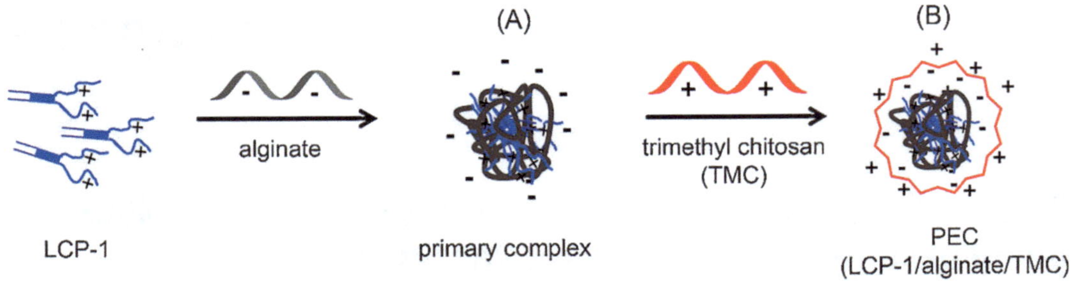

Fig. 1 Schematic illustration of the preparation of the LCP-1-loaded polyelectrolyte complex: LCP-1/alginate/TMC. (**a**) Primary complex (LCP-1/alginate), (**b**) Ternary PEC (LCP-1/alginate/TMC)

Polysaccharides composed of repeating monosaccharide units have been widely used for vaccine delivery due to their biodegradability, biocompatibility, and low toxicity [14, 15]. They can form polyelectrolyte complexes (PECs) with counterparts possessing opposite charges through electrostatic interactions [16, 17]. Among them, chitosan and its derivatives are the most common cationic nanocarriers used for the delivery of antigens [16, 18]. Since chitosan has very poor water solubility, its soluble derivative, trimethyl chitosan (TMC) is typically used as a component of self-adjuvanting delivery systems [19, 20]. TMC is partially quaternized, water-soluble, and mucoadhesive.

Various methods have been employed for the preparation of PECs, such as polyelectrolyte titration, jet mixing, and ionic gelation [21]. Titration is usually favored, as it is simple and proceeds under mild conditions [17]. This protocol describes the preparation of PECs comprised of peptide antigen (LCP-1), negatively charged alginate, and positively charged TMC (Fig. 1) [22]. The protocol includes two parts: (1) synthesis of TMC; and (2) formulation of LCP-1/alginate/TMC PECs.

2 Materials

2.1 Synthesis of TMC

1. Low molecular weight chitosan (75–85% deacetylated).
2. Sodium iodide.
3. *N*-Methyl-2-pyrrolidone (NMP).
4. Aqueous sodium hydroxide solution, 15% (w/w).
5. Diethyl ether and ethanol mixture (50:50 v/v).
6. Sodium hydroxide pellets.
7. Aqueous sodium chloride (NaCl) solution, 10% (w/v).
8. Deuterium water (D_2O).
9. Acetone.

2.2 Formulation and Characterization of PECs (LCP-1/Alginate/TMC)

1. Stock solution of alginate sodium in Milli-Q water (2 mg/mL).
2. Stock solution of trimethyl chitosan (TMC) in Milli-Q water (2 mg/mL).
3. Stock solution of LCP-1 in Milli-Q water (1 mg/mL).
4. Phosphotungstic acid (0.5% w/v).

2.3 Equipment

1. Magnetic stirrer with hot plate, magnetic stir bar, and ice bath.
2. Thermometer.
3. Condenser.
4. Freeze dryer.
5. Nuclear magnetic resonance (NMR) spectrometer (300 MHz or higher).
6. Probe-type sonicator.
7. Malvern Zetasizer Nano-ZS (Malvern Instruments Ltd., UK or equivalent) to measure dynamic light scattering (DLS).
8. Transmission electron microscope.
9. Dialysis bag (3500 Da).
10. Balance, microbalance.
11. Laboratory glassware and apparatus.

3 Methods

This titration method was developed for the preparation of LCP-1/alginate/TMC PECs; however, it can also be used for other PEC nanoparticles (e.g., LCP-1/chondroitin sulfate/TMC, LCP-1/dextran/TMC, LCP-1/hyaluronic acid/TMC, and LCP-1/heparin/TMC [22]). Although, different quantities of charged polymers are required to formulate the specific PEC nanoparticles.

3.1 Synthesis of TMC

1. To 250 mL round bottom flask add low molecular weight chitosan (1 g) and sodium iodide (2.4 g), add 40 mL of NMP.
2. Stir the mixture until fully dissolved.
3. Add 6 mL of 15% aqueous NaOH solution and stir the mixture for 20 min. The temperature should be kept at 60 °C (*see* **Note 1**).
4. Add 6 mL of the methylation agent, methyl iodide, to the reaction mixture. Attach a reflux condenser to the flask and stir the mixture for 1 h at 60 °C.
5. Slowly pour the reaction mixture into 50 mL of diethyl ether–ethanol solution to precipitate the product (*see* **Note 2**).

6. Filter the precipitate using a Buchner funnel under vacuum. Wash the product with diethyl ether (2 × 5 mL washes) (*see* **Note 3**).
7. Add 2.4 g of sodium iodide to the 250 mL round bottom flask containing the product from **step 6**, then add 40 mL of NMP. Attach a reflux condenser to the flask (*see* **Note 4**).
8. Stir the mixture at 60 °C until fully dissolved.
9. Add 5.5 mL of 15% aqueous NaOH to the solution and stir the mixture for 20 min.
10. Add 3.5 mL of methyl iodide to the reaction mixture and stir for 45 min at 60 °C.
11. Add another 1 mL of methyl iodide, and 0.6 g of NaOH pellets to the mixture (*see* **Note 5**).
12. Stir the mixture for 1 h at 60 °C.
13. Repeat **steps 5** and **6**.
14. Transfer the resultant precipitate into the 100 mL round bottom flask and add 50 mL of 10% (w/v) NaCl. Stir for at least 18 h to replace I^- ions with Cl^- ions.
15. Transfer the solution into the dialysis bag (3500 Da) and dialyze it at room temperature against deionized water for 3 days, changing the water at least two times a day.
16. Freeze the purified solution in a dry ice-acetone cold bath, then lyophilize the frozen compound using a freeze dryer.
17. Analyze the product by NMR spectroscopy. Typical spectra are as follows: (D_2O, 300 MHz): ppm 5.25–5.70 (m, $H_{(1)}$), 3.55–4.75 (m, $H_{(2)}$, $H_{(3)}$, $H_{(4)}$, $H_{(5)}$, $2H_{(6)}$), 3.43 (s, OCH_3), 3.35 (s, OCH_3), 3.25 (bs, N-$(CH_3)_3$), 2.39 (bs, N-$(CH_3)_2$), and 1.98 (bs, C(O)CH_3). The degree of quaternization (DQ), degree of demethylation (DM), degree of 3- and 6-*O*-methylation (DOM-3 and DOM-6, respectively), and degree of acetylation (DA) of TMC were calculated from ^1H NMR spectra, according to the following equations [23, 24]:

$$DQ\% = (CH_3)_3/(9 \times H_{(1)}) \times 100.$$

$$DM\% = (CH_3)_2/(6 \times H_{(1)}) \times 100.$$

$$DOM - 3\% = (OCH_3)/(3 \times H_{(1)}) \times 100.$$

$$DOM - 6\% = (OCH_3)/(3 \times H_{(1)}) \times 100.$$

$$DA\% = (C(O)CH_3)/(3 \times H_{(1)}) \times 100.$$

$(CH_3)_3$, $(CH_3)_2$, (OCH_3), and $(C(O)CH_3)$ are the integrals of the hydrogens of the trimethylated amino groups at 3.25 ppm, the demethylated amino groups at 2.39 ppm, the methylated hydroxyl groups at either 3.35 (DOM-6) or 3.43

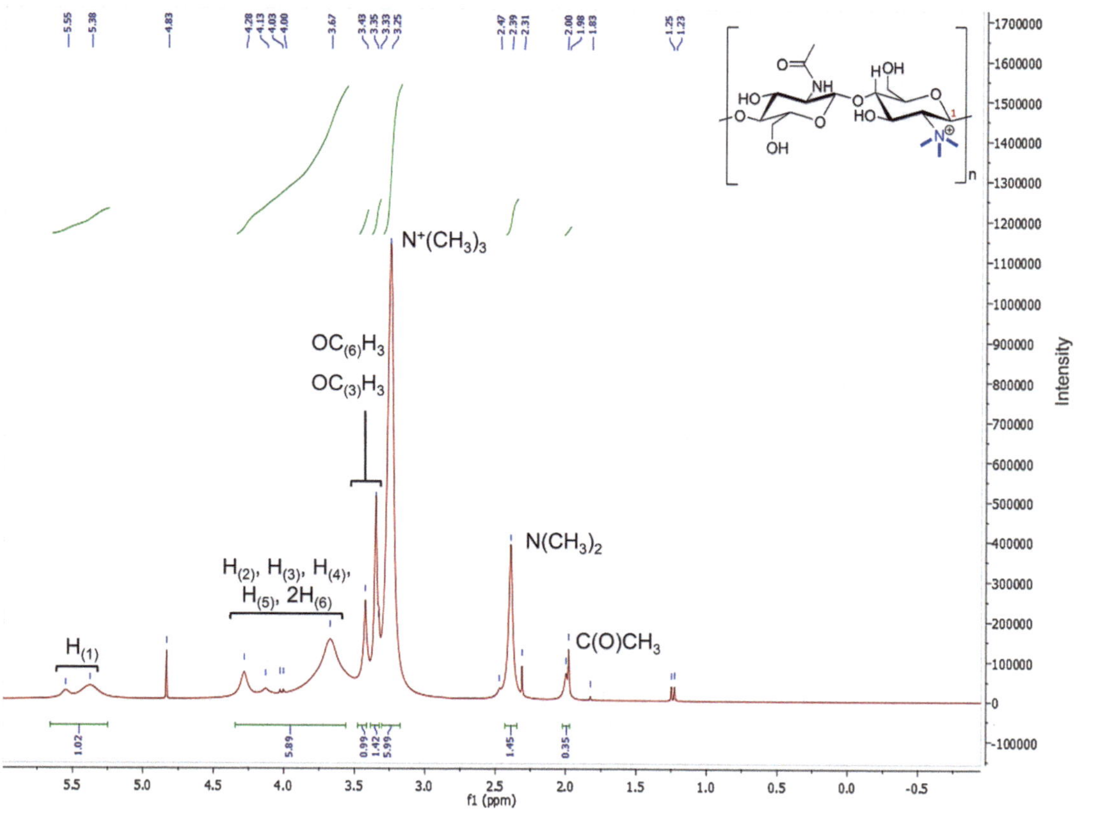

Fig. 2 1H NMR spectra of trimethyl chitosan

(DOM-3) ppm, and the acetylated groups at 1.98 ppm, respectively. $H_{(1)}$ is the integral of peaks between 5.25 and 5.70 ppm: the signal related to the carbohydrate hydrogen atom bound to the C-1's of the TMC molecule (Fig. 2).

3.2 Preparation of PECs (LCP-1/Alginate/TMC)

3.2.1 Prepare the Primary Complexes (LCP-1/Alginate)

1. Dilute the alginate stock solution (2 mg/mL) to 0.1, 0.2, 0.4, 0.6, 0.8, and 1.0 mg/mL, with Milli-Q water (*see* **Note 6**).
2. Transfer 100 μL of LCP-1 solution (100 μg) into a 2 mL centrifuge tube; dilute 10 times with Milli-Q water.
3. Add 50 μL of the diluted alginate solution (0.1 mg/mL, containing 5 μg alginate) into the centrifuge tube and mix thoroughly.
4. Sonicate the mixture for 4 min (2 × 2 min) using a probe-type sonicator at 120 W (duration 2 min, duty cycle 50%) in an ice bath.
5. Transfer the mixture to a scintillation vial and stir it continuously for 1 h at room temperature to stabilize the complexes.
6. Measure the particle size, polydispersity index (PDI), and zeta potential using dynamic light scattering (DLS).

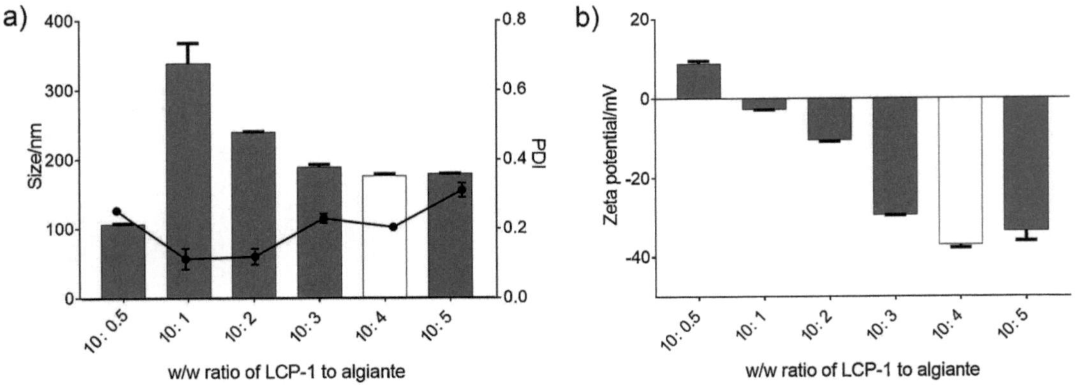

Fig. 3 Optimization of the ratio of LCP-1 to alginate for the preparation of primary complexes (LCP-1/alginate): the optimum ratio is indicated by white shading. (**a**) The bars show size (nm), while the line represents polydispersity index (PDI). (**b**) The bars show zeta potential (mV). LCP-1/alginate = 10:4, size 177 ± 2 nm, PDI 0.205 ± 0.01, zeta potential −37.1 ± 0.7 mV

7. Repeat **steps 3–6** five times, *except* in **step 3**, add 50 μL of alginate solution at concentrations of 0.2, 0.4, 0.6, 0.8, and 1.0 mg/mL (instead of 50 μL of the diluted alginate solution) to prepare primary complexes with the mixing ratios of LCP-1: alginate of 10:0.5, 10:1, 10:2, 10:3, 10:4, and 10:5, respectively (Fig. 3) (*see* **Note 7**).

3.2.2 Preparation of Ternary PECs (LCP-1/Alginate/TMC)

1. Dilute TMC stock solution (2 mg/mL) to 0.6, 1.0, 1.2, and 1.6 mg/mL with Milli-Q water.
2. Add 50 μL of TMC solution (0.6 mg/mL, containing 30 μg TMC) dropwise to freshly prepared LCP-1/alginate complex suspension (containing 100 μg of LCP-1, 40 μg of alginate).
3. Stir the solution for 1 h at room temperature to produce the final PEC nanoparticles.
4. Measure the particle size, PDI, and zeta potential using DLS.
5. Repeat **steps 2–4** five times, *except* at **step 2**, add 50 μL of TMC solution at concentrations of 0.6, 1.0, 1.2, 1.6, and 2.0 mg/mL (instead of 50 μL of the diluted TMC solution) to produce ternary PEC nanoparticles at ratios of 10:4:3, 10:4:5, 10:4:6, 10:4:8, and 10:4:10, respectively (*see* **Note 8**).
6. Measure the particle size, PDI, and zeta potential of the primary complexes and final ternary PEC nanoparticles using DLS. The expected particle size and zeta potential for different concentration are shown in Fig. 4 (*see* **Note 9**).
7. Examine the morphology of the PEC nanoparticles using transmission electron microscopy. Drop PEC nanoparticles (dilution might be required) onto a carbon-coated copper grid and leave it for 1 min to allow the nanoparticles to deposit

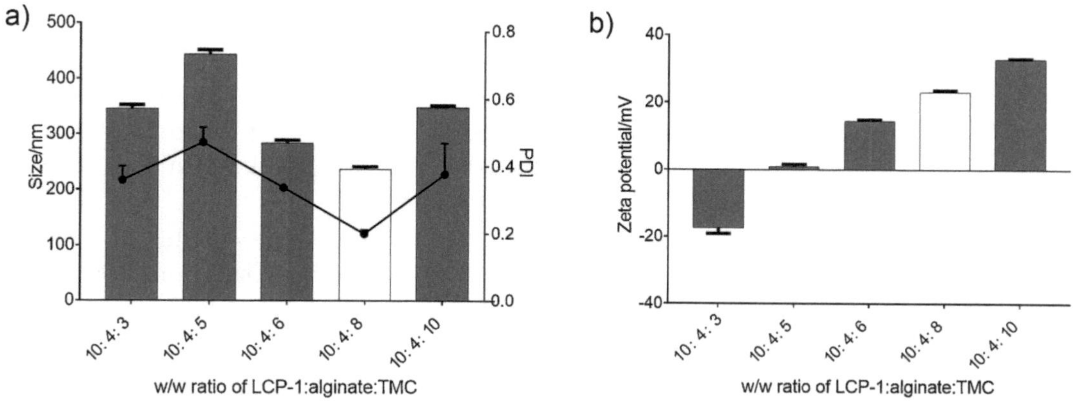

Fig. 4 Optimization of PEC nanoparticle (LCP-1/alginate/TMC) preparation using DLS monitoring: the optimum mixing ratio is indicated by white shading. (**a**) The bars show size (nm), while the line represents PDI. (**b**) The bars show zeta potential (mV). LCP-1/alginate/TMC = 10:4:8, size 237 ± 4 nm, PDI 0.199 ± 0.01, zeta potential 29.3 ± 1.5 mV

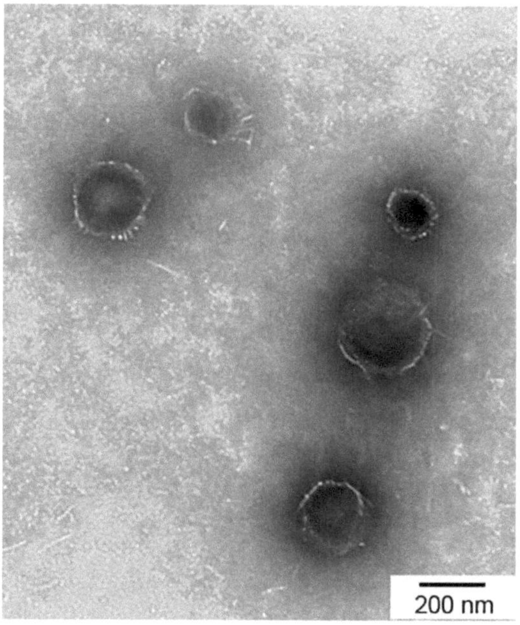

Fig. 5 Transmission electron microscopy images of LCP-1-loaded PEC nanoparticles (LCP-1/alginate/TMC)

onto the grid. Then, drop 0.5% (w/v) phosphotungstic acid onto the grid and leave it for 30 s. Air dry the grids before observation. The individual particles may be directly observed and measured by TEM (Fig. 5).

4 Notes

1. To increase the temperature of the solution to 60 °C, use a hot plate and silicone oil bath. Assemble the thermometer into the oil bath using the holder. Fix the round bottom flask onto the oil bath using a retort stand.
2. The temperature of the diethyl ether–methanol solution should be low (~4 °C) to improve precipitation/isolation of the product. The solution should be precooled in the refrigerator or immersed in an ice bath.
3. The chitosan derivative will precipitate in diethyl ether–ethanol solution, so should be filtered immediately as the precipitate tends to aggregate in larger lumps as time goes on. At this stage, the TMC degree of quaternization is around 20% [25].
4. Follow **steps 7–13** to synthesize TMC with a DQ of 60–90%.
5. This step helps to produce TMC with a higher DQ (80–90%).
6. It takes a few hours, to overnight, for polymers to swell and dissolve in water.
7. The optimum mixing ratio of LCP-1 to alginate is determined as the minimal amount of alginate required to produce uniform (PDI < 0.3), small-sized nanoparticles (<200 nm) with a surface charge of around −30 mV, and where further addition of alginate does not decrease the charge. The ratio of LCP-1: alginate of 10:4 was used to prepare the primary complexes.
8. PEC nanoparticles of higher concentration can be produced; however, they may require longer stirring time to allow polymer complexation and particle stabilization.
9. According to the change in charge, the ratio of LCP-1, alginate, and TMC should be 10:4:10. However, as particle size and PDI both increased at this ratio, we instead determined 10:4:8 to be the optimum ratio.

References

1. Nevagi RJ, Toth I, Skwarczynski M (2018) Peptide-based vaccines. In: Koutsopoulos S (ed) Peptide applications in biomedicine, biotechnology and bioengineering. Woodhead Publishing, Oxford
2. Skwarczynski M, Toth I (2016) Peptide-based synthetic vaccines. Chem Sci 7(2):842–854
3. Bartlett S, Skwarczynski M, Toth I (2020) Lipids as activators of innate immunity in peptide vaccine delivery. Curr Med Chem 27 (17):2887–2901
4. Nevagi RJ, Skwarczynski M, Toth I (2019) Polymers for subunit vaccine delivery. Eur Polym J 114:397–410
5. Fujita Y, Taguchi H (2011) Current status of multiple antigen-presenting peptide vaccine systems: application of organic and inorganic nanoparticles. Chem Cent J 5(1):48
6. Skwarczynski M, Toth I (2014) Recent advances in peptide-based subunit nanovaccines. Nanomedicine 9:2657–2669
7. Reddy ST, Van Der Vlies AJ, Simeoni E et al (2007) Exploiting lymphatic transport and complement activation in nanoparticle vaccines. Nat Biotechnol 25(10):1159–1164
8. Henriksen-Lacey M, Christensen D, Bramwell VW et al (2011) Comparison of the depot effect and immunogenicity of liposomes based

on dimethyldioctadecylammonium (DDA), 3β-[N-(N',N'-Dimethylaminoethane)carbomyl] cholesterol (DC-Chol), and 1,2-Dioleoyl-3-trimethylammonium propane (DOTAP): prolonged liposome retention. Mol Pharm 8(1):153–161

9. Yang JR, Luo YC, Shibu MA et al (2019) Cell-penetrating peptides: efficient vectors for vaccine delivery. Curr Drug Deliv 16(5):430–443
10. Hussein WM, Liu TY, Skwarczynski M et al (2014) Toll-like receptor agonists: a patent review (2011 - 2013). Expert Opin Ther Pat 24(4):453–470
11. Zheng J, Gao S, Cui X et al (2019) Adjuvants and delivery systems based on polymeric nanoparticles for mucosal vaccines. Int J Pharm 572:118731
12. Marasini N, Skwarczynski M, Toth I (2017) Intranasal delivery of nanoparticle-based vaccines. Ther Deliv 8(3):151–167
13. Malonis RJ, Lai JR, Vergnolle O (2020) Peptide-based vaccines: current progress and future challenges. Chem Rev 120 (6):3210–3229
14. Sedaghat B, Stephenson R, Toth I (2014) Targeting the mannose receptor with mannosylated subunit vaccines. Curr Med Chem 21 (30):3405–3418
15. Li P, Wang F (2015) Polysaccharides: candidates of promising vaccine adjuvants. Drug Discov Ther 9(2):88–93
16. Zhao L, Skwarczynski M, Toth I (2019) Polyelectrolyte-based platforms for the delivery of peptides and proteins. ACS Biomater Sci Eng 5(10):4937–4950
17. Meka VS, Sing MKG, Pichika MR et al (2017) A comprehensive review on polyelectrolyte complexes. Drug Discov Today 22 (11):1697–1706
18. Li X, Min M, Du N et al (2013) Chitin, chitosan, and glycated chitosan regulate immune responses: the novel adjuvants for cancer vaccine. Clin Dev Immunol 2013:387023
19. Nevagi RJ, Dai W, Khalil ZG et al (2019) Structure-activity relationship of group a streptococcus lipopeptide vaccine candidates in trimethyl chitosan-based self-adjuvanting delivery system. Eur J Med Chem 179:100–108
20. Marasini N, Ghaffar KA, Giddam AK et al (2017) Highly immunogenic trimethyl chitosan-based delivery system for intranasal lipopeptide vaccines against Group A *Streptococcus*. Curr Drug Deliv 14(5):701–708
21. Jeong Y, Lee D, Choe K et al (2017) Polypeptide-based polyelectrolyte complexes overcoming the biological barriers of oral insulin delivery. J Indust Eng Chem 48:79–87
22. Zhao L, Jin W, Cruz JG et al (2020) Development of polyelectrolyte complexes for the delivery of peptide-based subunit vaccines against group A *Streptococcus*. Nanomaterials (Basel) 10:823
23. Marasini N, Giddam AK, Ghaffar KA et al (2016) Multilayer engineered nanoliposomes as a novel tool for oral delivery of lipopeptide-based vaccines against group A *Streptococcus*. Nanomedicine 11(10):1223–1236
24. Polnok A, Borchard G, Verhoef JC et al (2004) Influence of methylation process on the degree of quaternization of N-trimethyl chitosan chloride. Eur J Pharm Biopharm 57(1):77–83
25. Verheul RJ, Amidi M, Van Der Wal S et al (2008) Synthesis, characterization and in vitro biological properties of O-methyl free N,N,N-trimethylated chitosan. Biomaterials 29 (27):3642–3649

Chapter 10

Multiepitope Fusion Antigen: MEFA, an Epitope- and Structure-Based Vaccinology Platform for Multivalent Vaccine Development

Siqi Li, Kuo Hao Lee, and Weiping Zhang

Abstract

Vaccines are regarded as the most cost-effective countermeasure against infectious diseases. One challenge often affecting vaccine development is antigenic diversity or pathogen heterogeneity. Different strains produce immunologically heterogeneous virulence factors, therefore an effective vaccine needs to induce broad-spectrum host immunity to provide cross-protection. Recent advances in genomics and proteomics, particularly computational biology and structural biology, establishes structural vaccinology and highlights the feasibility of developing effective and precision vaccines. Here, we introduce the epitope- and structure-based vaccinology platform multiepitope-fusion-antigen (MEFA), and provide instructions to generate polyvalent MEFA immunogens for vaccine development. Conceptually, MEFA combines epitope vaccinology and structural vaccinology to enable a protein immunogen to present heterogeneous antigenic domains (epitopes) and to induce broadly protective immunity against different virulence factors, strains or diseases. Methodologically, the MEFA platform first identifies a safe, structurally stable and strongly immunogenic backbone protein and immunodominant (ideally neutralizing or protective) epitopes from heterogeneous strains or virulence factors of interest. Then, assisted with protein modeling and molecule dynamic simulation, MEFA integrates heterogeneous epitopes into a backbone protein via epitope substitution for a polyvalent MEFA protein and mimics epitope native antigenicity. Finally, the MEFA protein is examined for broad immunogenicity in animal immunization, and assessed for potential application for multivalent vaccine development in preclinical studies.

Key words MEFA (multiepitope fusion antigen), Vaccinology platform, Multivalent vaccine, Epitopes, Structural vaccinology

1 Introduction

Infectious diseases are a leading cause of death and a major threat to global health [1, 2]. Vaccines are considered the most cost-effective countermeasure to reduce human mortality and morbidity, to eliminate pandemic or epidemic risk, and to lower antibiotic use. Success in vaccine and vaccination against infectious diseases includes the eradication of smallpox and the control of polio,

Fadil Bidmos et al. (eds.), *Bacterial Vaccines: Methods and Protocols*,
Methods in Molecular Biology, vol. 2414, https://doi.org/10.1007/978-1-0716-1900-1_10,

measles, mumps, and diphtheria. However, major challenges remain in developing vaccines for other diseases including human immunodeficiency virus (HIV), diarrhea caused by viral, bacterial, or parasitic pathogens, tuberculosis (TB) and malaria which continuously claim millions of human lives annually [3, 4]. While a lack of clear understanding of host pathogen interactions and difficulties in identifying immune biomarkers or immune correlates of protection are commonly major challenges in vaccine development, antigenic diversities and immunological heterogeneity among pathogen strains or virulence factors have also severely hampered development of effective vaccines against certain infectious diseases. Strategies including conserved antigen and cocktail vaccines have been attempted to overcome antigenic diversity or immunological heterogeneity, but they often encounter bottlenecks at improving protective efficacy. Different from conventional approaches, we introduce a new vaccinology platform called multiepitope fusion antigen, MEFA [5, 6], to construct polyvalent immunogens and to develop broadly protective vaccines. Benefiting from recent advance in genomics and proteomics, particularly computational biology and structural biology, MEFA combines epitope vaccinology and structural vaccinology concepts to generate epitope- and structure-based polyvalent protein immunogens for developing cross protective vaccines against heterogeneous strains, pathogens, or diseases [7].

The MEFA platform enables integration of foreign epitopes, ideally protective or neutralizing epitopes (epitopes induce protective immune responses) of various pathogenic strains or virulence factors, into a backbone protein and mimicking of epitope native antigenicity, thus constructing a polyvalent immunogen for broadly protective immunity. In principle, a protective epitope, if it retains native antigenicity after being integrated into a backbone protein, induces protective host immunity against the representing virulence factor or strain. Therefore, an MEFA immunogen carrying multiple protective epitopes from different strains serves as a polyvalent antigen for a cross protective vaccine. By substituting backbone protein epitopes with protective epitopes (of heterogeneous strains or virulence factors) of interest, we produce a chimeric polyvalent immunogen (Fig. 1), and by applying protein modeling and molecular dynamic simulation, we mimic epitope native antigenicity. Mechanistically, the MEFA vaccinology platform combines novel in silico analyses with conventional in vivo empirical studies to accelerate development of broadly protective multivalent vaccines. First, an MEFA backbone immunogen and epitopes of interest (from heterogeneous strains or virulence factors) are identified, and an MEFA immunogen is constructed in silico, followed by synthesis and cloning of an MEFA gene into an expression vector. Then, the MEFA protein is expressed, extracted and used for animal immunization studies. Finally, the MEFA protein-

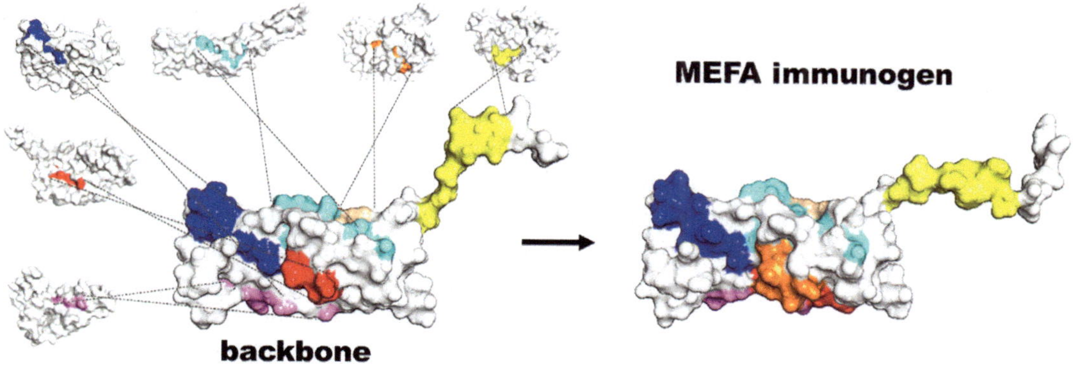

Fig. 1 Illustration of constructing a polyvalent MEFA immunogen by using the MEFA vaccinology platform. Surface-exposed epitopes of a backbone immunogen are substituted with protective or immunodominant epitopes from heterogeneous virulence factors or strains of interest to construct an MEFA immunogen

induced broad-spectrum immune responses are measured, and application of an MEFA immunogen for multivalent vaccine development is evaluated in preclinical investigation prior to human subject safety, immunogenicity, and efficacy studies.

2 Materials

2.1 Computation and Programs

MEFA backbone protein selection, epitope identification, and protein modeling are carried out on standard desktop or laptop computers through open-access servers. Atomic molecular dynamics simulation requires a local server and remote access to license-based CHARMM server or open-source GROMACS-5.0.7 MD server.

Programs or servers that are applied for in silico analyses are as follows.

1. IEDB (http://www.iedb.org/) or BepiPred (http://www.cbs.dtu.dk/services/BepiPred/).
2. ExPASy (https://www.expasy.org/).
3. PyMol (https://www.pymol.org/).
4. Phyre2 (http://www.sbg.bio.ic.ac.uk/~phyre2/html/page.cgi?id=index).
5. GROMACS (http://www.gromacs.org/).
6. Or CHARMM (https://www.charmm.org/), and MMTSB toolset (http://blue11.bch.msu.edu/mmtsb/Main_Page).

2.2 A Backbone Protein and Heterogeneous Epitopes of Interest

The MEFA platform requires a backbone protein and epitopes of heterogeneous strains or virulence factors of interest to construct an MEFA immunogen. A backbone protein should be safe and structurally stable, preferably a mutant key virulence factor which becomes avirulent after epitope substitution and induces protective immunity (housekeeping proteins may not be desirable since they potentially induce host immune tolerance or immunity against normal flora organisms). A backbone protein needs to be strongly immunogenic, possess multiple well-separated continuous epitopes, and ideally exhibits systemic and mucosal adjuvanticity. Additionally, a backbone protein can be easily expressed in an expression vector and an *Escherichia coli* strain and extracted at a great purity and in high yield. In the case there is no suitable backbone proteins available from a target pathogen, mutant cholera toxin (CT) of *Vibrio cholerae* and heat-labile toxin (LT) of enterotoxigenic *E. coli*, which have been demonstrated with little or low enterotoxicity but strong immunogenicity and mucosal and systemic adjuvanticity [8–11], can serve as alternative backbone proteins.

Epitopes from heterogeneous strains or virulence factors of interest are presented on an identified backbone protein. Preferably, one epitope represents one strain or one virulence factor and induces protective immune responses against its representing strain or virulence factor. With several protective epitopes integrated into a backbone immunogen, an MEFA immunogen induces broad immunity against multiple heterogeneous strains or virulence factors. Neutralizing or functionally protective epitopes which have been characterized are available at databases including Immune Epitope Database—IEDB (www.iedb.org). However, in many cases, protective epitopes likely have not been identified. Alternatively, immunodominant epitopes which are identified in silico can be used initially. Eventually, as we recently described [12–15], immunodominant epitopes are screened in empirical studies and protective epitopes are selected for MEFA immunogen construction.

2.3 MEFA Immunogen Construction, Expression, and Extraction

1. pET28a or other expression vectors, to clone an MEFA immunogen gene.
2. *E. coli* strain BL21-CodonPlus (DE3) or other *E. coli* strains, to host MEFA plasmids for MEFA protein expression.
3. LB agar plates and 2× YT medium broth.
4. Kanamycin, 30 μg/mL in final (for pET28a).
5. IPTG (isopropyl-β-D-thiogalactopyranoside).
6. B-PER, Bacterial Protein Extraction Reagent (in phosphate buffer; solution contains a proprietary, nonionic detergent in 20 mM Tris–HCl; pH 7.5).
7. Ni-NTA Agarose, Ni-NTA columns, wash buffer (50 mM NaH_2PO_4, 300 mM NaCl, 20 mM imidazole, 0.05% Tween 20, pH 8.0), elution buffer (50 mM NaH_2PO_4, 300 mM NaCl, 250 mM imidazole, 0.05% Tween 20, pH 8.0), to extract 6× His-tagged recombinant protein (*see* **Note 1**).

8. Spectrum Spectra/Por dialysis membrane tubing, at suitable pore sizes depending on MEFA protein molecular mass (5–10 kDa less than the MEFA protein molecular mass).
9. 1x IB solubilization buffer: 50 mM CAPS, pH 11.0, supplemented with 1% N-lauroylsarcosine and 1 mM DTT.
10. Protein refolding and dialysis buffer: 20 mM Tris–HCl, pH 8.5, supplemented with 0.1 mM DTT or without DTT.

2.4 Protein Biochemical Property and Immunogenicity Characterization

1. SDS-PAGE, for MEFA protein Coomassie blue stain or Western blot with specific antibodies.
2. Coomassie blue stain buffer (0.125% Coomassie blue R350, 50% methanol, 10% acetic acid), and destain buffer (20% methanol, 10% acetic acid, in ddH_2O).
3. Protein immunogenicity studies: MEFA protein-induced T cell and/or B cell immune responses are examined in animal immunization studies. Antigen specific adaptive immunity and in vitro protective activities, which vary and are specific to the pathogen of interest, are measured to assess MEFA immunogen broad immunity (B cell epitopes are used for MEFA immunogen construction in this protocol).

Animal immunization: 8-week-old mice, 25-gauge needle and 1-mL syringe, alum or incomplete Freund's adjuvant; mouse fecal and serum samples collected for measuring antigen-specific antibody responses.

Antibody titration: 96-well microtiter plates, ELISA coating antigens specific to the virulence factors targeted by MEFA protein, HRP-conjugated goat anti-mouse IgG and IgA secondary antibodies, 3,3′,5,5′-tetramethybenzidine (TMB) Microwell Peroxidase Substrate, a 96-well plate reader.

Antibody in vitro protection: assays to evaluate in vitro protection of MEFA-induced antibodies vary and are specific to pathogens or virulence factors of interest.

Preclinical challenge studies: if animal challenge models are available, immunizing animals with an MEFA protein to assess epitope-specific immunogenicity, then challenging the immunized animals with pathogen strains to measure cross protection and to evaluate potential application of an MEFA immunogen for multivalent vaccine development.

3 Methods

3.1 Backbone Protein Selection

Online programs that predict epitopes, locate epitope positions, image protein secondary structure, and examine protein stability are applied to select a suitable backbone protein. B cell epitopes on a backbone protein, including epitope amino acid sequence, position, and antigenic score, are predicted and analyzed by using

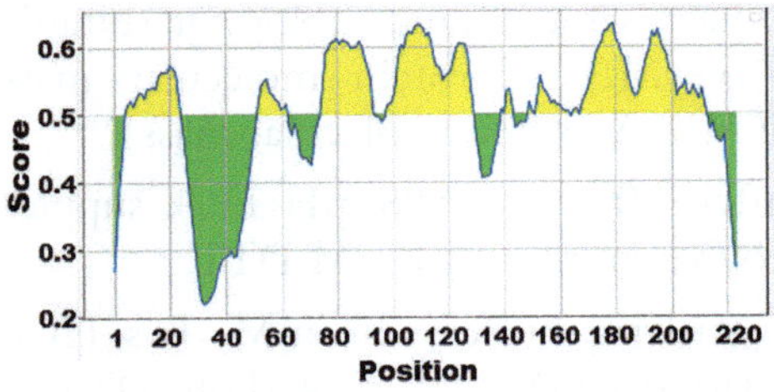

Predicted epitopes

No.	Start	End	Epitope	Length
1	5	24	KQLFKKKFVKEEHDKKTGQE	20
2	53	63	IDSQNMTKAAQ	11
3	76	94	YRSLGNYPATANASAATQL	19
4	100	120	SLGKVSADEAKNPFTGTAMIF	20
5	153	164	GDMFPFINVKEG	12

Fig. 2 Illustration of in silico identifying B cell immunodominant epitopes with IEDB program. Left: a diagram to show in silico predicted epitopes, with antigenic scores and locations, from a backbone protein. Right: a table to indicate epitope position, sequence and length

immune epitope database and analysis resource server IEDB [16]. Backbone protein half-life and stability are evaluated with ExPASy ProtParam [17]. Protein homology/analogy recognition engine version 2.0—Phyre^2 [18] and PyMol are used to generate backbone protein 3D structure images and to show epitope presentation on a backbone protein (*see* **Note 2**).

3.1.1 To Predict Backbone Immunogen Epitopes

1. Open IEDB www.iedb.org/.
2. Select B-cell Tools.
3. Click "Prediction of linear epitopes from protein sequence" for B cell epitope prediction.
4. Enter protein Swiss-Prot ID or input a backbone protein amino acid sequence, select "Bepipred Linear Epitope Prediction 2.0" for B-cell epitopes, and submit to proceed with epitope prediction.
5. Identified epitopes are presented in a graph and tables (Fig. 2) (*see* **Note 3**).

3.1.2 To Examine Backbone Protein Stability

1. Open ProtParam tool link at ExPASy (https://web.expasy.org/ProtParam/).
2. Enter backbone protein Swiss-Prot ID or accession number (if protein is deposited in ExPASy database); alternatively, input protein amino acid sequence as directed.
3. Press "computer parameters" command to proceed computation.
4. Retrieve data of protein molecular weight, extinction coefficients, estimated protein half-life, and protein instability index (*see* **Note 4**).

3.1.3 To Generate Backbone Protein Model and to Illustrate Protein 3D Images and Epitopes

1. Open Phyre^2 site http://www.sbg.bio.ic.ac.uk/~phyre2/html/page.cgi?id=index.
2. Type in your email address (to receive analysis updates) in a designated box shown on the window.

3. Input protein amino acid sequence and name your optional job description.
4. Select "Intensive" as the modeling and analysis mode.
5. Press "Phyre Search."
6. Receive email notices of computation updates, results in a "xxx. pdb file format" (xxx is the name of your optional job description; the confidence in the intensive model should be greater than 90%), and a link to view your protein model and 3D structure image.
7. Download your "xxx.pdb" file from Phyre^2 site, view protein structure, and further analyze "xxx.pdb" data for 3D structure with PyMol/ EdiPyMol.
8. Download PyMol (https://www.pymol.org/).
9. Open your Phyre^2 file "xxx.pdb" through PyMol program by using "File"→"Open"→"Downloads" by selecting your "xxx. pdb file.
10. Select the "Display" tab on the top to view protein amino acid sequence.
11. Select the "S" tab to show "surface" at the right for display type of the secondary structure in a 3D model, and the "C" for a uniform color for the backbone protein model (Fig. 3a).
12. Mark epitope amino acid sequences in different colors to image the location and surface exposure of each epitope (Fig. 3b) (*see* **Note 5**).

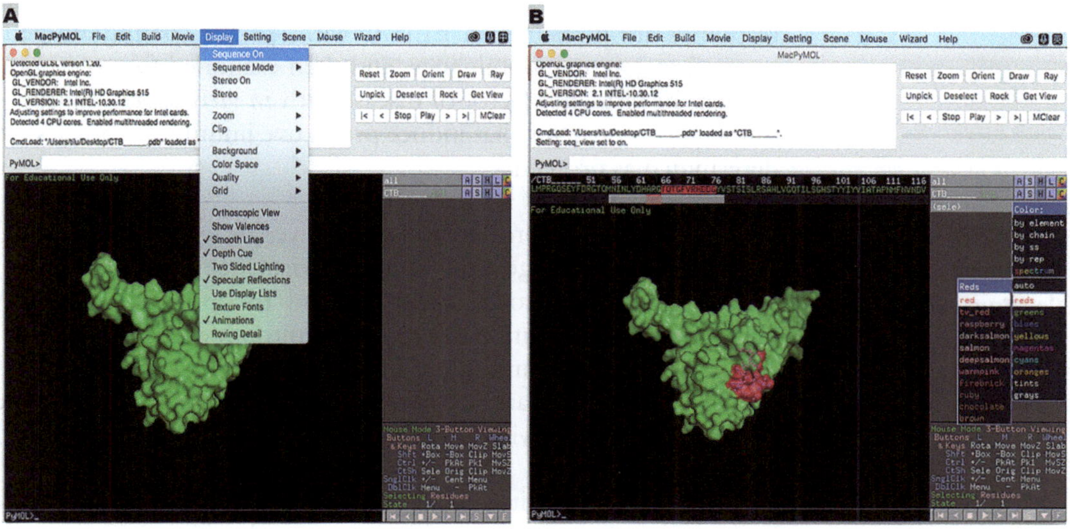

Fig. 3 Illustration of a backbone protein 3D model (**a**) and position of a backbone epitope (**b**), by PyMol program

3.2 Identification of Epitopes from Target Heterogeneous Strains or Virulence Factors

With a desirable backbone immunogen selected, epitopes from the heterogeneous strains or virulence factors of interest must be identified. Protective epitopes are ideal for MEFA immunogen construction. Having epitope native antigenicity retained, an MEFA immunogen carrying multiple protective epitopes of various strains or virulence factors induces broadly protective immune responses. As pointed out in Subheading 3.1, protective epitopes from well characterized pathogens are deposited at IEDB (www.iedb.org) and can be used directly for MEFA immunogen construction. For strains or virulence factors without protective epitopes being identified, immunodominant epitopes in silico identified can be used first; eventually protective epitopes selected by screening immunodominant ones in empirical studies are included in an MEFA immunogen.

3.2.1 To Predict Epitopes of Heterogeneous Strains or Virulence Factors of Interest

Immunodominant linear epitopes from each virulence factor or strain are predicted with IEDB, the same as backbone protein epitope prediction described in Subheading 3.1.1, with the only difference is the input of different amino acid sequences.

3.2.2 To Identify Protective Epitopes

Immunodominant epitopes identified in silico may not necessarily induce protective immune responses. Practically, empirical studies are followed to screen immunodominant epitopes and to identify the protective epitopes. A panel of immunodominant epitopes from a virulence factor or strain is selected (based on antigenic scores), and each epitope is genetically fused to a carrier protein for epitope fusions (to mimic epitope native antigenicity). With each epitope fusion gene cloned in an expression vector, individual epitope fusion proteins are extracted and used for mouse immunization. As we described [12–14], epitope fusion-induced immune responses are measured in ELISAs with specific coating antigens to confirm immune dominance, and then examined for protection against the virulence factor or disease with *in vitro* assays. Epitopes in the fusion proteins that induce protective immune responses are determined as protective.

3.3 MEFA Immunogen Construction

Having a backbone protein and protective (or immunodominant) epitopes from heterogeneous strains or virulence factors identified, we are ready in silico to construct and to optimize an MEFA immunogen. Substitution of backbone epitopes with selecting protective epitopes (or immunodominant epitopes if protective ones are not identified) is carried out either sequentially, substituting one epitope at a time, or with all epitopes replaced at once and modified afterward.

To substitute epitopes sequentially, we select an epitope of interest that possesses a similar length and antigenic score as a backbone epitope, proceed the substitution, construct a chimeric protein, and examine the resultant protein comparatively with the backbone protein for protein stability in ExPASy as in Subheading 3.1.2 and for epitope presentation with Phyre^2 and PyMol as in Subheading 3.1.3. If the top-ranked epitope from a virulence factor or strain does not match well with a backbone epitope, we select the second or the third ranked epitope whichever shows the better match with the backbone epitope. With assessment of no major alteration in protein stability and secondary structure, we proceed to substitution of the next epitope and continue the process until all epitopes are substituted. Alternatively, based on epitope length and antigenic score, we pairwise epitopes of interest with backbone epitopes and complete epitope substitution at once. The resultant MEFA protein is examined with ExPASy, Phyre^2, and PyMol for MEFA protein stability, protein secondary structure, and epitope presentation. If protein stability and structure are altered significantly, we adjust epitope pairing and optimize epitope substitution.

A MEFA protein is further characterized with molecular dynamic simulation programs from GROMACS or CHARMM as we described previously [6]. With comparative analyses of protein dynamic properties, including protein structure stability, conformational flexibility and solvent-accessible surface, of an MEFA protein with a backbone protein, MEFA protein overall structure and stability similarity is examined. With analyses of dynamic simulation focusing on an epitope on the MEFA versus its native format on a virulence factor protein, mimicking of epitope native antigenicity is evaluated.

3.3.1 GROMACS to Characterize an MEFA Protein

1. Download and install SSH client from link https://www.netsarang.com/en/all-downloads/ (Xshell for Windows OS); log in from your desktop as an SSH client to connect remote server (Linux OS) and execute commands (*note*: more information at https://netsarang.atlassian.net/wiki/spaces/ENSUP/pages/31555780/Getting+Started).
2. Download and install GROMACS (http://www.gromacs.org/) and FFTW (version 3.3.8) (http://www.fftw.org/download.html).
3. Input "MEFA protein (or backbone or each virulence factor).pdb" file into Xshell generated from Phyre^2 under your server account.
4. Generate an MEFA protein topology form the input pdb.file, with command **gmx pdb2gmx -f MEFA protein.pdb -o mefa_processed.gro -water spce**, with the selection of **GROMOS96 53a6 force field**.
5. Select periodic boundary conditions (PBC) at "cubic" shape and "1.0 nm" parameters with **gmx editconf -f mefa_processed.gro -o mefa_newbox.gro -c -d 1.0 -bt cubic**.

6. Enter solvent water into the simulation box with **gmx solvate -cp mefa_newbox.gro -cs spc216.gro -o mefa_solv.gro -p topol.top**.
7. Conduct system neutralization with command **gmx grompp -f ions.mdp -c mefa_solv.gro -p topol.top -o ions.tpr**, add counter ions through "genion" with **gmx genion -s ions.tpr -o mefa_solv_ions.gro -p topol.top -pname NA -nname CL -np 9**.
8. Minimize energy with commands **gmx grompp -f minim.mdp -c mefa_solv_ions.gro -p topol.top -o em.tpr** and **gmx mdrun -v -deffnm em**.
9. Equilibrate NVT with **gmx grompp -f nvt.mdp -c em.gro -p topol.top -o nvt.tpr** and **gmx mdrun -deffnm nvt**.
10. Equilibrate NPT with command **gmx grompp -f npt.mdp -c nvt.gro -t nvt.cpt -p topol.top -o npt.tpr** and **gmx mdrun -deffnm npt.**
11. Carry out molecule dynamics (MD) simulations, with **gmx grompp -f md.mdp -c npt.gro -t npt.cpt -p topol.top -o md_0_1.tpr** and **gmx mdrun -deffnm md_0_1**.
12. Calculate Cα root-mean-square deviation (RMSD) and root-mean-square fluctuations (RMSF) (Fig. 4), with commands **gmx trjconv -s md_0_1.tpr -f md_0_1.xtc -o md_0_1_noPBC.xtc -pbc mol -ur compact, gmx rms -s md_0_1.tpr -f md_0_1.trr -o rmsd.xvg**, and **gmx rmsf -s md_0_1.tpr -f md_0_1.xtc -o rmsf.xvg**, respectively.

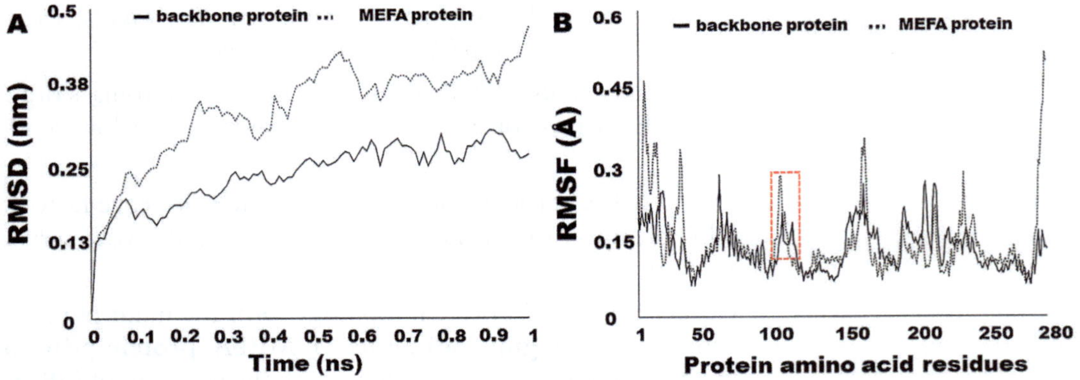

Fig. 4 GROMOS molecule dynamic simulation to show root-mean-square deviation—RMSD (**a**) and root-mean-square fluctuation—RMSF (**b**) of a backbone protein and an MEFA immunogen protein. RMSD and RMSF are used to assess structural and antigenic similarity between a backbone protein and an MEFA protein, or between an epitope on the MEFA and the epitope on the native virulence factor protein. Boxed is a backbone protein epitope and a substituting epitope on an MEFA protein

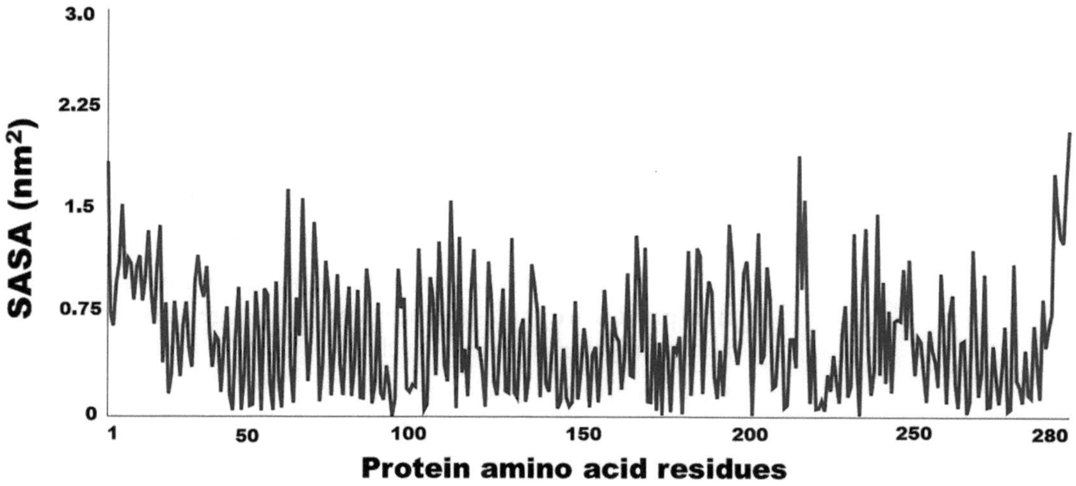

Fig. 5 GROMOS molecule dynamic simulation to show solvent average surface area (SASA) of a backbone protein (or an MEFA immunogen protein, or the epitope of interest on an MEFA protein or native virulence factor protein). SASA is used to assess structural and antigenic similarity between a backbone protein and an MEFA protein, or between an epitope on the MEFA and the epitope on the native virulence factor protein

13. Calculate solvent-accessible surface areas (SASA) (Fig. 5) with **gmx sasa -s md_0_1.tpr -f md_0_1.trr -o area.xvg -or -oa**.
14. Compare results of the MEFA versus backbone, and each epitope on the MEFA versus the native epitope on each virulence factor, to evaluate MEFA protein construction in silico (if overall protein structure or epitope native antigenicity is significantly altered, MEFA is optimized by modifying epitope position or selecting a different epitope from the representing virulence factor) (*see* **Note 6**).

3.3.2 CHARMM to Characterize MEFA Protein

Alternatively, an MEFA protein is examined with license based CHARMM, by constructing protein comparative modeling using Rosetta [19–21], carrying out molecular dynamics (MD) simulation, and analyzing protein secondary structure.

1. Download and install MMTSB toolset (http://blue11.bch.msu.edu/mmtsb/Main_Page).
2. Using "convpdb.pl" from MMTSB toolset to clean up the PDB file with **convpdb.pl -segnames model_pdb.pdb > ms.pdb**. Here ms.pdb is the final clean PDB file for CHARMM.
3. Download and install CHARMM (https://www.charmm.org/).
4. Prepare protein structure in CHARMM format **charmm pdbid=ms < mkprotein.inp**.
5. Prepare water box for protein solvation **charmm WaterDepth=8 target=protein cubic=1 < mkwaterbox.inp**.

6. Set protein solvation and neutralizing charge **charmm target=protein < solvation.inp**.
7. Apply 5.0 ns equilibrium in NVT, harmonic positional restrains on heavy atoms but gradually reduced **charmm < nvt_equil.5 ns.inp > npt_equil.5ns.log**.
8. Conduct the first 50 ns production simulation **charmm-gpu < nvt_prod.0.inp > nvt_prod.0.log** (*see* **Note** 7).
9. Get protein dcd out of water box **charmm < get-dcd-nvt_prod.0.inp** and **charmm cycle=1 < get-dcd-nvt_prod.restart.inp** for additional cycle.
10. Calculate protein backbone and Cα RMSD **charmm dcdin=0 < anal-rms-asa.inp**, and **charmm dcdin=1 < anal-rms-asa.inp** for additional dcd file.
11. Calculate accessible surface areas (ASA) **charmm dcdin=0 < anal-cs*.inp**, and **charmm dcdin=1 < anal-cs*.inp** for additional dcd file. Here the anal-cs*.inp files are predefined. It is based on different protein segments (i.e., a range of residue id for epitope region).
12. Normalize ASA into the relative ASA. The calculation using the ASA of each epitope region divided by total ASA (*see* **Note 8**).

3.4 MEFA Protein Expression, Extraction, and Immunogenicity Characterization

After protein structure stability and epitope antigenicity have been assessed and verified in silico, we synthesize an MEFA immunogen gene and clone the gene for protein expression and extraction. The MEFA gene can be inserted into the pET28a expression vector directly at synthesis, or with two restriction enzyme sites flanked at the MEFA gene for cloning into the pET28a vector, for expression of 6xHis-tagged or tag-less recombinant MEFA protein in *E. coli* strain BL21 (DE3), by following standard molecular biology protocols [22]. MEFA protein expression and extraction are verified by SDS PAGE with Coomassie blue staining and western blotting with specific antibodies; MEFA protein broad immunogenicity is confirmed in mouse immunization.

3.4.1 MEFA Protein Expression and Extraction

Recombinant 6xHis-tag MEFA protein is purified with Ni-NTA agarose in a column (www.qiagen.com), whereas tag-less MEFA protein expressed in inclusion body is extracted with B-PER™ Bacterial Protein Extraction Reagent (https://www.thermofisher.com).

1. Pick a colony from overnight growth on a LB/kan agar plate, culture in 5–10 ml 2× YT medium supplemented with kanamycin (30 μg/mL) at 37 °C overnight on a shaker (200 rpm).
2. Transfer 2–3 mL bacterial growth to 200 mL 2× YT medium supplemented with kanamycin (30 μg/mL) in a 500 mL flask, culture at 37 °C on a shaker (200 rpm) till the OD_{600} reaches 0.5–0.7.

3. Add IPTG (0–1 mM; optimal concentration varies for different proteins), culture for 4 more hours, and transfer into a 250 mL high-speed centrifuge bottle.
4. Harvest bacteria by centrifugation for 15 min at 17,000 × *g* (12,000 rpm with rotor F21-8x50Y), discard supernatant.
5. Freeze the pellet in a −80 °C freezer overnight.
6. Thaw the frozen pellet and add 10 mL B-PER reagent (in phosphate buffer to lyse bacteria. Vortex or pipet up and down until the suspension is homogenous, incubate for 30 min at room temperature (RT) on the shaker (100 rpm).
7. Sonicate bacterial lysate for 5 min (15 s on, 15 s off, at 30% amplitude; on ice).
8. Centrifuge bacterial lysate at 17,000 × *g* at 4 °C for 15 min, resuspend pellet in 10–20 mL 1× phosphate buffered saline (PBS, 137 mM NaCl, 2.7 mM KCl, 10 mM Na_2HPO_4, 1.8 mM KH_2PO_4).
9. To extract 6xHis tag MEFA protein, gently mix the suspension with 5–10 mL Ni-NTA agarose resin by rotating at 4 °C overnight.
10. Apply the suspension to a 100 mL column, flow through, wash 4× with 100 mL wash buffer, elute 4× with 100 mL elution buffer to collect 6 × His tag MEFA protein, run samples (20 μL) on SDS-PAGE. Use Amico or equivalent spin column (at a suitable pore size), or spectrum tubing with polyethylene glycol (PEG) at 4 °C, to concentrate the MEFA protein, and store at −80 °C (ready for immunization).
11. To extract tagless MEFA inclusion body protein, homogenize suspension from **step 8** by passing up and down with 20 mL syringe (use 18G needle).
12. Add 100 mL PBS, vortex, centrifuge at 17,000 × *g* at 4 °C for 15 min, collect pellet.
13. Repeat **step 12** twice to wash protein pellet.
14. Dissolve the purified inclusion body protein pellet in 5–10 mL PBS.
15. Transfer resuspended protein to a 1.5 mL micro centrifuge tube (1 mL per tube), centrifuge with a bench top microcentrifuge at 16,200 × *g* at RT for 10 min, discard supernatant.
16. (To refold purified inclusion body protein) Suspend and solubilize protein pellet with 1 mL 1× IB solubilization buffer (freshly made) by vigorously pipetting and vortex; incubate suspension on a shaker (100 rpm) at RT for 1–2 h.
17. Centrifuge at 16,200 × *g* at RT for 10 min and collect supernatant.

18. Check protein with SDS-PAGE, proceed to one-step dialysis and refolding.
19. To dialyze and refold protein, transfer all of the supernatant into a dialysis tubing membrane with a pore size 5–10 kDa less than the MEFA protein molecular mass.
20. Prepare dialysis buffer for a minimum of two buffer exchanges, for a volume of greater than 50 times of the protein suspension sample volume (for an example, 500 mL dialysis and refolding buffer for 10 mL protein suspension).
21. Dialyze for 4 h at 4 °C and continue with two additional buffer exchanges using dialysis buffer without DTT; If visible precipitation occurs following dialysis, centrifuge the dialysis protein solution at 16,200 × *g* for 10 min at 4 °C.
22. Collect clear supernatant containing refolded MEFA protein, aliquot, measure protein concentration, and store MEFA protein at −80 °C.
23. Characterize refolded protein in ELISAs, SDS-PAGE Coomassie blue stain (incubation for 1 h in stain buffer, and then in destain buffer overnight; room temperature (50 rpm shaker), and western blot with MEFA protein specific antibodies.

3.4.2 MEFA Protein Immunization and MEFA-Specific Antibody Titration

Eight-week-old female BALB/c mice, 8–10 per group, are commonly used for immunization with MEFA protein to examine broad immunogenicity (*see* **Note 9**).

1. Mix and emulsify 50 μg MEFA protein (in 50 μL PBS) with an equal amount of incomplete Freund's adjuvant (IFA); 50 μL PBS mixed with IFA as a negative control.
2. Inject each mouse with MEFA protein (or PBS for the control group) intraperitoneally (IP) [or intramuscularly (IM), subcutaneously (SC), or intradermally (ID)].
3. Two booster injections (of the same dose and route of the primary) are followed at the interval of 2 weeks.
4. Collect mouse serum and fecal samples 2 weeks after the final booster (*see* **Note 10**).
5. Store serum and fecal supernatant at −80 °C until the use for antibody titration and antibody neutralization studies.
6. To titrate antigen-specific antibody responses, coat 96-well microtiter 2HB ELISA plates with epitope specific antigen (50–100 ng of each virulence factor protein in 100 μL 0.05 M carbonate-bicarbonate buffer, pH 9.6; per well) overnight at 4 °C.
7. Discard coating reagent, wash wells three times with PBST (PBS with 0.05% Tween 20), 150 μL PBST per well.

8. Block uncoated sites with 5% nonfat milk (in PBST), 150 mL per well, for 1 h at 37 °C.
9. Discard blocking buffer, wash 3× with PBST.
10. Add twofold serial dilution of mouse serum samples (1:200 initial dilution) or fecal suspension samples (1:50 initial dilution), 100 μL per well, in triplicate, incubate 1 h at 37 °C.
11. Wash wells with PBST three times and one time with PBS, add HRP-conjugated goat anti-mouse IgG (1:5.000 dilution) or IgA (1:3000 dilution), 100 μL per well, incubate for 1 h at 37 °C.
12. Wash wells with PBST three times, add 3,3′,5,5-′-tetramethylbenzidine (TMB; 0.4 g/L in an organic base) Microwell Peroxidase Substrate System, 100 μL per well, incubate at room temperature for 25 min.
13. Read optical density at OD_{650} in a plate reader.
14. Calculate OD readings into $\log_{10}$, by multiplying the highest dilution that produces OD readings greater than 0.3 (after subtraction of the background readings) by the adjusted OD and converting to a $\log_{10}$ scale (*see* **Note 11**).

3.4.3 MEFA Protein-Induced Antibody In Vitro and In Vivo Protective Activities

MEFA-induced antibody protective activities are examined with in vitro assays, or in vivo if animal challenge models are available. These assays vary and are pathogen or disease specific.

4 Notes

The protocols for in silico epitope identification, protein stability assessment, protein 3D image creation and molecule dynamic simulation provided in this chapter are basic and simplified. Detailed instructions to run programs and to analyze data are available from individual program home pages. While most programs are user friendly, molecule dynamic simulation from CHARMM and GROMCS is comprehensive and requires specialized computer-related skills. Other online programs can be used alternatively or synergistically to identify consensus epitopes and to enhance protein structure and stability characterization. Additionally, programs in MacOS and WinOS versions can be slightly different. Native antigenicity of epitopes on MEFA are assessed comparatively with their original format or individual representing virulence factor proteins, based on epitope dynamic property, structure conformational flexibility of residues, accessible surface areas, epitope antigenic scores as well as visual assessment of epitope conformation based on 3D structure. Eventually, in silico constructed MEFA immunogens are validated in empirical studies for broad immunogenicity and potential application of multivalent vaccine development.

Mice are used here to initially confirm MEFA protein broad immunogenicity. Other animal species, if they are relevant to the disease of interest, will be preferred for immunogenicity studies. Since B cell epitopes are discussed, only MEFA-induced antibody responses, but not T cell epitopes and T cell responses, are included in this chapter. Additionally, in vitro antibody neutralization assays are specific to each virulence factor or disease, whereas suitable animal challenge models may not exist for many pathogens and diseases, thus no protocols for antibody functional assays and animal challenge studies are provided. A suitable animal challenge model uses the animal species that are naturally susceptible to the pathogen and mimics disease progression, and after infection develop similar disease outcomes and similar levels of immune responses as humans [23–26]. Therefore, vaccine efficacy can be evaluated preclinically by immunizing the animals and then infecting the immunized animals with a pathogen. Eventually, vaccine candidacy is examined in human subject studies. Control human infection models (CHIM) [27], which directly test immunogenicity, safety and efficacy of vaccine candidates in volunteers, however, enables the use of animal challenge models to be skipped and acceleration of vaccine development.

1. Other wash and elution buffers can be found at the QIAGEN home page.
2. Protein secondary structure can be generated with CHARMM and ExPASy as well.
3. IEDB predicts T-cell epitopes as well; instructional details for epitope prediction are at the IEDB webpage www.iedb.org/.
4. A protein is classified stable if an instability index is less than 40. Instructions for protein stability assessment are available at ExPASy home page.
5. Phyre2 and PyMol program updates and online assistance are at Phyre^2 and PyMol webpages.
6. A constructed MEFA protein is expected to show protein structure and stability similar to the backbone protein; epitopes of interest on the MEFA protein display similar structural conformation and solvent-accessible surface areas as they are on virulence factor proteins.
7. Depending on convergence, one may increase the simulation length, for another 50 ns production simulation with **charmm-gpu cycle=1 < nvt_prod.restart.inp > nvt_prod.1.log**. The index number can be changed to extend another 50 ns until the simulation is converged.

8. Backbone or MEFA model was first solvated in a cubic box of TIP3P water, with the box size was set at 69 Å. Sodium ions were added to neutralize the total charge of the system. Energy minimization was first performed to remove improper molecular arrangement. Followed by 5.0 nanosecond (ns) equilibrium, harmonic positional restrains were applied on heavy atoms but gradually reduced to equilibrate the protein structures and water orientation. Finally, the production simulation was performed for a total simulation length of 350 ns. In the production run, NPT simulation was performed in Langevin dynamics with the constant temperature of 298 K and the constant pressure at 1.0 atm. Timestep 2.0 fs was used with the SHAKE algorithm applied to all hydrogen-containing bonds. Particle mesh Ewald was utilized for electrostatics cutoff of 13 Å. The cutoff of van der Waals interactions is 13 Å, with a switching function between 12 and 13 Å was used. The time evolution of the backbone or MEFA RMSD was first calculated to benchmark the convergence of simulation. The solvent accessible surface area (ASA) was calculated using 1.4 Å as the size of water probe and was further normalized into the relative ASA by calculating ASA for each epitope region. The relative ASA is a normalized calculation using each epitope region divided by total ASA.
9. To avoid gender bias, both sexes, in equal distribution, are used.
10. Fecal pellets are suspended in fecal reconstitution buffer: 10 mM Tris, 100 mM NaCl, 0.05% Tween-20, 5 mM sodium azide, pH 7.4, supplemented with protease inhibitor phenylmethylsulfonyl fluoride (0.5 mM final concentration); 1 g fecal pellet in 5 ml buffer; spin at 13,000 × *g* at room temperature for 15 min to collect supernatant.
11. Other adjuvants including double mutant heat-labile toxin of enterotoxigenic *E. coli* (dmLT) can also be used for mouse immunization under parenteral or mucosal routes. The amount of each coating antigen and secondary dilutions are optimized in standard checkboard test.

Acknowledgments

This work is supported by NIH R01AI121067-01A1 and University of Illinois at Urbana-Champaign. We also thank Shuangqi Wang, Ti Lu, and Ipshita Upadhyay for technical assistance.

References

1. Collaborators GD, Collaborators GH (2017) Global, regional, and national disability-adjusted life-years (DALYs) for 333 diseases and injuries and healthy life expectancy (HALE) for 195 countries and territories, 1990-2016: a systematic analysis for the global burden of disease study 2016. Lancet 390 (10100):1260–1344
2. Global health estimates 2016: Deaths by cause, age, sex, by country and by region, 2000–2016 (2018). https://www.who.int/healthinfo/global_burden_disease/estimates/en/index1.html. Accessed 21 Jan 2021
3. Nabel GJ (2013) Designing tomorrow's vaccines. New Engl J Med 368(6):551–560. https://doi.org/10.1056/NEJMra1204186
4. Mascola JR, Ahmed R (2015) Vaccines: vaccines for cancer and infectious diseases. Curr Opin Immunol 35:V–Vii. https://doi.org/10.1016/j.coi.2015.07.009
5. Ruan X, Sack DA, Zhang W (2015) Genetic fusions of a CFA/I/II/IV MEFA (multiepitope fusion antigen) and a toxoid fusion of heat-stable toxin (STa) and heat-labile toxin (LT) of enterotoxigenic *Escherichia coli* (ETEC) retain broad anti-CFA and antitoxin antigenicity. PLoS One 10(3):e0121623
6. Duan Q, Lee KH, Nandre RM, Garcia C, Chen J, Zhang W (2017) MEFA (multiepitope fusion antigen)-novel technology for structural vaccinology, proof from computational and empirical immunogenicity characterization of an enterotoxigenic *Escherichia coli* (ETEC) Adhesin MEFA. J Vaccines Vaccin 8(4):367. https://doi.org/10.4172/2157-7560.1000367
7. Seo H, Duan Q, Zhang W (2020) Vaccines against gastroenteritis, current progress and challenges. Gut Microbes 11(6):1486–1517. https://doi.org/10.1080/19490976.2020.1770666
8. Lycke N, Holmgren J (1986) Strong adjuvant properties of cholera toxin on gut mucosal immune responses to orally presented antigens. Immunology 59(2):301–308
9. Clements JD, Hartzog NM, Lyon FL (1988) Adjuvant activity of *Escherichia coli* heat-labile enterotoxin and effect on the induction of oral tolerance in mice to unrelated protein antigens. Vaccine 6(3):269–277
10. Lycke N (2005) Targeted vaccine adjuvants based on modified cholera toxin. Curr Mol Med 5(6):591–597
11. Freytag LC, Clements JD (2005) Mucosal adjuvants. Vaccine 23(15):1804–1813. https://doi.org/10.1016/j.vaccine.2004.11.010
12. Huang JC, Duan QD, Zhang WP (2018) Significance of enterotoxigenic *Escherichia coli* (ETEC) heat-labile toxin (LT) enzymatic subunit epitopes in LT enterotoxicity and immunogenicity. Appl Environ Microbiol 84(15): e00849-00818. https://doi.org/10.1128/AEM.00849-18
13. Lu T, Moxley RA, Zhang W (2019) Mapping the neutralizing epitopes of enterotoxigenic *Escherichia coli* (ETEC) K88 (F4) fimbrial adhesin and major subunit FaeG. Appl Environ Microbiol 85(11):e00329–e00319. https://doi.org/10.1128/AEM.00329-19
14. Lu T, Seo H, Moxley RA, Zhang W (2019) Mapping the neutralizing epitopes of F18 fimbrial adhesin subunit FedF of enterotoxigenic *Escherichia coli* (ETEC). Vet Microbiol 230:171–177. https://doi.org/10.1016/j.vetmic.2019.02.015
15. Lu T, Moxley RA, Zhang W (2020) Application of a novel epitope and structure vaccinology-assisted fimbria-toxin multiepitope fusion antigen of enterotoxigenic *Escherichia coli* for multivalent vaccine development against porcine post-weaning diarrhea. Appl Environ Microbiol 86(24):e00274-20. https://doi.org/10.1128/AEM.00274-20
16. Vita R, Mahajan S, Overton JA, Dhanda SK, Martini S, Cantrell JR et al (2019) The immune epitope database (IEDB): 2018 update. Nucleic Acids Res 47(D1): D339–D343. https://doi.org/10.1093/nar/gky1006
17. Artimo P, Jonnalagedda M, Arnold K, Baratin D, Csardi G, de Castro E et al (2012) ExPASy: SIB bioinformatics resource portal. Nucleic Acids Res 40(W1):W597–W603. https://doi.org/10.1093/nar/gks400
18. Kelley LA, Mezulis S, Yates CM, Wass MN, Sternberg MJE (2015) The Phyre2 web portal for protein modeling, prediction and analysis. Nat Protoc 10(6):845–858. https://doi.org/10.1038/nprot.2015.053
19. Chivian D, Baker D (2006) Homology modeling using parametric alignment ensemble generation with consensus and energy-based model selection. Nucleic Acids Res 34(17): e112. https://doi.org/10.1093/nar/gkl480
20. Raman S, Vernon R, Thompson J, Tyka M, Sadreyev R, Pei JM et al (2009) Structure prediction for CASP8 with all-atom refinement using Rosetta. Proteins Struct Funct Genet

77:89–99. https://doi.org/10.1002/prot.22540

21. Song JS, Liu YQ, Liu CZ, Xie JP, Ma LX, Wang LP et al (2013) Cumulative analgesic effects of EA stimulation of sanyinjiao (SP 6) in primary dysmenorrhea patients: a multicenter randomized controlled clinical trial. Zhen Ci Yan Jiu 38(5):393–398
22. Ausubel FM, Brent R, Kingston RK, Moore DD, Seidman JG, Smith JA, Struhl K (1999) Short protocols in molecular biology, 4th edn. Wiley, New York, NY
23. Spira WM, Sack RB, Froehlich JL (1981) Simple adult rabbit model for *Vibrio cholerae* and enterotoxigenic *Escherichia coli* diarrhea. Infect Immun 32(2):739–747. https://doi.org/10.1128/IAI.32.2.739-747.1981
24. Gerdts V, Wilson HL, Meurens F, van Drunen Littel-van den Hurk S, Wilson D, Walker S et al (2015) Large animal models for vaccine development and testing. ILAR J 56(1):53–62. https://doi.org/10.1093/ilar/ilv009
25. Swearengen JR (2018) Choosing the right animal model for infectious disease research. Animal Model Exp Med 1:100–108. https://doi.org/10.1002/ame2.12020
26. Ramin Sedaghat Herati EJW (2018) What is the predictive value of animal models for vaccine efficacy in humans. Cold Spring Harb Perspect Biol 10:a031583. https://doi.org/10.1101/cshperspect.a031583
27. Gordon SB, Rylance J, Luck A, Jambo K, Ferreira DM, Manda-Taylor L et al (2017) A framework for controlled human infection model (CHIM) studies in Malawi: report of a Wellcome Trust workshop on CHIM in low income countries held in Blantyre, Malawi. Wellcome Open Res 2:70. https://doi.org/10.12688/wellcomeopenres.12256.1

Chapter 11

Production, Isolation, and Characterization of Bioengineered Bacterial Extracellular Membrane Vesicles Derived from *Bacteroides thetaiotaomicron* and Their Use in Vaccine Development

Régis Stentz, Ariadna Miquel-Clopés, and Simon R. Carding

Abstract

Bacterial extracellular vesicles (BEVs) possess features that make them well suited for the delivery of therapeutics and vaccines. This chapter describes methods for engineering the commensal human intestinal bacterium *Bacteroides thetaiotaomicron* (Bt) to produce BEVs carrying vaccine antigens and accompanying methods for isolating and purifying BEVs for mucosal vaccination regimens.

Key words *Bacteroides thetaiotaomicron*, Bacterial extracellular vesicles, Crossflow ultrafiltration, Vaccines, Immunization

1 Introduction

Conventional vaccines based on the use of attenuated or inactivated forms of the target pathogen have successfully eradicated smallpox and rinderpest as well as significantly reducing the burden of many other infectious diseases throughout the past century. However, the time needed to identify vaccine targets, the high cost of vaccine development and manufacture, and the limited production capacity, make these traditional approaches less than optimal in the rapid response to epidemics and pandemics [1]. Furthermore, these vaccines are usually delivered parenterally via injection, which makes mass immunization costly particularly in resource-poor developing countries [2]. There is, therefore, a need for the development of new vaccines that are versatile, cost-effective, safe, and enable global immunization. To this end, various new vaccination technologies have emerged including the use of synthetic protein and peptide antigens [3]. Protein subunit vaccines are attractive because of their inherent safety although they can suffer from poor

Fadil Bidmos et al. (eds.), *Bacterial Vaccines: Methods and Protocols*,
Methods in Molecular Biology, vol. 2414, https://doi.org/10.1007/978-1-0716-1900-1_11,

immunogenicity and high manufacturing costs [4]. To address these constraints, nanoparticle-based delivery technologies have been developed which includes nanoparticle-sized extracellular vesicles naturally produced by bacteria.

Bacterial extracellular vesicles (BEVs) are spherical nanostructures composed of membrane-derived lipid bilayers with a diameter of between 20 and 400 nm. BEVs generated by gram-negative bacteria primarily consist of vesicles derived from the outer membrane containing phospholipids, outer membrane proteins, lipopolysaccharides, and capsular polysaccharides with their lumen principally filled with periplasmic content [5]. These components, which include microbe-associated molecular pattern molecules, confer inherent and potent adjuvanticity on BEVs which together with their natural temperature, chemical resistance, and straightforward isolation [6–8] makes them well suited as vaccine delivery vehicles capable of enhancing the immunogenicity of protein/peptide antigens without the need for chemical adjuvants [9]. The ability of BEVs to interact with, and be acquired by, mucosal epithelial and immune cells [10–12] further enhances their suitability for mucosal administration and the generation of local and systemic immunity [13].

We have engineered the gram-negative bacterium *Bacteroides thetaiotaomicron* (Bt), a prominent member of the intestinal microbiota of all animals [14, 15], to incorporate virus-, bacteria-, and human-derived proteins into its BEVs [8, 16]. These engineered Bt BEVs have been used to protect the gastrointestinal or respiratory tracts against infection, tissue inflammation and injury. Here, we describe the methods to implement secretion of vaccine antigens and other proteins into Bt BEVs for mucosal delivery, which are outlined in Fig. 1.

2 Materials

Unless stated otherwise, all solutions are prepared with double-distilled water (ddH_2O) with all reagents being stored at ambient (room) temperature (~20 °C). Most methods require the use of piston pipettes and sterile single-use pipette tips.

2.1 Synthetic Gene Design

1. Protein sequence (e.g., stalk of the hemagglutinin antigen of influenza A virus strain H5N1 [8, 17]).
2. Secretion sequence signal (e.g., N-terminal sequence of Bt outer membrane protein A [OmpA] (BT_3852)).
3. Optional additional sequences (e.g., His-tag or FLAG-tag sequences).
4. *Escherichia coli/Bacteroides* shuttle expression vector sequence (e.g., pGH90 [18]) (*see* **Note 3**).

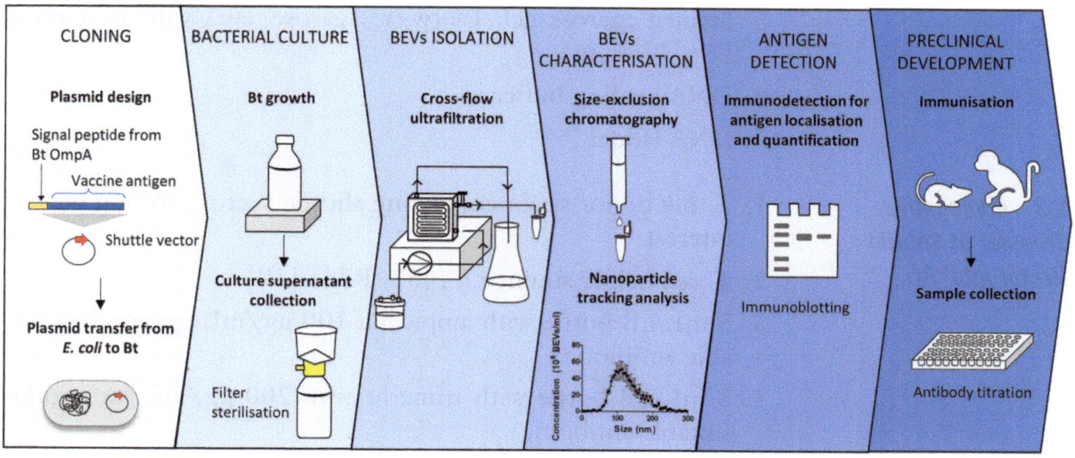

Fig. 1 Overview of the production and extraction of bioengineered bacterial extracellular vesicles (BEVs) derived from *B. thetaiotaomicron* for the development of novel vaccines. Immunization and sample collection for preclinical development and evaluation are described in **Notes 1** and **2**

2.2 Cloning of Synthetic Gene

The required materials for the cloning of the conjugative plasmid harboring the gene of interest are as follows:

2.2.1 Generation of Recombinant DNA

1. *E. coli/Bacteroides* shuttle expression vector.
2. Restriction enzymes.
3. Ligation buffer and enzyme.

2.2.2 Transformation of Competent Cells

1. Crushed ice for thawing the MAX Efficiency® DH10B competent cells.
2. S.O.C. medium: commercially available.
3. Luria-Bertani (LB) agar plates: prepare according to the manufacturer's instructions.
4. Shaking incubator.
5. LB broth for the dilution of the bacterial cell suspension: prepare according to the manufacturer's instructions.
6. Agar plates containing ampicillin (working concentration 100 μg/mL).
7. 1.5 mL centrifuge tubes.
8. Water bath.

2.2.3 Screening of Recombinant Bacteria

1. Shaking incubator.
2. Benchtop centrifuge.
3. Small-scale plasmid isolation kit.
4. Gel electrophoresis chamber and electrophoresis power supply.

5. Stained agarose gel: 1% (w/v) agarose, e.g., with EvaGreen® Dye, Biotium.
6. DNA loading buffer.
7. DNA ladder.

2.3 Conjugative Transfer of Shuttle Vector into Bt

1. *E. coli* donor strain containing shuttle vector carrying gene of interest.
2. *E. coli* helper strain, e.g., J53/R751 [19].
3. 5 mL LB bottle with ampicillin 100 μg/mL (or another suitable antibiotic).
4. 5 mL LB bottle with trimethoprim 200 μg/mL (or another suitable antibiotic)
5. Brain heart infusion (BHI) agar supplemented with hemin and antibiotics (gentamicin 200 μg/mL and/or erythromycin 5 μg/mL, or another suitable antibiotic): Dissolve 18.5 g of BHI powder and 7.5 g of agar in 0.5 liter of deionized water and add 0.75 μM of hemin (BHIH). Autoclave the medium and leave for a minimum of 24 h in an anaerobic cabinet to fully deoxygenate.
6. Filter disc 0.45 μm pore size, 25 mm.
7. 50 mL conical centrifuge tubes.
8. Tweezers.
9. Sterile 25 mL wide neck universal glass bottle.

2.4 Assessing Protein Expression and Secretion into BEV

2.4.1 Culture of Bt Transconjugants

1. BHIH agar plates containing gentamicin 200 μg/mL and erythromycin 5 μg/mL.
2. Sterile inoculation loop(s).
3. 20 mL BHIH containing erythromycin 5 μg/mL in Universal bottles.
4. Anaerobic cabinet.
5. 50 mL conical centrifuge tubes.
6. Refrigerated centrifuge.
7. −20 °C freezer.

2.4.2 Cell Total Protein Extraction

1. 0.2 M Tris–HCl, pH 7.2.
2. Sonicator.
3. Refrigerated centrifuge.
4. Bradford reagent.
5. 96-well microplate.
6. Bovine serum albumin (BSA).

2.4.3 BEV Total Protein Extraction

1. 20 mL syringe.
2. 0.22 μm pore-size polyethersulfone (PES) membranes (Sartorius).
3. Centrifugal concentrator, 100 kDa molecular weight cutoff.
4. Refrigerated centrifuge.
5. 0.2 M Tris–HCl, pH 7.2.
6. Sonicator.
7. Bradford reagent.
8. 96-well microplate.
9. BSA.

2.4.4 Protein Western Blotting/Antigen Immunodetection

1. Gel electrophoresis equipment.
2. Protein gel.
3. SDS sample loading buffer.
4. Reducing agent.
5. Antioxidant for protein electrophoresis.
6. Running buffer.
7. Blotting equipment.
8. Western blotting membrane.
9. Tweezers.
10. Tris-Glycine transfer buffer (25×): Dissolve 18.2 g Tris Base and 90.0 g of glycine in 450 mL of deionized water. Mix well and adjust the volume to 500 mL with deionized water. The pH of the buffer is 8.3. Store the buffer at 20 °C. The buffer is stable for 6 months at 25 °C.
11. Methanol.
12. Orbital shaker.
13. Tris-buffered saline (TBS) buffer: 50 mM Tris–HCl, 150 mM NaCl, pH 7.5.
14. TBST buffer: TBS buffer with 0.05% Tween 20.
15. Blocking buffer: TBST with 5% non-fat dry milk.
16. Chemiluminescent substrate.
17. Primary antibody (e.g., 6×-His Tag monoclonal antibody) used at a working concentration recommended by the manufacturer or by other providers if the antibody is not commercially available (*see* **Note 4**).
18. Secondary antibody labelled with horse radish peroxidase (HRP) (*see* **Note 5**).
19. Imaging System.

2.5 Bacteria Medium-Scale Culture

2.5.1 BHIH Bacterial Culture

1. BHIH medium.
2. Anaerobic cabinet.
3. Magnetic stirrer.
4. Sterile magnetic stirring bar.

2.5.2 BDM+ Bacterial Culture

1. *Bacteroides* defined medium Plus (BDM+): To prepare 500 ml, dissolve 2.61 g of KH_2PO_4 and 7.03 g of K_2HPO_4 * 3 H_2O into 481 mL of deionized water, add the following solutions to a final concentration of 15 mM NaCl and 8.5 mM $(NH4)_2SO_4$, adjust the pH to 7.4 using 5 M NaOH, autoclave and place in the anaerobic cabinet to equilibrate for a minimum of 24 h. Add the rest of the solutions to a final concentration of 30 mM of glucose, 0.2 mM L-histidine, 100 nM vitamin B12, 6 μM vitamin K3 (menadione), 0.1 mM $MgCl_2$, 50 μM $CaCl_2$, 4.1 mM L-cysteine and 1.4 μM $FeSO_4$ * 7 H_2O. Leave the medium for a minimum of 24 h in the anaerobic cabinet to fully deoxygenate. Add 2 μM of Protoporphyrin IX freshly made before using the media. All stock solutions in exception of Protoporphyrin IX are prepared in advance and autoclaved or filter sterilized, as below:
   - 1.5 M NaCl (autoclaved).
   - 0.85 M $(NH4)_2SO_4$ (autoclaved).
   - 3 M Glucose (filtered).
   - 0.2 M L-histidine, store at 4 °C (filtered).
   - 1 mM vitamin B12, store at 4 °C (filtered).
   - 6 mM vitamin K3 (menadione), dissolve in ethanol and store at −20 °C (filtered).
   - 0.1 M $MgCl_2$ * 6 H_2O, store in anaerobic cabinet (autoclaved).
   - 50 mM CaCl2 * 2 H_2O, store in anaerobic cabinet (autoclaved).
   - 0.41 M L-cysteine hydrochloride, store in anaerobic cabinet (filtered).
   - 2 mM Protoporphyrin IX.
   - 1.4 mM $FeSO_4$ * 7 H_2O, store in anaerobic cabinet (filtered).
2. Anaerobic cabinet.
3. Magnetic stirrer.
4. Sterile magnetic stirring bar.

2.5.3 Supernatant Collection

1. Spectrophotometer (wavelength: 600 nm) with cuvette holder and cuvettes.

2. Refrigerated high speed floor centrifuge, including appropriate rotor with 500 mL sealable centrifuge bottles.
3. 0.2 μm PES bottle top filter unit (500 mL).
4. Membrane vacuum pump.
5. Sterile 500 mL bottles.

2.6 EV Isolation

1. Filtration cassette Vivaflow 50 R (100,000 MWCO, Hydrostat, model VF05H4, Sartorius).
2. Peristaltic pump for running the Vivaflow-unit.
3. Sterile phosphate-buffered saline (PBS), pH 7.4.
4. 0.22 μm PES syringe filters.
5. 5 mL sterile syringes.
6. 15 mL conical centrifuge tubes.
7. 1.5 mL sterile low-bind microcentrifuge tubes.
8. Deionised water.
9. 0.5 M NaOH solution.
10. 10% (v/v) ethanol.

2.7 EVs Purification

In this part of the chapter, the use of size-exclusion chromatography is described for the removal of remaining proteins. Two options are proposed: 2.7.1 and 2.7.2 for an increased resolution.

2.7.1 Routine Purification

1. qEVoriginal/35 nm SEC columns (IZON).
2. Support to maintain column in a vertical position.
3. 1.5 mL lo-bind microcentrifuge tubes.
4. Amicon Ultra 0.5 mL centrifugal filters (RC, 10 kDa MWCO).
5. Sterile PBS, pH 7.4.
6. 0.22 μm PES membrane syringe filter.
7. 1 mL sterile syringes.
8. LB and BHIH agar plates.

2.7.2 High-Resolution Fractionation

1. CL2-B Sepharose.
2. Sterile PBS, pH 7.4.
3. Chromatography column (120 cm × 1 cm) (e.g., Econo-Column® Chromatography Column) in PBS.
4. Chromatography fraction collector.
5. UV spectrophotometer.
6. Vivaspin 20 centrifugal concentrator (100 kDa molecular weight cutoff).
7. 0.22 μm PES membrane syringe filter.

2.8 EVs Size and Concentration Analysis

1. Nanoparticle Analyzer (ZetaView TWIN Particle Tracking Analyzer instrument or equivalent).
2. Particle-free deionized water.
3. 1 or 10 mL sterile syringes.

2.9 Antigen Localization with Proteinase K Assay

1. Proteinase K.
2. Phenylmethanesulfonyl fluoride (PMSF).
3. Water bath.
4. Sodium Dodecyl Sulphate (SDS).
5. See also materials in Subheadings 2.4.3 and 2.4.4.

2.10 Antigen Quantification

1. Recombinant antigen.
2. See also materials in Subheadings 2.4.3 and 2.4.4.

3 Methods

3.1 Synthetic Gene Design

The gene can be synthesized *de novo* using commercial gene synthesis services. The N-terminus of the protein of interest is fused in frame to the signal peptide of the product of BT_3852 (OmpA of Bt); MKKILMLLAFAGVASVASA. The chimeric protein sequence is tested in silico for cleavage of the OmpA signal sequence using http://www.cbs.dtu.dk/services/SignalP/. If unsuccessful, change or add amino acids as appropriate to the N-terminus of the gene of interest, downstream from the signal peptide sequence. To facilitate immunodetection and/or purification of the protein in downstream applications, a fusion tag can be added to the gene. It is important that the coding sequence of the desired protein incorporates codon usage optimization for expression in Bt, which is usually provided as part of gene synthesis services. The desired target sequence is then integrated into an acceptor vector.

3.2 Cloning of Gene of Interest

3.2.1 Generation of Recombinant DNA

1. Digest the plasmid containing the synthetic gene with restriction enzymes to excise the gene from the vector carrying the synthetic gene (e.g., pEX-K168).
2. Digest the *E. coli/Bacteroides* shuttle vector pGH90 (*see* Subheading 2.1) (*see* **Note 6**).
3. Ligate the gene into the digested pGH090 expression vector to allow translational fusion (*see* **Note 7**).

3.2.2 Ransformation of Competent Cells

1. Prepare LB agar plates containing ampicillin.
2. Thaw one vial of competent MAX Efficiency® DH10B on ice.
3. Gently add 1–5 μL of ligation mixture to the MAX Efficiency® DH10B.

4. Place the vial with the bacteria suspension on ice for 30 min.
5. Induce a heat-shock at 42 °C for 30 s.
6. Incubate the bacteria on ice for an additional 5 min.
7. Add 950 μL S.O.C. medium.
8. Incubate at 37 °C with agitation (250 rpm).
9. Seed 100 μL of the bacterial suspension onto agar plates containing ampicillin.
10. Incubate the plates for 16–18 h at 37 °C.

3.2.3 Screening of Cloned Recombinant DNA

1. Pick individual colonies from agar plates and add each colony to a tube containing 5 mL LB medium and ampicillin.
2. Incubate the liquid cultures at 37 °C and 250 rpm for 16–18 h.
3. Isolate plasmid DNA from each culture using a small-scale isolation kit (e.g., QIAprep Spin Miniprep Kit) according to the manufacturer's instructions.
4. Digest the plasmid DNA using the appropriate restriction enzymes (*see* **Note 8**).
5. Resolve digested DNA on a TBE gel.
6. Confirm the identity of the plasmids containing the insert of the expected size by DNA sequencing using appropriate primers.

3.3 Conjugative Transfer of Shuttle Vector into Bt

3.3.1 Triparental Mating Procedure

1. Prepare BHIH agar plates either with or without gentamycin and erythromycin.
2. Grow cultures of the *E. coli* donor strain with ampicillin (containing the plasmid with the correct inserted sequence, *see* Subheading 3.2.3) and the *E. coli* helper strain with trimethoprim in 10 mL of LB, at 37 °C, under agitation for 16–18 h. In parallel, grow culture of the Bt recipient strain in BHIH in an anaerobic cabinet at 37 °C for 16–18 h.
3. Inoculate 10 mL of LB with 100 μL of the *E. coli* donor strain and the helper strain (no antibiotics added) cultures and incubate for 2 h at 37 °C with agitation (e.g., 200 rpm). In parallel, inoculate in 30 mL BHIH contained in a 50 mL tube with 800 μL of the Bt culture and incubate for 2 h at 37 °C in an anaerobic cabinet.
4. Add donor and helper cultures to the Bt recipient in the 50 mL tube, mix by vortexing briefly and centrifuge at 2000 × *g* for 15 min at 20 °C.
5. Remove the supernatant and resuspend cells in 100 μL of BHIH. Transfer cell suspension to the surface of a sterile 0.45 μm filter placed on a BHIH agar plate. Incubate the plate aerobically for 16–18 h at 37 °C.

6. Transfer the filter to a sterile wide-necked Universal bottle and add 1 mL of BHIH and resuspend the bacterial conjugation mixture by vortexing thoroughly.
7. Make serial dilutions and plate 100 μL of each dilution and the non-diluted cell suspensions onto BHIH agar plates containing gentamycin (to prevent E. coli growth) and erythromycin (selection of Bt transconjugants).

3.4 Assessing Protein Expression and Secretion into BEVs

3.4.1 Culture of Bt Transconjugants

1. Pick 4 individual colonies and restreak each one on separate BHIH agar plates containing gentamicin and erythromycin.
2. Incubate the plates anaerobically at 37 °C for 48 h.
3. Inoculate bottles containing 20 mL of BHIH with each of the 4 isolated clones.
4. Incubate the bottles anaerobically at 37 °C for 48 h.
5. Centrifuge the 20 mL of culture in 50 mL tubes, at 6000 × *g* for 15 min, at 4 °C.
6. Collect the supernatant.
7. Wash cell pellets once in PBS before storing at −20 °C prior to analysis.

3.4.2 Cell Total Protein Extraction

1. Resuspend thawed cell pellets in 250 μL of 0.2 M Tris–HCl (pH 7.2).
2. Disrupt the cells via sonication using eight 10-s pulses (amplitude, 6 μm), with 30-s pauses on ice between each pulse.
3. Cell extracts are obtained after centrifugation at 14,000 × *g* for 30 min at 4 °C and harvesting the supernatant.
4. Measure total protein concentration of each supernatant/sample using the Bio-Rad protein assay according to the manufacturer's instructions using BSA to generate a standard curve.

3.4.3 BEV Total Protein Extraction

1. Filter the 20 mL supernatants through 0.22 μm pore-size PES membranes to remove debris and cells.
2. Concentrate the supernatants by ultrafiltration (100 kDa molecular weight cutoff, Vivaspin 20) to a final volume of 250 μL.
3. Discard the filtrate.
4. Rinse the retentate with 20 mL of 0.2 M Tris–HCl, pH 7.2 and concentrated to 250 μL.
5. Collect the retentate and disrupt the vesicles via sonication using eight 10-s pulses (amplitude, 6 μm), with 30-s pauses on ice between each pulse.
6. Measure the total protein content and concentration using the Bio-Rad protein assay according to the manufacturer's instructions using BSA to generate a standard curve.

3.4.4 Protein Western Blotting/Antigen Immunodetection

1. Add BEV and cell extracts obtained in Subheading 3.4.3 to loading buffer containing 0.4 M freshly prepared dithiothreitol (DTT).
2. Load 7 μg of the total protein onto a 12% precast gel and separate by electrophoresis at 180 V for 40 min.
3. Transfer the proteins from the gel onto a polyvinylidene difluoride (PVDF) membrane using the XCell II™ Blot Module or equivalent (according to the manufacturer's instructions) at 25 V for 2 h in a solution containing Tris-Glycine transfer buffer and methanol 20% (v/v).
4. Incubate the membrane with blocking buffer by gently shaking for 30 min at 20 °C using an orbital shaker.
5. Discard the blocking solution and incubate the membrane for 16–18 h at 4 °C in blocking buffer containing primary antibody (Usually 1:1000 to 1:10000).
6. After washing 3 times with TBST, membranes are incubated with HRP-conjugated secondary antibody in blocking buffer for 1 h at 20 °C.
7. After 3 washes with TBST, Enhanced Chemiluminescent substrate (ECL) is added to detect bound antibody.

3.4.5 Medium-Scale Bacterial Culture and Harvesting Conditioned mediaBHIH Bacterial Culture

1. Inoculate 10 mL of BHIH with a frozen stock for ~16 h.
2. Inoculate 500 mL of BHIH with 0.5 mL of the preinoculum for 17 h (starting OD_{600} ~ 0.005) with mild stirring.

3.4.6 BDM+ Bacterial Culture

Initially BHIH was used as a standard medium for generating Bt BEVs [8, 16]. However, considering the need to reduce and even exclude animal-derived products from medical/therapeutic formulations to be used in humans, we have modified a chemically defined *Bacteroides* growth media (BDM) [20] (*see* Subheading 2.5.2) for Bt BEV production.

1. Inoculate 10 mL of BHIH with a frozen stock for ~16 h.
2. Inoculate 10 mL of BDM+ with 100 μL BHIH culture from **step 1** for ~8 h.
3. Inoculate 500 mL of BDM+ with 0.5 mL of the preinoculum for 17 h (starting OD_{600} ~ 0.005) with gentle stirring.

3.4.7 Supernatant Collection

1. Collect bacterial cultures at final OD_{600} 1.5–2.5.
2. Precool centrifuge and canisters for 5 min at 4 °C.
3. Decant the culture into two 2 Nalgene™ PPCO centrifuge bottles and centrifuge at 6037 × *g* for 30 min at 4 °C.

4. Filter-sterilize the supernatant with a 0.22 μm bottle-top filter unit and transfer the filtrate into a sterile 500 mL bottle (*see* **Note 9**).
5. Samples are stored at 4 °C prior to BEVs isolation for up to 24 h.

3.5 Isolation of BEVs

During the filtration process, the membrane will retain any molecules >100 kDa including BEVs (retentate) and concentrate them in the reservoir, whereas molecules <100 kDa will be removed in the flow-through (filtrate) directly to the waste. BEVs isolation should be performed at ambient (20–22 °C) temperature.

Three procedures are used in a step wise manner for isolating BEVs, which are illustrated in Fig. 2.

3.5.1 Module Rinsing

1. Set up the system as illustrated in Fig. 2 Option 2.
2. Place 400 mL of deionized water in the reservoir and pump the liquid through the system at an initial speed setting of 2 to remove air pockets and then increase the speed setting to 4–5 (200–300 mL/min) until 400 mL has been run through the system. Collect filtrate in waste bin. Check for any leaks.

3.5.2 Sample Concentration

1. Set up the module as illustrated in Fig. 2 Option 1.
2. Fill the reservoir with 500 mL filtrate.
3. Pump sample through the system. The initial recirculation speed setting is 2 for at least 1 cycle and then adjusted to setting 4–5 for sampling (200–300 mL/min). Maximum recirculation speed setting is 5. Reduce the speed for lower volumes to avoid foaming.
4. Concentrate the sample until there is ~5 mL left in the system.
5. Switch off the pump.
6. Pour 500 mL of PBS pH 7.4 into the reservoir and pump through the system at an initial speed setting of 2 to remove any air pockets and then increase the speed to setting 4–5.
7. Reduce the recirculation speed setting to 1–2 (20–40 mL/min) to avoid foaming.
8. Switch off the pump with 1–4 mL remaining in the system.
9. Set up the module as illustrated in Fig. 2 Option 3.
10. Start the pump at a speed setting of 1–2 and carefully collect the concentrated samples in a 15 mL tube or microcentrifuge tube (use a bigger tube than the volume to be collected to account for foam).
11. When no more concentrated sample emerges from the tubing, switch off the pump.

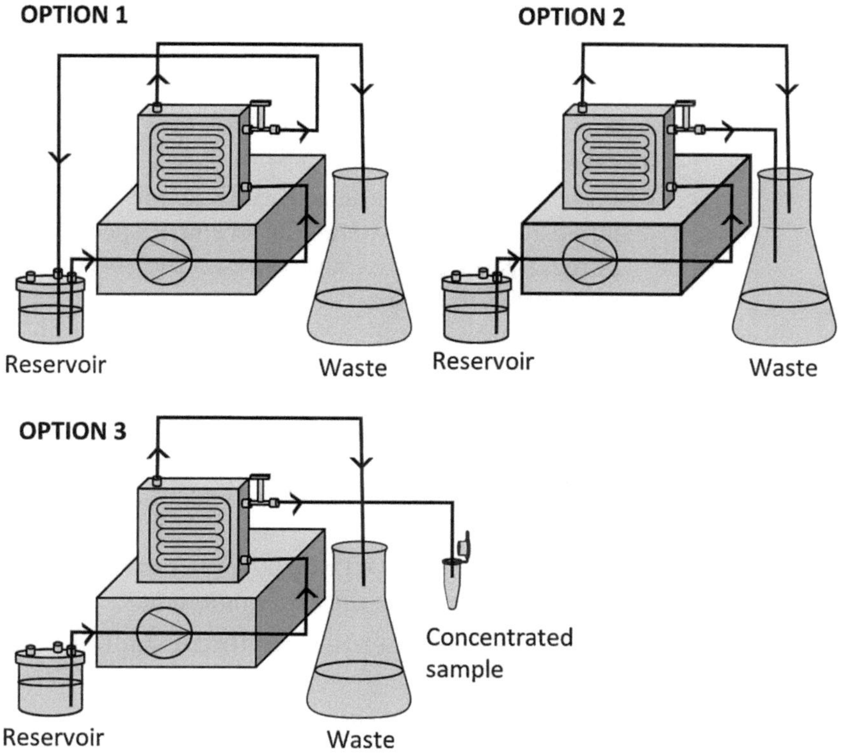

Fig. 2 BEV isolation module system. Option 1. Sample concentration. Option 2. Module rinsing/decontamination. Option 3. Collection of concentrated samples

12. Place 0.5–1 mL of PBS in the reservoir and pump through the system at speed setting 1–2 to flush out remaining BEVs. Collect in the original 15 mL tube or microcentrifuge tube. Final volume collected depends on concentration of BEVs needed.
13. Switch off the pump.
14. Centrifuge at 15,000 × *g* for 20 min at 4 °C to remove any precipitate.
15. Filter-sterilize the supernatant using a 0.22 μm syringe filter, collecting the filtrate in sterile 1.5 mL lo-bind microcentrifuge tubes or 15 mL tube.

3.5.3 Module Decontamination and Washing

1. Set up the system as illustrated in Fig. 2 Option 2.
2. Place 400 mL of deionized water in the reservoir and pump liquid through the system at an initial speed setting of 2 to remove any air pockets and increase the speed setting to 4–5 (200–400 mL/min). Collect filtrate in waste bin.
3. When all the water has been collected, switch off the pump.
4. Set up the system as illustrated in Fig. 2 Option 1.

5. Place 250 mL of decontamination solution (250 mL 0.5 M NaOH) and pump it through the system at a speed setting of 3–4 (50–100 mL/min). Allow to recirculate for a minimum of 20 min, then switch off the pump.
6. Set up the system as illustrated in Fig. 2 Option 2.
7. Place 400 mL of deionized water in the reservoir and pump liquid through the system at an initial speed setting of 2 to purge any air pockets and increase the speed setting to 4–5 (200–400 mL/min).
8. Switch off the pump.
9. Set up the system as illustrated in Fig. 2 Option 2.
10. Place 250 mL of 10% Ethanol in the reservoir and pump liquid through the system at an initial speed setting of 2 to purge any air pockets and increase the speed setting to 3–4 (50–100 mL/min).
11. When half of the 10% Ethanol has been filtered, switch off the pump and dismantle the system module.
12. Store cassette membrane in 10% Ethanol at 4 °C to avoid contamination.
13. Rinse the reservoir with water and let it air dry. Reservoir and tubing are left to dry at 20 °C.

    BEV isolation waste is discarded in the sink drain.

3.6 BEVs Purification

Contaminants (e.g., proteins) of BEV preparations can be removed by size exclusion chromatography (SEC). We recommend using the method described in Subheading 3.7, **step 1** for routine preparations and in Subheading 3.7, **step 2** if increased resolution is needed, for instance, to size fractionate BEVs.

3.6.1 Routine Purification

1. Bring the BEVs preparation, column buffer (PBS pH 7.4) and qEVoriginal/35 nm SEC column(s) to ~20 °C. Do not remove the column caps until operational temperature is reached.
2. Secure the column in a vertical position using a stand.
3. Carefully remove the column top-cap.
4. Attach a column reservoir (if available) and add 1.5× column volume of buffer (PBS 1×, 15 mL).
5. Remove the bottom column cap and allow the buffer to run under gravity to waste.
6. If any buffer other than PBS is used, flush with at least 3 column volumes of the buffer (>30 mL).
7. The column will stop flowing when the buffer has entered the loading frit.
8. Load 0.5–1 mL of isolated BEVs onto the loading frit.

9. Collect the void volume (3 mL) fraction(s) into 1.5 mL lo-bind microcentrifuge tubes.
10. Allow the sample to completely run into the column.
11. Top up the reservoir with 15 mL buffer (PBS 1×) and collect 0.5–1 mL elution fractions.
12. For a loading volume of 1 mL of BEVs and collecting 0.5 mL elution fractions, BEVs will elute in fractions 7–12 with proteins eluting in fractions 10–20.
13. After the eluted fractions have been collected, flush the column with 1.5 volumes of buffer (PBS 1×, 15 mL) before loading another sample.
14. If storing the column, flush with buffer containing 20% ethanol or 0.05% sodium azide.
15. Store the column at 4 °C to avoid contamination.
16. Filter-sterilize the BEVs using a 0.22 μm syringe filter, collecting the filtrate in sterile 1.5 mL lo-bind microcentrifuge tubes or 15 mL tube.
17. Pool BEVs elution fractions and concentrate to desired volume using Amicon Ultra 0.5 mL centrifugal filters (RC, 10 kDa MWCO).

3.6.2 High-Resolution Size Fractionation

See **Note 10**.

3.7 BEVs Size and Concentration Analysis

Size and concentration of isolated BEVs suspension is determined by nanoparticle tracking analysis (NTA) using a suitable NTA instrument. The protocol described below is for the ZetaView PMX-220 TWIN instrument from Particle Metrix GmbH.

1. Prepare instrument set up according to the manufacturer's instructions.
2. Inject 10–20 mL particle-free deionized water using a 10 mL syringe to perform cell quality check; avoid injecting air bubbles.
3. Inject 5–10 mL of 1:25,000 100 nm polystyrene standard bead suspension to perform focus auto-alignment.
4. Inject 10–20 mL particle-free deionized water using a 10 mL syringe to rinse instrument; avoid injecting air bubbles.
5. Dilute aliquots of BEVs suspension in 1:1000 to 1:20,000 in particle-free deionized water.
6. Inject 1 mL sample with a syringe; avoid injecting air bubbles.

7. Acquire size distribution video data using the following settings.
   - temperature: 25 °C;
   - frames: 60;
   - duration: 2 s;
   - cycles: 2;
   - positions: 11;
   - camera sensitivity: 80; and,
   - shutter value: 100.
8. Analyse data using the ZetaView NTA software (version 8.05.12) with the following post acquisition settings:
   - minimum brightness: 20;
   - max area: 2000;
   - min area: 5; and,
   - trace length: 30.

3.8 Proteinase K Assay

To establish if heterologous proteins are expressed in the lumen or at the surface of BEVs we use broad-spectrum proteinase K. Proteinase K digests proteins exposed at the surface of BEVs but not in the lumen. Extracts of BEVs obtained in the presence or absence of Proteinase K samples and analyzed by Western blotting using antibodies specific for the heterologous protein makes it possible to distinguish between surface- and lumen-expressed antigens (*see* **Note 11**).

1. Add 100 mg/L of proteinase K into intact 10^{11} BEVs/mL or solubilized (in 1% SDS) and incubate for 1 h in a water bath at 37 °C.
2. The activity of proteinase K is stopped by addition of 1 mM PMSF.
3. Load samples onto a 12% polyacrylamide gel and perform a Western blot following steps described in Subheading 3.4.4.

3.9 Antigen Quantification

The amount of antigen expressed in BEVs is readily determined by Western blotting using serial dilutions of the recombinant antigen and comparing the intensity of the bands visualized on the blot to estimate the concentration of antigen in the BEVs using Image Lab Software (Bio-Rad).

1. Prepare serial dilutions of the recombinant antigen and the isolated BEVs.
2. Load samples onto a 12% polyacrylamide gel and perform a Western blot (Subheading 3.4.4).
3. Analyse the image of the blot using the quantity tool of Image Lab software (Bio Rad).

4 Notes

1. Immunization: Details of immunization protocols are provided in Carvalho et al. [8, 16]. In short, animals receive a primary immunization of filter-sterilized BEVs (*see* **Note 12**) via the nasal or oral route with booster immunizations carried out 7–14 days late with an infectious challenge following after a further 7–10 days. At necrosis body fluids and tissues are harvested for downstream analyses of antibody and immune cell profiles and histopathology.
2. Sample collection and antibody profiles: Serum bronchoalveolar lavage and saliva samples are routinely used in ELISAs to identify and quantify antigen-specific IgA and IgG antibodies as described in Carvalho et al. [8, 16]. In short, the ELISAs include coating the plate with recombinant protein for 16 h at 4 °C. After washing and incubating with blocking solution, serial dilutions of samples are added and incubated for 16 h at 4 °C. After washing, a secondary antibody conjugated to HRP is added for 1 h at 20 °C. A chromogenic substrate is then added and absorbance at 450 nm is recorded using a spectrophotometer. Tissue homogenates (e.g., salivary glands and lungs) can also be used in this antibody detection assay.
3. One possibility is to use the plasmid-borne inducible gene expression system developed for Bt that is based upon a Bt endogenous mannan-inducible promoter [21]. This system allows to create translational fusions and generate protein products with the possibility of adding a C-terminal polyhistidine tag.
4. A 6x-His Tag antibody should be used if the chimeric protein contains a His-tag sequence. An alternative is to use antibodies raised against epitopes of the antigenic protein either commercially available or from other sources.
5. The secondary antibody can be conjugated to with various enzyme, fluorescent proteins, biotin or to polymers. The secondary anti-Ig antibody must have specificity for the immunoglobulins present in the species in which the primary antibody was raised.
6. The digestion of the *E. coli/Bacteroides* shuttle vector pGH090 with NcoI and EcoRI is provided as an example. In the case of DNA sequence constraints or if the EcoRI restriction site cannot be removed from the internal sequence of the synthetic gene, other restrictions sites can be used for the design of the 3′-end of the fragment (e.g., BamHI or SmaI) that are located downstream from the NcoI site of pGH090 [18].

7. If necessary, the DNA fragment containing the gene of interest can be extracted after excision of the band at the expected size from an agarose gel after electrophoresis using the QIAquick Gel Extraction Kit (Qiagen) or equivalent, following the manufacturer's instructions.
8. As an example, we describe the cloning of the gene encoding H5F the highly conserved stalk region of the hemagglutinin molecule of IAV strain H5N1 (VN/04:A/ VietNam/1203/04 [17]) into the *Bacteroides* expression vector pGH090 [8]. The synthetic gene fused at its 5′-end with a signal peptide sequence was designed to contain a BspHI restriction site at its 5′-end and an EcoRI restriction site at its 3′-end to enable DNA cloning. BspHI was chosen because a lysine residue follows the first amino acid (methionine) in the sequence of the signal peptide. Therefore, the lysine AAA/G codon which starts with an A and follows the ATG start codon is included in the BspHI (TCATGA) restriction sequence. BspHI restriction enzyme generates a cohesive end compatible with the NcoI cohesive end of the restricted pGH090 vector to allow translation fusion of the synthetic gene. As a result, the BspHI site was lost and a combination of EcoRI and a site located within the vector and outside of the gene of interest was used to digest recombinant plasmids.
9. BEV sterility is confirmed by checking for the growth of any contaminating bacterial cells. Spread 100 μL of the filter-sterilized solution on BHIH agar plates, incubate in an anaerobic cabinet for 48 h at 37 °C and confirm the absence of colonies.
10. To increase the resolution of the SEC procedure and obtain a better separation of vesicles of different sizes, the SEC can be performed using a 120 cm × 1 cm column (Econo-Column® Chromatography Columns, Bio-Rad) filled with 90 mL of CL2-B Sepharose (Sigma-Aldrich) [22]. The absorbance of the fractions is measured at 280 nm and the first fractions displaying an absorbance peak are pooled. Pooled fractions are concentrated to 1 mL with a Vivaspin 20 centrifugal concentrator (100 kDa molecular weight cutoff, Sartorius) and the retentate is filtered through a 0.22 μm PES membrane (Sartorius). The concentration of the vesicles can then be determined using nanoparticle tracking analysis as described in Subheading 3.7, **step 1**.
11. If the protein is expressed on the surface of the BEV Proteinase K will degrade it and the band will be absent on the immunoblot. If the protein is expressed in the lumen, the band will still be evident. SDS-treatment of vesicles makes their luminal contents accessible to Proteinase K and serves as control for enzyme activity and confirmation of the protein being expressed in the lumen of BEVs.

12. The sterility of BEV suspensions stored at 4 °C is examined by checking for growth of any contaminating bacterial cells prior to immunization. Add 90 μL of sterile BHIH broth to 10 μL of BEV suspension and spread the 100 μL onto BHIH agar plates. Incubate in both aerobic and anaerobic conditions for 48 h at 37 °C and check for the absence of colony.

Acknowledgments

This work was supported in part by the UK Biotechnology and Biological Sciences Research Council (BBSRC) under grant numbers BB/J004529/1, BB/R012490/1, and BBS/E/F000PR10355. We thank Dr. Emily Jones for her contribution in setting up methods for SEC purification and the use of the Zetaview NTA, and Dr. Rokas Juodeikis for contributing to the development of BDM+ medium.

References

1. Rauch S, Jasny E, Schmidt KE et al (2018) New vaccine technologies to combat outbreak situations. Front Immunol 9:1963. https://doi.org/10.3389/fimmu.2018.01963
2. Wallis J, Shenton DP, Carlisle RC (2019) Novel approaches for the design, delivery and administration of vaccine technologies. Clin Exp Immunol 196:189–204. https://doi.org/10.1111/cei.13287
3. Francis MJ (2018) Recent advances in vaccine technologies. Vet Clin North Am Small Anim Pract 48:231–241. https://doi.org/10.1016/j.cvsm.2017.10.002
4. Shin MD, Shukla S, Chung YH et al (2020) COVID-19 vaccine development and a potential nanomaterial path forward. Nat Nanotechnol 15:646–655. https://doi.org/10.1038/s41565-020-0737-y
5. Schwechheimer C, Kuehn MJ (2015) Outer-membrane vesicles from gram-negative bacteria: biogenesis and functions. Nat Rev Microbiol 13:605–619. https://doi.org/10.1038/nrmicro3525
6. Arigita C, Jiskoot W, Westdijk J et al (2004) Stability of mono- and trivalent meningococcal outer membrane vesicle vaccines. Vaccine 22:629–642. https://doi.org/10.1016/j.vaccine.2003.08.027
7. Kanojia G, Raeven RHM, van der Maas L et al (2018) Development of a thermostable spray dried outer membrane vesicle pertussis vaccine for pulmonary immunization. J Control Release 286:167–178. https://doi.org/10.1016/j.jconrel.2018.07.035
8. Carvalho AL, Fonseca S, Cross K et al (2019) Bioengineering commensal bacteria-derived outer membrane vesicles for delivery of biologics to the gastrointestinal and respiratory tract. J Extracell Vesicles 8:1632100. https://doi.org/10.1080/20013078.2019.1632100
9. Chen DJ, Osterrieder N, Metzger SM et al (2010) Delivery of foreign antigens by engineered outer membrane vesicle vaccines. Proc Natl Acad Sci U S A 107:3099–3104. https://doi.org/10.1073/pnas.0805532107
10. Stentz R, Carvalho AL, Jones EJ et al (2018) Fantastic voyage: the journey of intestinal microbiota-derived microvesicles through the body. Biochem Soc Trans 46:1021–1027. https://doi.org/10.1042/BST20180114
11. Cecil JD, Sirisaengtaksin N, O'Brien-Simpson NM et al (2019) Outer membrane vesicle-host cell interactions. Microbiol Spectr 7. https://doi.org/10.1128/microbiolspec.PSIB-0001-2018
12. Jones EJ, Booth C, Fonseca S et al (2020) The uptake, trafficking, and biodistribution of *Bacteroides thetaiotaomicron* generated outer membrane vesicles. Front Microbiol 11:57. https://doi.org/10.3389/fmicb.2020.00057
13. Miquel-Clopés A, Bentley EG, Stewart JP et al (2019) Mucosal vaccines and technology. Clin Exp Immunol 196:205–214. https://doi.org/10.1111/cei.13285

14. Smith CJ, Rocha ER, Paster BJ (2006) The medically important *Bacteroides* spp. in health and disease. In: Dworkin M, Falkow S, Rosenberg E, Schleifer KH, Stackebrandt E (eds) The prokaryotes. Springer, New York, NY. https://doi.org/10.1007/0-387-30747-8_14
15. Human Microbiome Project Consortium (2012) Structure, function and diversity of the healthy human microbiome. Nature 486:207–214. https://doi.org/10.1038/nature11234
16. Carvalho AL, Miquel-Clopés A, Wegmann U et al (2019) Use of bioengineered human commensal gut bacteria-derived microvesicles for mucosal plague vaccine delivery and immunisation. Clin Exp Immunol 196:287–304. https://doi.org/10.1111/cei.13301
17. Valkenburg SA, Mallajosyula VV, Li OT et al (2016) Stalking influenza by vaccination with pre-fusion headless HA mini-stem. Sci Rep 6:22666. https://doi.org/10.1038/srep22666
18. Wegmann U, Horn N, Carding SR (2013) Defining the *Bacteroides* ribosomal binding site. Appl Environ Microbiol 79:1980–1989. https://doi.org/10.1128/AEM.03086-12
19. Shoemaker NB, Getty C, Gardner JF et al (1986) Tn*4351* transposes in *Bacteroides* spp. and mediates the integration of plasmid R751 into the *Bacteroides* chromosome. J Bacteriol 165:929–936. https://doi.org/10.1128/jb.165.3.929-936.1986
20. Bryant WA, Stentz R, Le Gall G et al (2017) *In silico* analysis of the small molecule content of outer membrane vesicles produced by *Bacteroides thetaiotaomicron* indicates an extensive metabolic link between microbe and host. Front Microbiol 8:2440. https://doi.org/10.3389/fmicb.2017.02440
21. Horn N, Carvalho AL, Overweg K et al (2016) A novel tightly regulated gene expression system for the human intestinal symbiont *Bacteroides thetaiotaomicron*. Front Microbiol 7:1080. https://doi.org/10.3389/fmicb.2016.01080
22. Durant L, Stentz R, Noble A et al (2020) *Bacteroides thetaiotaomicron*-derived outer membrane vesicles promote regulatory dendritic cell responses in health but not in inflammatory bowel disease. Microbiome 8:88. https://doi.org/10.1186/s40168-020-00868-z

Chapter 12

Membrane Vesicles Produced by *Shewanella vesiculosa* HM13 as a Prospective Platform for Secretory Production of Heterologous Proteins at Low Temperatures

Jun Kawamoto and Tatsuo Kurihara

Abstract

Extracellular membrane vesicles (EMVs) produced by Gram-negative bacteria are useful as a vaccine platform. During growth in broth at 18 °C, *Shewanella vesiculosa* HM13 produces a large number of EMVs that contain a 49-kDa major cargo protein, named P49. Enhanced green fluorescent protein fused to the C-terminus of P49 is delivered to EMVs, suggesting that P49 is useful as a carrier to target foreign proteins to EMVs for production of artificial EMVs with desired functions. This method is potentially useful for the preparation of designed vaccines and is described in detail in this chapter.

Key words Extracellular membrane vesicles, Protein fusion, Recombinant protein production, Protein secretion, Protein transport

1 Introduction

Gram-negative bacteria produce extracellular membrane vesicles (EMVs), which play various physiological roles such as intercellular communication, defense, pathogenicity, and catalysis [1–3]. Given their high immunogenicity and nonreplicating nature, these vesicles are useful as vaccines [4, 5]. By modifying the composition of EMVs, it is possible to develop artificial EMVs with desired immunogenic properties [6, 7]. A method for heterologous protein transport to EMVs would contribute to the development of such engineered EMVs.

Shewanella vesiculosa HM13 is a Gram-negative bacterium isolated from the intestine of horse mackerel and is a prospective host for secretory production of heterologous proteins at low temperatures, which may be beneficial for improving stability and/or decreasing toxicity of some expressed proteins [8–11]. *S. vesiculosa* HM13 produces a large number of EMVs that contain a 49-kDa protein, named P49, of unknown function as

Fadil Bidmos et al. (eds.), *Bacterial Vaccines: Methods and Protocols*,
Methods in Molecular Biology, vol. 2414, https://doi.org/10.1007/978-1-0716-1900-1_12,

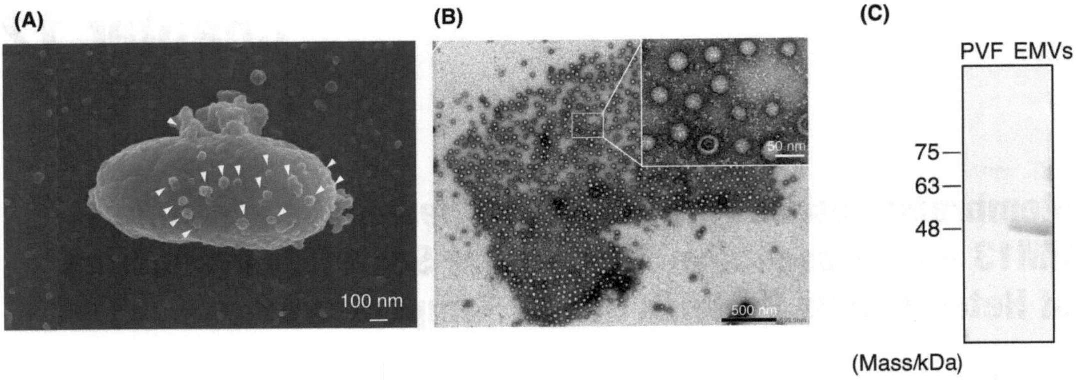

Fig. 1 EMVs and their cargo proteins from *S. vesiculosa* HM13. (**a**) Field emission-scanning electron microscopy image of *S. vesiculosa* HM13. Blebs on the cell surface and vesicles secreted into the extracellular milieu were observed. White triangles indicate blebbing of the outer membrane to produce EMVs. Bar indicates 100 nm. (**b**) Transmission electron microscopy (TEM) observation of EMVs. TEM images demonstrate that *S. vesiculosa* HM13 produces EMVs with a uniform diameter of approximately 50 nm. The inset represents a magnified image of the boxed area. Bar indicates 500 and 50 nm, respectively. (**c**) SDS-PAGE gel image of PVF and purified EMVs. The EMVs of *S. vesiculosa* HM13 carry a single major cargo protein, P49. (Parts of this figure have been reproduced/modified from [8], with permission from Frontiers Media SA)

their single major cargo protein (Fig. 1). P49 is expected to be useful as a carrier to transport heterologous proteins to EMVs to obtain engineered EMVs with desired properties. As a proof of concept, we fused enhanced green fluorescent protein (eGFP) to the C-terminus of P49 and expressed it in *S. vesiculosa* HM13 [8]. We found that the fusion protein was delivered to EMVs. Although there is room for improvement in yield and purity, this system may serve as a new platform for the preparation of functionalized EMVs. Here, we describe the experimental details of the construction of the recombinant *S. vesiculosa* HM13 strain producing the P49-eGFP fusion protein, preparation of EMVs, and analysis of the localization of the fusion protein.

2 Materials

2.1 Equipment

1. Cryopreservation beads (*see* **Note 1**).
2. 1.5 mL tubes.
3. 50 mL tubes.
4. 500 mL flasks.
5. 0.22-μm polyethersulfone (PES) syringe filters.
6. 0.45-μm PES syringe filters.
7. Syringes.
8. Temperature-controlled shaker with cooling function.

9. Reciprocal shaker incubator.
10. UV-visible spectrophotometer.
11. Ultracentrifuge with a 12 × 10 mL angle rotor Ti-50.
12. Polycarbonate ultracentrifuge tubes, 16 × 76 mm.
13. Low-speed centrifuge.
14. Ultrasonic probe sonicator.
15. Polyacrylamide gel electrophoresis (SDS-PAGE)-associated instruments.
16. Semidry blotter.
17. Filter paper.
18. Hydrophobic polyvinylidene fluoride (PVDF) transfer membrane.
19. Benchtop reciprocal shaker.
20. Chemiluminescence image scanner.
21. Polymerase chain reaction (PCR) machine.
22. Agarose gel electrophoresis apparatus.
23. Heat block incubator.

2.2 Solutions and Reagents

1. Rifampicin stock solution: 50 mg/mL in DMSO. Store at −30 °C under dark conditions.
2. Kanamycin stock solution: 50 mg/mL in water. Filter with a 0.22-μm PES syringe filter. Store at −30 °C.
3. EMV suspension buffer (DPBSS): Dulbecco's phosphate-buffered saline (2.7 mM KCl, 8.9 mM $Na_2HPO_4{\cdot}7H_2O$, 1.5 mM KH_2PO_4, and 135.9 mM NaCl, pH 7.2) with additional 0.2 M NaCl. Filter with a 0.22-μm syringe filter.
4. Trichloroacetic acid (TCA) solution (100% w/v).
5. Acetone (prechilled at −30 °C).
6. Precast 5–20% gradient polyacrylamide gels.
7. 2× SDS-sample buffer: 125 mM Tris–HCl (pH 6.8), 4% SDS, 20% glycerol, 0.04% bromophenol blue, and 10% 2-mercaptoethanol.
8. 10× SDS-running buffer: 250 mM Tris base, 1.9 M glycine, and 1% SDS.
9. Protein transfer anode buffer 1: 300 mM Tris-base, 0.05% SDS, and 20% methanol.
10. Protein transfer anode buffer 2: 25 mM Tris-base, 0.05% SDS, and 20% methanol.
11. Protein transfer cathode buffer: 25 mM Tri-base, 0.05% SDS, 40 mM 6-aminohexanoic acid, and 20% methanol.

12. Tris-buffered saline (TBS): 137 mM NaCl, 2.68 mM KCl, 25 mM Tris–HCl (pH 7.4).
13. TBS-T: 0.05% (w/v) Tween 20 in TBS.
14. Blocking solution: 5% (w/v) skim milk in TBS-T.
15. Primary antibody: rabbit anti-GFP.
16. Secondary antibody: goat anti-rabbit IgG (H + L)-HRP conjugate.
17. Chemiluminescent substrate for horseradish peroxidase (HRP) enzyme (*see* **Note 2**).

2.3 Bacterial Strains, Plasmid, and Culture Media

1. *S. vesiculosa* HM13-Rif^r (*see* **Note 3**).
2. *S. vesiculosa* P49-eGFP.
3. Competent *Escherichia coli* S17–1/λ*pir* (*see* **Note 4**).
4. pKNOCK plasmid (*see* **Note 5**).
5. pGreen plasmid (*see* **Note 6**).
6. Luria Bertani (LB) medium: Dissolve 10 g tryptone, 5 g yeast extract, and 10 g NaCl in 1 L water and adjust the pH to 7.0 with NaOH. Autoclave and store at room temperature (RT).
7. LB agar: Mix 15 g agar powder with 1 L of LB liquid medium, prepared as described above. Autoclave, allow to cool to 56 °C, supplement as required, and pour into sterile Petri plates. Store at 4 °C (*see* **Note 7**).

3 Methods

3.1 Construction of a P49-eGFP Fusion Protein Expression Strain

All PCRs are performed in 50 μL reactions using a high-fidelity proof reading polymerase (*see* **Note 8**), according to manufacturer's protocol. Details for PCR primers used, including indication of complementary sequences to allow fusion of the amplicons, are shown in Table 1.

3.1.1 Construction of pKP49eGFP and Transformation into Conjugal Donor Strain

1. Amplify a 523 bp-fragment of the 3′-terminal region of the P49 gene (without stop codon) using genomic DNA from *S. vesiculosa* HM13 and primers P49C-fwd/P49C-rev. With the addition of 15 bp overhangs on both primers to allow directional fusion to amplified eGFP and pKNOCK sequences, the total amplicon size is 553 bp.
2. Amplify the entire eGFP gene from the pGreen plasmid using primers GFP-fwd/GFP-rev. With the addition of a 15 bp overhang to the GFP-rev primer to allow fusion to the linear pKNOCK sequence, the total amplicon size is 732 bp. No additional overhang is added to the GFP-fwd primer, but the first 15 bp are complementary to the overhang on the P49C-rev primer.

Table 1
Primers used in PCR reactions

Primer name	Sequence (5′ to 3′)[a]
P49C-fwd	ACTAGTGGATCCCCCTCCTACAGGTACAACAGTACCAAACGAT
P49C-rev	TTCTCCTTTACTCATCTTAGAACCGTTAGAAGTGTTAGATACGAATCCA
GFP-fwd	ATGAGTAAAGGAGAAGAACTTTTCACTGG
GFP-rev	GAATTCCTGCAGCCCCTATTTGTATAGTTCATCCATGCCATGTG
pKNOCK-fwd	GGGCTGCAGGAATTC
pKNOCK-rev	GGGGGATCCACTAGT
pK-check-fwd	ATTCCCCTCCACCGCG
pK-check-rev	CCTCGAGGTCGACGGTATCG

[a]Complementary sequences used to fuse amplicons for construction of pKP49eGFP are indicated, where appropriate, by matched text color

3. Amplify the complete 2,096 bp linear sequence of the pKNOCK plasmid by inverse-PCR using primers pKNOCK-fwd/pKNOCK-rev.
4. Check the correct amplification of PCR products by analyzing 5 μL each on a 1% agarose gel, visualized with 0.5 mg/L ethidium bromide solution.
5. Purify each specific amplicon using a silica column PCR cleanup kit and elute in 25 μL RNase-free water (*see* **Note 9**).
6. Assemble the DNA fragments using a ligation-free DNA assembly kit (*see* **Note 10**) to generate pKP49eGFP (Fig. 2).
7. After fusion reaction, add 50 μL of competent *E. coli* S17–1/λ *pir* cells.
8. Incubate on ice for 30 min.
9. Incubate at 42 °C in a heat block incubator for 30 s to 1 min.
10. Immediately put the tube on ice and incubate for 3 min.
11. Add 300 μL LB medium and incubate at 37 °C and 200 rpm for 1 h on a reciprocal shaker.
12. Spread the 300 μL culture on LB agar containing 50 μg/mL kanamycin.
13. Incubate at 37 °C for overnight.
14. Isolate single colonies and confirm transformants by colony PCR using primers P49C-fwd/GFP-rev, with correct clones amplifying a 1,270 bp fragment.

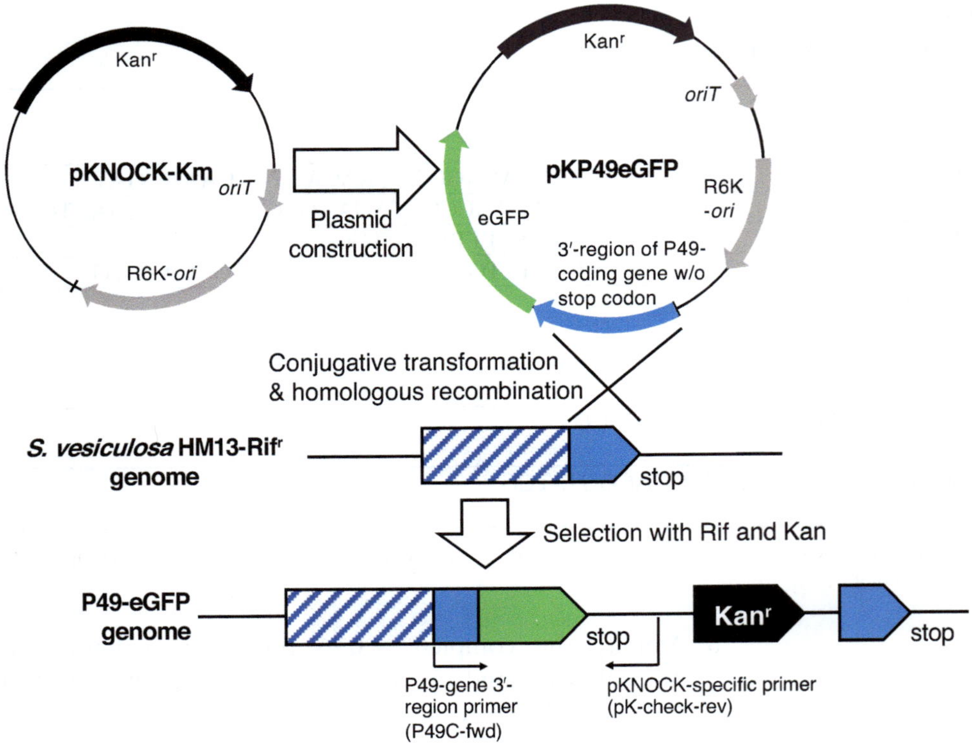

Fig. 2 Schematic illustration of the construction of the eGFP-fused P49 expression strain of *S. vesiculosa* HM13. A 523 bp sequence corresponding to the 3′ end of the P49 gene was fused to the 717 bp eGFP gene, and cloned into the *pir*-dependent suicide plasmid, pKNOCK. Integration of the plasmid into the genome by homologous recombination was confirmed by PCR with a set of primers annealing to the P49 gene and the pKNOCK-specific region

15. For further confirmation, the plasmids were extracted from the transformants and analyzed by sequencing using primers pK-check-fwd/pK-check-rev.
16. Store glycerol stocks of confirmed culture at −80 °C until use as plasmid-donor cells.

3.1.2 Conjugal Transfer of pKP49eGFP from E. coli *S17–1/λpir into* S. vesiculosa *HM13-Rif^r*

1. Inoculate a cryopreservation bead of *S. vesiculosa* HM13-Rif^r (plasmid-recipient) onto an LB agar plate containing 50 μg/mL rifampicin. Incubate at 18 °C until single colonies are formed.
2. At the same time, plate out the plasmid donor strain *E. coli* S17–1/*λpir* carrying pKP49eGFP, prepared in Subheading 3.1.1, onto LB agar containing 50 μg/mL kanamycin. Incubate overnight at 37 °C.
3. Inoculate a single colony of *S. vesiculosa* HM13-Rif^r into 5 mL LB containing 50 μg/mL rifampicin. Incubate overnight at 18 °C and 180 rpm until the OD_{600} reaches 1–2.

4. At the same time, inoculate a single colony of *E. coli* S17–1/λ *pir* carrying pKP49eGFP into 5 mL LB medium containing 50 μg/mL kanamycin. Incubate overnight at 37 °C and 185 rpm.
5. For both donor and recipient strains, subculture 50 μL of the overnight growth into fresh 5 mL aliquots of LB containing the appropriate antibiotic and continue to incubate at the appropriate temperature for each. For the culture of *E. coli* S17–1/λ *pir* carrying pKP49eGFP, incubate at 37 °C and 180 rpm until the OD_{600} reaches 0.5–1 (3–4 h). For the recipient *S. vesiculosa* HM13-Rif^r culture, incubate at 18 °C and 180 rpm until the OD_{600} reaches about 1 (approx. 12 h).
6. Mix 100 μL aliquots of donor and recipient cultures by pipetting in a 1.5 mL tube and centrifuge for 5 min at 5,000 × *g* and 18 °C to pellet cells.
7. Remove the supernatant and add 150 μL of fresh LB to gently resuspend the cell pellet.
8. Drop the mixture onto the center of an LB agar plate containing no antibiotics and allow it to dry in a Class II safety cabinet for about 30 min without spreading.
9. Incubate at 18 °C for overnight for conjugative plasmid transfer.
10. Transfer the cell lawn to a 1.5 mL tube containing 1 mL LB and suspend the cells.
11. Inoculate 100–200 μL of the cell suspension onto an LB agar plate containing 50 μg/mL rifampicin and 50 μg/mL kanamycin.
12. Incubate at 18 °C until single colonies are formed.
13. Isolate representative colonies for verification of correct clones.

3.1.3 Verification of Plasmid Integration into the *S. vesiculosa* HM13-Rif^r Chromosome

1. Prepare a standard Taq polymerase PCR reaction mixture containing a P49C-fwd and pK-check-rev (Fig. 2), allowing 10–20 μL per colony to be tested.
2. Aliquot 10–20 μL PCR reaction mixture, and mix in single colonies to be tested, in respective PCR tubes. Amplify 30 cycles using manufacturer's recommendations for times and temperatures for each step.
3. Analyze 10 μL of each PCR product by 1% agarose gel electrophoresis.
4. Check the size of amplified gene fragments, which should be 1,304 bp for correct clones.
5. Isolate single colonies for which targeted plasmid integration was detected and preserve confirmed *S. vesiculosa* P49-eGFP cultures using cryopreservation beads at −80 °C until use.

3.2 Bacterial Culture for EMV Preparation

1. Inoculate a cryopreservation bead of *S. vesiculosa* P49-eGFP onto an LB agar containing 50 μg/mL rifampicin and 50 μg/mL kanamycin.
2. Incubate at 18 °C until single colonies are formed.
3. Inoculate a single colony into 5 mL LB containing 50 μg/mL rifampicin and 50 μg/mL kanamycin.
4. Incubate at 18 °C and 180 rpm until the OD_{600} reaches 1–2 (about 16 h).
5. Inoculate 50 μL of the seed culture into fresh 5 mL LB containing 50 μg/mL rifampicin and 50 μg/mL kanamycin (*see* **Note 11**).
6. Incubate at 18 °C and 180 rpm until the OD_{600} reaches 2–3 (about 24 h) (*see* **Note 12**).
7. Subject the cell culture to the subsequent EMV preparation.

3.3 EMV Preparation

1. Transfer the 5 mL culture to 1.5 mL tubes and pellet the bacterial cells by centrifugation at 6,800 × *g* and 4 °C for 15 min.
2. Carefully transfer the supernatants to new 1.5 mL tubes.
3. Collect the cell pellets into a single 1.5 mL tube and wash the combined pellet three times with 1 mL of fresh LB medium, discarding the supernatant after each wash (*see* **Note 13**).
4. Keep cells on ice and use them for soluble and insoluble protein preparation described in Subheading 3.4.1, or store at −80 °C until use.
5. Centrifuge the supernatants from **step 2** at 13,000 × *g* and 4 °C for 15 min to remove possible contaminants such as broken cell debris in EMV-containing fraction.
6. Filter the supernatants through a 0.45-μm PES syringe filter to remove residual cellular materials.
7. Transfer the filtrates into a 16 × 76 mm polycarbonate ultracentrifuge tube, prepare a weight-matched balance tube, and ultracentrifuge at 100,000 × *g* and 4 °C for 2 h (*see* **Note 14**).
8. Transfer the supernatant (*see* **Note 15**) to new 1.5 mL tubes and store as a post-vesicle fraction (PVF) at −80 °C until use.
9. Add 1 mL DPBSS to the ultracentrifuge tubes and suspend the pellet (*see* **Note 16**).
10. Collect the EMV suspension into 1.5 mL tubes (*see* **Note 17**) for Western blotting, or store at −80 °C until use.

3.4 Localization Analysis of P49-Fusion Protein

3.4.1 Preparation of Cellular Protein Samples

1. Suspend the cells (prepared in Subheading 3.3, **step 4**) in 10× volume of DPBSS (typically 300–500 μL for cells from a 5 mL culture sample).
2. Sonicate cells on ice for 3 min (0.5 s ON, 1 s OFF, at 20% amplitude).
3. Transfer the suspension to a new 1.5 mL tube and pellet undisrupted cells at 20,500 × *g* and 4 °C for 15 min.
4. Transfer the supernatant into a 16 × 76 mm polycarbonate ultracentrifuge tube, add 5 mL DPBSS, and prepare a weight-matched balance tube (*see* **Note 14**).
5. Ultracentrifuge at 100,000 × *g* and 4 °C for 2 h.
6. Collect the supernatant as soluble protein fraction and store at −80 °C for Western blot analysis.
7. Add 50 μL DPBSS into the ultracentrifuge tube to resuspend the pellet by pipetting or gently vortexing.
8. Transfer the suspension to a new 1.5 mL tube.
9. Store the suspension as insoluble protein fraction at −80 °C for Western blot analysis.

3.4.2 Sample Preparation for SDS-PAGE and Western Blotting

1. Add 1/10 volume of 100% TCA to each of the four samples (i.e., soluble and insoluble cell protein fractions, PVF, and EMVs).
2. Vortex and incubate on ice for 30 min.
3. Centrifuge at 20,500 × *g* and RT for 30 min.
4. Discard supernatants.
5. Add 20 μL prechilled acetone to each tube and wash by pipetting.
6. Centrifuge at 20,500 × *g* and RT for 30 min.
7. Discard supernatants and repeat the wash step twice.
8. Air-dry the pellets for 5 min.
9. Add 40 μL of 1× SDS-sample buffer to each tube and suspend samples by pipetting.
10. Boil the suspensions at 100 °C for 5 min and cool at RT.

3.4.3 SDS-PAGE and Western Blotting

1. For each sample prepared in Subheading 3.4.2, mix 2.4 μL with 27.6 μL of 1× SDS-sample buffer and load 10 μL (corresponding to 100 μL culture) to respective lanes of a precast 5–20% gradient polyacrylamide gel.
2. Run the gel at 120 V in 1× SDS-running buffer until the dye front reaches the bottom of the gel.
3. For semidry Western-blot transfer of the proteins (*see* **Note 18**), cut the PVDF membrane to the size of the separation

gel, wash in methanol and soak in 30–50 mL Protein transfer cathode buffer for 10 min.

4. Soak the gel in 30–50 mL of Protein transfer cathode buffer for 10 min.
5. Prepare three sheets of gel-sized, thick absorbent filter paper (*see* **Note 19**) by soaking one in Protein transfer anode buffer 1, another in Protein transfer anode buffer 2, and the third in Protein transfer cathode buffer, each for 10 min.
6. Place the filter paper soaked in Protein transfer anode buffer 1 on a semidry blotter, followed by the filter paper soaked in Protein transfer anode buffer 2.
7. Place the PVDF membrane on the filter papers on the blotter, followed by the gel and then the filter paper incubated in Protein transfer cathode buffer.
8. Run the transfer at 0.23 A for 30 min.
9. Following protein transfer, immerse the membrane in 40 mL blocking solution for 1 h on a reciprocal benchtop shaker at RT (or overnight at 4 °C).
10. Rinse the membrane three times in 40 mL TBS-T for 10 min each at RT.
11. Incubate the membrane in 20 mL primary antibody solution, that is, rabbit anti-GFP antibody diluted 1:10,000 in blocking solution (*see* **Note 20**), for 1 h on a reciprocal benchtop shaker at RT.
12. Rinse the membrane three times in 40 mL TBS-T for 10 min each at RT.
13. Incubate the membrane in 50 mL secondary antibody solution, that is, goat anti-rabbit IgG (H + L)-HRP conjugate diluted 1:50,000 in blocking solution (*see* **Note 20**), for 1 h on a reciprocal benchtop shaker at RT.
14. Rinse the membrane three times in 40 mL TBS-T for 10 min each at RT.
15. Transfer the membrane onto a sheet of plastic wrap.
16. Spread 700 μL chemiluminescent HRP substrate, on the surface of the membrane and incubate for 5 min at RT.
17. Wrap the membrane in a plastic wrap and place it facing down on chemiluminescence image scanner.
18. Expose the membrane to obtain the band corresponding to the fusion protein (Fig. 3).

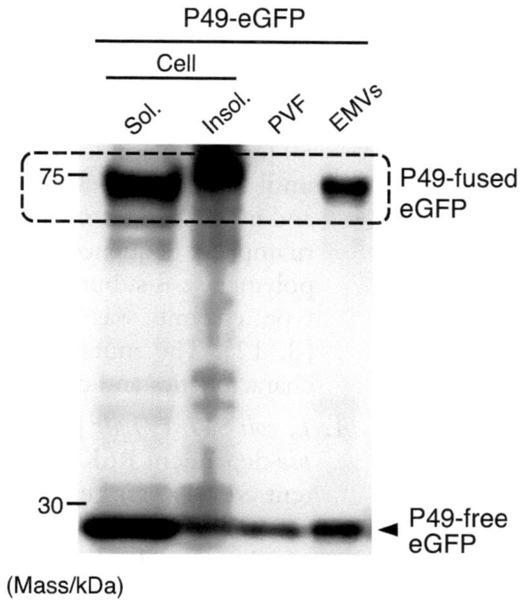

Fig. 3 Detection of eGFP fused to the C-terminus of P49 in *S. vesiculosa* HM13 EMVs. Soluble (Sol.) and insoluble (Insol.) cell fractions as well as PVF and EMVs were subjected to SDS-PAGE and Western blotting. The fusion protein was detected with an anti-GFP antibody. A fusion protein of about 75 kDa was detected in both cellular fractions and EMVs, but not PVF, indicating that P49 transports the fusion partner to EMVs. P49-free eGFP (~29 kDa) was also detected in these fractions. Proteolytic cleavage of the fusion protein probably occurs after transport of the fusion protein to EMVs (*see* **Note 21**). (This figure has been reproduced/modified from [8], with permission from Frontiers Media SA)

4 Notes

1. For long-term storage of *S. vesiculosa* HM13 and its derivatives, it is better to use cryopreservation beads rather than glycerol stocks because the latter method often causes inefficient recovery of the bacterial strains. For efficient recovery of the bacterial strains from cryopreservation beads, it is better to use new bacterial colonies (smooth pink colonies) for preparation of the stock.
2. Any commercial HRP conjugate substrate kit, such as Chemi-Lumi One Ultra (Nacalai Tesque, Kyoto, Japan), may be used according to manufacturer's protocol.
3. *S. vesiculosa* HM13-Rif^r is a spontaneous rifampicin-resistant mutant of a hypervesiculating psychrotrophic bacterium, *S. vesiculosa* HM13 (Fig. 1), and the parent strain of *S. vesiculosa* P49-eGFP. It is necessary to have a resistance marker in the plasmid recipient strain that is different from

that of the plasmid being transferred. This allows selection of transconjugants by growth in the presence of both antibiotics (Fig. 2). *S. vesiculosa* HM13-Rif^r was generated by spreading the wild-type bacterial suspension on an LB plate containing 50 μg/mL rifampicin and incubating at 18 °C for several days until single colonies were formed. The colonies were isolated and inoculated onto a new culture plate containing 50 μg/mL rifampicin. Sequence analysis of the *rpoB* gene coding for RNA polymerase ß-subunit demonstrated that Ser532 of the wild-type enzyme was replaced with Tyr in the Rif^r mutant [8, 12]. The mutation of *rpoB* does not affect the growth characteristics and cargo loading of P49 to the EMVs.

4. *E. coli* S17-1/λ*pir* [13] is used as a conjugal donor strain for *pir*-dependent R6K plasmids, such as pKNOCK [14]. Competent S17-1/λ*pir* cells, prepared as described by Inoue et al. [15], allow heat shock transformation with the desired plasmid, which can then be conjugally transferred to a target recipient cell.
5. pKNOCK [14] is a plasmid harboring an RP4 *oriT*, a *pir*-dependent R6K *ori*, and a kanamycin resistance (Kan^r) gene (Fig. 2), which has been shown to work as a suicide vector in *S. vesiculosa* [8].
6. pGreen is a plasmid carrying the eGFP-coding gene [16] and was used for construction of the eGFP-integration plasmid, pKP49eGFP.
7. Because rifampicin should be protected from light, rifampicin-containing plates must be stored under dark conditions at 4 °C.
8. A high-fidelity proofreading polymerase, such as Q5 High-Fidelity DNA Polymerase (New England Biolabs Japan Inc., Tokyo, Japan) or similar, is required to reduce the possible introduction of mutations during amplification of the sequences.
9. Any PCR clean up kit, such as Wizard SV Gel and PCR Clean-Up System (Promega, Madison, WI, USA) or similar, may be used. If required (i.e., more than one amplicon band visible), specific amplicons may be gel purified using an appropriate kit.
10. Different commercial kits are available for directional ligation-free DNA assembly. We used NEBuilder HiFi DNA Assembly Master Mix (New England Biolabs) according to the manufacturer's instruction.
11. Detailed characterization of EMVs such as proteomics, lipidomics, and analysis of surface carbohydrates [9, 11] requires higher amounts of purified EMVs. For these purposes, mid-scale (30 mL) or large-scale (1 L) cultures are required, with concentration using tangential flow filtration (TFF) with a

100-kD cut-off filter, as described by Chutkan et al. [17], rather than ultracentrifugation. After TFF concentration, EMVs can be collected by ultracentrifugation as described in Subheading 3.3.

12. The maximum yield of EMVs carrying high purity P49 is usually obtained from cultures in the mid- to late-stationary phase ($OD_{600} = 2–3$).
13. To remove EMVs remaining at the cell surface, the cell pellet should be washed at least three times. Avoid carrying over the supernatant. If the cell pellet is loose, leave some supernatant at the bottom of the tube, add 500 μL LB medium into the tube, centrifuge at 2,000 × *g* at 18 °C for 15 min and carefully pipet to remove all the supernatant including the remaining EMVs from the cell pellet.
14. The tubes must be balanced to within 0.01 g on a scale.
15. After ultracentrifugation, the EMV pellet was visible (Fig. 4). To prevent pellet dispersion, immediately collect the supernatant (PVF) from the ultracentrifuge tube using a syringe needle, without touching the pellet.
16. After ultracentrifugation, EMVs often adhere to the bottom of the centrifuge tube, and it is difficult to resuspend the EMV pellet with DPBSS. To avoid yield loss, add 100–200 μL DPBSS at a time, and gently pipet up and down without directly touching the pellet. Repeat resuspension with the remaining DPBSS until the pellet is completely dispersed.
17. Store purified EMVs of *S. vesiculosa* HM13 at −80 °C. Long-term storage (more than 4 days) of EMVs at 4 °C often affects the size of the EMVs and stability of the cargo protein, probably due to unwanted aggregation of EMVs and proteolysis of

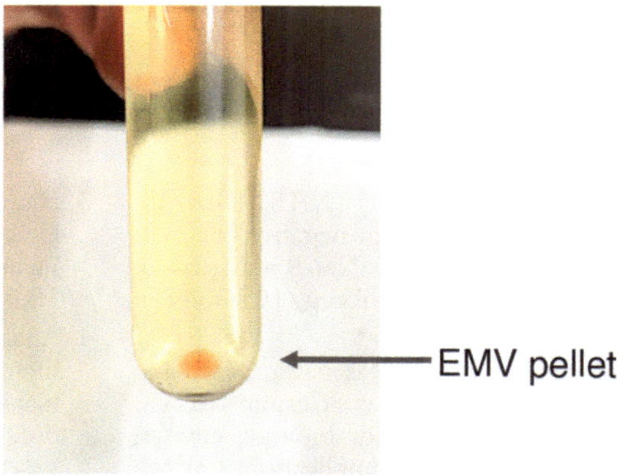

Fig. 4 EMV pellet obtained by ultracentrifugation. EMVs purified from 5 mL of the culture supernatant of *S. vesiculosa* HM13 were visible as a pale pink pellet after ultracentrifugation

the cargo. However, EMVs can be stored at 4 °C for use within 2–3 days for morphological studies by transmission electron microscopy (TEM), nanoparticle tracking analysis, and dynamic light scattering. For proteome analysis, fresh EMVs or EMVs stored at −80 °C should be used.

18. For Western blotting, we used a semidry method described by Komatsu [18] with modification.
19. Extra-thick blot filter paper can be purchased precut (e.g., Bio-Rad Blot-Absorbent Filter Paper 7.5 cm × 10 cm × 2.45 mm) to match the size of precast gels.
20. Dilution may be different depending on source of the antibodies.
21. Analysis of secretory production of eGFP fused to P49 demonstrated the loading of P49-free eGFP, in addition to the fusion protein, onto EMVs. The appearance of P49-free eGFP might be due to the proteolytic cleavage of the fusion protein after being transported to EMVs. Regulation of this putative proteolytic activity could expand the utility of this secretory protein production system, for example, for production of P49-free foreign protein by facilitating cleavage or production of fusion protein with higher purity by suppressing cleavage.

Acknowledgments

This work was supported in part by JSPS KAKENHI (JP17H04598, JP18K19178, and JP20K20570 to TK and JP16K14885 and JP20K05786 to JK) and a grant from the Institute for Fermentation, Osaka (L-2019-2-012 to TK). TEM observations were performed in collaboration with the Analysis and Development System for Advanced Materials (ADAM) at the Research Institute for Sustainable Humanosphere, Kyoto University.

References

1. Schwechheimer C, Kuehn MJ (2015) Outer-membrane vesicles from Gram-negative bacteria: biogenesis and functions. Nat Rev Microbiol 13:605–619. https://doi.org/10.1038/nrmicro3525
2. Toyofuku M, Tashiro Y, Hasegawa Y et al (2015) Bacterial membrane vesicles, an overlooked environmental colloid: biology, environmental perspectives and applications. Adv Colloid Interf Sci 226:65–77. https://doi.org/10.1016/j.cis.2015.08.013
3. Nagakubo T, Nomura N, Toyofuku M (2020) Cracking open bacterial membrane vesicles. Front Microbiol 10:3026. https://doi.org/10.3389/fmicb.2019.03026
4. Anand D, Chaudhuri A (2016) Bacterial outer membrane vesicles: new insights and applications. Mol Membr Biol 33:125–137. https://doi.org/10.1080/09687688.2017.1400602
5. McCarthy PC, Sharyan A, Sheikhi Moghaddam L (2018) Meningococcal vaccines: current status and emerging strategies. Vaccines (Basel)

6:12. https://doi.org/10.3390/vaccines6010012

6. Gerritzen MJH, Martens DE, Wijffels RH et al (2017) Bioengineering bacterial outer membrane vesicles as vaccine platform. Biotechnol Adv 35:565–574. https://doi.org/10.1016/j.biotechadv.2017.05.003
7. Irene C, Fantappiè L, Caproni E et al (2019) Bacterial outer membrane vesicles engineered with lipidated antigens as a platform for *Staphylococcus aureus* vaccine. Proc Natl Acad Sci U S A 116:21780–21788. https://doi.org/10.1073/pnas.1905112116
8. Chen C, Kawamoto J, Kawai S et al (2020) Isolation of a novel bacterial strain capable of producing abundant extracellular membrane vesicles carrying a single major cargo protein and analysis of its transport mechanism. Front Microbiol 10:3001. https://doi.org/10.3389/fmicb.2019.03001
9. Casillo A, Di Guida R, Carillo S et al (2019) Structural elucidation of a novel lipooligosaccharide from the cold-adapted bacterium OMVs producer *Shewanella* sp. HM13. Mar Drugs 17:34. https://doi.org/10.3390/md17010034
10. Kamasaka K, Kawamoto J, Chen C et al (2020) Genetic characterization and functional implications of the gene cluster for selective protein transport to extracellular membrane vesicles of *Shewanella vesiculosa* HM13. Biochem Biophys Res Commun 526:525–531. https://doi.org/10.1016/j.bbrc.2020.03.125
11. Di Guida R, Casillo A, Yokoyama F et al (2020) Detailed structural characterization of the lipooligosaccharide from the extracellular membrane vesicles of *Shewanella vesiculosa* HM13. Mar Drugs 18:231. https://doi.org/10.3390/md18050231
12. Taniguchi H, Aramaki H, Nikaido Y et al (1996) Rifampicin resistance and mutation of the *rpoB* gene in *Mycobacterium tuberculosis*. FEMS Microbiol Lett 144:103–108. https://doi.org/10.1111/j.1574-6968.1996.tb08515.x
13. Hoffmann A, Thimm T, Dröge M et al (1998) Intergeneric transfer of conjugative and mobilizable plasmids harbored by *Escherichia coli* in the gut of the soil microarthropod *Folsomia candida* (Collembola). Appl Environ Microbiol 64:2652–2659. https://doi.org/10.1128/AEM.64.7.2652-2659.1998
14. Alexeyev MF (1999) The pKNOCK series of broad-host-range mobilizable suicide vectors for gene knockout and targeted DNA insertion into the chromosome of Gram-negative bacteria. Biotechniques 26:824–826, 828. https://doi.org/10.2144/99265bm05
15. Inoue H, Nojima H, Okayama H (1990) High efficiency transformation of *Escherichia coli* with plasmids. Gene 96:23–28. https://doi.org/10.1016/0378-1119(90)90336-P
16. Hellens RP, Edwards EA, Leyland NR et al (2000) pGreen: a versatile and flexible binary Ti vector for *Agrobacterium*-mediated plant transformation. Plant Mol Biol 42:819–832. https://doi.org/10.1023/a:1006496308160
17. Chutkan H, MacDonald I, Manning A, Kuehn MJ (2013) Quantitative and qualitative preparations of bacterial outer membrane vesicles. Methods Mol Biol 966:259–272. https://doi.org/10.1007/978-1-62703-245-2_16
18. Komatsu S (2015) Western blotting using PVDF membranes and its downstream applications. Methods Mol Biol 1312:227–236. https://doi.org/10.1007/978-1-4939-2694-7_24

Chapter 13

Glycine Induction Method: Effective Production of Immunoactive Bacterial Membrane Vesicles with Low Endotoxin Content

Satoru Hirayama and Ryoma Nakao

Abstract

Bacteria are known to release nanometer scale proteoliposomes termed bacterial membrane vesicles (MVs), and it is considered that native and bioengineered MVs would be applicable for development of acellular vaccines and novel drug delivery systems in medical settings. However, important considerations for manufacturing purposes include the varied productivity of MV among bacterial species and strains, as well as endotoxicity levels due to the lipopolysaccharide component. The method for MV induction using glycine described here is simple and provides a solution to these problems. Glycine weakens bacterial peptidoglycans and significantly increases bacterial MV formation, while the relative endotoxin activity of glycine-induced MVs is extremely reduced as compared to that of noninduced MVs. Nevertheless, glycine-induced MVs elicit strong immune responses at levels nearly equivalent to those of noninduced MVs. Taken together, the present method for induction by glycine is convenient for research studies of bacterial MVs and has potential for use in medical applications including vaccine development.

Key words Membrane vesicles, Glycine, *Escherichia coli*, Endotoxin, Adjuvant

1 Introduction

Bacterial membrane vesicles (MVs) are spherical nanostructural bodies, released from the cells under various growth conditions. Bacterial MVs generally contain various antigens and have adjuvant activities; thus, their application to vaccines has been widely studied, though amounts produced vary among species and strains. When targeting bacteria with a small amount of MV formation, the resultant low yield is a major obstacle for development of medical applications, and no general method for massive production of MVs has been established.

Various MV formation mechanism models have been proposed, some of which involve bacterial peptidoglycans (PGs). Investigation of a model utilizing PG endopeptidases found that

Fadil Bidmos et al. (eds.), *Bacterial Vaccines: Methods and Protocols*,
Methods in Molecular Biology, vol. 2414, https://doi.org/10.1007/978-1-0716-1900-1_13,

MV production is likely enhanced due to accumulation of substances such as PG fragments in the periplasm [1], while another study showed that MV formation in *Escherichia coli* was increased by deletion of genes involved in PG biosynthesis [2]. Furthermore, it was recently reported that the action of endolysin, a PG-degrading enzyme derived from phage, which causes explosive cell lysis, was shown to induce formation of MVs [3, 4]. Given this background, it is considered that a mechanism that acts on bacterial PGs is likely an effective means to induce MVs.

Very recently, we reported a simple method for adding glycine during bacterial culture to increase bacterial MV production [5]. In *E. coli*, PGs consist of alternating repeating units of two amino sugars, *N*-acetylmuramic acid and *N*-acetylglucosamine, with L-alanine, D-glutamic acid, meso-2,6-diaminopimelic acid, and D-alanine internally bridging between them [6] (Fig. 1). Glycine has a function to weaken PGs by competing for and replacing DL-alanine contained in them [7]. Addition of a large amount of glycine (e.g., ~2.0%) was found to severely suppress bacterial growth, whereas bacterial MV production was significantly increased by adjusting the added amount. In studies of an *E. coli* probiotic strain (Nissle 1917) [8–10], addition of 1.0% glycine increased the yield of MVs by approximately 70-fold regarding the protein amount and approximately 50-fold regarding the lipid amount as compared to no addition [5]. Our recent findings

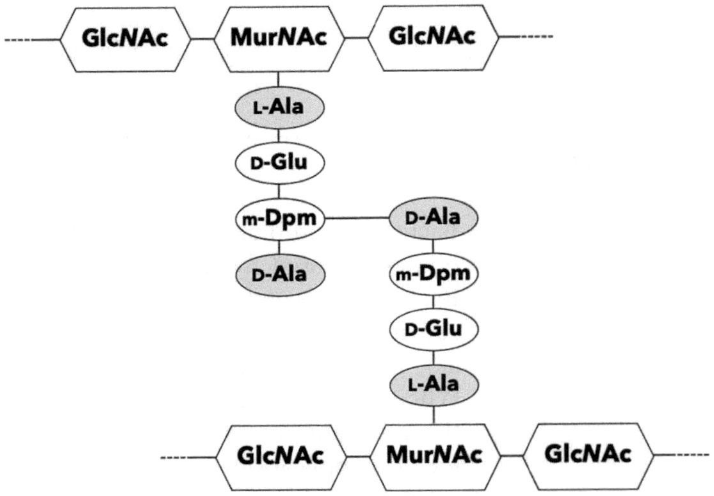

Fig. 1 Structure of *E. coli* peptidoglycan. The peptidoglycan (PG) is formed from two alternating amino sugars, a straight chain of *N*-acetylglucosamine (Glc*N*Ac) and *N*-acetylmuramic acid (Mur*N*Ac) with L-alanine (L-Ala), D-glutamic acid (D-Glu), meso-2,6-diaminopimelic acid (m-Dpm), and D-alanine (D-Ala) internally bridging between them. Glycine weakens PG by competing for and replacing DL-alanine contained within

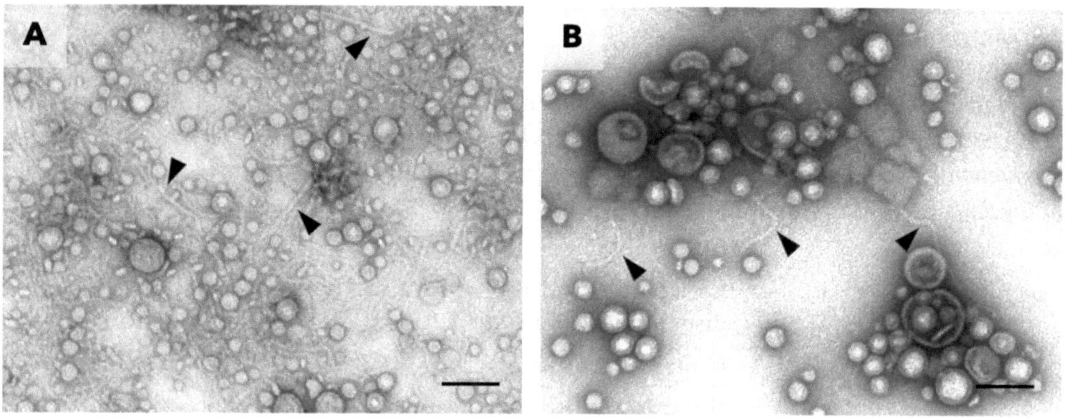

Fig. 2 Noninduced and glycine-induced MVs of *E. coli*. Noninduced MVs (**a**) and glycine-induced MVs (**b**) of a flagellar-deficient derivative of *E. coli* Nissle 1917 were purified by ultracentrifugation, then observed by transmission electron microscopy. The mean of diameter of glycine-induced MVs was apparently greater than that of noninduced MVs. No flagella were observed as the flagella master regulator gene *flhD* was deleted in both preparations [5], while fimbrial structures (black arrow heads) associated with MVs were noted in both. Scale bars represent 100 nm

showed that glycine-induced MVs had increased average particle size (Fig. 2) and altered protein composition. Furthermore, endotoxin activity was reduced in glycine-induced MVs by approximately eightfold, whereas the cytokine-inducing activity in mouse macrophage-like cells J774.1 in vitro and mucosal adjuvant activity in vivo were comparable to those of noninduced MVs [5]. These differences in properties are thought to be related to the fact that the mode of MV formation is also changed. As a result, it is considered that glycine induction can be used as a method to easily and significantly induce formation of bacterial MVs that would be suitable for application to vaccines as well as adjuvants. The most effective amount of added glycine may vary depending on bacterial species and target strain. Nevertheless, a glycine induction method is considered to be at least practically applicable for increased MV yield in *E. coli* [5] and *Acinetobacter baumannii* (unpublished findings by Nakao R et al.).

In the present chapter, a glycine induction method for preparation of large amounts of MVs with low endotoxin content is described (using the *E. coli* Nissle 1917 strain as an exemplar), along with methods for assessing cytokine-inducing activity and mucosal adjuvanticity of the MVs. Because both MV productivity and safety are important considerations in the context of potential future clinical applications, it is proposed that the present method will be useful for various medical research fields, especially those involved in drug delivery research and vaccinology.

2 Materials

2.1 Induction of E. coli MV Production by Addition of Glycine

1. A flagella-deficient derivative of *E. coli* Nissle 1917 strain (DSM 6601, serotype O6:K5:H1) (*see* **Note 1**).
2. LB broth: 10 g/L tryptone, 5 g/L yeast extract, and 10 g/L NaCl in distilled water. Autoclave and store at room temperature or 4 °C.
3. LB agar plates: LB broth added 15 g/L agar. After autoclaving, dispense into Petri dishes. Store at room temperature or 4 °C.
4. Culture tubes and flasks.
5. Incubators with and without shaker.
6. 20% glycine solution: Dissolve 20 g of glycine in 80 mL distilled water then bring the volume up to 100 mL (*see* **Note 2**). Sterilize using a 0.22-μm filter and store at room temperature (*see* **Note 3**).
7. Centrifuge.
8. 400-mL polypropylene centrifuge bottles with screw caps.
9. Bottle top PVDF membrane filter units with pore size of 0.45 μm (*see* **Note 4**).
10. Aspirator with a vacuum pump (*see* **Note 4**).
11. Ultracentrifuge with a fixed-angle rotor.
12. 70-mL polycarbonate ultracentrifuge bottles with aluminum screw caps.
13. Phosphate-buffered saline (PBS), 10 × stock: NaCl 80 g/L, KCl 2 g/L, Na_2HPO_4 14.4 g/L, KH_2PO_4 2.4 g/L in distilled water. Mix solid with ca. 900 mL distilled water, adjust pH to 7.4, then bring final volume to 1 L and autoclave to sterilize. For 1× PBS, dilute 1/10 with distilled water, and sterilize by filtration through 0.22-μm filter (*see* **Notes 5** and **6**).
14. 20 mM Tris–HCl buffer (pH 8.0): Prepare 1 M Tris–HCl buffer (pH 8.0) by weighing 12.1 g of Tris base and dissolving it in 80 mL distilled water. Adjust pH to 8.0 with concentrated HCl then bring the volume up to 100 mL and autoclave to sterilize. Dilute the 1 M Tris–HCl (pH 8.0) 1/50 with distilled water, and sterilize by filtration through 0.22-μm filter (*see* **Notes 5** and **6**).

2.2 Quantification of MV Components

2.2.1 Quantification of the Amount of Total Protein in MVs

1. Bovine serum albumin (BSA) solution: Dissolve BSA in distilled water to obtain 10 mg/mL stock solution. Dispense aliquots into microtubes and store at −20 °C.
2. 96-well microtiter plates (*see* **Note 7**).

3. Coomassie brilliant blue (CBB) G-250 dye-based reagent (i.e., Bradford assay reagent, e.g., Bio-Rad Protein Assay, Bio-Rad, Hercules, CA, USA).
4. 96-well microplate reader.

2.2.2 Quantification of the Amount of Total Lipid in MVs

1. Linoleic acid solution: Dissolve water-soluble linoleic acid in distilled water to obtain 1 mg/mL stock solution. Dispense aliquots into microtubes and store at −20 °C.
2. FM4-64 dye solution: Dissolve FM4-64 dye (*N*-(3-triethylammoniumpropyl)-4-(6-(4-(diethylamino) phenyl)hexatrienyl) pyridinium dibromide) in distilled water to prepare a 0.5 mg/mL stock solution. Dispense aliquots into brown microtubes and store at −20 °C protected from light.
3. 96-well black microtiter plate.
4. Fluorescence microplate reader.

2.2.3 Quantification of Endotoxin in MVs

1. Lipopolysaccharide (LPS) solution: Dissolve commercial LPS in distilled water to obtain 1 mg/mL solution (*see* **Note 8**). Dispense aliquots into microtubes and store at −20 °C.
2. Endotoxin-free water.
3. Endotoxin-free pipette tips.
4. Endotoxin-free 96-well microtiter plate.
5. Limulus amebocyte lysate (LAL) assay kit (e.g., Endospecy ES-50M Set, Seikagaku Co, Tokyo, Japan).
6. Incubator.
7. 0.6 M acetic acid in endotoxin-free water (*see* **Note 9**).
8. Fluorescence microplate reader.

2.3 Evaluation In Vitro: Cytokine-Inducing Activity of MVs

2.3.1 Addition of MVs to Macrophage-like Cells and Sample Preparation

1. Mouse macrophage-like cell line J774.1 (*see* **Note 10**).
2. Supplemented RPMI 1640 medium (S-RPMI): RPMI 1640 supplemented with 10% heat-inactivated fetal calf serum, 100 units/mL penicillin G, and 0.1 mg/mL streptomycin sulfate.
3. 90-mm nontreated culture dishes (*see* **Note 10**).
4. CO_2 incubator.
5. Microscope.
6. Centrifuge.
7. 50-mL polypropylene centrifuge tubes.
8. Trypan blue solution: 0.4% (w/v) trypan blue dissolved in distilled water.
9. Cell counting system.

10. 24-well microtiter plates.
11. 1.5-mL microtubes.

2.3.2 Quantitative Real-Time PCR

1. RNA purification kit suitable for use with cultured cells (e.g., RNeasy Mini Kit, QIAGEN, Hilden, Germany).
2. Centrifuge.
3. 15-mL or 50-mL polypropylene centrifuge tubes.
4. Real-time PCR cDNA synthesis kit (e.g., ReverTra Ace qPCR RT Master Mix with gDNA Remover, TOYOBO, Tokyo, Japan).
5. 96-well PCR plates with plate seals.
6. Thermal cycler.
7. Forward and reverse primers and TaqMan probes for detection of the expression of the target cytokines and β-actin internal control (Table 1).
8. Enzyme and buffer for real-time PCR (e.g., Premix Ex Taq [Probe qPCR], TaKaRa Bio, Shiga, Japan).
9. TE buffer: 10 mmol/L Tris-HCl (pH 8.0), 1 mmol/L EDTA (pH 8.0), autoclaved.
10. DNA standard for absolute quantification: Prepare template DNA of interest (e.g., plasmid cloned target sequences or synthesized gene). Adjust to 0.5×10^{10} copies/μL with TE buffer.
11. Real-time PCR instrument.

2.3.3 Sandwich ELISA for Cytokines

1. 96-well ELISA plate.
2. Plastic wrap.
3. Capture antibody for target cytokine molecules (e.g., anti-murine IL-6 antibody).
4. PBS (*see* Subheading 2.1).
5. PBST: 0.05% Tween-20 in PBS.
6. Blocking buffer: 1% BSA in PBS.
7. Recombinant protein of target cytokine molecules (e.g., IL-6).
8. Diluent: 0.1% BSA in PBST.
9. Detection antibody for target cytokine molecules (e.g., biotinylated anti-murine IL-6 antibody).
10. Conjugated avidin: Avidin-alkaline phosphatase (AP) conjugate or avidin-horseradish peroxidase (HRP) conjugate.

Table 1
Primers and probes for quantitative real-time PCR

Target		Sequence (5′–3′)
IL-4	Forward	CGCCATGCACGGAGATG
	Reverse	CGAGCTCACTCTCTGTGGTGTT
	Probe[a]	TGCCAAACGTCCTCACAGCAACGA
IL-6	Forward	CCAGAAACCGCTATGAAGTTCCT
	Reverse	CACCAGCATCAGTCCCAAGA
	Probe[a]	TCTGCAAGAGACTTCCATCCAGTTGCC
IL-12p40	Forward	AGCTCGCAGCAAAGCAAGAT
	Reverse	TGGAGACACCAGCAAAACGA
	Probe[a]	TGTCCTCAGAAGCTAACCATCTCCTG
TNF-α	Forward	AGACCCTCACACTCAGATCATCTTC
	Reverse	CCTCCACTTGGTGGTTTGCTA
	Probe[a]	CAAAATTCGAGTGACAAGCCTGTAGCCC
IFN-α	Forward	CTGCTAGTGATGAGCTACTGGTCAA
	Reverse	GGGTCAAGGCTCTCTTGTTCCT
	Probe[a]	CTGCTCCCTAGGATGTGACCTGCCTCA
IFN-β	Forward	GCTCCTGGAGCAGCTGAATG
	Reverse	TCCGTCATCTCCATAGGGATCT
	Probe[a]	TCAACCTCACCTACAGGGCGGACTTC
IFN-γ	Forward	AGCCAGATTATCTCTTTCTACCTCAGA
	Reverse	GCAATACTCATGAATGCATCCTTT
	Probe[a]	CAGGCCATCAGCAACAACATAAGGGTC
β-Actin	Forward	CACCGATCCACACAGAGTACTTG
	Reverse	CAGTGCTGTCTGGTGGTACCA
	Probe[a]	CAGTAATCTCCTTCTGCATCCTGTCAGCAA

Sequences other than those for TNF-α have been previously published [11]
[a]Probes are labeled with FAM (6-carboxyfluorescein) at 5′-end and TAMRA (6-carboxytetramethylrhodamine) at 3′-end of the sequences

11. Substrate solution: For AP reaction, prepare 3 g/L *para*-nitrophenyl phosphate in diethanolamine buffer (9.7 mL/L diethanolamine, 0.1 g/L $MgCl_2 \cdot 6H_2O$, 0.2 g/L NaN_3 in distilled water, pH adjusted to 9.6). For HRP reaction, prepare 1.25 mmol/L 3,3′,5,5′-tetramethylbenzidine and 2.21 mmol/L hydrogen peroxide, and less than 1% dimethyl sulfoxide in a 0.08 mol/L acetate buffer at pH 4.9.
12. 96-well microplate reader.

2.4 Evaluation In Vivo: Mucosal Adjuvanticity of MVs

2.4.1 Intranasal Immunization of Mice

1. 5-week-old female BALB/c mice (*see* **Note 11**).
2. PBS (*see* Subheading 2.1).
3. Ovalbumin (OVA) stock solution: Dissolve OVA in PBS to obtain 2 mg/mL solution. Dispense aliquots into microtubes and store at −20 °C.
4. Isoflurane.
5. Anesthesia inhalation device (*see* **Note 12**).

2.4.2 Sample Collection from Mice

1. PBS (*see* Subheading 2.1).
2. Parasympathetic stimulant solution: Prepare 0.8 mg/mL isoproterenol in PBS. Separately, prepare 0.2 mg/mL pilocarpine in PBS. Mix equal volumes of the two solutions and filter the solution through a 0.22-μm filter (*see* **Note 13**).
3. 1 mL syringes.
4. 26G needles.
5. 21G nonbeveled needles (*see* **Note 14**).
6. Scissors.
7. Tweezers.
8. PBS containing 0.1% BSA: Dissolve BSA at 0.1% concentration (w/v) in PBS.
9. 1.5-mL microtubes.

2.4.3 ELISA for Evaluation of Mucosal Adjuvanticity of MVs

1. 96-well ELISA plate.
2. Plastic wrap.
3. OVA solution (*see* Subheading 2.4.1).
4. ELISA coating buffer: 1.59 g/L Na_2CO_3, 2.93 g/L $NaHCO_3$, 0.2 g/L NaN_3 in distilled water, pH adjusted to 9.6.
5. PBST (*see* Subheading 2.3.3).
6. 1% skim milk in PBST.
7. 0.5% skim milk in PBST.
8. Incubator.
9. Enzyme-linked detection antibody (e.g., rabbit AP-labeled anti-murine IgG (H + L) antibody, goat HRP-labeled anti-murine IgA (alpha) antibody).
10. Substrate solution (*see* Subheading 2.3.3).
11. 96-well microplate reader.

3 Methods

Perform all procedures at room temperature, unless otherwise specified.

3.1 Induction of E. coli MV Production by Addition of Glycine

1. For seed culture, pick a single colony of *E. coli* from an overnight LB agar plate culture and inoculate a small amount (e.g., 2 mL) of LB broth. Incubate at 37 °C for 8 h with shaking at ca. 150 rpm.
2. Inoculate 1/200 volume of *E. coli* culture into 80 mL of LB broth without glycine (noninduction control) (*see* **Note 15**), and LB broth with glycine added to a final concentration of ~2.0% (*see* **Note 16**). Incubate at 37 °C for 16 h with shaking at ca. 150 rpm (*see* **Note 17**).
3. Centrifuge bacterial cultures at 7,200 × *g* for 30 min at 4 °C and collect the culture supernatants.
4. Filter culture supernatants using 0.45-μm PVDF filters to completely remove remaining bacterial cells (*see* **Note 4**).
5. Ultracentrifuge culture supernatants at 103,800 × *g* for 2 h at 4 °C to obtain MV pellets.
6. Suspend MV pellets in small amount (e.g., 1.5 mL) of PBS or 20 mM Tris-HCl (pH 8.0) (*see* **Notes 6**, **18** and **19**).

3.2 Quantification of MV Components

The amounts of total protein and total lipid in MVs can be quantified by the Bradford and FM4-64-based methods, respectively. We routinely perform the Bradford method for quality control of MVs, and sometimes use the FM4-64-based method as an alternative quality control technique. The endotoxin activity of MVs can be also measured by the Limulus assay and normalized by the amount of protein or lipid contained in MVs.

3.2.1 Quantification of the Amount of Total Protein in MVs

1. Make a two-fold serially diluted solution (e.g., 0.1–1000 μg/mL) of standard BSA with distilled water.
2. Dilute MV samples two-fold serially with distilled water (*see* **Note 20**).
3. Pipette 10 μL each of standard BSA and MV samples into appropriate wells of a 96-well microtiter plate (*see* **Notes 7** and **21**).
4. Add appropriate volume (e.g., 200 μL/well) of dye reagent for Bradford protein assay (*see* **Note 22**).
5. Incubate the plate for 5–15 min at room temperature.
6. Measure absorbance at 595 nm.
7. Estimate the amount of protein contained in the MV samples from the calibration curve created by the BSA standards.

3.2.2 Quantification of the Amount of Total Lipid in MVs

Using FM4-64, which is a lipophilic dye, and water-soluble linoleic acid, the amount of lipid contained in MVs is calculated as the linoleic acid equivalent.

1. Make a twofold serially diluted solution (e.g., 0.5–500 μg/mL) of standard linoleic acid with distilled water.
2. Dilute MV samples twofold serially with distilled water (*see* **Note 23**).
3. Dilute FM4-64 dye solution to 1/200 (i.e., 2.5 μg/mL) with distilled water.
4. Pipette 5 μL each of standard linoleic acid and MV samples into appropriate wells of a 96-well black microtiter plate (*see* **Note 21**).
5. Add 100 μL/well of 2.5 μg/mL FM4-64 dye.
6. Incubate the plate for 10 min at room temperature protected from light.
7. Detect fluorescence from FM4-64 dye with a fluorescence plate reader with the excitation and emission wavelengths at 535 nm and 625 nm, respectively.
8. Estimate the amount of lipid contained in the MV samples from the calibration curve created by the linoleic acid standards.

3.2.3 Quantification of Endotoxin in MVs

It is recommended to first make a rough estimate of the required dilutions for standards and samples, to determine the readable range, and then narrow the dilution range to quantify samples.

1. Make a tenfold serially diluted solution of standard LPS and MV samples with endotoxin-free water.
2. Pipette 25 μL each of standard LPS and MV samples into appropriate wells of an endotoxin-free 96-well microtiter plate (*see* **Note 21**).
3. Add 25 μL/well of LAL reagent, incubate at 37 °C for 30 min.
4. Add 100 μL/well of 0.6 M acetic acid.
5. Measure absorbance at 405 nm.
6. Estimate the required dilution rates of LPS standards and MV samples.
7. Make a twofold serially diluted solution of standard LPS and MV samples within measurable range using endotoxin-free water.
8. Repeat **steps 2** through **5**.
9. Estimate the endotoxin unit (EU) of the MV samples from the calibration curve created by the LPS standards (*see* **Note 8**).

3.3 Evaluation In Vitro: Cytokine-Inducing Activity of MVs

This section presents two methods for assessing MV immunoactivity using a mouse macrophage-like cell line; quantification of intracellular cytokine mRNA expression levels and determination of the amount of cytokines into the medium. Different cytokines (and/or their specific mRNA expression profiles) may be evaluated, affecting the choice of reagents used for the real-time PCR and ELISA methods described. We normally assay for IL-6, IL-12 and TNF-α, and β-actin for internal control.

3.3.1 Addition of MVs to Macrophage-like Cells and Sample Preparation

1. Incubate J774.1 cells in 50 mL of S-RPMI in 5 separate non-treated culture dishes (10 mL/dish) at 37 °C in a humidified atmosphere containing 5% CO_2 until cells are confluent (*see* **Notes 10** and **24**).
2. Harvest the cells from dishes and transfer them to a centrifuge tube.
3. Disperse cells well, then remove an aliquot of cell suspension (e.g., 20 μL) and mix with 0.4% (w/v) trypan blue solution 1:1.
4. Count live cells (cells not stained blue) and estimate total number of live cells in the suspension.
5. Collect cells by centrifugation at 300 × *g* for 10 min at room temperature and discard the supernatant.
6. Resuspend cells in an appropriate volume S-RPMI 1640 prewarmed at 37 °C to adjust number of live cells to 7.3×10^5/mL.
7. Dispense 900 μL of cell suspension (i.e., 6.6×10^5 cells) into each of 21 wells of a 24-well microtiter plate (*see* **Note 24**).
8. Standardize glycine-induced and noninduced MVs to desired concentrations (e.g., 0.1, 1, and 10 ng-protein/mL) with PBS.
9. Add 100 μL of each MV sample to appropriate wells of the 24-well plate containing 900 μL of cell suspension and mix. For the negative control wells, add and mix 100 μL of PBS alone.
10. Incubate for 2 h (for samples used in quantitative real-time PCR) or 12 h (for samples used in cytokine ELISA) at 37 °C in humidified atmosphere containing 5% CO_2.
11. Pipette cells and MV mixture into a 1.5-mL microtube and centrifuge at 300 × *g* for 10 min at room temperature. Collected cells can be used for analysis of cytokine mRNA expression by quantitative real-time PCR (*see* Subheading 3.3.2), while the supernatant can also be used for determining the amounts of secreted cytokines by ELISA (*see* Subheading 3.3.3).

3.3.2 Quantitative Real-Time PCR

1. Prepare total RNA from collected cells prepared in Subheading 3.3.1 using an RNA purification kit appropriate for culture cells, according to the manufacturer's instructions.
2. Synthesize cDNA from total RNA using the cDNA synthesis kit for real-time PCR, according to the manufacturer's instructions.
3. Prepare a master mixture for quantitative real-time PCR containing 0.04 μM each of forward and reverse primers and probe, specific for the cytokine-encoding mRNA to be quantified, and the appropriate concentration of enzyme and buffer (20 μL/PCR reaction).
4. Dispense 18 μL/well of the master mixture into a 96-well real-time PCR plate.
5. Dilute the appropriate standard plasmid tenfold serially (e.g., 10^0–10^9 copies per 2 μL) with sterile water or TE buffer.
6. Add 2 μL each of the serially diluted standard plasmid and the cDNA samples to appropriate wells (Fig. 3) containing the PCR mixture dispensed in **step 4**.
7. Seal the plate and perform real-time PCR reaction (1 cycle of 95 °C for 30 s, 40 cycles consisting of 95 °C for 5 s and 60 °C for 34 s). Set to measure the fluorescence of the sample at 60 °C.
8. Create a calibration curve using C_t (threshold cycle) values of standards and plasmid copy numbers. Use the calibration curve to determine the copy number of the target gene contained in the cDNA sample (*see* **Note 25**).

3.3.3 Cytokine ELISA

1. To coat an ELISA plate with a capture antibody specific for the cytokine to be quantified, dilute the antibody with PBS and dispense 100 μL/well into a 96-well microtiter plate.
2. Seal the plate with plastic wrap and incubate overnight at room temperature.
3. Remove contents and wash the plate with 300 μL/well PBST four times.
4. Dispense 300 μL/well blocking buffer into the plate.
5. Seal the plate with plastic wrap and incubate for 1 h at room temperature.
6. Remove contents and wash the plate with 300 μL/well PBST four times.
7. Twofold serially dilute the recombinant cytokine protein standard (e.g., ranging from 2^{-8} to 2^1 ng/mL) with PBST. Also, dilute culture supernatants prepared in Subheading 3.3.1 with PBST (*see* **Note 26**).

	1	2	3	4	5	6	7	8	9	10	11	12
A	10^9	10^8	10^7	10^6	10^5	10^4	10^3	10^2	10^1	10^0	NC	
B	10^9	10^8	10^7	10^6	10^5	10^4	10^3	10^2	10^1	10^0	NC	
C	1	2	3	4	5	6	7	8	9	10		
D	11	12	13	14	15	16	17	18	19	20	21	
E	1	2	3	4	5	6	7	8	9	10		
F	11	12	13	14	15	16	17	18	19	20	21	
G	1	2	3	4	5	6	7	8	9	10		
H	11	12	13	14	15	16	17	18	19	20	21	

Standards (1×10^0~1×10^9 copies)

Samples no. 1~21 (Triplicate)

Fig. 3 Example 96-well plate layout for real-time PCR detection of cytokine expression. Dilutions of standards are tested in duplicate, whereas each of the 21 cDNA samples from stimulated J774.1 cells are tested in triplicate. NC = negative control

	1	2	3	4	5	6	7	8	9	10	11	12
A	2^1	2^0	2^{-1}	2^{-2}	2^{-3}	2^{-4}	2^{-5}	2^{-6}	2^{-7}	2^{-8}	NC	
B	2^1	2^0	2^{-1}	2^{-2}	2^{-3}	2^{-4}	2^{-5}	2^{-6}	2^{-7}	2^{-8}	NC	
C	1	2	3	4	5	6	7	8	9	10		
D	11	12	13	14	15	16	17	18	19	20	21	
E	1	2	3	4	5	6	7	8	9	10		
F	11	12	13	14	15	16	17	18	19	20	21	
G	1	2	3	4	5	6	7	8	9	10		
H	11	12	13	14	15	16	17	18	19	20	21	

Standards (2^{-8}~2^1 ng/mL)

Samples no. 1~21 (Triplicate or 2-fold dilutions)

Fig. 4 Example 96-well plate layout for detection of cytokine concentration by ELISA. Dilutions of recombinant cytokine standards are tested in duplicate, whereas each of the 21 culture supernatant samples from stimulated J774.1 cells may be tested either in triplicate for a single dilution, or in three different concentrations (serial twofold dilutions). NC = negative control

8. Add 100 μL/well each of the diluted standards and samples to appropriate wells (Fig. 4) of the plate.
9. Seal the plate with plastic wrap and incubate at room temperature for 2 h.
10. Remove contents and wash the plate with 300 μL/well PBST four times.
11. Add 100 μL/well of the appropriate biotinylated cytokine-specific detection antibody, diluted with diluent to the concentration recommended by the manufacturer.
12. Seal the plate with plastic wrap and incubate at room temperature for 2 h.

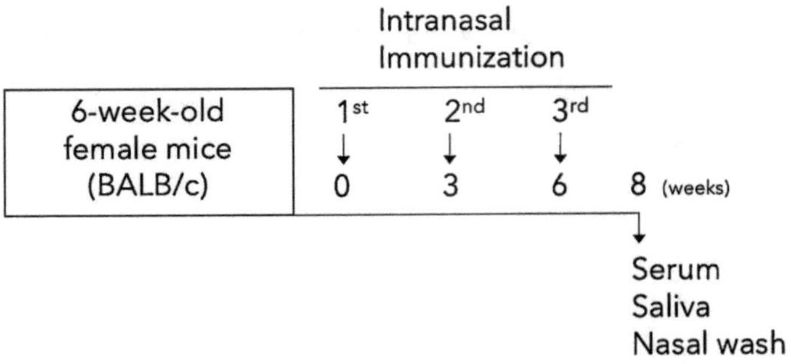

Fig. 5 Immunization timeline. Immunize mice initially at 6 weeks of age, and again 3 and 6 weeks later. Collect samples (saliva, serum, and nasal wash) at 2 weeks after final immunization

13. Remove contents and wash the plate with 300 μL/well PBST four times.
14. Add 100 μL/well of enzyme-conjugated avidin, appropriately diluted with diluent, as recommended by the manufacturer.
15. Seal the plate with plastic wrap and incubate at room temperature for 30 min.
16. Remove contents and wash the plate with 300 μL/well PBST four times.
17. Add 100 μL/well of substrate solution into the plate.
18. Sequentially (e.g., 10 min, 30 min, 60 min, and 120 min) measure the absorbance of 405 nm (for AP) or 650 nm (for HRP) (*see* **Note 27**).

3.4 Evaluation In Vivo: Mucosal Adjuvanticity of MVs

Intranasally immunize mice with OVA as an antigen and bacterial MVs as an adjuvant to evaluate the adjuvant activity of MVs. The immunization schedule is shown in Fig. 5.

3.4.1 Nasal Immunization to Mice

1. Purchase 4 mice of 5-week-old female BALB/c per experimental group (*see* **Note 28**). In this case, 16 mice were divided into the following 4 groups: (1) PBS (mock, negative control); (2) OVA (antigen alone); (3) OVA + glycine-induced MVs; (4) OVA + noninduced MVs.
2. Maintain the mice for 1 week to acclimate to environment.
3. On inoculation day (i.e., day 0 when mice are 6 weeks old), prepare the following samples for administration to each group: (1) 50 μL of PBS; (2) 50 μL of OVA (0.5 μg/μL) in PBS; (3) 50 μL of mixture of OVA (0.5 μg/μL) + glycine-induced MVs (0.1 μg/μL) in PBS; (4) 50 μL of mixture of OVA (0.5 μg/μL) + noninduced MVs (0.1 μg/μL) in PBS.
4. Anesthetize mice with isoflurane by inhalation (*see* **Note 29**).

5. Using a pipette, inoculate 5 μL per nostril (total 10 μL per mouse) with the appropriate sample for each group. For test samples, this equates to 5 μg per mouse OVA +/− 1 μg per mouse of either glycine-induced or noninduced MVs (*see* **Note 30**).
6. At 21 days after first immunization (9 weeks old), nasal immunization is performed again by repeating **steps 3–5**.
7. At 21 days after second immunization (12 weeks old), nasal immunization is performed again by repeating **steps 3–5**.
8. At 14 days after the third immunization (14 weeks old), collect saliva, serum, and nasal wash samples, as described in Subheading 3.4.2.

3.4.2 Sample Collection from Mice

1. To collect saliva samples, inject 200 μL of the parasympathetic stimulant solution into the abdominal cavity. When saliva secretion is induced, collect saliva from the mouth with a pipette. Store at −80 °C (*see* **Note 31**).
2. For serum samples, collect whole blood from anesthetized mice by cardiac puncture through the diaphragm with 1-mL syringe and 26G needle (*see* **Notes 32** and **33**). Centrifuge blood samples at 300 × *g* for 10 min to precipitate blood cells, collect the serum in new microtubes. Store at −20 °C.
3. For nasal wash samples, remove the head of the deceased mouse after exsanguination and dissect out the lower jaw. Insert a syringe needle (21G nonbeveled) into the nasal cavity from the posterior opening, and flush with 1 mL of PBS containing 0.1% BSA. Collect the outflow from the nostrils in 1.5-mL microtube. Repeat the flushing step three times, pool the replicates and centrifuge at 300 × *g* for 10 min to remove cell debris. Store at −20 °C.

3.4.3 ELISA for Evaluation of Mucosal Adjuvanticity of MVs

The samples collected in Subheading 3.4.2 may be assayed for OVA-specific IgG, IgA, IgM, and IgE antibodies, as desired, by using the appropriate enzyme-linked detection antibody. Each type of sample (i.e., saliva, serum, nasal wash) is tested separately, but a single ELISA plate can be used to assay for two types of OVA-specific antibodies, for example, IgG and IgA (Fig. 6).

1. To coat ELISA plates, dispense 100 μL/well of 0.1 μg/μL OVA in ELISA coating buffer (i.e., 10-μg OVA per well).
2. Seal the plate with plastic wrap and incubate overnight at 4 °C.
3. Remove contents and wash the plate with 300 μL/well PBST three times.
4. Dispense 150 μL/well of 1% skim milk in PBST to block.

	1	2	3	4	5	6	7	8	9	10	11	12
A		1	2	3	4	5	6	7	8	NC		
B		1	2	3	4	5	6	7	8	NC		
C		9	10	11	12	13	14	15	16	NC		
D		9	10	11	12	13	14	15	16	NC		
E		1	2	3	4	5	6	7	8	NC		
F		1	2	3	4	5	6	7	8	NC		
G		9	10	11	12	13	14	15	16	NC		
H		9	10	11	12	13	14	15	16	NC		

Samples no. 1~16 IgG, duplicate (rows A–D)

Samples no. 1~16 IgA, duplicate (rows E–H)

Fig. 6 Example 96-well plate layout for samples assayed for OVA-specific IgG and IgA antibodies. Sets of samples (saliva, serum, or nasal wash) are tested in duplicate (each at a single dilution) for OVA-specific IgG (top half of the plate) and IgA (bottom half of the plate) by using the appropriate enzyme-linked detection antibody

5. Seal the plate with plastic wrap and incubate for 2 h at 37 °C.
6. Remove contents and wash the plate with 300 μL/well PBST three times.
7. Appropriately dilute the samples prepared in Subheading 3.4.2 with 0.5% skim milk in PBST (*see* **Note 34**).
8. Add 100 μL/well of each of diluted sample to appropriate wells on the plate (Fig. 6).
9. Seal the plate with plastic wrap and incubate at 37 °C for 1 h.
10. Remove contents and wash the plate with 300 μL/well PBST three times.
11. Add 100 μL/well of appropriately enzyme-linked detection antibody, diluted in 0.5% skim milk in PBST to the concentration recommended by the manufacturer.
12. Seal the plate with plastic wrap and incubate at 37 °C for 1 h.
13. Remove contents and wash the plate with 300 μL/well PBST three times.
14. Add 100 μL/well of substrate solution into the plate.
15. Seal the plate with plastic wrap and incubate at 37 °C.
16. Sequentially (e.g., 10 min, 30 min, 60 min) measure the absorbance of 405 nm (for AP) or 650 nm (for HRP) (*see* **Note 27**).

4 Notes

1. Nissle 1917 is a representative probiotic *E. coli* strain. The flagella-deficient derivative is used as an MV producer, in order to completely exclude flagella from the MV preparation without any sample loss [5].
2. If it is difficult to dissolve, heat the solution at 50 °C.
3. If crystals form during storage, they will dissolve when heated at 50 °C.
4. Filtration using bottle top filter unit and aspirator is convenient for rapid filtration of bacterial supernatant. Depending on the bacterial species and strains, a 0.22-μm PVDF filter may also be required. Check for sterility of flowthrough by plating on LB agar plate. Resterilize using 0.45- or 0.22-μm filters as needed.
5. Filtration is recommended to remove contaminants in buffer.
6. As a solvent for suspending MV pellets, PBS, Tris–HCl, distilled water, or another buffer can be used, depending on the purpose of the individual study.
7. It is preferable to use a 96-well plate with low protein adsorption.
8. Refer to the product standard certificate for each lot to calculate the EU of LPS.
9. Different solutions may be required depending on the LAL assay kit used.
10. We recommend that J774.1 cells are cultured on nontreated culture surfaces and passaged by simple dilution method (e.g., 1:10 split), because J774.1 cells are very sticky on tissue culture treated surface. Even if treated with trypsin-EDTA, it is difficult to efficiently detach J774.1 cells from the tissue culture treated surface. Cell scrapping (e.g., by cell scraper) is not recommended as it is known to affect cell viability.
11. Purchase animals at least 1 week before starting experiments.
12. An anesthesia bottle can be used instead.
13. Prepare and mix the solutions of isoproterenol and pilocarpine just prior to intraperitoneally administrating the mixture to mice.
14. It is also possible to cut the tip of the 21G beveled needle to make it nonbeveled.
15. Due to the low MV yield of the noninduced control, multiple bacterial culture supernatants (e.g., from 5–8 flasks of 80-mL cultures) are required to measure the amount of protein and lipid in later step.

16. Effective glycine concentrations may vary by bacterial species and strain. It is important to consider the optimum glycine concentration for the target strain.
17. If necessary, monitor the degree of increase in turbidity of the bacterial culture over time.
18. Glycine-induced MV pellets are large. Therefore, be careful to not create foam when pipetting to uniformly mix them in the buffer. In this case, it is useful to control flow speed/volume of pipetting by cutting the end of pipet-tips. On the other hand, due to the low MV yield from noninduced culture (*see* **Note 15**), the MV pellets must be collected together from several flask cultures and suspended with a small amount of buffer. If necessary, to prepare MVs with greater purity, perform further purification by repetitive washing or density gradient centrifugation [12].
19. In general, antigens carried by MVs are more stable against degradation by heat or enzymes, as compared with those in free-form. Nevertheless, the suitable storage conditions of MVs must be determined by preliminary experiments, because stability of MVs may not only differ among individual strains, but also vary due to the preparation method. We usually prepare MVs in an appropriate solution, store at −20 °C, and use for testing within 1 year after isolation. For longer storage (e.g., ~2 years), MVs prepared in an appropriate solvent can be also lyophilized by freeze-dryer apparatus then stored at 4 °C. Prior to use for experiment, lyophilized MVs must be reconstituted with the same volume of distilled water as before lyophilizing.
20. In most cases, optimal results will be obtained by testing each sample without dilution or by testing different serial dilutions (e.g., 2, 4 and eightfold).
21. We recommend measuring standards and samples in duplicate or triplicate.
22. Depending on the MV sample, it may aggregate, making protein measurement difficult. In that case, use dye reagent for BCA method [13] instead.
23. The number of twofold serial dilutions required depends on the yield of MVs and the amount of buffer used for suspension of pellet. We suggest that 1/5 diluted sample is used as starting material, and further make six twofold serial dilutions for testing.
24. In this case, 5 culture dishes of 10-mL S-RPMI (i.e., 50 mL total) will yield more than 2×10^7 J774.1 cells. This is sufficient for performing an assay using 21 wells in a 24-well plate, accommodating triplicate samples of the following 7 groups:

glycine-induced MVs at three different concentrations (0.01, 0.1, and 1 ng-protein/mL), noninduced MVs at three different concentrations (0.01, 0.1, and 1 ng-protein/mL), and a negative control (PBS alone).

25. The copy number of the target gene should be normalized by dividing it by the copy number of the internal control (e.g., β-actin).
26. A single dilution of each supernatant sample may be tested in triplicate, or each sample may be tested in three different dilutions (serial twofold dilutions). How much the sample needs to be diluted depends on the cytokine being tested and may need to be determined empirically. Depending on the cytokine being tested, a dilution of ca. ~16 fold may be required.
27. When using HRP-linked detection antibody and the corresponding substrate, 100 μL/well of 1 N HCl can be added at the appropriate time to stop the HRP reaction and measure the absorbance at 450 nm. As the solution changes from blue to yellow, its absorbance increases.
28. It is mandatory that approval by the appropriate ethical committee as well as related government agency be obtained before performing any animal experiments. All mice must be housed and maintained in a suitable animal facility according to all regulations/laws applicable to animal welfare.
29. Mouse inhalation anesthesia can be performed with isoflurane at a concentration of 2% and flow rate of 2 L/min.
30. Insufficient anesthesia may cause mice to sneeze. It is best to wait 1 min after inoculation of one nasal cavity before inoculating the other so that the first dose does not block the airway, thus lowering the possibility of sneezing.
31. Storage at −80 °C reduces viscosity. The storage is recommended to easily handle saliva samples.
32. Slowly drain the blood to prevent heart collapse.
33. Approximately 1 mL can be collected.
34. In most cases, optimal results will be obtained by testing 1000-fold dilution for serum IgG, and 100-fold dilutions for serum IgA, IgM, IgE, salivary S-IgA, and nasal wash S-IgA.

Acknowledgments

The authors wish to thank Michiyo Kataoka for technical assistance with the transmission electron microscope observations. This study was supported by JSPS KAKENHI (JP18K15160, JP19K22644, JP20K18492, JP20H03861) and the Japan Agency for Medical Research and Development (AMED) (JP18fk0108124).

References

1. Schwechheimer C, Kuehn MJ (2015) Outer-membrane vesicles from gram-negative bacteria: biogenesis and functions. Nat Rev Microbiol 13(10):605–619
2. McBroom AJ, Johnson AP, Vemulapalli S, Kuehn MJ (2006) Outer membrane vesicle production by *Escherichia coli* is independent of membrane instability. J Bacteriol 188 (15):5385–5392
3. Turnbull L, Toyofuku M, Hynen AL, Kurosawa M, Pessi G, Petty NK, Osvath SR, Cárcamo-Oyarce G, Gloag ES, Shimoni R (2016) Explosive cell lysis as a mechanism for the biogenesis of bacterial membrane vesicles and biofilms. Nat Commun 7(1):1–13
4. Toyofuku M (2019) Bacterial communication through membrane vesicles. Biosci Biotechnol Biochem 83(9):1599–1605
5. Hirayama S, Nakao R (2020) Glycine significantly enhances bacterial membrane vesicle production: a powerful approach for isolation of LPS-reduced membrane vesicles of probiotic *Escherichia coli*. Microb Biotechnol 13 (4):1162–1178
6. Vollmer W, Blanot D, De Pedro MA (2008) Peptidoglycan structure and architecture. FEMS Microbiol Rev 32(2):149–167
7. Hammes W, Schleifer K, Kandler O (1973) Mode of action of glycine on the biosynthesis of peptidoglycan. J Bacteriol 116 (2):1029–1053
8. Rembacken B, Snelling A, Hawkey P, Chalmers D, Axon A (1999) Non-pathogenic *Escherichia coli* versus mesalazine for the treatment of ulcerative colitis: a randomised trial. Lancet 354(9179):635–639
9. Kruis W, Frič P, Pokrotnieks J, Lukáš M, Fixa B, Kaščák M, Kamm M, Weismueller J, Beglinger C, Stolte M (2004) Maintaining remission of ulcerative colitis with the probiotic *Escherichia coli* Nissle 1917 is as effective as with standard mesalazine. Gut 53 (11):1617–1623
10. Behnsen J, Deriu E, Sassone-Corsi M, Raffatellu M (2013) Probiotics: properties, examples, and specific applications. Cold Spring Harb Perspect Med 3(3):a010074
11. Ichinohe T, Watanabe I, Ito S et al (2005) Synthetic double-stranded RNA poly(I:C) combined with mucosal vaccine protects against influenza virus infection. J Virol 79 (5):2910–2919
12. Hirayama S, Nakao R (2021) Intranasal vaccine study using *Porphyromonas gingivalis* membrane vesicles: isolation method and application to a mouse model. Methods Mol Biol 2210:157–166. https://doi.org/10.1007/978-1-0716-0939-2_15
13. Smith PK, Krohn RI, Hermanson GT, Mallia AK, Gartner FH, Provenzano M, Fujimoto EK, Goeke NM, Olson BJ, Klenk D (1985) Measurement of protein using bicinchoninic acid. Anal Biochem 150(1):76–85

Chapter 14

Methods for Assessment of OMV/GMMA Quality and Stability

Francesca Micoli, Renzo Alfini, and Carlo Giannelli

Abstract

Outer membrane vesicles (OMV) represent a promising platform for the development of vaccines against bacterial pathogens. More recently, bacteria have been genetically modified to increase OMV yield and modulate the design of resulting particles, also named generalized modules for membrane antigens (GMMA). OMV/GMMA resemble the bacterial surface of the pathogen, where key antigens to elicit a protective immune response are and contain pathogen-associated molecular patterns (e.g., lipopolysaccharides, lipoproteins) conferring self-adjuvanticity. On the other hand, OMV/GMMA are quite complex molecules and a comprehensive panel of analytical methods is needed to ensure quality, consistency of manufacture and to follow their stability over time. Here, we describe several procedures that can be used for OMV/GMMA characterization as particles and for analysis of key antigens displayed on their surface.

Key words Outer membrane vesicles, OMV, GMMA, Vaccines, Analytical methods

1 Introduction

During the last years outer membrane vesicles (OMV) have received great attention as platform for the development of vaccines against bacterial pathogens [1]. Gram-negative bacteria, during growth, naturally release small bilayered membrane structures from the cell surface with formation of native OMV (nOMV), a process that has been associated with several biological functions [2]. OMV release often happens at levels that are too low to support vaccine manufacture and different ways have been identified to increase OMV shedding. In particular, vesicle-like aggregates of insoluble outer membrane proteins can be chemically extracted from whole bacteria using detergents (e.g., deoxycholate) resulting in detergent-extracted OMV (dOMV), or bacteria can be genetically manipulated to increase blebbing resulting in the so-called mutant-derived OMV (mdOMV) or GMMA (generalized modules for membrane antigens) [3, 4]. By this way, vesicle integrity is preserved and GMMA present outer membrane antigens in their native environment and conformation. Additional mutations

Fadil Bidmos et al. (eds.), *Bacterial Vaccines: Methods and Protocols*,
Methods in Molecular Biology, vol. 2414, https://doi.org/10.1007/978-1-0716-1900-1_14,

can be introduced to reduce OMV endotoxicity, more often through modification of the lipid A structure [5]. OMV producer strains can also be engineered for expression of specific protein or polysaccharide antigens, eventually supporting the development of multicomponent vaccines [6]. Simplicity of manufacture make OMV an attractive technology for affordable vaccines [7]. Furthermore, OMV resemble the bacterial surface of the pathogen, where key antigens to elicit a protective immune response are and contain pathogen-associated molecular patterns, such as lipopolysaccharides (LPS) or lipoproteins, that confer self-adjuvanticity [5]. OMV have optimal size for cellular uptake and display antigens in multiplicity, favoring B cells activation, all factors contributing to make OMV a promising vaccine platform [1, 6, 8]. On the other hand, OMV are quite complex molecules (Fig. 1) and development of a comprehensive panel of analytical methods is needed to ensure quality, consistency of manufacture and to follow OMV stability over time. Here, we describe a number of procedures (Table 1) that can be used for OMV characterization as particles and for analysis of key antigens displayed on their surface.

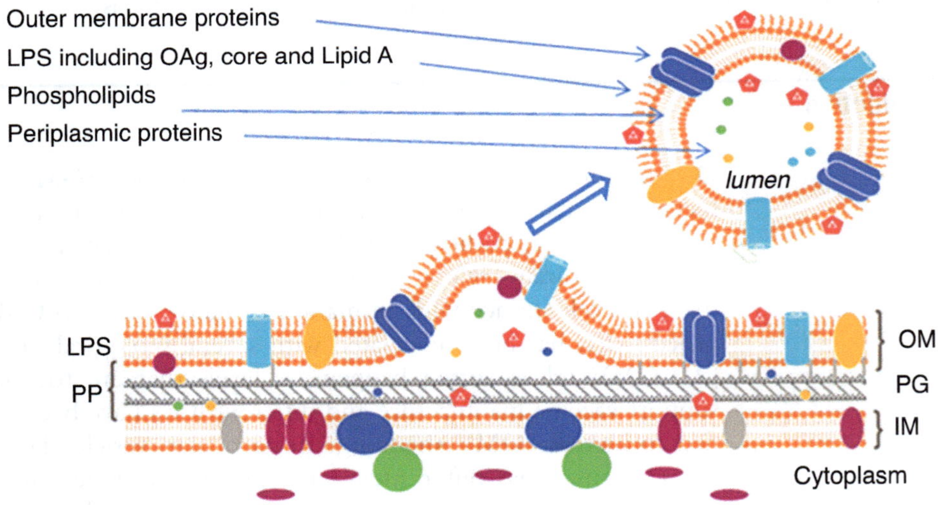

Abbreviations

LPS: Lipopolysaccharides

OAg: O-antigen

OM: Outer membrane

PG: Peptidoglycan

PP: Periplasm

IM: Inner membrane

Fig. 1 Schematic representation of OMV formation and their main components

Table 1
Methods for OMV/GMMA characterization

Attribute	Method	Performed on	References
Identity	Dot blot/western blot	GMMA	–
Particle size	DLS/NTA/HPLC-SEC MALS	GMMA	[9]
Purity: Soluble proteins	HPLC-SEC (fluorescence emission)	GMMA	[9]
Purity: DNA content	HPLC-SEC (ABS260/ABS280)	GMMA	[9]
Protein pattern/composition	SDS-PAGE/MS	GMMA	[10, 11]
Total protein content	Micro BCA/Lowry/amino acidic analysis	GMMA	[12]
Total sugar content	HPAEC-PAD/Dische/cELISA	GMMA or extracted OAg/core	[13–15]
Lipid A structure	MALDI-MS	Isolated lipid A	[16]
Lipid A amount	HPLC RP-QqQ	GMMA	–
OAg molecular size distribution and molar ratio of OAg of different length	HPLC-SEC/semicarbazide	Extracted OAg/core	[14]
OAg O-acetylation level	^{1}H NMR/Hestrin	Extracted OAg	[17]
Key proteins quantification	SRM/cELISA	GMMA	[10, 18]

OAg: O-antigen portion of lipopolysaccharide (LPS) molecules

2 Materials

Prepare all solutions using ultrapure water (grade 1, >18 MΩ-cm at 25 °C; prepared by purifying deionized water) and analytical grade reagents. Prepare all reagents at room temperature (RT), unless otherwise specified.

2.1 Dot Blot/Western Blot

1. Phosphate buffered saline with 0.05% Tween 20 (PBS/T): weigh 0.5 g of Tween 20 in a weighing boat and transfer into a 1000 mL graduated cylinder by washing weighting boat with PBS. Add PBS up to 1000 mL and mix on a magnetic stirrer until the solution is homogeneous.
2. PBS/T with 3% bovine serum albumin (PBS/T + 3% BSA): weigh in a weighing boat 3 g of BSA and transfer in a 100 mL graduated cylinder; add PBS/T solution up to 100 mL. Mix on a magnetic stirrer.
3. PBS/T with 0.1% BSA (PBS/T + 0.1% BSA): Transfer 87 mL of PBS/T in a 100 mL glass bottle and add 3 mL of PBS/T+ 3% BSA. Mix on a magnetic stirrer.

4. IgG Mouse anti-*Salmonella* Typhimurium primary antibody 1:3000 in PBS/T + 0.1% BSA: Transfer 9 mL of PBS/T + 0.1% BSA in a 15 mL Falcon tube; add 3 μL of mouse anti-*S.* Typhimurium primary antibody (Abcam, code AB8274) and mix by vortex. Prepare this solution just before use.
5. Goat anti-mouse IgG secondary antibody (AP conjugate) 1:6000 in PBS/T+ 0.1% BSA: Transfer 18 mL of PBS/T+ 0.1% BSA in a 50 mL Falcon tube; add 3 μL of goat anti-mouse IgG AP conjugate secondary antibody (Sigma, code A3438) and mix by vortex. Prepare this solution just before use.
6. Developing solution: In a 50 mL Falcon tube, solubilize 2 tablets of SIGMA Fast BCIP with 20 mL of water by vigorously vortexing and keep it at 4 °C protected from light until its use (to be prepared just before use).

2.2 Dynamic Light Scattering (DLS)

1. DLS instrument (i.e., Zetasizer Nano—Malvern Panalytical).

2.3 Size Exclusion High-Performance Liquid Chromatography with Multiangle Light Scattering (HPLC-SEC/MALS)

1. Tosoh TSK gel 6000 PW column (30 cm × 7.5 mm; cat. 805765) connected in series with a Tosoh TSK gel 4000 PW column (30 cm × 7.5 mm; cat. 805763) and with Tosoh TSK gel PWH guard column (7.5 cm × 7.5 mm; cat. 806732).
2. HPLC system equipped with UV, fluorimeter, and multiangle static light scattering (MALS) detectors (e.g., DAWN HELEOS II, Wyatt).
3. Filter PBS eluent through 0.22 μm filter before use.

2.4 Nanoparticle Tracking Analysis (NTA)

1. NS300 NanoSight instrument (Malvern) equipped with a CMOS camera and a 488 nm monochromatic laser. Data acquisition and processing with NTA software.
2. Low-bind tubes.
3. Low-bind pipette tips.

2.5 SDS-Page

1. Prestained molecular weight standard.
2. 1 M (1,4-Dithioerythritol) (DTT) solution.
3. 10% acrylamide gels, Bis-Tris buffer, for SDS-PAGE.
4. Sample loading buffer (4×) containing lithium dodecyl sulfate and Coomassie G250.
5. Brilliant Blue G—Colloidal Concentrate stain: Dilute the content to the working concentration in the original container according to the dye datasheet. Shake the working concentration dye container before use.
6. 3-(N-morpholino)propanesulfonic acid (MOPS) running buffer: Dilute 25 mL of 20× MOPS running buffer with 475 mL of water in a graduate glass cylinder.

7. Fixing solution: Under chemical hood mix in a 100 mL graduated cylinder 20 mL methanol (MeOH), 2 mL acetic acid and then add water to reach 50 mL.

2.6 Micro BCA

1. Micro BCA Protein Assay Kit (Thermo Fisher, cat. 23235).

2.7 Ultracentrifugation

1. Ultracentrifuge equipment with appropriate rotor (*see* **Note 1**).

2.8 Lowry

1. 0.8 N sodium hydroxide (NaOH): Weigh 1.6 g of NaOH, transfer it in a 50 mL Falcon tube and dissolve into a final volume of 50 mL water.
2. 20% sodium carbonate: Weigh 4 g of sodium carbonate, transfer it in a 100 mL cylinder and dissolve into a final volume of 20 mL water.
3. 0.4% Potassium Tartrate/0.2% cupric sulfate: Weigh 200 mg of $(CHOHCOOK)_2$ ½ H_2O, transfer it in a 50 mL Falcon tube and add 25 mL of water. Weigh 100 mg of $CuSO_4$ 5 H_2O, transfer it in the same Falcon tube. Dissolve it by vortex and add 25 mL of water to reach a final volume of 50 mL.
4. 10% sodium dodecyl sulfate (SDS): Weigh 5 g of SDS, transfer it in a 100 mL cylinder and dissolve into a final volume of 50 mL water.
5. Copper–tartrate–carbonate (CTC) solution: In a 15 mL Falcon tube, add 3 mL of 0.4% potassium tartrate–0.2% cupric sulfate solution and 3 ml of 20% sodium carbonate. Vortex the solution, wait 10 min and vortex the solution again before use.
6. Reagent A: In 50 mL Falcon tube, add 6 mL CTC Solution, 6 mL 10% SDS, 6 mL 0.8 N NaOH, 6 mL water and mix well with the vortex mixer (*see* **Note 2**).
7. Folin reagent: In a 50 mL Falcon tube covered with aluminum foil (to protect from light) add 5 mL water and 1 mL Folin–Ciocalteu's solution (Merck, 1.09001) (*see* **Note 2**).
8. 120 μg/mL BSA standard: Weigh 120 μL of 2 mg/mL BSA in a Falcon tube. Calculate the amount of water to add using the following formula:

$$\text{Water (mg)} = 16.67 \times 2\ \text{mg/mL BSA weight (mg)} - 2\ \text{mg/mL BSA weight (mg)}$$

2.9 Amino Acid Analysis

1. Hydrolysis tubes (Waters, cat. WAT079007 or WAT007571).
2. Protein low-bind tubes, 2 mL.
3. Total Recovery HPLC vials (Waters, cat. 186000384c).
4. Kit AccQTag Ultra (Waters, cat. 186003836).

5. Hydrolyzed Amino Acids Standard (Waters, cat. WAT088122).
6. UPLC BEH C18 130 Å, 1.7 μm, 2.1 mm 100 mm column (Waters, cat. 186,003,837).
7. UHPLC system equivalent to Waters ACQUITY UPLC H-CLASS, equipped with UV detector.
8. Column preheater (Waters, cat. 205000730).
9. Post Column connector PEEK id 0.0025 (Waters, cat. 430001783).
10. AccQTag Ultra Eluent B/10 (eluent line B): Filter 0.2 μm 1 L water. With a 1 L graduated cylinder measure accurately 900 mL of filtered water and transfer it in an appropriate bottle for eluent. With a 100 mL graduated cylinder measure accurately 100 mL of AccQTag Ultra Eluent B and transfer in the eluent bottle previously filled with water. Mix accurately.
11. HPLC Needle/seal washing solution: Prepare in a bottle a solution with 50% water/acetonitrile (ACN).
12. 6 N Hydrochloric acid (HCl) in single-use ampules.
13. Device to perform dry hydrolysis in HCl vapor phase: Alfatech Hydrosmart 2 with hydrolysis vessels, cat. 20001001.
14. 1% Phenol in 6 N HCl: Under the chemical hood transfer a maximum of 40 mg of phenol powder inside a weighed vial, close the vial and weigh it (this procedure is needed to avoid phenol breathing). Determine the amount of 6 N HCl for phenol dissolution using the following formula:

    $$\mu\text{L } 6\text{ N HCl} = 100 \times \text{mg Phenol}$$

    Transfer the required volume of 6 N HCl into the vial containing the weighed phenol. Close the screw cap vial and vortex until complete dissolution.
15. 100 mM HCl: In a 25 mL graduated cylinder place 418 μL 6 N HCl and add water up to reach 25 mL total volume.
16. 6 N HCl, 0.1% phenol solution: In a 2 mL glass vial mix 630 μL of 6 N HCl and 70 μL of 1% Phenol solution in 6 N HCl. Prepare it just before use.

2.10 High-Pression Anion Exchange Chromatography Coupled with Pulsed Amperometric Detection (HPAEC-PAD) Analysis (Neutral Sugars)

1. 11.25 μg/mL Rhamnose (Rha), Galactose (Gal), Glucose (Glc), Mannose (Man) sugars mix (neutral sugar) standard solution: In different 2 mL vials, weigh accurately 9 mg of each standard and dissolve them in a suitable amount of water added by weighing on balance, in order to obtain solutions at 4.5 mg/mL. To calculate the amount of water to be added to each vial, use the following formulas:

μg sugar = weight μg × purity% × (sugar MW without hydration water)/(sugar MW with hydration water),

(If the sugar has no hydration water, the last ratio present in the formula is equal to 1).

μL water = (μg sugar)/4.5

Weigh an empty 500 mL glass bottle and transfer 1 mL of each sugar solution prepared above; then add water to a total weight of 400 g in order to obtain a solution with a concentration of 11.25 μg/mL for each standard.

2. 4 × 250 mm CarboPac PA10 column (Thermo Fisher, cat. 046110) coupled with a 4 × 50 mm PA10 guard column (Thermo Fisher, cat. 046115).
3. 8 M TFA solution: Under chemical hood, with the glass cylinder, measure 30 mL of TFA and transfer it into a glass bottle. Add in the bottle 20.5 mL of water and gently mix.
4. 4 M TFA solution: Dilute 10 mL of 8 M TFA adding 10 mL of water.
5. 50 mM NaOH: Fill the eluent bottle with 2 L of water using a volumetric flask. Degas for 15 min by bubbling helium. Add 5.2 mL of 50% NaOH to the bottle. Degas for 10 min more.
6. 500 mM NaOH: Fill the eluent bottle with 2 L of water using the volumetric flask. With a pipette, remove 26 mL of water from the bottle. Degas for 15 min by bubbling helium. Add 26 mL of 50% NaOH to the bottle. Degas for 10 min more.
7. 1 M sodium acetate (AcONa) 100 mM NaOH: Degas 1.5 L of water for 15 min by bubbling helium. Dissolve the whole content of preweighed AcONa (Thermo Fisher, cat. 059326) by adding 500 mL of degassed water directly into the AcONa bottle. Transfer the solution to a 1 L volumetric flask. Wash twice the AcONa bottle with about 100 mL more of degassed water and add to the solution in the 1 L volumetric flask. Fill the 1 L volumetric flask with degassed water, up to reach 1 L of volume for the AcONa solution. Place a 0.22 μm vacuum-cup filter on an empty eluent 2 L bottle, connect it to the vacuum system and filter the prepared 1 L solution. Degas the solution for 10 min by bubbling Helium. Add 5.2 mL of 50% NaOH to the bottle. Degas with Helium for 10 min more.

2.11 HPAEC-PAD Analysis (Amino Uronic Acid)

1. 4 × 50 mm CarboPac PA1 guard column (Thermo Fisher, cat. 043096), 4 × 250 mm CarboPac PA1 column (Thermo Fisher, cat. 035391).
2. TFA-HCl Mixture: According to the volume needed (3 mL for each sample and 14 mL for the calibration curve), prepare the quantity of solution needed in a glass bottle according to the volumes reported in Table 2.

Table 2
Preparation of acid solution "TFA-HCl Mixture"

Final volume mL	TFA mL	HCl mL
20	2.6	17.4
25	3.3	21.7
30	3.9	26.1
35	4.6	30.4
40	5.2	34.8
45	5.9	39.1
50	6.5	43.5
55	7.2	47.8
60	7.8	52.2
65	8.5	56.5

2.12 Dische Colorimetric Method

1. Disposable polystyrene cuvette with polypropylene sealing cap.
2. Sulfuric acid: In 50 mL glass bottle add 40 mL of sulfuric acid. Close the bottle and cool in ice for at least 1 h before use.
3. 1 M cysteine: inside an Eppendorf tube, weigh cysteine hydrochloride to achieve a weight of 157.6 mg. Add water and mix by vortexing. Calculate the volume of water to add by using the following formula:

$$\text{water to be added (mL)} = \frac{\text{mg cysteine hydrochloride}}{157.6}$$

   Keep cooled in ice before usage. This solution has to be prepared and used the same day.

2.13 High-Performance Liquid Chromatography Mass Spectrometry (HPLC-MS) (See Note 3)

1. Luna 3u C8(2), 50 × 2 mm, 100A (Phenomenex, cat. 00B-4248-B0).
2. SPE online cartridge holder, MercuryMS 20 mm cartridge holder (Phenomenex, cat. CH0–5845).
3. Strata-X 25 μm On-Line Extraction, Cartridge 20 × 2.0 mm (Phenomenex, cat. 00M-S033-B0-CB).
4. 50% Isopropyl alcohol (IPA) solution (v/v): Using a 100 mL graduated cylinder, measure 40 mL of IPA and transfer it in a bottle. Measure then 40 mL of water and add to the bottle, mixing with the IPA.
5. 5 nmol/mL (1220 ng/mL) 3-hydroxymyristic acid (3OH-My) in IPA/Water 1:1 (v/v): Weigh 10 mg of 3OH-My in a low-bind tube and dissolve them in a suitable amount of IPA added by weight on balance, in order to obtain

a 10 mg/mL solution. To calculate the amount of IPA to be added use the following formula:

$$\text{mg IPA} = \frac{\text{weight } \mu\text{g}}{10} \times \text{purity\%} \times 0.785$$

Then, to obtain a 1.22 mg/mL 3OH-My solution in IPA, in a low-bind tube, weigh accurately 150 mg of the solution prepared and add by weight the quantity of IPA calculated with the following formula:

$$\text{mg IPA} = \left(\frac{\text{weight mg}}{150} \times 1230\right) - \text{weight mg}$$

Weigh (tare) then an empty 1 L glass bottle and transfer 1000 μL of this solution, add IPA up to achieve a total weight of 392.5 g (500 mL× density 0.785 g/mL). Add then 500 g of water to achieve a total weight of 892.5 g (500 mL) and mix. 1 mL aliquots of the standard solution in suitable cryotubes can be stored at −80 °C.

6. 5 nmol/mL (1080 ng/mL) 3-hydroxylauric acid (3OH-La) IPA/Water 1:1 (v/v): Weigh about 10 mg of 3OH-La in a low-bind tube and dissolve them in a suitable amount of IPA added by weight on balance, in order to obtain a 10 mg/mL solution. To calculate the amount of IPA to be added use the following formula:

$$\text{mg IPA} = \frac{\text{weight } \mu\text{g}}{10} \times \text{purity\%} \times 0.785$$

Then, to obtain a 1.08 mg/mL 3OH-La solution in IPA, in a low-bind tube, weigh accurately 150 mg of the solution prepared and add by weight the quantity of IPA calculated with the following formula:

$$\text{mg IPA} = \left(\frac{\text{weight mg}}{150} \times 1389\right) - \text{weigh mg}$$

Weigh (tare) an empty 1 L glass bottle and transfer 1000 μL of this solution, add then IPA up to achieve a total weigh of 392.5 g (500 mL× density 0.785 g/mL). Add then 500 g of water to achieve a total weight of 892.5 g (500 mL) and mix. 1 mL aliquots of the standard solution in suitable cryotubes can be stored at −80 °C.

7. 1.5 M NaOH: In a 50 mL Falcon tube add 500 μL of 50% NaOH to 5.9 mL of water.

8. Eluent A (40%MeOH, 0.05% formic acid (FA)): In a 200 mL graduated cylinder measure 80 mL MeOH, add 100 μL FA and finally add water up to 200 mL. Transfer into a 250 mL glass bottle.

9. Eluent B (70% ACN, 0.05% FA): In a 200 mL graduated cylinder measure 140 mL of ACN, add 100 μL FA and finally add up to 200 mL of water. Transfer into a 250 mL glass bottle.
10. Eluent C (0.1% TFA): With a graduated cylinder measure 100 mL of water, transfer into a glass bottle and add 100 μL of TFA.
11. Eluent D (33% ACN in ISP, 0.1% TFA): With a graduated cylinder measure 33 mL of ACN, transfer into a glass bottle. Measure then with a graduated cylinder 67 mL of isopropyl alcohol and mix it with the ACN in the glass bottle. Add 100 μL of TFA and mix.
12. Eluent E (65% ACN): In a 100 mL graduated cylinder measure 65 mL of ACN and add water to reach 100 mL. Transfer in a 100 mL glass bottle.

2.14 Matrix-Assisted Laser Desorption/Ionization Time-of-Flight Mass Spectrometry (MALDI-TOF MS)

1. 50% ACN: In a 15 mL Falcon tube, add 1 mL of water, then 1 mL of ACN.
2. 200 mg/mL Super-DHB matrix solution: Weigh 5 mg of Super-DHB matrix in a 500 μL Eppendorf tube, then dissolve in 25 μL of 50% Water/ACN and vortex. The solution has to be used the same day of preparation (mass spectrum acquisition day).
3. Standard MW peptide: Fill an Eppendorf tube with 100 μL of water and add an aliquot (5 μL) of Peptide calibration standard, mix by vortexing.
4. 5% acetic acid (v/v): In a 15 mL Falcon tube, put 1.9 mL of water and add 100 μL of acetic acid.
5. 4:1 chloroform–MeOH (v/v): In a 2 mL glass screw cap vial, mix 200 μL of MeOH and 800 μL of chloroform.

2.15 Competitive-ELISA (cELISA)

1. 10× Sample Dilution buffer (SDB) (10× PBS containing 1.0% BSA and 0.5% Tween 20): Add 10× PBS into a graduated cylinder up to 1.8 L. Add gently 10.0 mL of Tween 20 into the cylinder followed by 20.0 g BSA. Stir for 1 h and if needed continue stirring until the BSA is completely dissolved. When the BSA is completely dissolved, add 10× PBS up to 2 L and stir again the solution for few min. Check the pH of the solution (which should be in the range 6.5 ± 0.5) using a pH meter or a pH paper. Filter-sterilize the solution into two 1 L bottles using two bottle-top 0.22 μm filter units.
2. SDB working solution: In a graduated cylinder add water up to the volume needed to have a 1:10 dilution, according to the total volume to be prepared. Add 10× SDB (e.g., one 50 mL aliquot 10× SDB in 450 mL water). Stir for few min with a magnetic stirrer. Filter-sterilize the solution into 1 L bottles using a bottle-top 0.22 μm filter unit.

3. Blocking buffer (5% fat-free milk in PBS): Add 800 mL of 1× PBS into a graduated cylinder. Add 50 g of skim milk powder to the cylinder containing 1× PBS. Stir for 30 min. Check to ensure that no lumps of powdered milk remain, especially on the top surface. If so, break up the lumps and continue stirring for 30 min. When the milk is dissolved, add 1× PBS after checking pH (in the range of 7.4 ± 0.3) up to 1 L total volume and stir again for few min. Transfer the solution in a sterile glass bottle.
4. 20× Washing Buffer (20× PBS containing 1% Tween 20): Add into a graduated cylinder around 1.8 L of 20× PBS and 20 mL of Tween 20. Stir for 1 h. When the Tween 20 is completely mixed, add 20× PBS up to 2 L and stir again for few min. Check the pH of the solution using a pH meter or a pH paper, that needs to be in the range of 6.5 ± 0.5. Transfer the solution in a sterile glass bottle.
5. Washing Buffer solution: Take one 2 L cylinder, put 100 mL of neat solution and add Water up to 2.0 L total volume. Stir for few min with a magnetic stirrer and transfer in a bottle.
6. Carbonate coating buffer solution (0.05 M Carbonate buffer, pH 9.6): Use the Carbonate Buffer capsules from Sigma-Aldrich C3041-100CAP to prepare the volume of coating buffer as needed. Put inside the bottle a magnetic bar and pour the contents of one capsule for every 100 mL of buffer to be prepared. Add water up to the final volume. Stir until the tablets are dissolved. When the salts are dissolved, filter the solution and check its pH using a pH meter or a pH paper. The pH needs to be in the range 9.6 ± 0.3.
7. 10x Phosphate coating buffer (500 mM Sodium Phosphate Buffer, pH 7.0): dissolve 29.2 g of $NaH_2PO_4 \cdot H_2O$ and 77.3 g $Na_2HPO_4 \cdot 7H_2O$ in 1 L of water.
8. Phosphate coating buffer: dilute with water 1:10 the 10× Phosphate coating buffer, stir for few min with a magnetic stirrer and filter 0.22 μm the solution.

2.16 OAg Extraction

1. 10× Acetate buffer (final 100 mM pH 3.9 after ten-fold dilution): Weigh 1099 mg of AcONa and transfer it in a 100 mL graduated cylinder. Add, in sequence, 50 mL of water and 4950 μL of acetic acid. Add water up to 100 mL, insert a magnetic stirring bar and homogenize the solution.

2.17 Size Exclusion High-Performance Liquid Chromatography, Derivatization with Semicarbazide (HPLC-SEC SCA)

1. Tosoh TSK gel G3000 PWXL column (30 cm × 7.8 mm; cat. 808021) with a Tosoh TSK gel PWXL guard column (4.0 cm × 6.0 mm; cat. 808033).
2. Eluent 0.1 M NaCl, 0.1 M NaH_2PO_4, 5% ACN, pH 7.2: Weigh 5.85 g of NaCl, 3.22 g of NaH_2PO_4 and 10.37 g of Na_2HPO_4 and transfer the powders inside a 1 L glass graduated cylinder. Add water up to 850 mL in the 1 L

graduated cylinder. Put a magnetic stir bar inside the cylinder and solubilize the powder placing the cylinder on a magnetic stirrer. Measure, in a 50 mL graduated cylinder, 50 mL ACN and add them in the 1 L cylinder with the buffer solution. Add water to bring the total volume to 1 L. Leave the cylinder on the stirrer until the solution is completely homogeneous. Filter-sterilize with a 0.22 μm filter.

3. λ-DNA Molecular Weight Marker III 0.12–21.2 kb (Roche; cat. 10528552001): dilute 20× with HPLC eluent before injection.
4. 200 mg/L Sodium azide (NaN_3) solution.
5. Semicarbazide solution: dissolve 100 mg Semicarbazide hydrochloride and 90.5 mg of AcONa anhydrous in 10 mL of water.
6. 40 μg/mL 3-Deoxy-D-manno-oct-2-ulosonic acid (KDO) ammonium salt solution (the concentration value is referred to ammonium salt).

2.18 Nuclear Magnetic Resonance (NMR)

1. 4 M Sodium deuteroxide (NaOD): Weigh 160 mg of NaOH and transfer in a tube. Calculate with the following formula the quantity of deuterium hydroxide (D_2O) needed to achieve 4 M solution and add to the powder.

$$\text{Total mL of water} = \frac{\text{sodium hydroxide (mg)}}{160}$$

Close the tube and mix on a vortex. Keep the tube closed until use in order to limit carbonation.

2.19 Hestrin Colorimetric Method

To prepare solutions from powders, use magnetic stir bars and a magnetic stirrer to accelerate the dissolution.

1. 1 M Acetic acid: Dilute in a 100 mL glass graduate cylinder 5.7 mL of acetic acid up to 100 mL with water.
2. 1 mM AcONa solution, pH 4.5: Weigh 8.2 mg of AcONa anhydrous and dissolve it in a 100 mL glass graduated cylinder with 80 mL of water. The pH is adjusted to 4.5 with acetic acid 1 M solution. Calculate the total water volume of the solution needed to get the correct concentration using the following equation:

$$\text{Total mL of water} = \frac{\text{sodium acetate weight (mg)}}{0.082}$$

Add water to the cylinder in order to reach the calculated volume.

3. 30 mg/mL acetylcholine: Resuspend the contents of a 150 mg vial of acetylcholine chloride using 5 mL of AcONa solution 1 mM, pH 4.5. Prepare just prior use.

4. 2 M Hydroxylamine hydrochloride: Weigh 6.9 g of hydroxylamine hydrochloride and dissolve it in a 50 mL glass gruated cylinder with water up to 50 mL.
5. 3.5 M NaOH solution: Weigh 7.0 g of NaOH and dissolve it in a 50 mL glass graduated cylinder with water up to 50 mL.
6. 4 M HCl: In a 50 mL Falcon tube dilute 10 mL of 37% HCl by adding 20 mL of water.
7. 0.1 M HCl: In a 100 mL glass graduate cylinder, dilute 830 μL of 37% HCl up to 100 mL with water. Transfer and store it in a glass bottle.
8. 0.37 M Iron chloride solution in 0.1 M HCl: Weigh 3.0 g of iron (III) chloride and dissolve it in a 50 mL glass graduated cylinder with 0.1 M HCl up to 50 mL. Transfer the solution in a glass bottle together with a stir bar that will be useful to resuspend iron precipitate (if present) before use.
9. Basic hydroxylamine solution: In a 50 mL Falcon tube, dilute 10 mL of 2 M hydroxylamine solution with 10 mL of 3.5 M NaOH solution. Mix by vortex.

3 Methods

3.1 Dot Blot/Western Blot

Dot blot/western blot are used for verifying OMV/GMMA identity. Specific primary antibodies are used according to the key antigen(s) present on the samples. In the Materials section we have reported as an example the use of a primary antibody (and corresponding secondary antibody) specific for *S.* Typhimurium OAg, key component of *S.* Typhimurium GMMA [14].

1. Prepare a positive control (+ CTRL) by diluting the OMV/GMMA sample to use as positive control to 8 μg/mL (protein concentration) with PBS. Use PBS as negative control (– CTRL).
2. Dilute each OMV/GMMA sample to be analyzed to 8 μg/mL (protein concentration) with PBS.
3. Cut the 0.2 μm polyvinylidene difluoride (PVDF) membrane (other membrane materials could be more suitable, depending on the product type) in order to have a resulting area large enough to cover the Dot-blot apparatus area.
4. Wet the PVDF membrane in a plastic case with a suitable volume of methanol for about 1 min, manually shaking the case in order to make the membrane entirely wet.
5. Transfer the membrane in another case filled with PBS and leave the membrane with PBS for 3 min, shaking manually the case in order to facilitate the entire wetting of the

membrane with PBS. Perform all this preparation immediately before placing the PVDF membrane in the Dot-Blot apparatus (*see* **Note 4**).

6. Place the PBS-wet PVDF membrane in the Dot-Blot apparatus. Cut a small portion of the PVDF right bottom corner in order to recognize the layout order at the end of the procedure. Quickly close the Dot-Blot apparatus.
7. Spot 100 μL of samples/+ CTRL/− CTRL in each well (each sample and controls are analyzed in duplicate) (*see* **Note 5**).
8. After deposition, switch on the vacuum system at about −150 mBar and keep it turned on until all the loaded samples are dried (typically 1 min). Wait about 30 s more and then switch off the vacuum. Slowly open the Dot-Blot apparatus and use tweezers to get the membrane.
9. Transfer the membrane to a plastic case and block with PBS/T + 3% BSA for 30 min under moderate agitation.
10. While blocking, prepare the primary antibody solution. At the end of 30 min, transfer the membrane to a container containing 5 mL of primary antibody solution. Place the container on a rocker at 4 °C and leave under moderate agitation overnight
11. After the overnight incubation, place the container at RT for 10–15 min to let the temperature rise.
12. Remove the solution from the container and wash the membrane three times with 5 mL of PBS/T for 10 min/wash, under agitation.
13. Prepare the secondary antibody solution and incubate the membrane with 5 mL of this solution for 1 h under moderate agitation.
14. Wash the membrane three times with PBS/T for 10 min/wash, under agitation, as before.
15. Prepare the developing solution during the washing time.
16. Incubate the membrane with 6 mL of the freshly prepared developing solution and leave under moderate agitation until the spots are clearly visible (typically 1–2 min).
17. Stop the reaction by washing the membrane three times with 5 mL of water for 5 min/wash, under agitation. Leave the membrane in water. For the assay to be considered valid, there should be no visible spot for the - CTRL (PBS), whereas spots should be seen for the OMV/GMMA + CTRL, and both replicates of the tested samples (*see* **Note 6**).

3.2 Dynamic Light Scattering

DLS is one of the methods that can be used for OMV/GMMA particle size determination [9]. This technique allows determination of a mean hydrodynamic diameter (Z-average diameter) and a

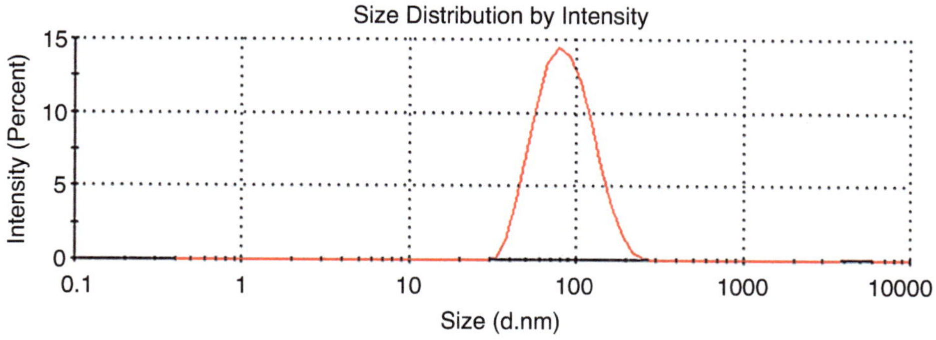

Fig. 2 Z-average diameter determination by DLS for a GMMA sample as an example

polydispersity index (PDI), describing the amplitude of the distribution. Hydrodynamic diameter is calculated using the Stokes–Einstein equation, obtaining the diffusion coefficient by measuring intensity fluctuations of scattered light produced by particles as they undergo Brownian motion.

The hydrodynamic diameter of the particles is expressed by a Z-average value (*see* **Note 7**), providing also a polydispersity index (PDI) of the size values calculated.

1. Adjust the settings on the Zetasizer Nano ZS as follows: Measurement type = Size, Material = Protein, Dispersant = buffer in which the sample is diluted (*see* **Note 8**), General options = Mark-Houwink parameters (set as default), Temperature = 25 °C, Equilibration time = 120 s (*see* **Note 9**), Measurement angle = 173° Backscatter (NIBS default), Measurement duration = Automatic, Measurements = 3, Delay between measurement = 30 s, Data processing = Analysis model General Purpose (normal resolution).
2. Dilute the sample to 50 μg/mL protein with an appropriate buffer and transfer to a DLS cuvette.
3. Insert the cuvette in the cell holder, place the thermal cap on top of the cuvette and close the lid.
4. Start the measurement. Figure 2 reports Z-average diameter determination for a GMMA sample as an example.

3.3 HPLC-SEC/MALS

HPLC-SEC analysis can be used to assess OMV/GMMA purity: presence of soluble proteins and relevant amounts of DNA can be detected. Use of MALS allows particle size determination too.

Presence of soluble proteins is assessed using the fluorescence channel detecting presence of eventual peaks at different (higher) elution times with respect to particles. For a quantitative estimate of protein impurities, the ratio of the area of the OMV/GMMA peak and of the area of soluble proteins can be used.

Presence of nucleic acids is assessed based on the ABS260/ABS280 ratio on the UV channels. DNA content can be accurately quantified by other methods like Threshold.

GMMA radii are evaluated with the MALS detector. The radius is calculated for each point of the particle peak and the average values, obtained using averaging formulas with different weights (Rn, Rw and Rz), are reported. The ratio Rw/Rn indicates whether the peak is homogenous with respect to radius: for a monodisperse sample the average radius is independent of the averaging method and the ratio is equal to 1.

1. Set the HPLC instrument as follows: column compartment temperature = 30 °C, autosampler compartment temperature = 4 °C, UV detector as first detector after column with acquisition channels 260 nm and 280 nm, fluorimeter detector as second detector after column with excitation wavelength 280 nm/emission wavelength 336 nm, multi angle light scattering detector as third detector after column.
2. Gently mix the sample in order to homogenize the content, dilute with PBS or with an appropriate buffer to 150 μg/mL of protein (sample concentration may be adapted depending on sample type).
3. Run the sample injecting 80 μL eluting with PBS at 0.5 mL/min flow rate and with 70 min run time (*see* **Note 10**). In Fig. 3a the chromatogram (fluorescence channel) of a purified *S.* Enteritidis GMMA sample (as an example) containing neglectable quantity of soluble proteins is reported. In Fig. 3b the chromatogram of a protein standards mixture (BioRad cod 151-1901: 670-1.35 kDa) is reported, showing the retention time range in which soluble proteins are expected. Figure 3c represents the sample of GMMA spiked with the protein standard mixture, clearly indicating ability of the method to distinguish OMV/GMMA particles from soluble proteins of different size. In Fig. 4 UV ABS chromatograms at 260 and 280 nm are overlapped for a GMMA sample to estimate the presence of nucleic acids. For sample containing very low/no DNA impurities, the ratio of the areas of the two chromatograms relative to the OMV/GMMA peak (at 260 and 280 nm respectively) is expected to be close to 1. Ratio > 1 reflects presence of DNA impurities. DNA impurities can be detected also at lower size compared to OMV/GMMA with the typical A260/A280 ratio.
4. Elaborate the MALS data using the following setting in the Wyatt Astra software: Despiking = heavy, Baseline = auto find baselines (check, selecting each detector, if the corresponding found baseline is appropriate. Correct the starting/ending points if needed (*see* **Note 11**), LS analysis = Sphere as model

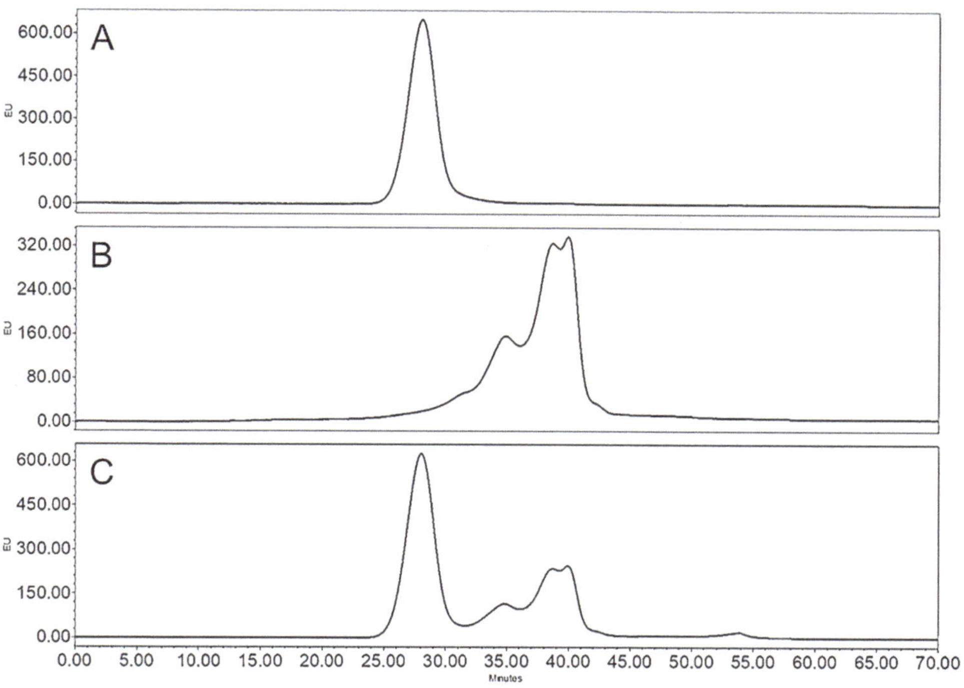

Fig. 3 HPLC-SEC analyses (fluorescence emission profiles) of a purified *S.* Enteritidis GMMA sample (**a**), of a protein standards mixture (**b**) and of the GMMA sample spiked with such standard (**c**), showing ability of this method to separate GMMA particles from soluble proteins of different size

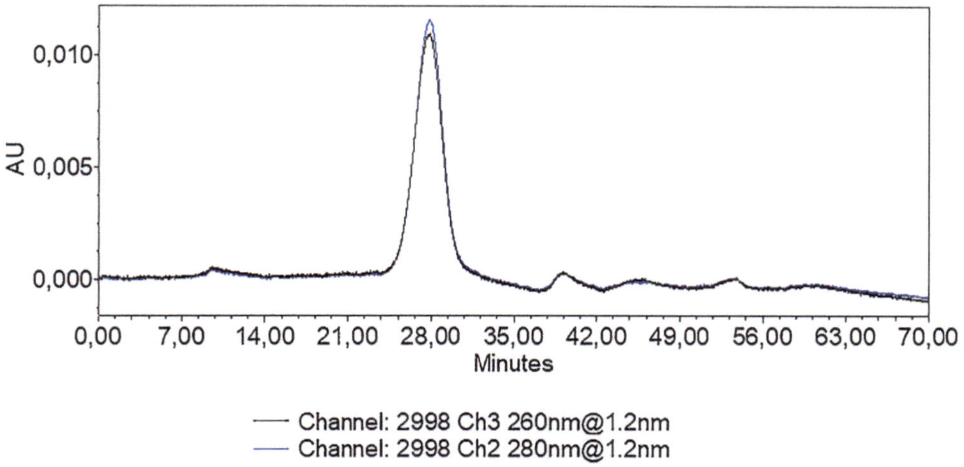

Fig. 4 HPLC-SEC analysis (UV detection at 260 and 280 nm), purified *S.* Enteritidis GMMA

with fit degree 1 (*see* **Note 12**). On the 90° detector (LS11) chromatogram, select as peak for the average radius calculation the time range corresponding to 20% height in OMV/GMMA peak leading and tailing (*see* **Note 13**). Check the fitting of the

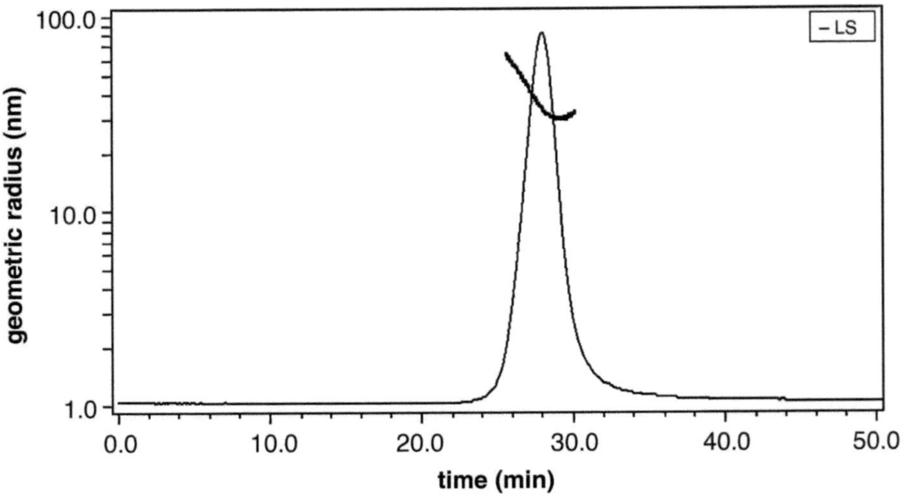

Fig. 5 LS Detector 11 (90°) with radius, purified *S.* Enteritidis GMMA

data and eventually disable some detectors removing them from the fitting. Chromatogram in Fig. 5 is reported as an example.

3.4 Nanoparticle Tracking Analysis (NTA)

1. Set the instrument as follows: minimum track length = automatic settings, minimal expected particle size and blur setting = applied, viscosity settings for water with automatic correction for the temperature, syringe pump speed 20.
2. Dilute the sample just before the analysis to 10–100 ng/mL of protein (the concentration needs to be adjusted depending on the sample) using low-bind material (pipette tips and tubes) and acquire five replicate videos of 30 s at 25 frames per second, generating five replicate histograms to be averaged.
3. Analyze particles movement with camera level = 16, slider shutter = 1300 and slider gain = 512. Test and adjust detection threshold value for the sample appearance. Figure 6 shows a typical NTA graph for a GMMA sample.

3.5 SDS-Page

SDS-PAGE analysis is used to look at protein pattern profile of OMV/GMMA preparations. Total protein composition can be also determined by MS [10, 11].

1. Dilute a GMMA sample to 300 μg/mL protein concentration.
2. In 1.5 mL Eppendorf tubes, add for each sample the following quantities: 4.5 μL of LDS sample buffer (4×), 1.5 μL of 1 M DTT and 10 μL of Sample solution.
3. Spin down the liquids, vortex and spin down again.
4. Place the tubes in a block heater at 100 °C for 10 min; then let them cool at RT and spin down the liquid.

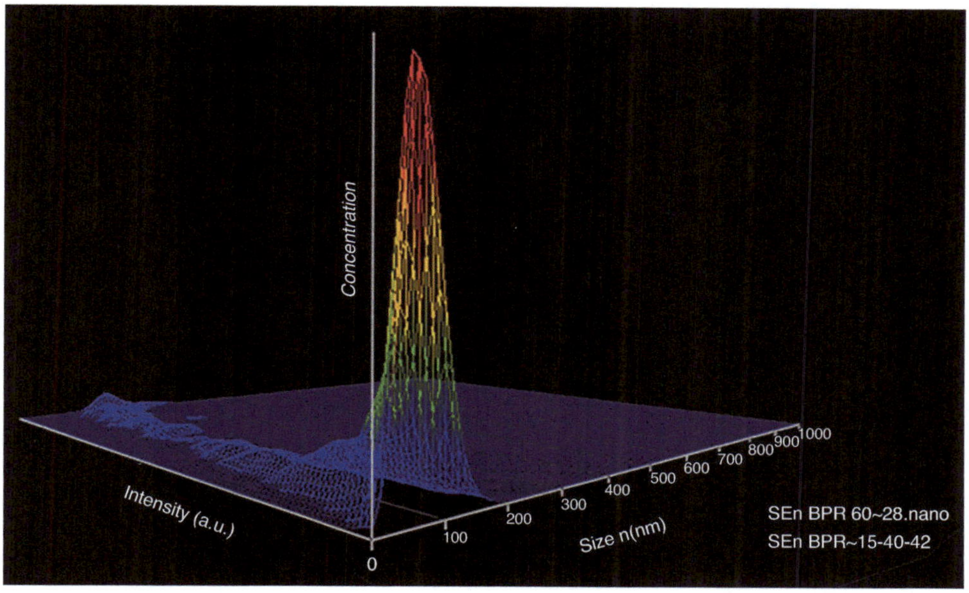

Fig. 6 NTA graph of *S.* Enteritidis GMMA (as an example), in which particle size, particle concentration, and relative intensity are plotted together

5. Insert the precast gel in the electrophoresis cell after having removed the strip present in the bottom of the gel.
6. Transfer MOPS buffer into the middle part of the cell up to cover the gel wells and more up to 2 cm from the front bottom of the cell.
7. Remove the wells protection from the gel, use the first well to load 10 μL of the marker, load the samples in the remaining wells.
8. Close the cell and connect it to the power supply.
9. Set up the power supply at 40 mA (constant) with 200 V as maximum limit and run the gel (typically 75 min). Wait until the blue marker of the wells reaches the bottom of the gel. Open the cell and remove the gel from its holder using the gel knife.
10. To fix the gel, place 50 mL of water into the staining container and place the gel in it.
11. Remove the water and add 50 mL of fixing solution. Leave the container closed on a rocker (set at 35 rpm) in the chemical hood for 30 min at RT.
12. Remove the fixing solution from the container and discard into the appropriate waste container.
13. Mix 40 mL of the dye solution and 10 mL of MeOH in a 50 mL Falcon tube, then add the staining solution in the container with the gel and leave it (closed) on a rocker (set at 35 rpm) overnight at RT.

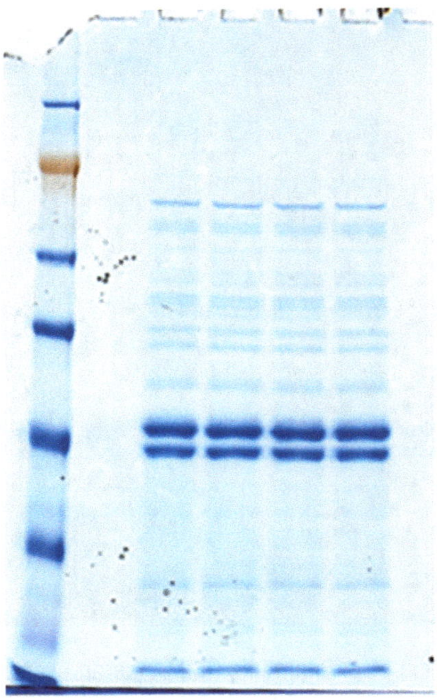

Fig. 7 SDS-PAGE analysis of *S. flexneri* 2a GMMA

14. Remove the staining solution from the container.
15. Wash the gel by adding to the container 50 mL of water and leave it on a rocker (set at 35 rpm) for 10 min. Repeat this operation twice.
16. Add 35 mL water and then 15 mL MeOH to the container with the gel and leave it (closed) on a rocker (set at 35 rpm) under chemical hood 1 h at RT.
17. Remove the solution from the container, then add 50 mL water and leave it (closed) on a rocker (set at 35 rpm) under chemical hood for 2 h at RT.
18. Collect a picture of the gel with a camera or scanner, as example in Fig. 7 a gel of *S. flexneri* 2a GMMA is reported.

3.6 Micro BCA

Micro BCA analysis is used to quantify OMV/GMMA total protein content.

1. Set the spectrophotometer wavelength at 562 nm.
2. Set a thermostatic bath to 60 °C.
3. To prepare 20 μg/mL BSA solution, open a BSA Standard ampoule from the Micro BCA kit, mix gently. Transfer 100 μL of BSA 2 mg/mL in a Falcon tube and weigh. Calculate the amount of water to add using the following formula:

$$\text{Water (mg)} = 100 \times 2\ \text{mg/mL BSA weight (mg)} - 2\ \text{mg/mL BSA weight (mg)}$$

Table 3
Dilutions for preparing calibration curve for Micro BCA analysis

Std conc. µg/mL	20µg/mL BSA µL	Water µL
0	0	500
1.8	45	455
2.4	60	440
5.0	125.0	375.0
10.0	250.0	250.0
15.0	375.0	125.0
20.0	500.0	0

4. In 5 mL tubes, prepare in duplicate the diluted solutions from the 20.0 µg/mL BSA solution, using the volumes reported in Table 3.
5. For each OMV/GMMA sample, first gently mix in order to homogenize, then prepare three replicates for analysis. The volume of sample, or diluted sample, required for each replicate is 500 µL (*see* **Notes 14** and **15**).
6. Prepare a reagent mixture by combining the reagents from the Micro BCA kit as follows: in a 50 mL Falcon tube, add 0.5 mL of Reagent C and 12.0 mL of Reagent B, and mix well with the vortex. Then add 12.5 mL of Reagent A and mix well with vortex.
7. To every tube containing standards and samples, add 500 µL of the freshly prepared reagent mixture.
8. Vortex each tube, cover with aluminum foil or Parafilm, and incubate in the preheated thermostatic bath for 1 h.
9. Allow the tubes to cool at RT for 10 min before starting sample readings.
10. Read each sample as per the following sequence: First read the "blank" cuvette (i.e., containing water only), if a single beam instrument is used (in order to subtract its absorbance from samples/standards absorbances). If a double beam instrument is used, leave the "blank" containing cuvette in the reference slot during all sample/standard readings. Next, read the BSA standard solutions (each in duplicate), in order of increasing concentration, and finally the samples (each in triplicate).
11. Calculate a quadratic fitting regression between the ABS measured and BSA concentrations (in the range 1.8–20 µg/mL) and calculate on it the concentration of the sample

analyte. If the concentration of a sample is outside the range of the calibration curve, dilute the sample with water and repeat the analysis.

3.7 Ultracentrifugation

The procedure is applied to separate OMV/GMMA particles from the solution, that is, to quantify soluble proteins eventually present in the sample. By ultracentrifugation, OMV/GMMA (centrifuged pellet) are separated from soluble proteins (centrifuge supernatant) (*see* **Note 16**).

1. Dilute the sample to a maximum concentration of 1 mg/mL (protein based).
2. Pellet OMV/GMMA by ultracentrifugation 4 °C, 30′, 110 k rpm, using a rotor with K-factor 15 (*see* **Note 1**).
3. For each sample tube, immediately (*see* **Note 17**) after ultracentrifugation ends, carefully transfer the supernatant with pipette, without touching nor resuspending the OMV/GMMA pellet, to a fresh tube for further analysis in **step 6** below.
4. Resuspend the pelleted OMV/GMMA fraction, after supernatant removal (*see* **Note 18**) and store the tube closed overnight at 4 °C.
5. The following day, pipet several times the solution in the tube, resuspending and homogenizing the pellet.
6. Analyze the protein content of the supernatant and resuspended pellet samples, using either the Micro BCA (*see* Subheading 3.6) or Lowry (*see* Subheading 3.8) method. The percentage of soluble proteins can be expressed as ratio between the protein content determined in the ultracentrifugation supernatant and the total protein content of the sample prior to ultracentrifugation.

3.8 Lowry

Lowry is a colorimetric method that can be used to quantify OMV/GMMA total protein content.

1. Set the spectrophotometer wavelength at 750 nm.
2. In 5 mL tubes, prepare (in duplicate) diluted solutions from the 120 μg/mL BSA standard solution, using the volumes reported in Table 4.
3. For each sample prepare three replicates for the analysis; the volume of sample, or diluted sample, required for each replicate is 400 μL (*see* **Notes 14** and **19**).
4. Add 400 μL of Reagent A to each tube containing standards/samples and mix well with the vortex. Add 200 μL of Folin reagent mixture to each tube containing standards/samples

Table 4
Dilutions for preparing calibration curve for Lowry analysis

Std conc. μg/mL	120 μg/mL BSA μL	PBS μL
0	0	400
9.0	30.0	370.0
21.0	70.0	330.0
42.0	140.0	260.0
60.0	200.0	200.0

and 400 μL of Reagent A. Vortex each tube. Cover the tubes with aluminum foil or Parafilm, incubate them at RT and start a 75 min countdown.

5. Read samples/standards following the sequence: First read the "blank" containing cuvette, if a single beam instrument is used (in order to subtract its absorbance from samples/standards absorbances). If a double beam instrument is used, leave the "blank" containing cuvette in the reference slot during all sample/standard readings. Next, read the BSA standard solutions (each in duplicate) in order of increasing concentration, and finally the samples (each in triplicate). For each tube, transfer its content in a quartz cuvette, place it in the spectrophotometer, read the ABS value and discard the cuvette content. Wash the cuvette with PBS after calibration curve readings (before starting the first sample reading) and between readings of different samples.
6. Calculate a linear regression between the ABS and BSA concentrations (in the range 9–60 μg/mL) and calculate on it the concentration of the sample analyte. If the concentration of a sample is outside the range of the calibration curve, dilute the sample with PBS and repeat the analysis.

3.9 Amino Acid Analysis

Amino acid analysis is another method to quantify total protein content of OMV/GMMA samples. In contrast to the Micro BCA and Lowry described above, Amino Acid Analysis quantifies actual content of each amino acid present in the sample, instead of assuming similar behavior of OMV/GMMA sample as a BSA standard in the colorimetric assays.

MS analysis can also be used to quantify specific proteins in OMV/GMMA that are relevant for vaccine efficacy. Such methods need to be set up and optimized specifically for the protein of interest. Examples can be found in [10, 18]. Before amino acid content determination, the time required to hydrolyze the sample has to be defined with a hydrolysis kinetic study (i.e., hydrolysis for

16, 24, 48, 72 h) in order to optimize conditions to achieve best amino acids recovery for the specific sample being analyzed.

1. For the HCl vapor phase hydrolysis (*see* **Notes 20** and **21**), before starting operations, treat glass hydrolysis tubes at 400 °C for 4 h, to avoid sample contamination (i.e., Gly, Ser Pro from skin, Asp, Glu from paper).
2. Switch on the Hydrosmart 2 at 114 °C in order to preheat the oven prior to inserting hydrolysis vessels.
3. For each sample, prepare hydrolysis tubes in triplicate by weight. Place the hydrolysis tube on the analytical balance, tare and add the sample to hydrolyze (*see* **Notes 20–22**).
4. Dry the hydrolysis tubes filled with sample solution in the centrifugal evaporator.
5. With the help of forceps, put the dried hydrolysis tubes inside the hydrolysis vessels (up to 12 hydrolysis tubes for each hydrolysis vessel; the oven can contain up to 3 hydrolysis vessels for a total of 36 tubes corresponding to 12 samples in triplicate).
6. Add at the bottom of the vessel (outside the hydrolysis tubes) 200 μL of 6 N HCl, 0.1% Phenol solution.
7. Cap the vessel, perform 4–5 vacuum–nitrogen cycles and close the cap valve leaving the vessels under nitrogen.
8. Place the vessels in their oven compartments, cover them with oven's cap and start the hydrolysis.
9. After the hydrolysis time is complete, extract carefully the vessels from the oven and let them cool for about 10 min under chemical hood, maintaining the samples under inert atmosphere.
10. Under a chemical hood, release the cap valve, unscrew the cap and take out with forceps hydrolysis tubes from the vessels.
11. To remove the eventual HCl residue drops, place sample hydrolysis tubes in a Speedvac concentrator for about 30 min.
12. Resuspend each sample in 100 μL 100 mM HCl and vortex well, as partial redissolution of the hydrolyzed sample is one of the main causes of low recovery (*see* **Note 22**).
13. Transfer the entire content of the hydrolysis tube in low-bind Eppendorf tube.
14. Prepare, in the low-bind tubes, two replicates of each dilution of the Standard Curve Solution starting from the 2.5 mM (2500 nmol/mL) standard solution of each amino acid (except for Cystine that has a concentration of 1.25 mM) as reported in Table 5, vortex all the standard dilutions tubes (*see* **Note 23**).
15. To derivatize the samples and standards directly in HPLC vials, transfer 70 μL of borate buffer AccQTag Ultra in each vial.

Table 5
Dilutions for preparing curve for amino acidic analysis

Amino acid nmol/mL	250 nmol/mL amino acid standard solution (μL)	Water (μL)	2500 nmol/mL amino acid standard solution (μL)
0	0	500	0
15	30	470	0
25	50	450	0
62.5	125	375	0
125	250	250	0
187.5	375	125	0
250	0	1800	200

Then to appropriate vials, add 10 μL of sample solution from **step 13** and each standard solution by inserting the tip in the buffer and pipetting 3–4 times. Vortex each vial for 10 s. Add 20 μL of reconstituted Reagent AccQTag Ultra (*see* **Note 24**) to each vial, releasing the reagent from the pipette by inserting the tip in the sample buffered solution and pipetting 3–4 times. Place all the vials at 55 °C for 10 min (*see* **Note 25**).

16. To run the amino acid analysis by HPLC-RP, set the HPLC instrument configuration as follows: column compartment = 49 °C, autosampler compartment = 20 °C, UV detector acquisition at 260 nm ABS, Flow rate = 0.7 mL/min, Run time = 10.2 min, Sample injection volume = 1 μL, Gradient Program (Fig. 8) (*see* **Note 26**).

    A representative chromatogram is reported in Fig. 9.

17. Calculate a linear regression for each amino acid standard (peak area vs concentration in the range 15–250 nmol/mL) and calculate on it the concentration of each amino acid in the sample analyte. For each standard/sample injection, check if the chromatogram 6-aminoquinolone (AMQ) area is higher with respect to the 80% of the average AMQ in the blanks. This check is performed to ensure that the derivatizing agent quantity is enough for the reaction (at least 5 times the amino groups). If the test fails, the corresponding standard/sample is not suitable for the analysis (*see* **Note 27**).

    Calculate the total protein of the sample by sum of the weight of each quantified amino acid, considering the condensation water using the following formula:

$$\text{Total protein} = \sum_{n=\text{Amino acid}} (\text{amino acid}_n \text{ molecular weight} - 18) \times \text{amino acid}_n \text{ concentration}(\text{nmol/mL})$$

	Time	Flow (mL/min)	%A	%B	%C	%D	Curve
1	Initial	0.700	2.0	0.0	98.0	0.0	Initial
2	0.29	0.700	2.0	0.0	98.0	0.0	11
3	5.49	0.700	9.0	80.0	11.0	0.0	7
4	7.10	0.700	8.0	15.6	57.9	18.5	6
5	7.30	0.700	8.0	15.6	57.9	18.5	6
6	7.69	0.700	7.8	0.0	70.9	21.3	6
7	7.99	0.700	4.0	0.0	36.3	59.7	6
8	8.59	0.700	4.0	0.0	36.3	59.7	6
9	8.68	0.700	2.0	0.0	98.0	0.0	6
10	10.20	0.700	2.0	0.0	98.0	0.0	6

Fig. 8 Gradient program for amino acid analysis by HPLC-RP. A -AccQTag Ultra Eluent A; B -AccQTag Ultra Eluent B/10; C -Water (Filtered 0.22 μm); D -AccQTag Ultra Eluent B

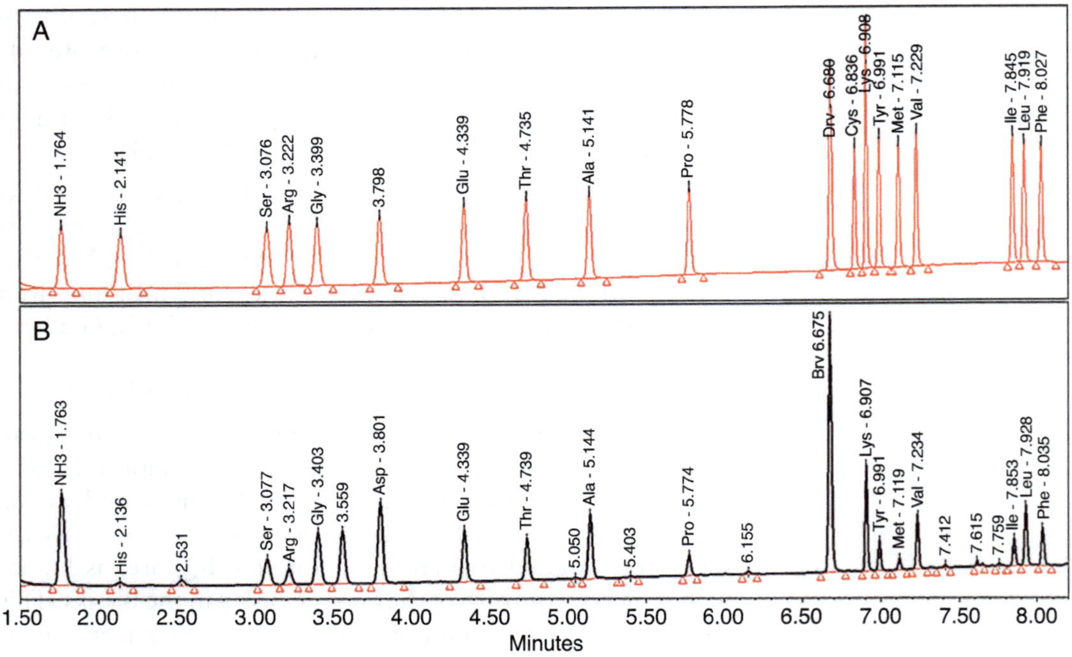

Fig. 9 Chromatograms (260 nm profiles) of standard amino acids mixture (**a**) and of a purified *S.* Typhimurium GMMA sample after hydrolysis and derivatization for amino acid analysis (**b**)

3.10 Total Sugar Quantification

LPS is one of the main constituents of OMV (naturally released)/GMMA membranes (Fig. 1). LPS molecules are constituted of three main portions: Lipid A is linked to the 3-deoxy-D-manno-octulosonic acid (KDO) terminus of the core region, which in turn is attached to the OAg chain (Fig. 10). Capsular polysaccharides can also be present on OMV/GMMA surface depending on the

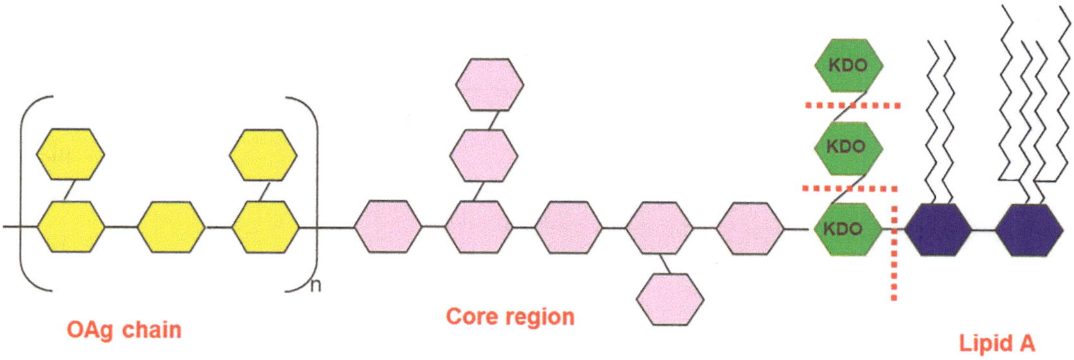

Fig. 10 Schematic representation of LPS molecules, constituted by lipid A, core region and repeating units of the OAg chain

specific type of OMV/GMMA [19]. Such sugars can be key target for protection as it is the case of *Salmonella* [20] and *Shigella* [21], and in-depth analysis of LPS/capsular polysaccharides can be needed.

According to the specific sugars that constitute the LPS/capsular polysaccharide molecules present, different analytical methods can be used. Here we report details on some methods that can be performed at this scope, starting from methods for total sugar determination. Such methods can be performed directly on OMV/GMMA particles avoiding any step of sugar chains isolation before measurement.

3.11 HPAEC-PAD (Neutral Sugars) [14]

This method allows quantification of neutral sugars, that are usually found as constituents of LPS core regions or OAg chains (such as in *S.* Typhimurium, *S.* Enteritidis, *Salmonella* Paratyphi A, *Shigella flexneri*). Concentrations of Rha, Gal, Glc, Man can be estimated through this analysis.

1. In 2 mL screw cap vials, prepare in duplicate the dilutions of calibration curve standard solutions, starting from the 11.25 μg/mL neutral sugar standard mix, as indicated in Table 6.
2. For triplicate analysis, prepare three 2 mL screw cap vials containing 450 μL of the diluted sample.
3. To each vial containing standards or samples, add 150 μL 8 M TFA (final concentration 2 M), close and vortex each vial for few seconds.
4. Place all vials in a rigid cardboard rack and incubate in a preheated oven at 100 °C for 4 h.
5. After this time, allow all the vials to cool at 2–8 °C for 30 min.
6. Remove the caps and dry the standards and samples overnight in centrifugal evaporator at RT in order to remove the TFA.

Table 6
Dilutions for preparing calibration curve for HPAEC-PAD neutral sugars analysis

μg/mL (each sugar)	11.25 μg/mL Neutral sugar standard solution μL	Water μL
0	0	450
0.5	20	430
1.0	40	410
2.5	100	350
5.0	200	250
7.5	300	150
10	400	50

7. Redissolve the content of each vial adding 450 μL of water and vortex for few seconds.
8. Filter the content of each vial into the sampler polypropylene vial using 0.45 μm nylon syringe filters (4 mm diameter).
9. Set up the instrument with CarboPac PA10 guard column, CarboPac PA10 column connected in series, Column/detector compartment set at 35 °C, Autosampler compartment set at 10 °C, Electrochemical detector equipped with gold working electrode and potential sets with standard carbohydrate 4-potential waveform.
10. Run the analysis of standards and samples with an injection volume of 25 μL in full loop mode with 50 min run time and 1 mL/min eluent flow rate using the following eluent program:

    20 min 18 mM NaOH (36% of eluent 50 mM NaOH), 10 min 28 mM NaOH 100 mM AcONa (36% 50 mM NaOH; 10% 1 M AcONa with 100 mM NaOH), 20 min 18 mM NaOH (36% of eluent 50 mM NaOH) (*see* **Note 28**).
11. For each sugar standard, calculate a linear regression between the peak areas and monomer concentrations (in the range 0.5–10 μg/mL) and calculate on it the concentration of each sugar monomer in the sample (*see* Fig. 11). If the concentration of a sample is outside the range of the calibration curve, dilute the sample with water and repeat the analysis.

3.12 HPAEC-PAD (Amino Uronic Acid) [13]

This method allows quantification of amino uronic acids, present for example in *Shigella sonnei* OAg and capsular polysaccharide displayed on OMV/GMMA surface.

1. In 2 mL screw cap vials, prepare in duplicate the dilutions of calibration curve standard solutions, starting from the 2.56

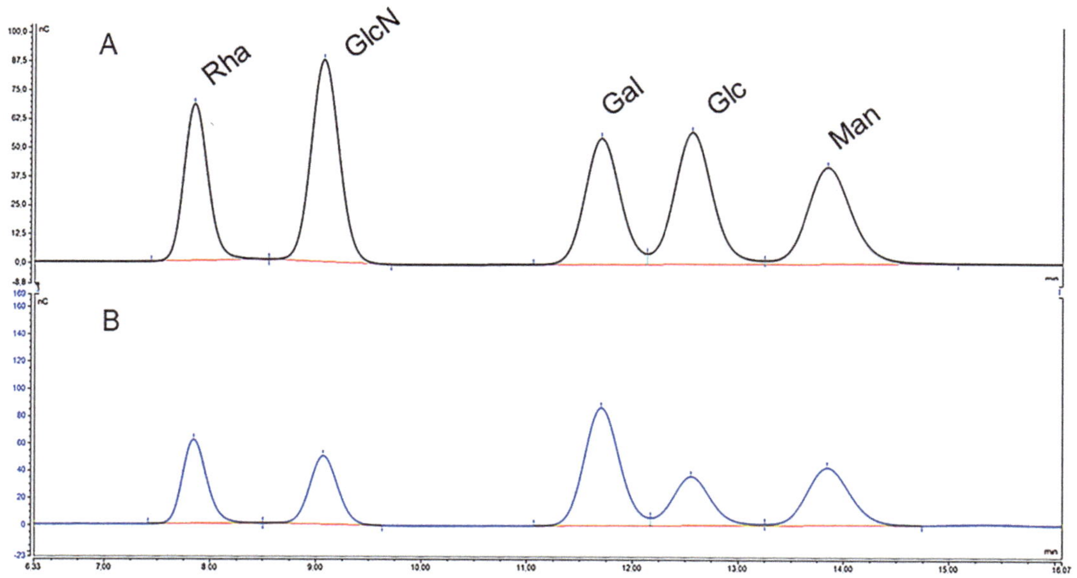

Fig. 11 HPAEC-PAD profile of standard neutral sugar mixture (**a**) and of a purified *S.* Typhimurium GMMA sample (**b**)

µg/mL and 0.512 µg/mL OAg standard solutions, as indicated in Table 7 (*see* **Note 29**) .

2. For triplicate analysis, prepare three 2 mL screw cap vials containing 300 µL of the diluted sample.
3. To each vial containing standard or sample, add 1 mL of TFA-HCl Mixture, close and vortex each vial for few seconds.
4. Place all vials in a preheated block heater at 80 °C for 4.5 h.
5. After this time, allow all the vials to cool at 2–8 °C for 15 min.
6. Remove the caps and dry the standards and samples under nitrogen flux in order to remove the major part of solvent/ HCl.
7. Finish drying the standards and samples overnight in centrifugal evaporator at RT.
8. Redissolve the content of each vial adding 300 µL of water and vortex accurately for few seconds.
9. Filter the content of each vial into the sampler polypropylene vial using 0.45 µm nylon syringe filters (4 mm diameter).
10. Set up the instrument with CarboPac PA1 guard column, CarboPac PA1 column connected in series, Column/detector compartment set at 25 °C, Autosampler compartment set at 10 °C, Electrochemical detector equipped with gold working electrode and potential sets with standard carbohydrate 4-potential waveform.

Table 7
Dilutions for preparing calibration curve for HPAEC-PAD amino uronic acid analysis

S. sonnei OAg µg/mL	2.56 µg/mL standard solution (µL)	0.512 µg/mL standard solution (µL)	Water (µL)
0	–	–	300
0.160	–	94	206
0.321	–	188	112
0.649	76	–	224
1.280	150	–	150
2.560	300	–	0

11. Run the analysis of samples and standards with an injection volume of 25 µL in full loop mode with 15 min run time and 1.5 mL/min eluent flow rate using 400 mM NaOH (*see* **Note 30**).
12. Calculate a linear regression between the areas of the peak of interest and the concentrations of the OAg standard (in the range 0.16–2.56 µg/mL) and calculate on it the concentration of the OAg in the sample (*see* Fig. 12, analyte peak at 8.9 min). If the concentration of a sample is outside the range of the calibration curve, dilute the sample with water and repeat the analysis.

3.13 Dische Colorimetric Method

Dische is a colorimetric method, that allows quantification of methyl pentose (6-deoxyhexose) content in samples (i.e., rhamnose and fucose) and can be often used as a simple and rapid method for sugar quantification in OMV/GMMA samples.

This method is based on the following steps: in the first step concentrated sulfuric acid at 100 °C hydrolyzes the polysaccharide and allows the monosaccharides to form methyl furfural (from methyl pentoses) or hydroxymethyl furfural (from hexoses). Both species react with sulfur present in cysteine. Compounds that derive from methyl furfural have a characteristic maximum in absorption spectrum at 396 nm, while compounds that derive from hydroxymethyl furfural have a maximum at 415 nm. In the single compound absorption spectra, considering the bell shape close to the maximum, the absorption at 396 nm due to chromophores deriving from hexoses is numerically equal to their absorption at 427 nm; the absorption at 427 nm due to chromophores deriving from methyl pentoses is close to zero. Consequently, performing the whole reaction on a mixture of hexoses and methyl pentoses, it is possible to calculate the methyl pentose contribution to 396 nm ABS by subtracting the 427 nm ABS.

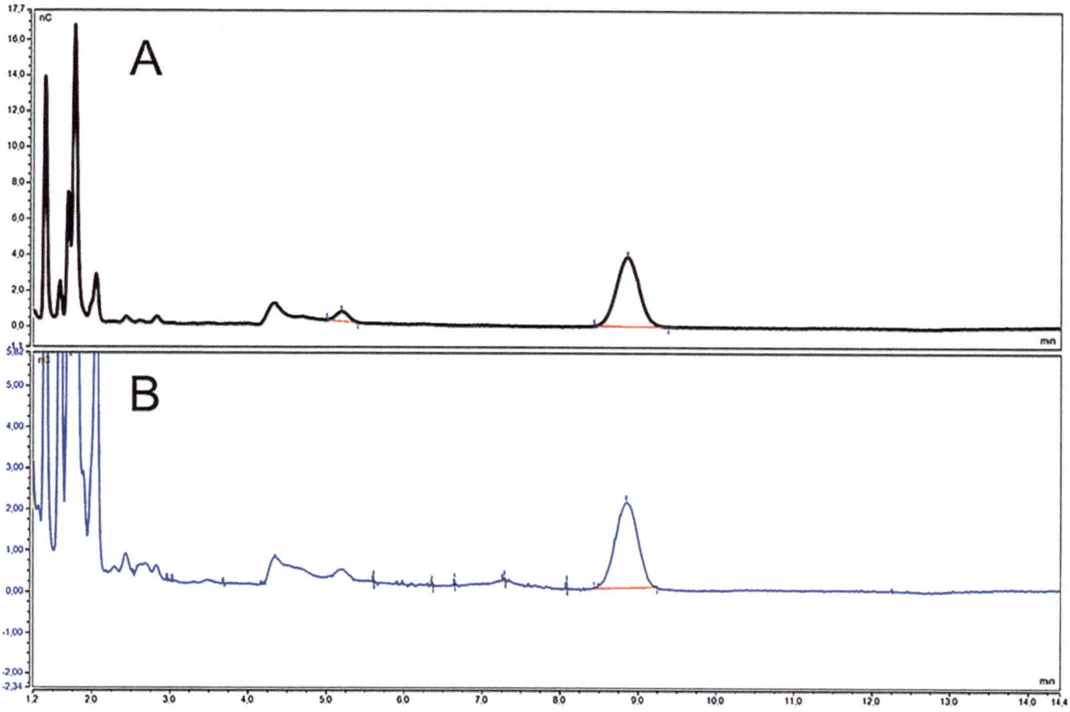

Fig. 12 HPAEC-PAD profile of *S. sonnei* standard OAg (**a**) and of a *S. sonnei* GMMA sample (**b**) after acid hydrolysis resulting in the formation of 2-amino-2-deoxy-α-L-altropyranuronic acid quantified in the analysis

The chromophore species is the same regardless the stereochemistry of the methylpentose sugars, that is, with a calibration curve of fucose standard it is possible to perform an accurate determination of sample containing rhamnose.

The procedure has been tested for the following interferences in methylpentose determination: hexoses (mannose, galactose, glucose), 2-aminohexoses: N-acetylglucosamine, uronic acid (glucuronic acid), 3,6-dideoxy hexoses (tyvelose), KDO (3-Deoxy-D-manno-oct-2-ulosonic acid), proteins (BSA up to 100 μg/mL with Rha concentration at 5 μg/mL), and DNA.

All substances above have no interference with methyl pentose determination, provided the ABS values of the sample at 396 and 427 nm remain under 1.2. (i.e., Rha can be quantified at 5 μg/mL in presence of 500 μg/mL glucose but Rha at 30 μg/mL in presence of 500 μg/mL Glc results in an ABS that exceeds this limit and needs a predilution).

1. Prepare a single set of the diluted solutions for a calibration curve from a fucose standard solution (50 μg/mL) directly in 2 mL Eppendorf tubes, following volumes reported in Table 8.
2. Gently mix samples before dilution in order to homogenize the vial content. Samples are analyzed in duplicate. The (diluted) sample quantity needed for each replicate is 500 μL (*see* **Note 31**).

Table 8
Dilutions for preparing calibration curve for Dische determination

Fucose µg/mL	50 µg/mL fucose µL	Water µL
2	20	480
4	40	460
8	80	420
16	160	340
32	320	180

3. To each standard and sample (500 µL), add 1050 µL of ice cooled sulfuric acid.
4. Vortex each tube.
5. Place the tubes in a thermoblock and heat for 5 min at 100 °C.
6. Cool the tubes in ice for 10 min.
7. Use an empty cuvette to set the spectrophotometer ABS to zero reading at 427 nm and 396 nm.
8. For each standard and sample, transfer 1 mL from the cooled tube into a cuvette and read the ABS 427 nm and 396 nm (to be used as "pre cys" values for the calculations).
9. Without removing the contents from the cuvettes, add then 32 µL of cysteine 1 M into each cuvette.
10. Close each cuvette with a cap and mix well by vortexing.
11. Wait 10 min at RT.
12. Read the ABS 427 nm and 396 nm of each cuvette (to be used as "post cys" values for the calculations). (*see* **Notes 32** and **33**).
13. For each standard and sample, calculate ΔABS using the following formula:

$$\mathrm{ABS} = \left(\mathrm{ABS}_{396}^{\mathrm{Post\ Cys}} - \mathrm{ABS}_{396}^{\mathrm{Pre\ Cys}}\right) - \left(\mathrm{ABS}_{427}^{\mathrm{Post\ Cys}} - \mathrm{ABS}_{427}^{\mathrm{Pre\ Cys}}\right)$$

Calculate a linear regression between the ΔABS and fucose standard concentrations (in the range 2–32 µg/mL) and calculate on it the concentration of methyl pentoses in the sample. If the concentration of a sample is outside the range of the calibration curve, dilute the sample with water and repeat the analysis.

3.14 HPLC-MS

The lipid A present in the sample is quantified through the 3-hydroxymyristic or 3-hydroxylauric acids present as primary esters in its structure (in Fig. 13 the lipid A structures of *S. sonnei*

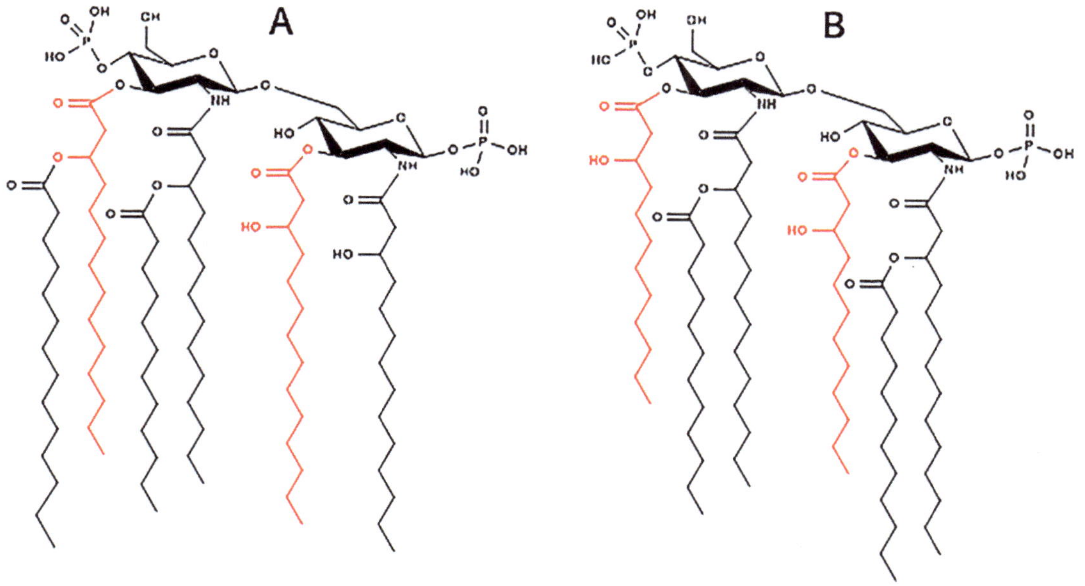

Fig. 13 Lipid A structures of *S. sonnei* (**a**) and *N. meningitidis* (**b**) reported as examples with the 3-hydroxy fatty acids (in red) quantified by HPLC-MS after the hydrolysis of ester bonds

(A) and *Neisseria meningitidis* (B) are represented as examples with the 3-hydroxy fatty acids quantified in this assay after the ester bonds hydrolysis evidenced in red). The procedure consists in an initial hydrolysis of the sample to release the 3-hydroxy-fatty acids, cleaving quantitatively the ester bonds from the glucosamine O-3 present in the lipid A.

The 3-hydroxy-fatty acid is then separated by RP-HPLC and quantified by MS detector, that is, triple quadrupole by selected reaction monitoring (SRM) or Orbitrap mass spectrometer by parallel reaction monitoring (PRM). Both samples/standards are assayed using an on-line SPE hydrophobic cartridge that allows to concentrate and desalt the analytes. The analysis, depending on the hydroxy-fatty acid type present as ester in the lipid A structure of OMV/GMMA, is performed to quantify only 3-hydroxymyristic acid (3OH-My) or 3-hydroxylauric acid (3OH-La) (hydroxy-fatty acids present as amide are not released in the hydrolysis conditions used).

An alternative method for quantification is reported in literature by Lyngby et al. [22], while Limulus amebocyte lysate (LAL) based assays are not deemed appropriate for this type of sample (*see* **Note 34**).

The parameters reported in the following paragraphs are for a system equipped with Thermo Quantum Access triple quadrupole.

1. Perform a *M/z* accuracy check in the region used (*see* **Note 35**).
2. Set up the MS instrument with the parameters as follows: Mass polarity = negative, ESI gas nitrogen, ESI spray voltage = 4000,

Table 9
Summary of the method time events

Time	Flow mL/min	Eluent A %	Eluent B %	Divert Valve position	Comment
0.00	1.00	100	0	Load	Injection, SPE wash
1.00	1.00	100	0	Load	End SPE wash
1.01	0.50	0	100	Load	Start dead volume compensation
1.75	0.50	0	100	Load	End dead volume compensation
1.76	0.25	0	100	Inject	Injection in column/elution
7.00	0.25	0	100	Inject	End elution
7.01	1.00	100	0	Load	SPE reequilibration
8.00	1.00	100	0	Load	SPE reequilibration

ESI Sheath gas pressure = 25, ESI Aux gas pressure = 10, Capillary temperature = 300 °C, Tube lens offset = −70, Skimmer Offset = 10, q cell collision gas Argon = 1 mTorr, Collision energy = 16 V, Data type = centroid.

3. Set up the MS scan event (time segment 0–10 min) with the following parameters: Q1 with 215 (for 3OH-La or 243 for 3OH-My) ± 0.02 *M/z* as precursor, Q3 with 59 ± 0.02 *M/z* as product using scan time 0.5 s (*see* **Note 36**).
4. Program the time events of the HPLC instrumental method as reported in Table 9.
5. Set up the instrument for the analysis connecting MS Divert valve, SPE cartridge, RP-column and tubes as shown in Fig. 14 closing valve position 4 with a screw cap fitting. Position the ESI cone as shown in Fig. 15 (B level).
6. To calculate the dead volume (needed only for first time setup; *see* **Note 37**) assemble the HPLC-QqQ equipment as reported in Fig. 14 connecting the ESI-MS directly to position 5 without the C_8 column. Program the HPLC pump, divert valve and QqQ spectrometer as reported in Table 10. Program the MS spectrometer to measure 3OH-La or 3OH-My according to the specific sample to be analyzed. Inject 1 μL of a standard dilution (no subjected to hydrolysis treatment) and collect the chromatogram. In Fig. 16 examples of chromatograms acquired by this way are reported.

   Consider the time needed for starting the leading of the first peak (i.e., 2.5 min in the Fig. 16) and calculate the eluent B volume pumped at that time with its flow rate (in the example: 2.5–1 = 1.5 min; 1.5 min * 0.25 mL/min = 375 μL). This is the dead volume of eluent B to be pumped after SPE wash step with eluent A before commute the divert valve. In terms of

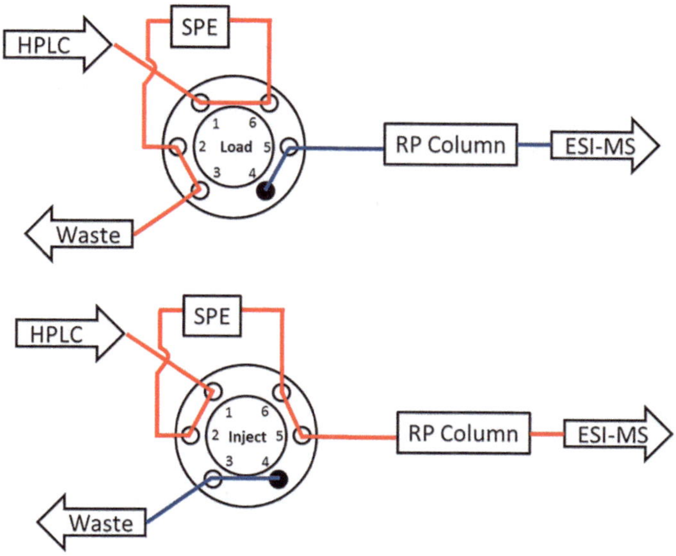

Fig. 14 SPE, RP-column and tubes connections in the two different valve positions

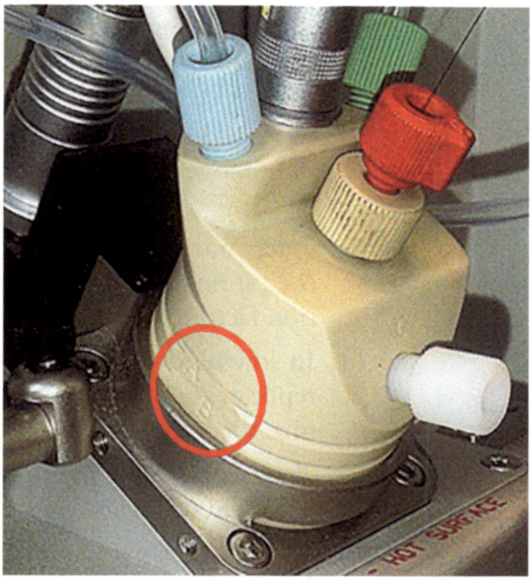

Fig. 15 Detail of ESI cone setup

time, it has to be divided by the flow rate 0.5 mL/min of the dead volume compensation step (0.75 min in the example). Adjust the time of the dead volume compensation step reported in Table 9 according to this value.

$$\text{Lasting (min) of dead volume compensation step at flow rate of } 0.5\ \text{mL/min} = \frac{(\text{first peak leading time (min)} - 1) \times 0.25}{0.5}$$

Table 10
Instrument setting for dead volume calculation

Time	Flow mL/min	Eluent A %	Eluent B %	Divert Valve position	ESI-QqQ settings
0.00	1.00	100	0	Load	–
1.00	1.00	100	0	Load	–
1.01	0.25	0	100	Inject	3OH-La/3OH-my
7.00	0.25	0	100	Inject	–
7.01	1.00	100	0	Load	–
8.00	1.00	100	0	Load	–

7. Perform a system equilibration setting the divert valve in inject position and fluxing the SPE cartridge and the column with 20 column volumes (about 3.2 mL) of eluent B. Then set the divert valve in load position and flux the SPE cartridge with 4 mL of eluent A.
8. Using low-bind tubes, prepare duplicate samples for a standard curve, starting from a vial containing 3-hydroxy-fatty acid standard (5 nmol/mL), following the scheme reported in Table 11.
9. For each sample to be analyzed, gently mix the sample in order to homogenize the content (*see* **Note 38**).
10. Using low-bind tubes, dilute the sample with water, in three independent replicates up to half of the desired dilution factor (i.e., if final dilution factor needed is 6, dilute only with dilution factor 3).
11. In low-bind tubes, transfer 200 μL of the previous dilution replicates and add 200 μL of IPA (dilution factor 2), homogenize the samples.
12. Transfer 200 μL of the previous replicates (in 50% IPA) in low-bind tubes and proceed with the following treatments.
13. To each tube containing standards and samples, add 40 μL of 1.5 M NaOH, close it, and mix by vortex.
14. Warm the tubes at 40 °C in a preequilibrated water bath or oven for 2 h.
15. Chill the tubes at 4 °C for 15 min.
16. Transfer the standards and samples in the HPLC low-bind vials.
17. Analyze standards and samples with 1 μL injections creating a sample list with the following order: six injections of the 5 nmol/mL standard (to be used just as system check; usually

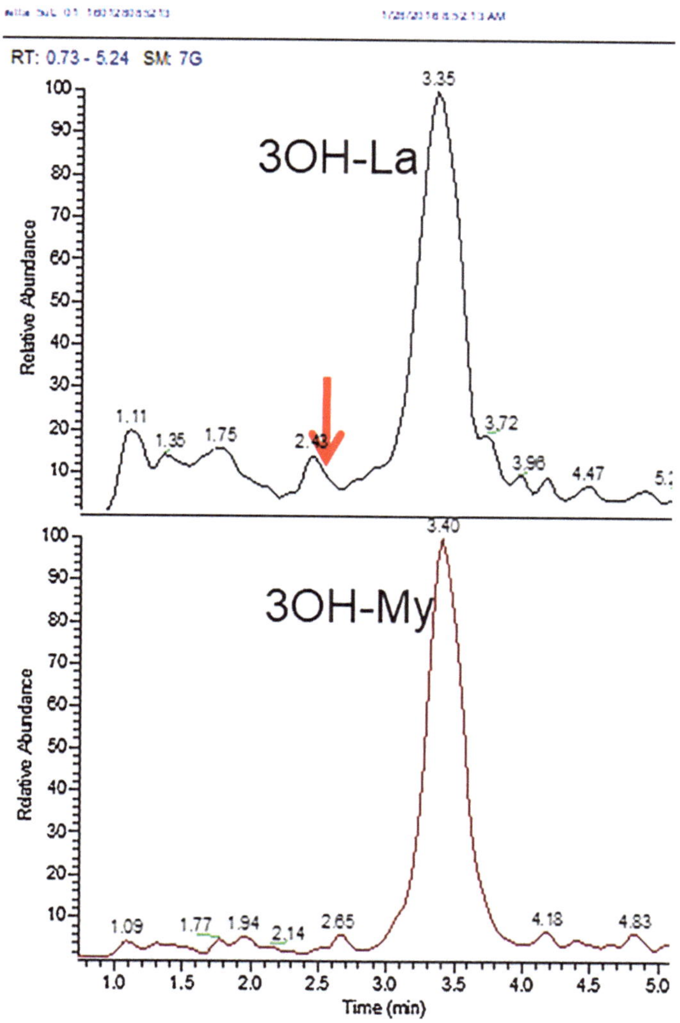

Fig. 16 Examples of chromatograms (HPLC-MS for lipid A quantification) for dead volume calculation

Table 11
Dilutions for preparing calibration curve for HPLC-MS lipid A quantification analysis

nmol/mL	5 nmol/mL STD μL	50% IPA μL
0.5	20	180
1.0	40	160
1.5	60	140
2.5	100	100
5.0	200	0

the area of the first injection is lower than the other replicates), one injection for each calibration standard in order of increasing concentration, one injection for each sample replicate, one injection for each calibration standard in order of increasing concentration. At the end of sample and standard list runs, after the last chromatographic analysis is completed, store the column with 10 column volumes (1.6 mL) of eluent E. After several analyses, Column Cleaning may be needed (*see* **Note 39**). In Fig. 17 are reported chromatograms related to 3OH-La (RT 4.71 min) and 3OH-My (RT 5.57 min).

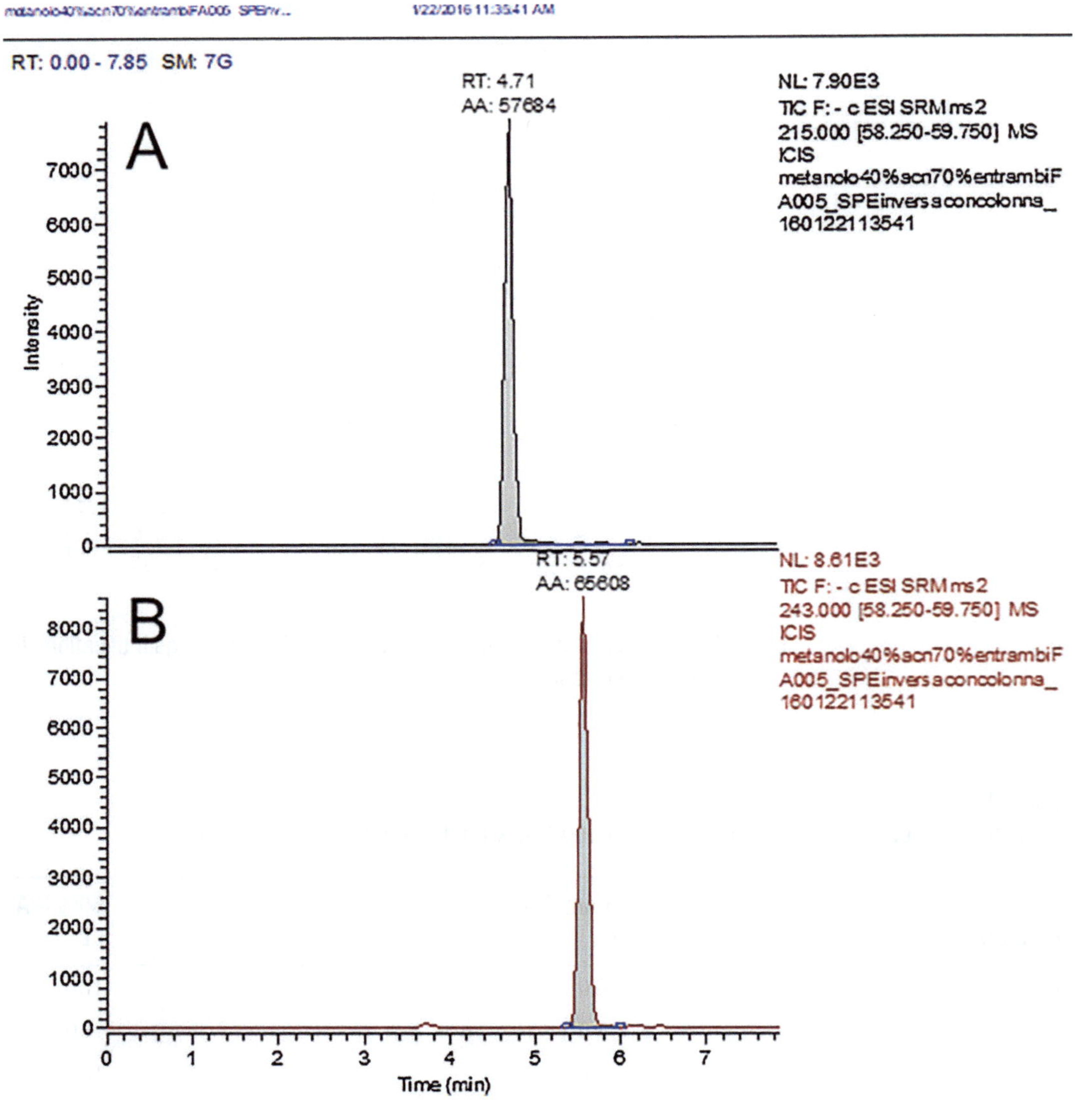

Fig. 17 Examples of chromatograms related to 3OH-La (RT 4.71 min) (**a**) and 3OH-My (RT 5.57 min) (**b**) quantification by HPLC-MS.

18. Calculate a linear regression for the 3-hydroxy-fatty acid standard between peak areas and concentrations and calculate on it the concentration of the sample analyte. Considering the structure of lipid A (that contains 2x 3OH-FA esters per molecule), lipid A amount is the half of 3OH-FA molar amount measured in the analysis. The result is expressed as nmol/mL of lipid A.

3.15 MALDI-TOF MS

1. Dilute the OMV/GMMA sample with PBS to a final concentration close to 1 mg/mL of protein.
2. In a Wheaton 2 mL screw cap vial, add 200 μL of sample and 50 μL of acetic acid 5% solution.
3. Close the container and keep it at 100 °C in a preheated oven for 2 h (*see* **Note 40**).
4. Cool the vial, vortex it, and transfer the whole content in a 2 mL Eppendorf tube.
5. Centrifuge at 14,000 × *g*, 10 °C, for 15 min and discard the supernatant.
6. Add 1.5 mL of water to the pellet and resuspend it.
7. Centrifuge at 14,000 × *g*, 10 °C, for 15 min, discard the supernatant.
8. Add 1.5 mL of water to the pellet and resuspend it.
9. Centrifuge at 14,000 × *g*, 10 °C, for 15 min, discard the supernatant.
10. Dry the pellet overnight in a centrifugal evaporator.
11. Just before deposition on MALDI plate, dissolve the whole dried pellet in 100 μL of 4:1 chloroform–MeOH by vortexing.
12. For each standard and sample to be assayed, pipet 2 μL into a 0.5 mL tube, add 2 μL of Super-DHB solution, and mix them carefully pipetting.
13. Withdraw 2 μL of the solution and load it on the target plate.
14. Let the spot dry at RT, possibly with gentle air flux (i.e., plate left under chemical hood near the front border).
15. To acquire MS Spectra, set up the spectrometer in order to work in negative, reflectron mode and acquire MS in the range 1000–3000 Da.
16. Calibrate the spectrometer by shooting with laser at the peptide standard spot coordinate.
17. For each sample, record the mass of the sample by shooting with the laser at the sample spot coordinate (*see* **Note 41**). Shoot with a laser intensity suitable for obtaining resolution of isotopic cluster peaks (usually less than 50% of its maximum), in different positions within the sample spot, summing the result spectrum until clear peaks appear over the baseline noise (Fig. 18) (*see* **Note 42**).

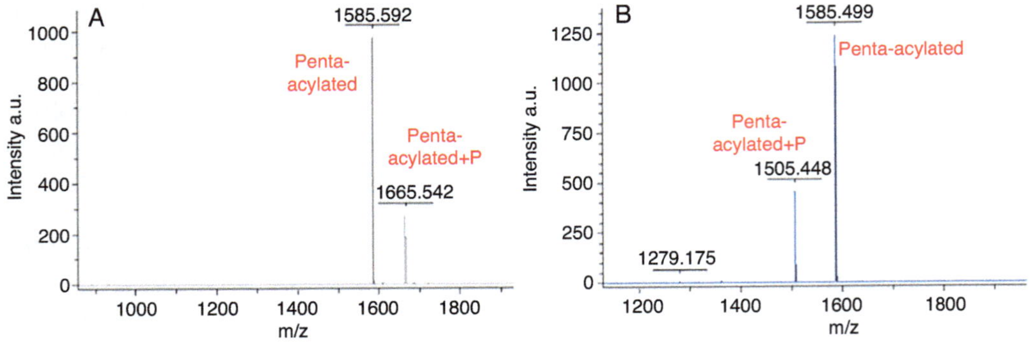

Fig. 18 MALDI-MS spectra of pentaacylated lipid A of *S.* Enteritidis GMMA isolated by mild hydrolysis in acetate buffer pH 4.5 with 3% N-Octyl-β-D-glucopyranoside preserving the lipid A structure (**a**), compared to acetic acid hydrolysis (**b**)

3.16 cELISA [15]

The cELISA working principle is based on the competition between the coating antigen and the specific antigen to quantify in OMV/GMMA samples for the antigen-specific primary antibody. The more antigen is present on OMV/GMMA, the less primary antibody can bind to the coating antigen, and the less signal can be detected by ELISA. The antigen of interest on OMV/GMMA is quantified by comparing the ELISA signal obtained with a standard curve. The standard curve is built by spiking the primary antibody with known amounts of OMV/GMMA displaying the antigen of interest.

MS analysis can also be used for quantifying specific protein antigens, as well as to determine entire OMV/GMMA protein composition [10].

1. Prepare 10 sequential dilutions (2- or 3-fold), named as Std01-Std10, of a freshly prepared OMV/GMMA sample solution, starting from a concentration previously set, based on the specific OMV/GMMA sample and antibodies used. To prepare 350 μL of each standard (sufficient for 50 μL each, plated in duplicate on each of three replicate plates), begin by weighing the appropriate volume of undiluted standard (Std01) in a tube and add the appropriate volume of dilution buffer. To prepare Std02 to Std10, weigh in the corresponding tubes the appropriate volume of previous standard and add the appropriate volume of dilution buffer.
2. Transfer 350 μL of each standard dilution point into successive wells in a row of a 1 mL deep-well plate. Use the diluted standards within 1 day.
3. Add 350 μL dilution buffer to each of two blank wells in the same row as the standards.
4. For positive controls, prepare three test dilutions of the sample used to prepare the standards, each a total of 0.8 mL in SDB in 2 mL tubes.

5. For each sample to be tested, prepare three test dilutions, each a total of 0.8 mL in SDB in 2 mL tubes.
6. Prepare 32 mL of coating solution by diluting the antigen to be used for coating the plate. Concentration depends on the specific antigen used, usually in the range 1–10 μg/mL.
7. Dispense 100 μL of coating solution in each well of 3 Nunc Maxisorp ELISA plates.
8. Seal the plates and incubate overnight at 4 °C.
9. For each of the three ELISA plates, aspirate the coating solution using a plate washer and add 200 μL of blocking buffer in each well. Wait 3 min between each of the three plates.
10. Incubate the plates at 25 °C for 1 h.
11. Before the end of the blocking step, prepare a 2× concentration primary antibody solution in SDB (dilution depends on the specific antibody used) which will be spiked with either the standard curve or the positive control or sample(s).
12. Add 350 μL of the 2× concentration of primary antibody solution to each standard dilution point into the 1 mL deep-well plate.
13. Pipette four/five times to mix the suspension.
14. Add to each of the three dilutions of the positive control, and each of the three dilutions of sample(s) to be tested, 0.8 mL of the 2× concentration of primary antibody solution and vortex the 2 mL tubes.
15. For each positive control and sample tube, dispense 400 μL/well into each of 4 wells of the deep-well plate containing the standards.
16. Incubate the deep-well plate at 4 °C in a Mixmate plate shaker. Minimum incubation time depends on the specific antigen being measured. If an overnight incubation is needed, prepare the standard, the positive control and sample dilutions the day before running the assay.
17. At the end of the blocking step (i.e., **step 10**), wash the ELISA plates.
18. Transfer 100 μL of each dilution point of the standard curve and the positive control and sample(s) from the deep-well plate to each of the three the ELISA plates. Maintain the same order in processing the three ELISA plates and allow an interval of execution time of at least 1.5 min.
19. Incubate the plates at 25 °C for 2 h.
20. Remove primary antibody solution and wash the ELISA plates with washing buffer solution.

21. Add 100 μL of secondary antibody solution to each well following the correct timing.
22. Incubate the plates at 25 °C for 1 h.
23. Approximately 10 min before the secondary antibody step ends, prepare the substrate solution by dissolving 2 tablets of Tris and p-nitrophenyl phosphate (pNPP) in 40 mL of water (substrate can be different according to the secondary antibody used).
24. As soon as the incubation of the secondary antibody times up, wash the plates and add 100 μL of substrate solution in each well, maintaining the timing between the three plates.
25. Incubate plates at 25 °C for 1 h.
26. Read the absorbances at 405 and 490 nm and calculate the difference between them (OD405nm–OD490nm).
27. Interpolate the standard curve using a four-parameter nonlinear (4PL) regression (*see* **Note 43**); the OD values on which the 4PL curve is linear represent the quantification range of the standard curve. For each plate, check if the positive control sample is within 30% of the expected value, otherwise exclude the entire plate data from the analysis. Consider the assay valid if at least two of the three plates are valid.
28. Calculate the amount of test samples by the average of all the wells (four replicates at each of the three dilutions tested and from the three different plates) with OD within the acceptable range of the curve.

3.17 OAg Extraction

OMV/GMMA treatment at low pH and high temperature results in cleavage of the labile linkage between KDO, at the proximal end of the core oligosaccharide, and lipid A, releasing the OAg-core chains or core alone in the supernatant (Fig. 10) and causing OMV/GMMA proteins and lipids to precipitate. The extracted sugars reflect the composition of LPS molecules on OMV/GMMA. They can be constituted by OAg chains attached to the core region (from LPS) and/or core region only (from lipooligosaccharides). OAg-core (simply indicated as OAg from here after) and core chains can be easily isolated (e.g., by size exclusion chromatography) and characterized in depth. Hydrolysis can be performed with acetate buffer at pH 3.9, or in 1% acetic acid, based on the OAg stability. The advantage to use 1% acetic acid is that the supernatant can be directly dried for further analysis without any desalting step (that can be eventually performed by the use a Cytiva PD10).

1. For polysaccharide extraction with acetate buffer, in a screw cap vial, add 0.9 mL of OMV/GMMA sample and 100 μL of 10× acetate buffer.

2. Keep the vial in an oven preheated at 100 °C for 100 min.
3. Transfer the whole vial content in an Eppendorf tube and centrifuge at 14,000 × *g* for 10 min.
4. Recover the supernatant that contains extracted polysaccharides (i.e., OAg, core).
5. For polysaccharide extraction with 1% acetic acid, dilute the OMV/GMMA sample with water to an appropriate concentration.
6. In a screw cap vial add 1 mL of diluted OMV/GMMA sample and 10 μL of acetic acid.
7. Keep the vial in an oven preheated at 100 °C for 100 min.
8. Transfer the whole vial content in an Eppendorf tube and centrifuge at 14,000 × *g* for 10 min.
9. Recover the supernatant that contains extracted polysaccharides (i.e., OAg, core).

3.18 HPLC-SEC SCA

OAg/core samples are analyzed by HPLC-SEC. A refractive index detector can be used to estimate apparent molecular size using dextrans (in the range 5–150 kDa) to run a calibration curve and GPC software. The samples are analyzed after derivatization with Semicarbazide to quantify the KDO present at the reducing end of OAg/core chains. This reaction is performed as a slight modification of the Semicarbazide assay for α-ketoacid determination [23] and allows calculation of molar concentration of the chains and quantification of molar ratio of populations at different size when present.

1. Using water, prepare in tubes, the dilutions of KDO ammonium salt 40 μg/mL standard solution in duplicate (as indicated in Table 12). Vortex all standard dilution tubes for few seconds.
2. Mix the OMV/GMMA sample in order to homogenize the content.

Table 12
Dilutions for preparing calibration curve for HPLC-SCA analysis

KDO (nmol/mL)	40 μg/mL $KDONH_4$ (μL)	Water (μL)
15.7	10	90
31.4	20	80
54.9	35	65
78.4	50	50
156.8	100	0

3. Transfer 100 μL of the sample into each of four different Eppendorf tubes (two to be derivatized with SCA and two to be used as blank replicates).
4. To each blank sample tube, add 100 μL of water. Vortex all tubes for few seconds.
5. To each standard tube and remaining sample tubes add 100 μL of Semicarbazide solution. Vortex all tubes for few seconds.
6. Keep all the tubes in a 50 °C preheated bath for 50 min.
7. Chill the tubes in a 2–8 °C fridge for 15 min.
8. Vortex all tubes for few seconds.
9. Transfer samples, blank samples and standards in HPLC vials.
10. Set up the instrument with G3000 PWXL guard columns connected in series, column compartment temperature set at 30 °C, autosampler compartment set at 4 °C and UV detector acquisition channel at 252 nm ABS.
11. Run the sample/standard analysis with 80 μL injection volume, 0.5 mL/min flow rate, 35 min run time, using isocratic conditions with eluent 0.1 M NaCl, 0.1 M NaH_2PO_4, 5% ACN, pH 7.2 (*see* **Note 44**). After the last chromatographic analysis is completed, store the system in NaN_3 0.02% preservative solution.
12. To quantify the KDO (nmol/mL) in the samples, correct the peak area corresponding to OAg/core in samples derivatized with Semicarbazide by subtracting the area of the corresponding blank. Quantify the amount of KDO using the calibration curve built with the peak areas of derivatized KDO standard at 252 nm (*see* Fig. 19). If the concentration of a sample is outside the range of the calibration curve, dilute the sample with water and repeat the analysis.

3.19 NMR

^{1}H NMR analysis is performed to confirm the identity of the OAg/core samples by detecting typical signals of the OAg chain and/or core region, confirming the presence of the characteristic sugars and to quantify the O-acetylation level of the polysaccharide, if present.

1. Record a ^{1}H NMR spectrum with the sample polysaccharide dissolved in 650 μL D_2O (*see* **Note 45**).
2. Add directly in the tube, after having recorded the first spectrum, 35 μL 4 M NaOD in order to obtain a final 200 mM concentration. Keep the sample at 37 °C for 2 h to achieve a complete de-O-acetylation.
3. Record a ^{1}H NMR spectrum of the de-O-acetylated sample. Figure 20 shows ^{1}H NMR spectra of *S. flexneri* 6 capsular polysaccharide isolated from GMMA.

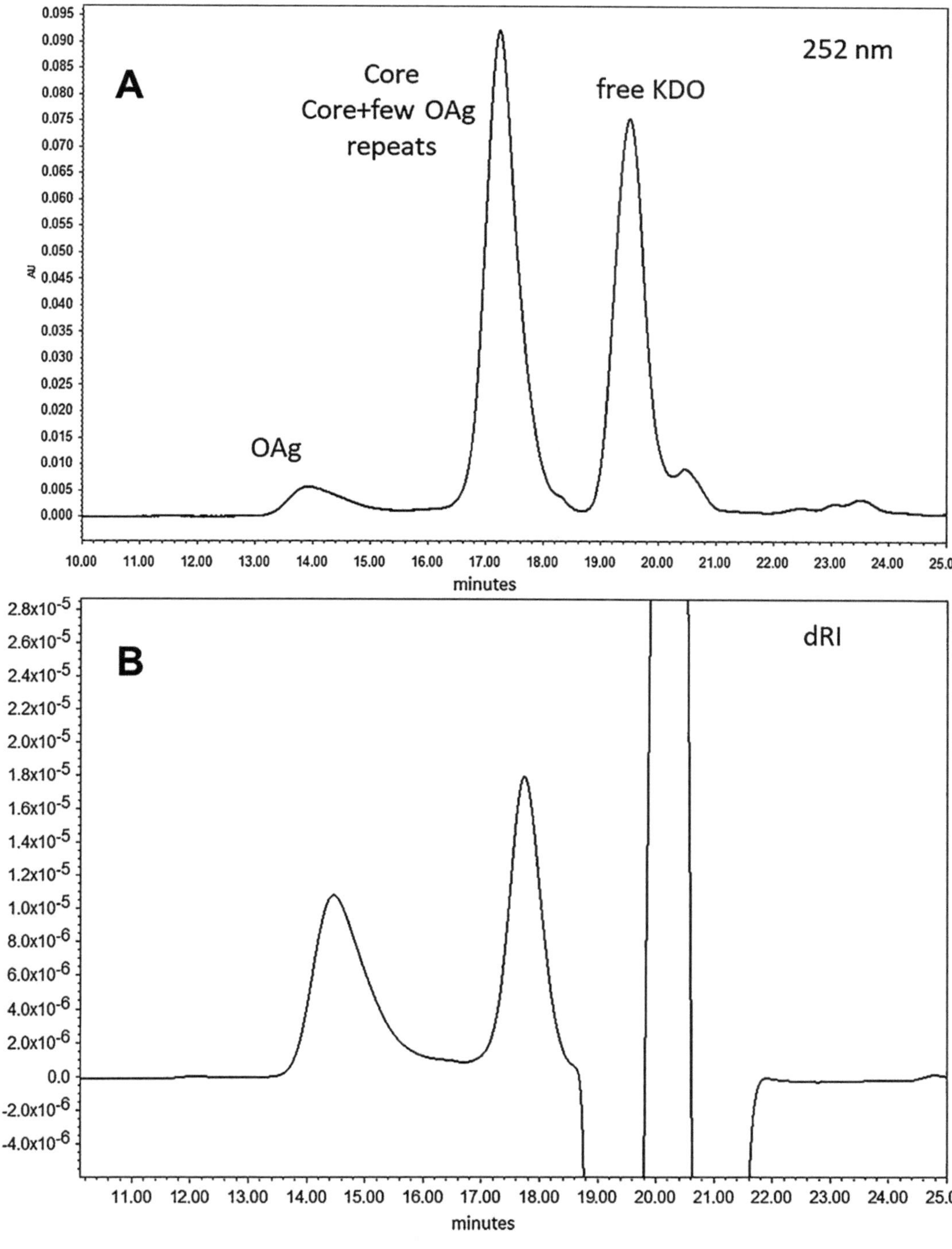

Fig. 19 HPLC-SEC chromatogram of OAg/core extracted from *S.* Typhimurium GMMA (Fig. 10): detection at 252 nm after derivatization with semicarbazide (**a**); detection in differential refractive index (dRI) of the underivatized sample (**b**)

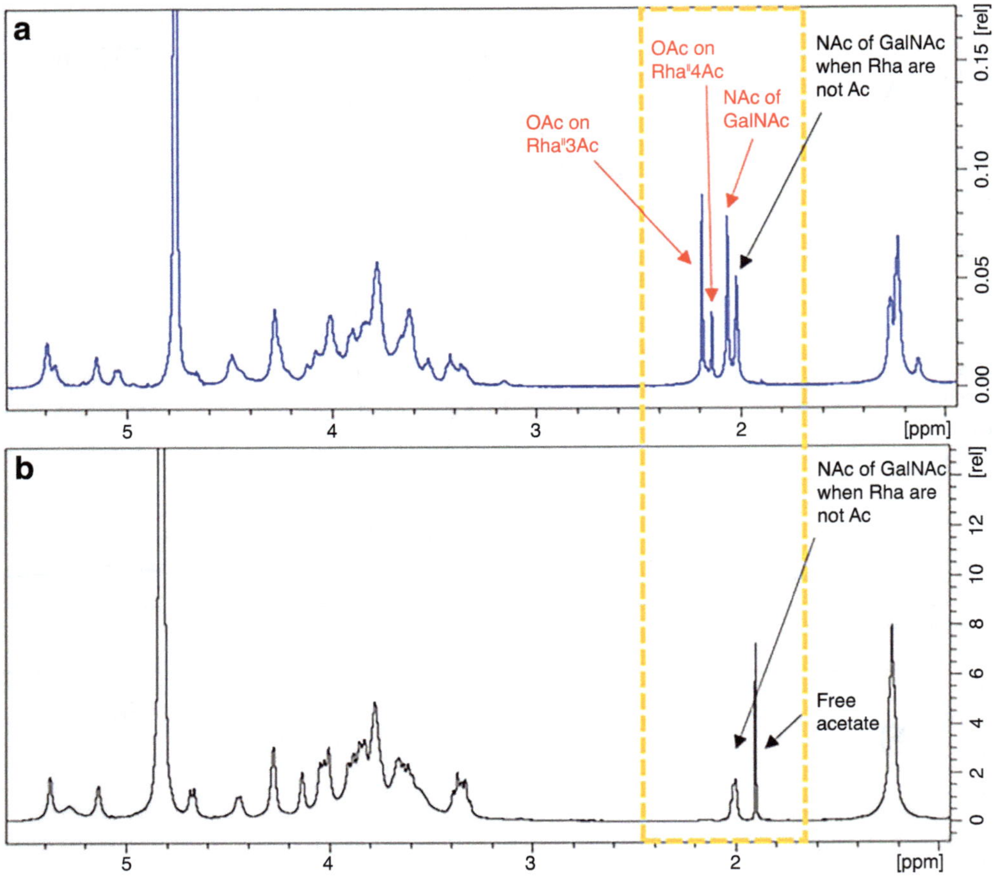

Fig. 20 ^{1}H NMR spectra of *S. flexneri* 6 capsular polysaccharide isolated from GMMA pre (**a**) and post (**b**) de-O-acetylation in NaOD

4. Calculate the O-acetylation level by comparing acetate anion signal (released after treatment with NaOD, at 1.91 ppm) and a known signal of the OAg repeating unit structure. Eventually correct the value for the acetate anion already present before the basic treatment.

3.20 Hestrin Colorimetric Method

The O-acetyl ester content can also be measured by the Hestrin colorimetric method. This method is based on the reaction of the ester groups with hydroxylamine in a basic media to form hydroxamic acid which, at low pH, generates a complex with Fe^{3+} with a maximum absorbance at 540 nm. The Hestrin method suffers from interference by common salts like phosphates; for this reason, OAg extracted samples often need to be desalted for it to work properly. O-acetylation is expressed as molar ratio (%) between acetic esters and OAg repeating units (*see* **Note 46**).

Table 13
Dilutions for preparing calibration curve for O-acetyl determination by Hestrin colorimetric method

Acetylcholine (μmol/mL)	3 mg/mL std. acetylcholine (μL)	Water (μL)
0 (blank)	0	1000
0.413	25	975
0.826	50	950
1.65	100	900
2.48	150	850
3.30	200	800
4.13	250	750

1. In a 15 mL Falcon tube, dilute 0.5 mL of 30 mg/mL acetylcholine adding 4.5 mL of water in order to obtain the standard solution at 3 mg/mL.
2. In 5 mL glass tubes prepare, in duplicate, the dilutions of acetylcholine standard solution (3.0 mg/mL) using volumes reported in Table 13.
3. In 5 mL glass tubes, add (in sequence and vortexing each time) 200 μL of diluted samples/standard curve solutions and 400 μL of the basic hydroxylamine solution.
4. Wait 3 min before proceeding to the next step.
5. Add 200 μL of 4 M HCl, 200 μL of the 0.37 M iron chloride solution and 100 μL of isopropanol.
6. Transfer all standards/samples to disposable cuvettes.
7. Read ABS of standards/samples in the following sequence with spectrophotometer wavelength set at 540 nm: first read the "blank" containing cuvette (if a single beam instrument is used this step is necessary in order to subtract the blank absorbance from samples/standards absorbances. If a double beam instrument is used, leave the "blank" containing cuvette in the reference slot during all sample/standard readings). Next, read the acetylcholine standard curve solutions (each in duplicate) from lower to higher concentrations. Finally, read the samples (each in triplicate).
8. Calculate a linear regression between the ABS and acetylcholine standard concentrations (in the range 0.413–4.13 nmol/mL) and calculate on it the concentration of the sample ester groups. If the concentration of a sample is outside the range of the calibration curve, dilute the sample with water and repeat the analysis.

4 Notes

1. With respect to the conditions reported in the text, calculate the centrifuge run time using the K-factor of the rotor in use and the maximum speed achievable (calculation tools are available in each rotor manufacturer website).
2. The reagent has to be prepared just before use.
3. All samples containing lipid A and fatty acid standards dilutions have to be placed and pipetted into low-bind tubes with low-bind pipette tips and HPLC low-bind vials have to be used.
4. Be sure that no air bubbles are present between paper and gasket and between paper and membrane and avoid complete membrane drying in all steps from this point.
5. Avoid loading border wells.
6. As an alternative to the dot blot, western blot analysis can be used. The OMV/GMMA sample is run by SDS-PAGE as described in Subheading 3.5. The gel is transferred on a membrane by using the appropriate apparatus and the recognition with a specific primary antibody is performed as described for dot blot.
7. Size distribution by intensity is preferred to measurements by number or by volume to have more reproducible results (Z-average) and because the refractive index (RI) values of particles are not known.
8. For data analysis, the viscosity and RI of the sample buffer solution (at 25 °C) needs to be indicated, the parameters for the main buffers are present in the software and a tool to calculate them for new buffer composition is present.
9. It is essential that the sample reaches the correct temperature before reading. For optimal measurement, temperatures of sample, cuvette and cuvette holder must be stable and homogenous during the reading.
10. Whenever the 260 or 280 baselines show a trend during equilibration of the column, wash for 200 min the stationary phase with a PBS solution containing 1 M NaCl. If needed, a stronger column wash (with a suitable titanium pump instrument) can be performed, with 0.1 M NaOH for 200 min. Be careful, after cleaning, to lower the pH with a phosphate buffer before storage of the column in the preservative solution.
11. MALS elaboration, for each point of the chromatogram, uses the 18 detector intensities with respect to the respective baselines to calculate the radius (peak area is not used). Be careful to choose the baseline parallel with an eventual signal evident drift.

12. Use a configuration with up-to-date light scattering detector normalization parameters (BSA monomer is injected as recommended by the instrument manufacturer to calculate these parameters).
13. This is needed, as the LS signals of many detectors are too noisy under that threshold.
14. For concentrated samples (i.e., >1 mg/mL protein), to avoid dilution accuracy errors, the sample is diluted by weight on analytical balance. Using as tare the empty dilution tube, register the weight of the sample and of added water in order to calculate the real dilution factor for each sample analyzed.
15. Check sample matrix compatibility as reported on the Thermo kit instructions.
16. Use the centrifugation conditions reported as starting point, to define the optimal conditions for a specific product a centrifugation kinetic study needs to be performed finding the conditions that allow to pellet quantitatively OMV/GMMA and not soluble proteins. The ultracentrifugation supernatant of the kinetic study can be for example analyzed by HPLC-SEC with a fluorescence detector as reported in Subheading 3.3. It is important to standardize in the centrifugation protocol also the tube filling volume.
17. OMV/GMMA tend to spontaneously resuspend with time.
18. This step can be used to change the buffer of the OMV/GMMA and to concentrate the sample by resuspension in a lower volume with respect to the starting one.
19. In case of samples containing TBS, dilute them with PBS in order to have a TBS final concentration of at maximum 1 mM to avoid interference in the assay.
20. In general, to run an amino acid determination, for each hydrolysis tube to match the amino acid calibration curves ranges, about 40 μg protein for GMMA samples are required.
21. During acid hydrolysis, tryptophan normally decomposes; cysteine and cystine, tyrosine, threonine, and serine partially decompose; methionine is partially oxidized; isoleucine and valine have lower hydrolysis rate, and therefore the recovery might be lower; asparagine and glutamine are hydrolyzed to aspartic acid and glutamic acid.
22. Once the total amino acid concentration has been obtained from the analysis, calculation of the amount of protein in the original sample is done using the following formula (to correlate the 100 μL resuspension volume to the initial dried sample volume):

$$\text{Sample protein } \mu\text{g/mL} = \frac{\text{protein conc. found in amino acid analysis} \times 100}{\text{Dried Sample initial volume } (\mu\text{L})}$$

23. Before taking up the standard solution for transfer, rinse the tip with it by pipetting up and down multiple times.
24. Before taking up the reagent for transfer, rinse the tip with it by pipetting up and down multiple times.
25. Do not add derivatizing agent to several vials and vortex them all at the end but proceed to capping and vortexing immediately for each vial.
26. If a binary UPLC system is used for the analysis, refer to the Waters Kit instruction material for the appropriate eluent scheme and gradient program.
27. Being the AMQ the product of derivatizing reagent reaction with water, if a degraded (i.e. hydrolyzed for humidity) reagent is used this test doesn't work.
28. After the last chromatographic analysis is completed, store the column in 18 mM NaOH solution.
29. In this case, a standard OAg sample is used for building the calibration curve, as the monomer (2-amino-2-deoxy-α-L-altropyranuronic acid) resulting from sample hydrolysis and quantified by the chromatographic analysis is not commercially available.
30. After the last chromatographic analysis is completed, store the column in 100 mM NaOH solution.
31. It is advisable, due to the short incubation time, to proceed in parallel with less than 15 standard/sample tubes. If the number of samples is higher, proceed with a cluster of 10 tubes for each 5 min incubation.
32. Both ABS have to be lower than 1.2 in order to be considered acceptable.
33. Do not leave concentrated sulfuric acid solution in polystyrene cuvettes but appropriately discard the solution just after usage (cuvettes are slowly corroded).
34. LAL is not deemed appropriate, as samples contain such high endotoxin units that they typically require a dilution factor of million-billion.
35. *M/z* accuracy check needs to be done at first implementation of the procedure on a new instrument or after a calibration of the MS instrument. This is required, as accuracy issues in *M/z*, especially for 59 Da fragment, were experienced. The check is performed by direct infusion of 3-OH-lauric acid/3-OH-myristic acid standard diluted in 70% ACN 0.05% FA to about 0.5–1 nmol/mL. Perform direct infusion of the standards and acquire/visualize in real time with the following parameters: First set Q1 in full scan mode and note the 215 *M/z* closest peak and/or the 243 *M/z* closest peak: each of these

two *M/z* values has to be set for the corresponding analyte precursor ion. Set QqQ with Q1 in order to filter the 3-OH-lauric acid or 3-OH-myristic acid ion using *M/z* value found in previous step, Q2 as collision cell with Argon pressure at 1 mTorr and a collision energy of 16 V, Q3 in full scan mode. Look at the 59 *M/z* closest peak: this *M/z* value has to be set for the analyte product ion (acetate). In the described method wherever the values 243, 215 and 59 Da are reported consider the actual *M/z* values found for analyte precursor ions and product ion.

36. The ESI and collision cell settings depend on the MS instrument, in this chapter are reported settings for the Thermo Quantum Access instrument. In the case of a different MS instrument, automatic optimization of ESI and collision cell parameters is needed: perform it by direct infusion of the standard diluted in eluent B.

37. The dead volume calculation is needed for this analysis because if the diverter valve commutes just after the SPE washing step, a considerable amount of eluent A is injected into the C8 RP column that results in no more equilibration in eluent B.

38. Samples and reference standards need to be in 50% IPA (v/v) to ensure complete hydrolysis of ester bonds. For this reason, the last dilution of the sample needs to be a two-fold dilution by volume in IPA.

39. Cleaning should be performed after every 15–30 samples assayed in triplicate. Run twice the program reported in Table 14 for the binary pump at 250 μL/min to wash the column. At the end of washing step, store the column with 10 column volumes (1.6 mL) of eluent E (65% ACN).

40. Depending on the purpose, a milder hydrolysis can be performed to preserve the lipid A structure (i.e., pyrophosphate groups): OMV/GMMA at 1 mg/mL in protein are kept in a screw-cap vial at 100 °C for 1 h in presence of 40 mM acetate buffer pH 4.5 and 3% N-Octyl-β-D-gluco-pyranoside.

Table 14
Column cleaning procedure

Min	Eluent C %	Eluent D %
0	75	25
2	75	25
32	0	100
35	75	25

41. The highest intensity spectra are usually recovered far from big crystal formations.
42. Using the MALDI instrument in reflectron mode, it is likely to find as highest peak exact $m/z + 1$ or exact $m/z + 2$ for the isotopic abundance of the elements present in the structure. During hydrolysis or matrix desorption, lipid A can lose one phosphate (-80 m/z) or other substituents so more than one peak can be observed for one product. Note that +22 m/z/ +38 m/z peaks, respectively, represent +Na and + K adducts.
43. $Y = ((b/(a - \text{Log10}(X)))^{\wedge}(d) - 1)/c$, where a, b, c, and d represent values of the curve parameters (i.e., respectively: a = minimum asymptote, b = Hill's slope, c = inflection point, and d = maximum asymptote), Y the amount determined, and X the absorbance reading.
44. Perform system equilibration with elution buffer before starting the analysis. This equilibration step lasts 70 min if the column system has to be equilibrated with a different eluent; 60 min or more are needed also to warm up the PDA lamps before the analysis.
45. The first ^{1}H NMR spectrum is recorded to ensure the absence of impurities or other signals at the same chemical shift of the acetate anion released after de-O-acetylation of the sample that would interfere with the quantification of the O-acetyl content.
46. Considering the multiple steps to prepare the sample for the O-acetylation measurement (OAg extraction, desalting), prior to calculating the O-acetyl ester/repeating unit ratio, it is recommended to measure the OAg quantity on the same treated sample prepared for ester measurements and use this value for the further calculation.

Acknowledgments

All authors were involved in drafting the book chapter and approved the final version. The development of this article was sponsored by GlaxoSmithKline Biologicals SA. The authors declare the following interests: FM, CG, and RAL are employees of the GSK group of companies.

References

1. Micoli F, MacLennan CA (2020) Outer membrane vesicle vaccines. Semin Immunol 50:101433. https://doi.org/10.1016/j.smim.2020.101433
2. Ellis TN, Kuehn MJ (2010) Virulence and immunomodulatory roles of bacterial outer membrane vesicles. Microbiol Mol Biol Rev 74(1):81–94. https://doi.org/10.1128/mmbr.00031-09
3. Gerke C, Colucci AM, Giannelli C et al (2015) Production of a *Shigella sonnei* vaccine based on generalized modules for membrane antigens (GMMA), 1790GAHB. PLoS One 10

(8):e0134478. https://doi.org/10.1371/journal.pone.0134478

4. van de Waterbeemd B, Streefland M, van der Ley P et al (2010) Improved OMV vaccine against *Neisseria meningitidis* using genetically engineered strains and a detergent-free purification process. Vaccine 28(30):4810–4816. https://doi.org/10.1016/j.vaccine.2010.04.082
5. Mancini F, Rossi O, Necchi F, Micoli F (2020) OMV vaccines and the role of TLR agonists in immune response. Int J Mol Sci 21(12):4416. https://doi.org/10.3390/ijms21124416
6. van der Pol L, Stork M, van der Ley P (2015) Outer membrane vesicles as platform vaccine technology. Biotechnol J 10(11):1689–1706. https://doi.org/10.1002/biot.201400395
7. Kis Z, Shattock R, Shah N, Kontoravdi C (2019) Emerging technologies for low-cost, rapid vaccine manufacture. Biotechnol J 14 (1):e1800376. https://doi.org/10.1002/biot.201800376
8. Li M, Zhou H, Yang C et al (2020) Bacterial outer membrane vesicles as a platform for biomedical applications: an update. J Control Release 323:253–268. https://doi.org/10.1016/j.jconrel.2020.04.031
9. De Benedetto G, Cescutti P, Giannelli C, Rizzo R, Micoli F (2017) Multiple techniques for size determination of generalized modules for membrane antigens from *Salmonella typhimurium* and *Salmonella enteritidis*. ACS Omega 2(11):8282–8289. https://doi.org/10.1021/acsomega.7b01173
10. Tani C, Stella M, Donnarumma D et al (2014) Quantification by LC-MS(E) of outer membrane vesicle proteins of the Bexsero® vaccine. Vaccine 32(11):1273–1279. https://doi.org/10.1016/j.vaccine.2014.01.011
11. Maggiore L, Yu L, Omasits U et al (2016) Quantitative proteomic analysis of *Shigella flexneri* and *Shigella sonnei* generalized modules for membrane antigens (GMMA) reveals highly pure preparations. Int J Med Microbiol 306(2):99–108. https://doi.org/10.1016/j.ijmm.2015.12.003
12. Micoli F, Alfini R, Di Benedetto R et al (2020) GMMA is a versatile platform to design effective multivalent combination vaccines. Vaccine 8 (3):540. https://doi.org/10.3390/vaccines8030540
13. Giannelli C, Raso MM, Palmieri E et al (2020) Development of a specific and sensitive HPAEC-PAD method for quantification of vi polysaccharide applicable to other polysaccharides containing amino uronic acids. Anal Chem 92(9):6304–6311. https://doi.org/10.1021/acs.analchem.9b05107
14. De Benedetto G, Alfini R, Cescutti P et al (2017) Characterization of O-antigen delivered by generalized modules for membrane antigens (GMMA) vaccine candidates against nontyphoidal *Salmonella*. Vaccine 35 (3):419–426. https://doi.org/10.1016/j.vaccine.2016.11.089
15. Rossi O, Aruta MG, Acquaviva A et al (2020) Characterization of competitive ELISA and formulated alhydrogel competitive ELISA (FAcE) for direct quantification of active ingredients in GMMA-based vaccines. Methods Protoc 3(3):62. https://doi.org/10.3390/mps3030062
16. Rossi O, Caboni M, Negrea A et al (2016) Toll-like receptor activation by generalized modules for membrane antigens from lipid A mutants of *Salmonella enterica* serovars typhimurium and enteritidis. Clin Vaccine Immunol 23(4):304–314. https://doi.org/10.1128/CVI.00023-16
17. Micoli F, Ravenscroft N, Cescutti P et al (2014) Structural analysis of O-polysaccharide chains extracted from different *Salmonella* typhimurium strains. Carbohydr Res 385:1–8. https://doi.org/10.1016/j.carres.2013.12.003
18. Biagini M, Spinsanti M, De Angelis G et al (2016) Expression of factor H binding protein in meningococcal strains can vary at least 15-fold and is genetically determined. Proc Natl Acad Sci U S A 113(10):2714–2719. https://doi.org/10.1073/pnas.1521142113
19. Raso MM, Gasperini G, Alfini R et al (2020) GMMA and glycoconjugate approaches compared in mice for the development of a vaccine against *Shigella flexneri* serotype 6. Vaccine 8 (2):160. https://doi.org/10.3390/vaccines8020160
20. MacLennan CA, Martin LB, Micoli F (2014) Vaccines against invasive *Salmonella* disease: current status and future directions. Hum Vaccin Immunother 10(6):1478–1493. https://doi.org/10.4161/hv.29054
21. Mani S, Wierzba T, Walker RI (2016) Status of vaccine research and development for *Shigella*. Vaccine 34(26):2887–2894. https://doi.org/10.1016/j.vaccine.2016.02.075
22. Lyngby J, Olsen LH, Eidem T, Lundanes E, Jantzen E (2002) Quantification of lipopolysaccharides in outer membrane vesicle vaccines against meningococcal disease. High-performance liquid chromatographic determination of the constituent 3-hydroxy-lauric acid. Biologicals 30(1):7–13. https://doi.org/10.1006/biol.2001.0285
23. Macgee J, Doudoroff M (1954) A new phosphorylated intermediate in glucose oxidation. J Biol Chem 210(2):617–626

Chapter 15

Production of Vaccines Using Biological Conjugation

Emily J. Kay and Vanessa S. Terra

Abstract

The production of conjugate vaccines within an *E. coli* (*Escherichia coli*) host provides an inexhaustible supply without the need for culture of pathogenic organisms. The machinery for expression of glycan and acceptor protein, as well as the coupling enzyme, are all housed within the *E. coli* chassis, meaning that there are no additional steps required for individual purification and chemical conjugation of components. In addition, there are far fewer purification steps necessary to obtain a purified glycoconjugate for use in vaccine testing. Here we describe production and purification of a HIS-tagged *Campylobacter jejuni* AcrA protein conjugated to *Streptococcus pneumoniae* serotype 4 capsule.

Key words Vaccine, Biological conjugation, *Streptococcus pneumoniae*, PGCT, PglB

1 Introduction

The use of biological conjugation, or protein glycan coupling technology (PGCT) was first exploited after the discovery of the PglB enzyme encoded within the genome sequence of *Campylobacter jejuni* [1] and its subsequent expression in *Escherichia coli* nearly 20 years ago [2]. In the intervening years, multiple improvements and refinements have been made to the initial conjugation protocols. These improvements include: introduction of a PglB consensus sequence (sequon) into the protein to be glycosylated [3], chromosomal integration of the various cloned components required for the reaction [4, 5], optimized culture conditions [6, 7], and genetic engineering of PglB to increase efficiency of glycan transfer [8]. In addition, alternative conjugation enzymes have been discovered such as PglL from *Neisseria meningitidis* [9, 10], PglS from *Acinetobacter* species [11] and Ngt from *Actinobacillus pleuropneumoniae* [12]. PglL and PglS are both O-linked oligosaccharyltransferases that are more promiscuous in terms of substrate specificity than PglB, although PglL has never

Fadil Bidmos et al. (eds.), *Bacterial Vaccines: Methods and Protocols*,
Methods in Molecular Biology, vol. 2414, https://doi.org/10.1007/978-1-0716-1900-1_15,

been demonstrated to transfer sugars with glucose at the reducing end, whereas PglS has [11]. PglL glycosylates a handful of proteins including PilE [9] and more recently an extended glycosylation recognition motif (MOOR) has been defined and used as a fusion to other proteins [10]. PglS glycosylates a single protein, ComP, and although the glycosylation site is known, the minimum section of protein required for glycosylation has yet to be defined [11]. Ngt is an N-linked oligosaccharyltransferase which acts in the cytoplasm, adding glucose directly from nucleotide activated sugar to protein [12].

There are many ways to achieve biological conjugation but some basic principles apply. In the example presented in this chapter, conjugation using CjPglB will be considered. The requirements for this enzyme are a glycan with an *N*-acetylated reducing end sugar that is built on undecaprenol pyrophosphate, and a host strain that is free of competing, compatible glycans, such as endogenous O-antigen. In addition, the host strain must possess all the necessary sugar precursors to assemble the glycan or extra biosynthetic pathways need to be added. A modified acceptor protein must be co-expressed, containing at least one accessible consensus sequon of (D/E-X-N-X-S/T; where X is any amino acid except for proline) [13], as well as the introduction of a signal peptide that targets the protein to the periplasm where the transfers occurs. The consensus sequon can either be engineered within the protein in an accessible loop, providing there is structural information for the protein in question, or added to the *N*- and/or *C*- terminus. The addition of a signal peptide sequence targets the protein to the periplasm. Though optimization is often required to find the best leader sequence, common options are those from proteins PelB and DsbA [14]. Finally, the conjugating enzyme, CjPglB, must be expressed concurrently with the glycan and acceptor protein [15]. Any of the components can be expressed from compatible plasmids or integrated into the chromosome of the host strain. Again, optimization of both plasmid backbone and inducer systems is often required.

Once all components are assembled and expressed within *E. coli*, the optimum growth and induction conditions must be determined [6]. Here we demonstrate the steps following large-scale growth, where the glycoconjugate must be harvested, purified via an affinity tag on the acceptor protein, and subjected to additional clean-up steps to remove contaminating proteins and endotoxin (Fig. 1).

In this example, AcrA, a *C. jejuni* protein with an added PelB leader sequence and two native internal glycosylation sequons, is conjugated to *Streptococcus pneumoniae* serotype 4 capsule (SP4), recombinantly expressed in *E. coli*, via the oligosaccharyl transferase enzyme, CjPglB [16]. The conjugate is purified first using a His-Trap column on an AKTA chromatography system (Fig. 2),

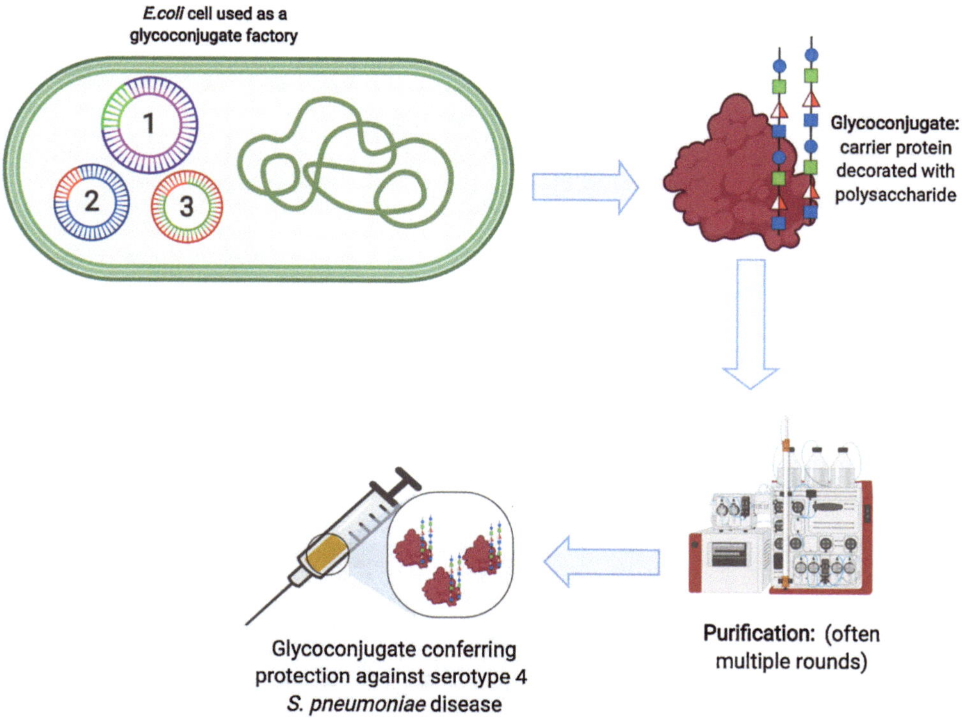

Fig. 1 Glycoengineering approach to the production and purification of glycoconjugate vaccines. An *E. coli* cell is transformed with the components needed to generate the glycoconjugate protein, shown here encoded on plasmids for (1) glycan expression; (2) target protein with leader sequence and glycosylation sequons; (3) the enzyme responsible for coupling the glycan to the protein. Any of the components could be inserted into the chromosome, as in the protocol described in this chapter where PglB (the coupling enzyme is inserted onto the chromosome. The resulting glycoconjugate must then undergo several rounds of purification to remove contaminants and unconjugated glycan. (This figure was created with BioRender.com)

followed by subsequent steps to remove endotoxin, and finally anion exchange chromatography to further purify the glycosylated protein (Fig. 3).

2 Materials

All buffers used for chromatography should be passed through 0.2 μm filter and degassed by spinning with a magnetic stir bar whilst applying vacuum for at least 30 min.

2.1 Cell Growth

1. *E. coli* W311B containing plasmids pWA2 and pB4 (*see* **Note 1**).
2. Growth medium SSOB: 2% tryptone, 0.5% yeast extract, 171 mM NaCl, 2.5 mM KCl, 10 mM $MgCl_2$, 10 mM $MgSO_4$. Weigh 40 g tryptone, 10 g yeast extract, 20 g NaCl, 372 mg KCl, 1.904 g $MgCl_2$ and 4.928 g $MgSO_4 \cdot 7H_2O$ and

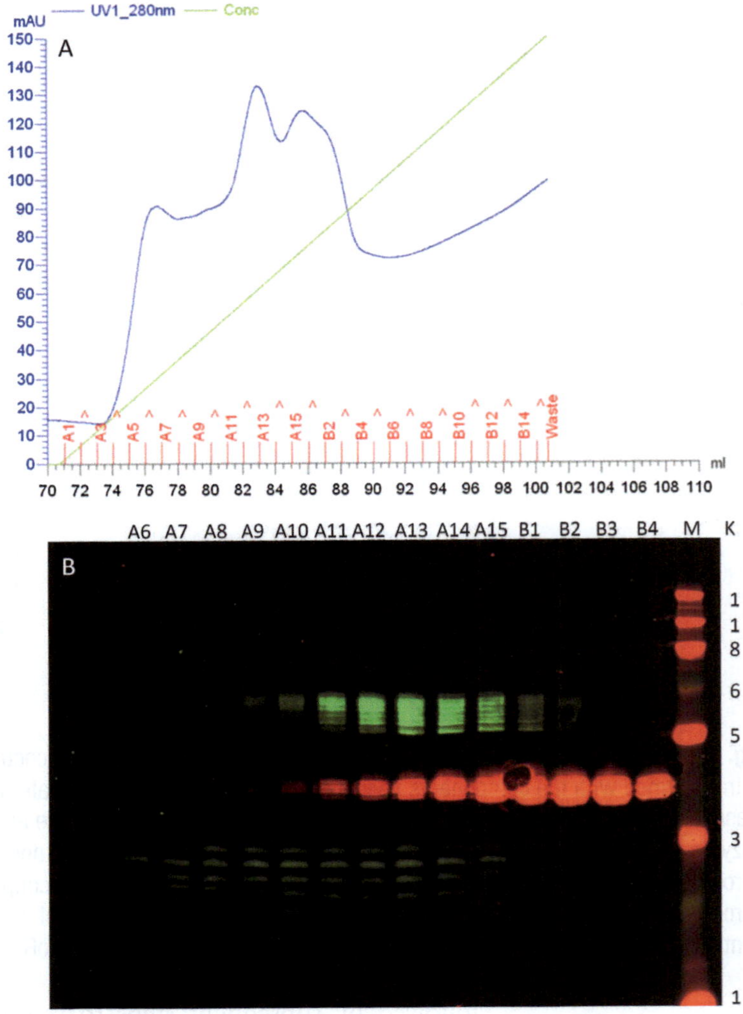

Fig. 2 (**a**) AKTA trace of AcrA-SP4 glycoconjugate eluted from the His-Trap column with imidazole. The green line denotes increasing concentration of imidazole, the *x* axis denotes the volume of liquid passed through the column, and the *Y* axis denotes the UV absorbance at 280 nm, as an indication of eluted protein concentration. Fractions are labeled by their position in the AKTA collection tray, which accommodates 15 tubes per row. The red lines indicate the 1 mL fractions collected between A1 and B14. (**b**) Selected eluate fractions separated on SDS-PAGE gel followed by Western blot with anti-His tag fluorescent antibody (red) and anti-SP4 capsule fluorescent antibody (green), as visualized by a LiCor Odyssey digital imager. *M* molecular weight marker PageRuler Plus

dissolve in 1.6 L distilled water. Once dissolved, make the volume up to 2 L with distilled water and autoclave at 121 °C with 15 psi of steam for 20 min.

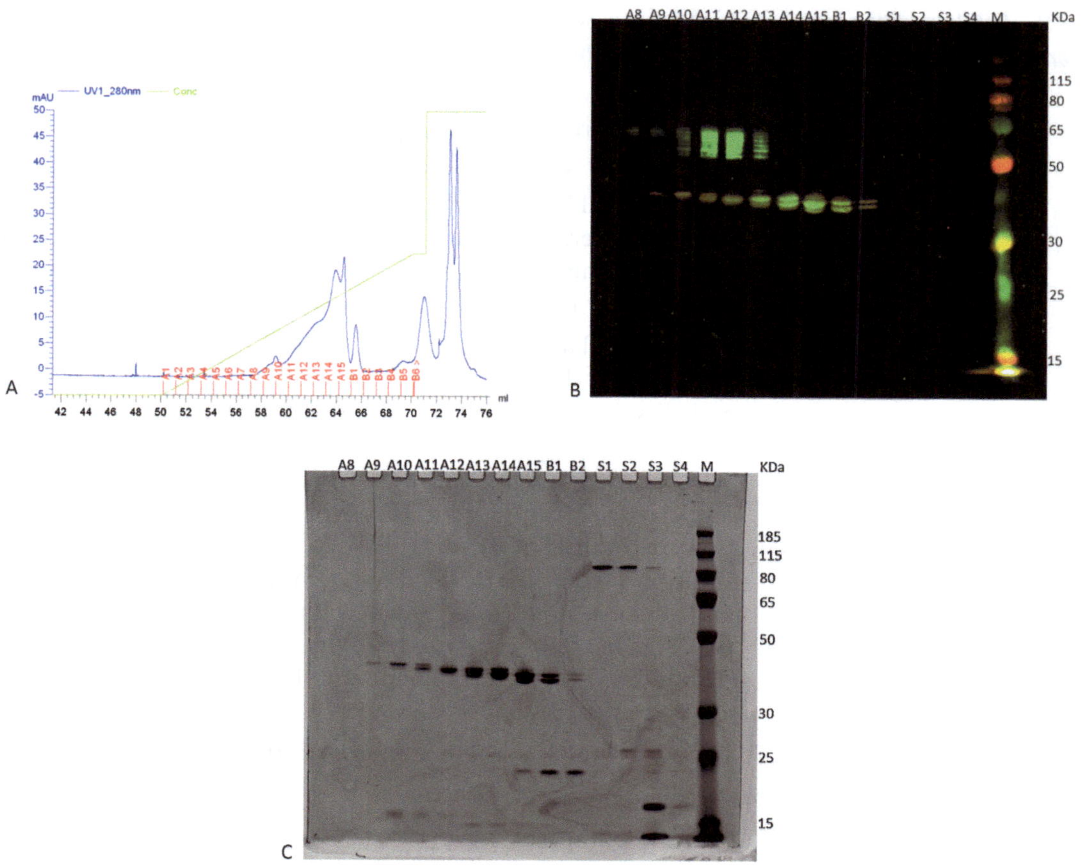

Fig. 3 (**a**) AKTA trace of AcrA-SP4 glycoconjugate eluted from the Resource S column with NaCl. The green line denotes increasing concentration of NaCl, the *x* axis denotes the volume of liquid passed through the column, and the *Y* axis denotes the UV absorbance at 280 nm, as an indication of eluted protein concentration. Fractions are labeled by their position in the AKTA collection tray, which accommodates 15 tubes per row. The red lines indicate the 1 mL fractions collected between A1 and B6. Selected eluate fractions were separated on SDS-PAGE gel followed by either (**b**) Western blot with fluorescent anti-His tag antibody (red) and fluorescent anti-SP4 capsule antibody (green), or (**c**) Coomassie staining, both visualized by a LiCor Odyssey digital imager. S1–S4 denote high salt fractions collected after the salt concentration was increased to 1 M NaCl. *M* molecular weight marker PageRuler Plus

3. Tetracycline: 10 mg/mL in 70% ethanol, filter-sterilized and stored in aliquots at −20 °C.
4. Ampicillin: 100 mg/mL in distilled water, filter-sterilized and stored in aliquots at −20 °C.
5. Inducer: 1 M IPTG, in distilled water, filter-sterilized and stored in aliquots at −20 °C.
6. PglB cofactor: 1 M $MnCl_2$, in distilled water and filter-sterilized (*see* **Note 2**).

2.2 Cell Harvesting and Lysis

1. Centrifuge with rotors capable of accommodating 500 mL containers and spinning at 6000 × *g* and 30 mL containers at 10,000 × *g*.
2. High pressure cell homogenizer (French press or similar).
3. Lysis buffer: 50 mM NaH_2PO_4, 0.3 M NaCl, pH 7.4. Weigh 3.45 g of NaH_2PO_4 and 8.77 g of NaCl and dissolve in 400 mL distilled water. Adjust the pH to 7.4 with concentrated NaOH before making the volume up to 500 mL and passing through a 0.2 μm filter membrane (*see* **Note 3**).
4. Lysozyme: 100 mg/mL in 10 mM Tris–HCl, pH 8. First prepare 1 M Tris–HCl, pH 8 by weighing 2.11 g of Tris base and dissolving it in 80 mL distilled water. Adjust the pH to 8 with concentrated HCl before making the volume up to 100 mL and passing through a 0.2 μm filter membrane, or autoclave at 121 °C with 15 psi of steam for 20 min. Next, prepare 10 mL of 10 mM Tris–HCl pH 8 by diluting 1 M solution 1:100 with distilled water. Weigh out 1 g of lysozyme and dissolve in a 10 mL final volume of 10 mM Tris–HCl, pH 8. The 100 mg/mL lysozyme solution is then passed through a 0.2 μm filter membrane before storage in 500 μL aliquots at −20 °C.
5. Benzonase nuclease high strength (250 U/μL).
6. 0.2 μm filter units.
7. Vacuum pump.

2.3 AKTA High-Performance Liquid Chromatography

1. AKTA Chromatography system.
2. 20% ethanol (1 L).
3. 0.5 N NaOH (500 mL).
4. Distilled, degassed water (1 L).
5. Peristaltic pump.
6. 0.2 μm filter units.
7. Magnetic stirrer, stir bar, and vacuum pump for degassing buffers.

2.4 HisTrap Purification

1. HisTrap 1 mL column (e.g., HisTrap FF column, Cytiva, or similar product).
2. Wash buffer I: 50 mM NaH_2PO_4, 0.3 M NaCl, 0.1% Triton X-114, pH 7.4. Weigh 1.725 g of NaH_2PO_4 and 4.385 g of NaCl and dissolve in 200 mL distilled water. Adjust the pH to 7.4 with NaOH before making the volume up to 249.75 mL and adding 250 μL of Triton X-114. Pass through a 0.2 μm filter membrane and store at 4 °C.
3. Wash buffer II: 50 mM NaH_2PO_4, 0.3 M NaCl, 20 mM imidazole, pH 7.4. Weigh 3.45 g of NaH_2PO_4, 8.77 g of

NaCl and 0.68 g of imidazole, and dissolve in 400 mL distilled water. Adjust the pH to 7.4 with NaOH before making the volume up to 500 mL and passing through a 0.2 μm filter membrane.

4. Elution buffer: 50 mM NaH_2PO_4, 0.3 M NaCl, 300 mM imidazole, pH 7.4. Weigh 1.725 g of NaH_2PO_4, 4.385 g of NaCl and 5.1 g of imidazole, and dissolve in 200 mL distilled water. Adjust the pH to 7.4 with HCl before making the volume up to 250 mL and passing through a 0.2 μm filter membrane.

2.5 Resource S Purification

1. Resource S 1 mL columns prepacked with ion exchange resin (e.g., SOURCE15S resin, Cytiva) (*see* **Note 4**).
2. Start buffer: 20 mM MES, pH 6. Weigh 1.952 g MES and dissolve in 400 mL distilled water. Adjust pH to 6 with NaOH and pass through a 0.2 μm filter membrane.
3. Elution buffer: 20 mM MES, 1 M NaCl, pH 6. Weigh 976 mg MES and 14.61 g NaCl in 250 mL. Adjust with NaOH to pH 6 before passing through a 0.2 μm filter membrane.
4. 2 M NaCl (250 mL).
5. 1 M NaOH (250 mL).
6. Storage buffer: 20% ethanol–0.2 M sodium acetate. 50 mL ethanol, 4.1 g sodium acetate, to 250 mL with water.

2.6 PD-10 Desalting and Endotoxin Removal

1. PD-10 desalting columns—containing Sephadex G-25 resin.
2. High-capacity endotoxin removal spin columns.
3. Centrifugal concentrator columns, such as Vivaspin, with a 10 kDa molecular weight cut off (MWCO).
4. PD-10 equilibration buffer: 50 mM NaH_2PO_4, 25 mM NaCl, pH 7.4. Weigh 1.725 g of NaH_2PO_4 and 0.365 g of NaCl and dissolve in 200 mL distilled water. Adjust the pH to 7.4 before making the volume up to 250 mL and passing through a 0.2 μm filter membrane.
5. Endotoxin removal buffer: 50 mM NaH_2PO_4, 0.1 M NaCl, pH 7.4. Weigh 1.725 g of NaH_2PO_4 and 1.461 g of NaCl and dissolve in 200 mL distilled water. Adjust the pH to 7.4 before making the volume up to 250 mL and passing through a 0.2 μm filter membrane.
6. Tube rotator with "Ferris Wheel" disk for end over end rotation.
7. Lyophilizer.

2.7 SDS-Page

1. Bis-Tris acrylamide gels.
2. Power pack.

3. Electrophoresis chamber and corresponding gel holder.
4. 10× MOPS running buffer: To 800 mL of distilled water add 41.86 g of MOPS free acid, 4.1 g of sodium acetate and 3.72 g of Na_2EDTA. Adjust solution to pH 7.7 using NaOH and make the total volume up to 1 L with distilled water.
5. 4× Laemmli sample buffer: 240 mM Tris–HCl, pH 6.8, 40% glycerol, 8% SDS, 0.02% Bromophenol blue. To 12 mL of distilled water add 12 mL of 1 M Tris, pH 6.8, 20 mL 100% glycerol, 4 g of SDS, 10 mg of Bromophenol blue, before making the final volume up to 50 mL. Add 100 μL of 2-mercaptoethanol per 900 μL (*see* **Note 5**).
6. Heat block that can accommodate 1 mL tubes and heat to 100 °C.

2.8 Western Blot

1. Nitrocellulose membrane.
2. Semidry transfer apparatus.
3. Power pack.
4. Tweezers.
5. Filter paper.
6. Towbin buffer: 25 mM Tris, 192 mM glycine, pH 8.3, 20% (v/v) methanol. To 50 mL distilled water add 303 mg Tris base, 1.44 g glycine and 20 mL methanol before adjusting the volume to 100 mL with distilled water (*see* **Note 6**).
7. Phosphate buffer saline (PBS): NaCl: 137 mM KCl: 2.7 mM Na_2HPO_4: 10 mM KH_2PO_4: 1.8 mM. Using a fine balance measure 8 g of NaCl, 200 mg of KCl, 1.44 g of Na_2HPO_4 and 245 mg of KH_2PO_4. Add all salts to 800 mL of distilled water. Adjust the pH to 7.4 with concentrated HCl before making the final volume up to 1 L with distilled water.
8. PBS-T (PBS 0.1% Tween): To 500 mL of PBS add 500 μL of Tween-20.
9. PBS-M: 2% milk powder in PBS, prepared freshly before use.
10. Primary antibodies: mouse anti-6×-His antibody and rabbit anti-SP4 antibody (Statens serum Institute, Denmark).
11. Secondary antibodies: IRDye® 680RD goat anti-mouse and IRDye® 800CW goat anti-rabbit conjugates.
12. Rocker or orbital shaking platform.
13. Digital imaging system for detection of fluorescence at 680 nm and 800 nm (*see* **Note 7**).

2.9 Coomassie Staining

1. Staining tray big enough to hold SDS-PAGE gel and around 30 mL liquid.

2. Coomassie staining solution: 50% MeOH, 10% HoAC, 40% H_2O, 0.25% Coomassie Blue R-250. To 80 mL distilled water add 100 mL methanol, 20 mL acetic acid, and 500 mg Coomassie Blue R-250. Stir the solution for 3–4 h and then filter through Whatman filter paper. Store at room temperature.
3. De-stain solution: 5% methanol, 7.5% acetic acid, 87.5% distilled water. To 875 mL distilled water add 50 mL methanol and 75 mL acetic acid.

3 Methods

All equipment should be appropriately cleaned according to the manufacturer's specifications and local equipment usage regulations.

3.1 Cell Growth and Expression

1. In a 5 L flask, inoculate 2 L SSOB containing appropriate supplements (20 μg/mL tetracycline; 100 μg/mL ampicillin, 1 mM IPTG, and 4 mM $MnCl_2$) with a preinduced starter culture of *E. coli* W311B+ pWA2 + pB4 (i.e., 50 mL of SSOB with 0.5 mM IPTG, 20 μg/mL tetracycline, and 100 μg/mL ampicillin, inoculated with a single colony from a plate, incubated overnight at 28 °C) to an OD_{600nm} 0.03 (*see* **Note 8**).
2. Incubate at 28 °C, shaking at 110 rpm for a total of 24 h.
3. Harvest cells by centrifugation at 5400 × *g* for 1 h at 4 °C in 500 mL centrifuge bottles.
4. Discard the supernatant and freeze pellets from each of the centrifuge buckets at −80 °C after weighing them (*see* **Note 9**).

3.2 Cell Lysis

1. Thaw pellet by submerging storage tube in cold water for 15 min.
2. Resuspend cell pellet (3–4 g wet weight) in 100 mL lysis buffer + 1 mg/mL lysozyme (1.2 mL 100 mg/mL) + 10 μL benzonase (2.5 KU) (*see* **Note 10**).
3. Lyse with a Stansted High Pressure cell homogenizer or similar high-pressure cell homogenizer. Prerinse the homogenizer with 70% ethanol, distilled water and lysis buffer before use. Sample should be passed through the high-pressure cell homogenizer as many times as required for the resuspended pellet suspension to significantly clear.
4. After lysis, centrifuge sample at 3000 × *g* for 15 min at 4 °C to pellet large debris.
5. Supernatant should be decanted into fresh centrifuge tubes and insoluble cell debris removed by centrifugation at 7800 × *g* for 1 h at 4 °C.

6. At this stage, the supernatant should be passed through a 0.2 μm filter into a suitable storage container for loading onto the HisTrap column (*see* **Note 11**).

3.3 Chromatography Purification with HisTrap Columns

1. Prepare a 1 mL HisTrap-FF column by flushing through with: 5 column volumes (CV) distilled water, followed by 5 CV lysis buffer (*see* **Note 12**).
2. Cleared and filtered cell lysate is loaded onto the prepared 1 mL HisTrap-FF column using a peristaltic pump at a rate of 1 mL/min.
3. After loading, wash the column for 1 h (~60 mL) at 4 °C with Wash I (contains 0.1% Triton-X114) (*see* **Note 13**).
4. Prepare AKTA by washing the pumps and feed lines with around 30 mL 20% ethanol and then with water, followed by 5 mL each of wash buffer II and elution buffer for pump A and B, respectively, until baseline levels.
5. Load the sample-containing HisTrap-FF column onto the AKTA, with wash buffer II running at 1 mL/min through the feed line so that you can attach the column drop-to-drop to avoid introducing air into the column.
6. Wash the column with 30 mL wash buffer II, until UV baseline is steady (collect wash fraction of UV peaks to analyze by SDS-PAGE/Western blot).
7. Elute protein from column using an imidazole gradient at 1 mL/min (feed line A = wash buffer II, feed line B = elution buffer. Change from 0–100% B over 30 mL, with 1 mL fractions collected) (*see* **Note 14**).
8. Analyze selected fractions showing increased OD_{280} (Fig. 2a) by SDS-PAGE with Coomassie staining and by Western blot for detection of His and glycan (Fig. 2b) (*see* **Note 15**).
9. Stored HisTrap columns in 20% ethanol (*see* **Note 16**).
10. Clean the AKTA with 30 mL of 0.5 M NaOH, water, and 20% ethanol before storage.

3.4 SDS-Page

1. Prepare each of the selected eluted fractions separately in Laemmli buffer (i.e., 5 μL sample, 10 μL PBS, 5 μL 4× Laemmli buffer containing mercaptoethanol) in 1 mL tubes and heat at 95 °C for 10 min. If both Coomassie staining and Western blot are desired, double the quantity of sample can be prepared and run on two identical gels.
2. Centrifuge heated samples for 3 min at maximum speed in a benchtop centrifuge to pellet any debris.
3. Dilute the 10× MOPS by adding 100–900 mL distilled water.

4. Assemble the gel in the electrophoresis chamber, fill the inner portion between the gel(s) and the gel holder with 1× MOPS running buffer. Pour the remaining 1× MOPS running buffer into the outer chamber.
5. Load 18 μL of each sample supernatant into appropriate wells of a NuPAGE 10% Bis-Tris Gel Novex® (or similar SDS-PAGE gel).
6. Run a protein ladder alongside your samples. Confirm that the protein ladder is compatible with the imaging system used.
7. Attach the electrodes to the power pack and run the gel at 100 V, surrounded by ice, for 2 h 25 min or until the dye front has just run off the bottom of the gel.
8. Remove the gel from the electrophoresis chamber and rinse with distilled water before proceeding with either Western blot or Coomassie staining.

3.5 Western Blot

1. Cut nitrocellulose membrane to the size of the gel and soak in MilliQ water for 10 min, then transfer to Towbin buffer (*see* **Note 17**).
2. Cut six pieces of filter paper to the size of the gel and soak them in Towbin buffer.
3. Assemble the transfer stack in the semi-dry transfer unit as follows, using tweezers to handle filter paper and membrane: on the anode, place three Towbin-soaked filter papers followed by the nitrocellulose membrane. Submerge the rinsed gel in Towbin buffer, then carefully place it on the membrane being careful not to move it once it has touched the membrane. At this point air bubbles should be removed by gently rolling a clean object over the top. The remaining three Towbin-soaked filter papers should be placed on top of the stack and air bubbles removed. The cathode is then placed on top and the electrodes connected to the power pack.
4. The transfer is run at 1 mA/cm^2 for 1 h before removing the membrane to a suitable clean washing tray.
5. Wash the membrane once with PBS to remove residual Towbin buffer and then incubate with PBS-M for 1 h at room temperature.
6. Pour off the PBS-M and wash the membrane three times with around 30 mL PBS-T. Gentle agitation should be applied during incubation of the wash steps by using a rocker or orbital shaker at room temperature.
7. Incubate the membrane with a mix of the two primary antibodies diluted in 10 mL PBS-M (1:10000 mouse anti-His antibody and 1:1000 rabbit anti-SP4 antibody) for 1 h at room temperature for simultaneous detection of protein (anti-His) and glycan (anti-SP4) (*see* **Note 18**).

8. Wash the membrane as in **step 6**.
9. Incubate the membrane with a mix of the two secondary antibodies, both at 1:10000 in 10 mL PBS-M, for 45 min at room temperature. The IRDye® 680RD goat anti-mouse allows for detection of recombinant proteins and IRDye® 800CW Goat anti-rabbit detection of glycosylation (*see* **Note 19**).
10. Wash the membrane as in **step 6** with a final wash in PBS only to wash the tween off the membrane before imaging.
11. Detect the fluorescent signal emitted by the secondary antibodies using a digital imager capable of fluorescence detection at 680 nm and 800 nm.

3.6 Coomassie Stain

1. Place the rinsed gel in a staining tray with around 30 mL of Coomassie stain. Place on a rocker or orbital shaker for for 2–4 h, until the gel is a uniform blue colour.
2. Pour off the Coomassie stain and de-stain using de-staining solution for 4–24 h, changing the de-stain solution periodically until background is clear and protein bands are prominent.
3. Image the gel using any image detection equipment or camera against a white background.

3.7 Desalting and Endotoxin Removal

1. Prepare an endotoxin removal column the day before use. Twist off bottom closure, loosen cap and place in 50 mL centrifuge tube. Spin at 500 × *g* for 1 min. Discard solution. Plug the bottom with the plug provided with the column and add 8 mL 0.2 N NaOH, replace cap and invert several times. Incubate overnight at room temperature.
2. Pool selected HisTrap eluate fractions from Subheading 3.3 that contain target protein, as determined by Western blot and Coomassie-stained gel in Subheadings 3.4–3.6 (Fig. 2) (*see* **Note 20**).
3. Prewash a 10 kDa MWCO centrifuge filter column (e.g., Vivaspin) twice by adding the maximum fill volume of MilliQ water at centrifuging at 4000 × *g* for 1 min, then load pooled sample onto the column and concentrate until sample volume is no more than 2.5 mL (*see* **Note 21**).
4. Use PD-10 desalting column gravity protocol. Remove storage buffer by pouring it off the column. Drip through 25 mL PD-10 equilibration buffer, then apply a maximum of 2.5 mL sample (*see* **Note 22**).
5. Discard flow through and then elute sample with 3.5 mL equilibration buffer, collecting the eluate in 1 mL fractions.
6. Clean the PD-10 column with 25 mL equilibration buffer, then store in 20% ethanol (*see* **Note 23**).

7. Pool the first 2.375 mL of eluate from PD-10 column and add 125 μL 2 M NaCl to increase the salt concentration to 0.1 M.
8. To prepare an endotoxin removal column, remove bottom plug and cap and centrifuge column at 500 × *g* 1 min, then wash with 8 mL each of 2 M NaCl, followed by water, and 3 rounds of endotoxin removal buffer, with centrifugation at 500 × *g* 1 min for each wash.
9. Plug the endotoxin removal column and add 2.5 mL sample (salt adjusted PD-10 eluate from **step 6**), replace cap and invert several times. Incubate the column at 4 °C with end-over-end mixing, on a tube rotator fitted with a "Ferris Wheel" disk, overnight.
10. Remove bottom plug and cap and centrifuge the column at 500 × *g* 1 min to elute sample off the endotoxin removal column into a fresh collection tube (*see* **Note 24**).
11. Remove salt before anion exchange by using PD-10 desalting columns as in **steps 4–6**.

3.8 Cation Exchange Chromatography

1. Prepare the Resource S column (*see* **Note 25**) by running through 5 CV each of MilliQ water, start buffer, elution buffer, and start buffer before loading sample.
2. The desalted sample in PD-10 equilibration buffer must be diluted in start buffer before loading onto the column (e.g., 3 mL sample diluted up to 35 mL with start buffer).
3. After loading, the column is attached to the AKTA and washed with 10–20 mL start buffer until the UV baseline is steady (*see* **Note 26**).
4. The sample is eluted from the column using a salt gradient up to 0.5 M NaCl (feed line A = start buffer, feed line B = elution buffer. Change from 100% A to 50% B over 20 mL, with 1 mL fractions collected).
5. The eluted fractions can be analyzed at this stage by SDS-PAGE with Coomassie staining and by Western blot as in Subheadings 3.4–3.6 (Fig. 3).
6. Clean the Resource S column on the AKTA using a high salt wash of 10 mL 1 M NaCl (this can be achieved by using only feed line B, that is, increase from 50% B to 100% B for 10 min) (*see* **Note 27**).
7. The AKTA is cleaned with 30 mL 0.5 M NaOH, distilled water, and 20% EtOH.

3.9 Buffer Exchange and Storage

1. Pool the desired fractions from Resource S column (*see* **Note 28**), concentrate to around 2.5 mL and buffer exchange using Vivaspin 10 kDa MWCO centrifuge filter column, as described in Subheading 3.7, **step 3** (again, prewash filter twice with MilliQ water before loading sample).

2. Exchange the buffer using Vivaspin 10 kDa MWCO centrifuge column by washing three times with PBS (*see* **Note 29**).
3. Determine protein concentration using any acceptable protein quantification method such as BCA (Pierce). It is also advisable to quantify the glycan (*see* **Note 30**).
4. Lyophilize in ready to use aliquots for long term storage (*see* **Note 31**).

4 Notes

1. pWA2 [17] is a derivative of pBR322 [18], which has a pMB1 origin of replication and an ampicillin resistance cassette, encoding AcrA, the protein to be glycosylated. pB4 [19] encodes the SP4 polysaccharide locus cloned into pBBR1MCS-3, which contains a tetracycline resistance cassette and a pBBR1 origin of replication, compatible with IncC, IncP, IncQ, and IncW group plasmids [20]. W311B is *Escherichia coli* W3110 [21] with PglB under an IPTG inducible promoter, integrated into the chromosome using a transposon [16]. It is essential that the plasmids are from different incompatibility groups [22, 23]. Even with different antibiotic resistance selection, incompatible plasmids will not be maintained stably within *E. coli* leading to variable copy number and even plasmid loss. This is a major issue for consistent and reproducible glycoconjugate production.
2. $MnCl_2$ must be made fresh each time of use or stored in ready to use 1 mL aliquots at −20 °C for up to 6 months as it is unstable. $MnCl_2$ is a cofactor for PglB and was shown to have a binding pocket in the active site of PglB [24, 25]. Addition of $MnCl_2$ was shown to improve glycosylation at a concentration of 4 mM [26].
3. Lysis buffer, wash buffer II and elution buffer are used at RT, wash buffer I is used at 4 °C, but all His-purification buffers may be stored at 4 °C for up to 1 month.
4. Other cation or anion exchange resin may be used but the buffers described in this protocol are specific for the Resource S columns (Cytiva). Other resins may require different buffers for binding and elution.
5. Instead of mercaptoethanol, dithiothreitol (DTT or Cleland's reagent) may be added to a final 1× concentration of 50 mM as a reducing agent. Sample buffer should not be stored with reducing agent; instead, it should be added to a working aliquot just before use.

6. Towbin buffer with methanol should not be stored. 100 mL is enough for transfer of 2 midi gels of size 8.7 cm × 13.3 cm. 10× Towbin buffer without methanol can be made in larger quantities and stored at room temperature, the buffer should be discarded once the salts visibly precipitate from solution.
7. A LI-COR Odyssey® (LI-COR Biosciences, UK) image detection system was used, but any comparable system capable of detecting the secondary antibodies may be used as appropriate.
8. If facility for incubating 2 L flasks is unavailable, then the culture may be split into smaller flasks as long as the ratio of culture to flask capacity does not exceed 2:5.
9. At this stage the pellet can be transferred to an appropriate tube and stored at −80 °C until purification.
10. If the pellet weighs more than 3–4 g wet weight, then add in a proportional amount of lysis buffer with lysozyme and benzonase.
11. It is best to load the sample onto the column as soon as possible rather than storing, as the protein is liable to degrade without the addition of protease inhibitors, which may then affect downstream applications.
12. It is advisable to keep a separate column for each glycoconjugate to be purified, but each column can be used multiple times. If the column has been previously used and stored stripped, then it must be recharged before use as follows: 5 CV distilled water, recharge with 1 mL 0.1 M $NiSO_4$, 5 CV water, 5 CV lysis buffer.
13. An on-column Triton wash is used to significantly reduce the amount of contaminating endotoxin [27]. It is necessary to perform the wash at 4 °C due to the cloud point of Triton X-114. Endotoxin removal is necessary if a functional O-antigen ligase is present and a further anionic exchange purification step is needed. The O-antigen ligase will attach recombinant glycan to the lipid A core which will then co-purify with the glycoconjugate protein. If the endotoxin removal step is not taken, the lipid-linked glycan will block the anion exchange column. Following the purification, a Limulus amebocyte lysate (LAL) assay may be used to determine how much endotoxin remains in the sample. There are many commercial kits available for this.
14. The gradient can be altered according to the protein to be eluted. It is best to collect at least 5 mL of eluate before the protein elutes. This allows for contaminating proteins to be run off the column before elution of the protein of interest. The gradient is achieved by increasing the % of elution buffer running through the column up to 100%.

15. As the protein AcrA has two glycosylation sites there should be two bands visible above the unglycosylated protein, denoting short repeat units attached at one or both glycosylation sites. In addition, glycan bands should be visible above the protein, showing polymerized glycan attached to the protein. Control samples of protein only and glycan only can be included in Western blots for comparison. Any glycan visible at the bottom of the gel will be glycan that is not attached to the target protein.
16. It is recommended to store columns stripped and only recharge with Ni^{2+} before use. A full cleaning of columns is done using 5–10 CV water, 5–10 CV stripping buffer, 60 CV 1 M NaOH (contact time at least 1 h), 5–10 CV wash II buffer and 5–10 CV 20% ethanol (protocol can be found in the manufacturer's manual). Stripping Buffer: 20 mM NaH_2PO_4, 0.5 M NaCl, and 50 mM EDTA at pH 7.4. To 200 mL of distilled water add 0.69 g of NaH_2PO_4, 7.305 g of NaCl, and 3.653 g of EDTA. Adjust to pH 7.4 with NaOH and make up to volume to 250 mL with distilled water. The EDTA will not dissolve until the pH is adjusted.
17. We recommend use of Hybond™-C Extra nitrocellulose membrane (Amersham Biosciences, UK) for better results.
18. A 10 mL volume of primary antibody solution in PBS-M is generally enough to completely cover a membrane in a tray 11 × 14 cm. If antibody is scarce then a heat-sealable bag, the size of the membrane, may be used to reduce the volume of the primary antibody solution required to completely cover the membrane. The aim is to cover the membrane with solution at all times without air bubbles crossing the surface. In the experiment described in this chapter, both primary antibodies are added for simultaneous detection, which requires a detection system that can detect two different secondary signals at once (e.g., fluorescence). If this is not available, the primary antibodies should be added separately.
19. There are different detection systems that can be used for Western blotting. For example, horseradish peroxidase, HRP-conjugated secondary antibodies may be used, followed by an enhanced chemiluminescence (ECL) detection kit. The membrane can then be exposed to X-ray film in the dark and then developed. The disadvantage is that only one primary antibody can be used at a time, so either two gels for each sample could be run, or the membrane stripped and reprobed.
20. Different fractions may be pooled according to downstream application or desired ratio of glycosylated to unglycosylated protein. As the sample will be further purified by ion exchange chromatography, it is preferable to pool as many fractions as

possible that contain visible polymerized glycan (i.e., visible green banding above the protein). Where a shift in protein is seen this denotes addition of smaller repeats added at either one or two of the available glycosylation sites. With reference to Fig. 2b, fractions A10-B2 would be pooled.

21. If the sample volume exceeds the maximum fill volume for the column then it may be added after the initial spin has reduced the volume, however, it is preferable to use a column with a larger capacity to start with. The sample will remain in the concentrator filter so the eluate may be discarded or kept for analysis as desired. A Vivaspin column at 4 °C will concentrate very slowly. If concentration time exceeds 3 h, the temperature can be increased from 4 °C to 10 °C. If the protein concentration is high as determined by Coomassie staining of the fractions before concentration, then a larger capacity concentrator column may be used and concentration stopped at a higher volume. If the sample is over concentrated then aggregation and precipitation of the protein may occur.

22. If the sample volume is less than 2.5 mL, first add the sample (noting the volume) to the column and wait until it has entered the resin bed, then add an amount of equilibration buffer so that the total volume of sample plus buffer added to the column is 2.5 mL. This avoids diluting the sample.

23. The PD-10 columns can be reused several times but it is recommended to use them with the same sample to avoid cross-contamination.

24. The endotoxin removal columns can be reused at least five times. To clean them, regenerate with 8 mL 0.2 N NaOH overnight. Elute by centrifugation at 500 × *g* for 1 min, then wash with 8 mL 2 M NaCl, followed by 8 mL water, centrifuging as indicated after each wash. Columns are stored at 4 °C, after adding 8 mL 20% ethanol to the column and plugging the column with the provided caps for the top and bottom.

25. Choice of ion exchange column is dependent on the pI of the protein to be purified, use Resource Q (anion) for proteins with a pI of 1–11 and Resource S (cation) for proteins with a pI of 3–14, according to the manufacturer's instructions. Additionally, the pH of the start and elution buffers will vary depending on protein. In the case of AcrA, the pI is 7, so Resource S was used with buffers at pH 6. AcrA was not stable and precipitated in Resource Q buffer at pH 8. As a starting point, Resource S buffers are used at 1 pH unit below the pI of the protein and Resource Q buffers at 1 pH unit above the pI of the protein. From the starting point, the pH can be further optimized to achieve the best yield versus purity of your sample, as appropriate.

26. The maximum pressure for the Resource S column is 1.5 MPa, therefore flow rates should be adjusted so as not to exceed this pressure.

27. The Resource S column is cleaned with 5–10 CV each of 2 M NaCl, 1 M NaOH, 2 M NaCl, distilled water, start buffer, distilled water, and 20% ethanol–0.2 M sodium acetate. If the column is blocked, as indicated by no flow through the column at the maximum pressure tolerance for the column, then inject with 1 mg/mL pepsin in a solution of 0.5 M NaCl and 0.1 M acetic acid, and leave overnight before flushing through with 5–10 CV of distilled water and cleaning as above. Once the column is in 20% ethanol–0.2 M sodium acetate it may be stored at 4 °C until next use.

28. As with the initial His-purified sample, different fractions may be pooled according to downstream application or desired ratio of glycosylated to unglycosylated protein. In this example, fractions A9–A15 (Fig. 3) were pooled. This was to capture as much protein as possible that is glycosylated while minimizing the proportion of contaminating protein (i.e., protein bands that appear on the Coomassie stained gel that are absent from the Western blot).

29. Vivaspin columns can be used for buffer exchange by diluting the concentrated sample up to the original volume with PBS and concentrating again. Repeating the process three times. If preferred, alternative buffer exchange methods may be used (e.g., dialysis cassettes).

30. Sugar content can be quantified using, for example, hot phenol–sulfuric acid method [28] or high-performance anion-exchange chromatography with pulsed amperometric detection (HPAEC–PAD) on a Dionex system [29].

31. Lyophilizing glycoconjugates suspended in PBS will result in salts being retained in the sample, therefore water should be used for reconstitution. Buffer exchange into 50 mM ammonium bicarbonate buffer may be preferable, because it sublimes into ammonia, water, and carbon dioxide upon lyophilization, often without causing measurable harm to the protein. A 50 mM ammonium bicarbonate buffer, pH 7.8, is made by dissolving 4 g NH_4HCO_3 in 800 mL distilled water, then making the final volume up to 1 L. If ammonium bicarbonate buffer is used then the sample should be reconstituted in PBS. If a lyophilizer is not available, then the purified glycoconjugate may be stored at −80 °C, but freeze–thaw cycles should be avoided.

References

1. Parkhill J, Wren BW, Mungall K et al (2000) The genome sequence of the food-borne pathogen *Campylobacter jejuni* reveals hypervariable sequences. Nature 403(6770):665–668. https://doi.org/10.1038/35001088
2. Wacker M, Linton D, Hitchen PG et al (2002) N-linked glycosylation in *Campylobacter jejuni* and its functional transfer into *E. coli*. Science 298(5599):1790–1793. https://doi.org/10.1126/science.298.5599.1790
3. Fisher AC, Haitjema CH, Guarino C et al (2011) Production of secretory and extracellular N-linked glycoproteins in *Escherichia coli*. Appl Environ Microbiol 77(3):871–881. https://doi.org/10.1128/aem.01901-10
4. Yates LE, Natarajan A, Li M et al (2019) Glyco-recoded *Escherichia coli*: Recombineering-based genome editing of native polysaccharide biosynthesis gene clusters. Metab Eng 53:59–68. https://doi.org/10.1016/j.ymben.2019.02.002
5. Strutton B, Jaffé SRP, Pandhal J, Wright PC (2018) Producing a glycosylating *Escherichia coli* cell factory: the placement of the bacterial oligosaccharyl transferase pglB onto the genome. Biochem Biophys Res Commun 495 (1):686–692. https://doi.org/10.1016/j.bbrc.2017.11.023
6. Kämpf MM, Braun M, Sirena D et al (2015) In vivo production of a novel glycoconjugate vaccine against *Shigella flexneri* 2a in recombinant *Escherichia coli*: identification of stimulating factors for in vivo glycosylation. Microb Cell Factories 14(1):12. https://doi.org/10.1186/s12934-015-0195-7
7. Caillava AJ, Ortiz GE, Melli LJ et al (2019) Improving bioreactor production of a recombinant glycoprotein in *Escherichia coli*: effect of specific growth rate on protein glycosylation and specific productivity. Biotechnol Bioeng 116(6):1427–1438. https://doi.org/10.1002/bit.26953
8. Ihssen J, Haas J, Kowarik M et al (2015) Increased efficiency of *Campylobacter jejuni* N-oligosaccharyltransferase PglB by structure-guided engineering. Open Biol 5 (4):140227–140227. https://doi.org/10.1098/rsob.140227
9. Faridmoayer A, Fentabil MA, Haurat MF et al (2008) Extreme substrate promiscuity of the *Neisseria* oligosaccharyl transferase involved in protein O-glycosylation. J Biol Chem 283 (50):34596–34604. https://doi.org/10.1074/jbc.M807113200
10. Pan C, Sun P, Liu B et al (2016) Biosynthesis of conjugate vaccines using an O-linked glycosylation system. mBio 7(2):e00443–e00416. https://doi.org/10.1128/mBio.00443-16
11. Harding CM, Nasr MA, Scott NE et al (2019) A platform for glycoengineering a polyvalent pneumococcal bioconjugate vaccine using *E. coli* as a host. Nat Commun 10(1):891. https://doi.org/10.1038/s41467-019-08869-9
12. Naegeli A, Neupert C, Fan Y-Y et al (2014) Molecular analysis of an alternative N-glycosylation machinery by functional transfer from *Actinobacillus pleuropneumoniae* to *Escherichia coli*. J Biol Chem 289 (4):2170–2179. https://doi.org/10.1074/jbc.M113.524462
13. Kowarik M, Young NM, Numao S et al (2006) Definition of the bacterial *N*-glycosylation site consensus sequence. EMBO J 25 (9):1957–1966. https://doi.org/10.1038/sj.emboj.7601087
14. Georgiou G, Segatori L (2005) Preparative expression of secreted proteins in bacteria: status report and future prospects. Curr Opin Biotechnol 16(5):538–545. https://doi.org/10.1016/j.copbio.2005.07.008
15. Cuccui J, Wren B (2015) Hijacking bacterial glycosylation for the production of glycoconjugates, from vaccines to humanised glycoproteins. J Pharm Pharmacol 67(3):338–350. https://doi.org/10.1111/jphp.12321
16. Herbert JA, Kay EJ, Faustini SE et al (2018) Production and efficacy of a low-cost recombinant pneumococcal protein polysaccharide conjugate vaccine. Vaccine 36 (26):3809–3819. https://doi.org/10.1016/j.vaccine.2018.05.036
17. Feldman MF, Wacker M, Hernandez M et al (2005) Engineering N-linked protein glycosylation with diverse O antigen lipopolysaccharide structures in *Escherichia coli*. Proc Natl Acad Sci U S A 102(8):3016–3021. https://doi.org/10.1073/pnas.0500044102
18. Sutcliffe JG (1978) Nucleotide sequence of the ampicillin resistance gene of *Escherichia coli* plasmid pBR322. Proc Natl Acad Sci U S A 75(8):3737–3741. https://doi.org/10.1073/pnas.75.8.3737
19. Kay EJ, Yates LE, Terra VS et al (2016) Recombinant expression of *Streptococcus pneumoniae* capsular polysaccharides in *Escherichia coli*. Open Biol 6(4):150243. https://doi.org/10.1098/rsob.150243

20. Kovach ME, Elzer PH, Hill DS et al (1995) Four new derivatives of the broad-host-range cloning vector pBBR1MCS, carrying different antibiotic-resistance cassettes. Gene 166 (1):175–176. https://doi.org/10.1016/0378-1119(95)00584-1
21. Bachmann BJ (1996) Derivations and genotypes of some mutant derivatives of *Escherichia coli* K-12. In: Neidhardt FC (ed) *Escherichia coli* and *Salmonella*: cellular and molecular biology, 2nd Ed edn. ASM Press, Washington, DC, pp 2460–2488
22. Thomas CM (2014) Plasmid incompatibility. In: Bell E (ed) Molecular life sciences: an encyclopedic reference. Springer New York, New York, NY, pp 1–3. https://doi.org/10.1007/978-1-4614-6436-5_565-2
23. Carattoli A (2009) Resistance plasmid families in *Enterobacteriaceae*. Antimicrob Agents Chemother 53(6):2227–2238. https://doi.org/10.1128/AAC.01707-08
24. Lizak C, Gerber S, Numao S et al (2011) X-ray structure of a bacterial oligosaccharyltransferase. Nature 474(7351):350–355. https://doi.org/10.1038/nature10151
25. Napiórkowska M, Boilevin J, Darbre T et al (2018) Structure of bacterial oligosaccharyltransferase PglB bound to a reactive LLO and an inhibitory peptide. Sci Rep 8(1):16297. https://doi.org/10.1038/s41598-018-34534-0
26. Garcia-Quintanilla F, Iwashkiw JA, Price NL et al (2014) Production of a recombinant vaccine candidate against *Burkholderia pseudomallei* exploiting the bacterial *N*-glycosylation machinery. Front Microbiol 5:381. https://doi.org/10.3389/fmicb.2014.00381
27. Zimmerman T, Petit Frère C, Satzger M et al (2006) Simultaneous metal chelate affinity purification and endotoxin clearance of recombinant antibody fragments. J Immunol Methods 314(1):67–73. https://doi.org/10.1016/j.jim.2006.05.012
28. Masuko T, Minami A, Iwasaki N et al (2005) Carbohydrate analysis by a phenol–sulfuric acid method in microplate format. Anal Biochem 339(1):69–72. https://doi.org/10.1016/j.ab.2004.12.001
29. Talaga P, Vialle S, Moreau M (2002) Development of a high-performance anion-exchange chromatography with pulsed-amperometric detection based quantification assay for pneumococcal polysaccharides and conjugates. Vaccine 20(19):2474–2484. https://doi.org/10.1016/S0264-410X(02)00183-4

Chapter 16

Immunological Assessment of Lung Responses to Inhalational Lipoprotein Vaccines Against Bacterial Pathogens

Anneliese S. Ashhurst, Cameron C. Hanna, Richard J. Payne, and Warwick J. Britton

Abstract

Lipopeptides or lipoproteins show potential as safe and effective subunit vaccines for protection against bacterial pathogens. Provided suitable adjuvants are selected, such as the TLR2-stimulating molecules Pam_2Cys and Pam_3Cys, these may be formulated as inhalational vaccines to optimize localized pulmonary immune responses. Here, we present methods to assess antigen-specific memory lymphocyte responses to novel vaccines, with a focus on immune responses in the lung tissue and bronchoalveolar space. We describe detection of T-cell responses via leukocyte restimulation, followed by intracellular cytokine staining and flow cytometry, enzyme-linked immunosorbent spot assay (ELISpot), and sustained leukocyte restimulation for detection of antigen-specific memory responses. We also detail assessment of antibody responses to vaccine antigens, via enzyme-linked immunosorbent assay (ELISA)-based detection. These methods are suitable for testing a wide range of pulmonary vaccines.

Key words Tuberculosis, Inhalational vaccination, Lungs, Bronchoalveolar lavage, Lipoprotein, ELISpot, ELISA, Intracellular cytokine staining, Flow cytometry, Leukocyte restimulation

1 Introduction

The concept of vaccination has been one of the most transformative and impactful innovations in the history of medicine. Vaccines have eliminated the devastating smallpox virus from our communities and have drastically reduced the incidence of many other bacterial and viral diseases that previously caused mass morbidity and mortality. The global pandemic caused by the novel severe acute respiratory syndrome coronavirus 2 (SARS-CoV-2), that emerged in late 2019, has for many brought into perspective our critical reliance on vaccines to control infectious disease [1]. There remain, however, many infectious diseases for which we do not yet have safe and highly effective vaccines. This includes the bacterial pathogen

Fadil Bidmos et al. (eds.), *Bacterial Vaccines: Methods and Protocols*,
Methods in Molecular Biology, vol. 2414, https://doi.org/10.1007/978-1-0716-1900-1_16,

Mycobacterium tuberculosis, the etiological agent of tuberculosis (TB), which remains the single biggest infectious disease killer throughout history [2].

To date, there is only one licensed vaccine for TB, *Mycobacterium bovis* Bacille Calmette-Guérin (BCG), which has been widely used in TB control programs since 1921. While it has proven efficacy in reducing severe childhood forms of TB, including extrapulmonary manifestations such as TB meningitis, it fails to provide reliable protection against pulmonary TB in adolescents and adults. In addition, as a live attenuated mycobacterial vaccine, BCG is not suitable for use in immunocompromised individuals, limiting its use [3, 4]. There is an urgent need to continue vaccine development to provide safe and more effective immunizations against this dangerous pathogen.

One promising approach that offers improved safety compared to live attenuated or viral-vectored vaccines is subunit vaccination, including vaccines that utilize antigenic peptide epitopes or proteins as antigens. These require coadministration of an immune-activating adjuvant in order to provoke a sufficient protective immune response. In order to optimize the effect of antigen–adjuvant coadministration, one strategy involves covalent conjugation of these components to form a self-adjuvanting vaccine. Conceptually, a self-adjuvanting vaccine optimizes the timing of antigen–adjuvant delivery to antigen presenting cells (APCs). This is beneficial for providing APCs the requisite secondary costimulation at the same moment they are receiving and processing antigen for presentation to T-cells. Moreover, this strategy is thought to enhance phagocytic uptake of vaccine antigens, as pattern recognition receptor (PRR) ligands can induce receptor-mediated phagocytosis [5]. This strategy has been demonstrated extensively by us and others for indications such as cancer [6–12], allergy [13], and bacterial infections [14–17]. For example, Rai et al. reported the use of a multistage biepitope vaccine conjugated to the TLR2 agonist, Pam_2Cys, which elicited significant $CD4^+$ and $CD8^+$ T-cell immunity and reduced *M. tuberculosis* burden in mice to a greater extent than the BCG vaccine [16].

We have recently reported the synthesis, immunological and efficacy testing of lipopeptide and lipoprotein-based self-adjuvanting vaccines, containing *M. tuberculosis* proteins or T cell epitopes, in preclinical murine models. The antigens were synthetically produced by solid-phase peptide synthesis and conjugated covalently to the TLR2-targeting adjuvants Pam_2Cys and Pam_3Cys using a peptide ligation strategy, providing high purity self-adjuvanted vaccine constructs [18, 19]. We opted for a totally synthetic strategy as opposed to recombinant protein expression in order to avoid contaminants and enable access to highly pure products. Moreover, most adjuvants are hydrophobic which creates difficulties when fusing them to highly hydrophilic peptides and proteins. A chemical synthesis approach can alleviate these issues as conjugation of the adjuvant to the peptide can take place in organic solvent.

We have also demonstrated immunological advantages of delivering vaccines such as these to the pulmonary mucosa via inhalation to provide a greater degree of localized protection [18, 20]. Growing evidence indicates that pulmonary or mucosal vaccination strategies will enhance localized lung immune responses against *M. tuberculosis* and other respiratory bacterial or viral infections [15, 21–24], and this is a rapidly developing field of vaccine research. We encourage investigators to explore different routes of immunization when developing novel vaccines, and consider matching the route of immunization to the route of transmission of the pathogen. However, not all types of vaccines are suitable for delivery by all routes, so careful selection is required. Lipopeptide or protein-based vaccines may be particularly suitable for mucosal delivery, as they do not carry the risks associated with live vaccines and can be used in individuals with compromised immune systems. Importantly, unlike live or viral vectored vaccines, protein-based vaccines may be given repeatedly to boost immunity and may also be readily formulated as dry powders to facilitate delivery as an inhalational vaccination [20].

In this chapter, we detail commonly used methods to perform assessment of vaccine-induced peptide or protein-specific memory immune responses, focusing on responses in the lung. These must be examined in an experimental animal model, with mice being the most frequently used. As an exemplar, we present assessment of peptide or protein-specific responses in murine models of pulmonary vaccination against *M. tuberculosis*. The materials and methods may, however, be adapted to assess the immune response against any peptide or protein vaccine antigen of interest. We recommend analyzing both localized and systemic antigen-specific immune responses to vaccination, that is, the immune response at the site of immunization, as well as the immune response in blood and secondary or peripheral lymphoid organs. Isolation of leukocytes from nonlymphoid tissue presents more difficulty than the standard methods used to process blood and lymphoid tissue. Therefore, we present here methods to isolate cells from bronchoalveolar lavage (BAL) and lung tissue to assess localized responses to inhalational vaccines.

To understand the antigen-specific memory immune response induced by a vaccine, it is of value to assess both cell-mediated and humoral responses. For intracellular bacterial pathogens such as *M. tuberculosis*, the critical protective role of T-cells is well established [25] and assessing these is the focus of the protocols detailed here. We describe detection of vaccine-induced antigen-specific T-cell responses via several methods: leukocyte restimulation followed by intracellular cytokine staining (ICS) and flow cytometry, enzyme-linked immunosorbent spot (ELISpot) assay, and sustained leukocyte restimulation for detection of antigen-specific memory responses. The different methods provide a means to obtain a wide range of data on the T-cell responses to vaccine antigens, and also take into consideration that the investigator may be limited by the equipment or reagents available. Consideration should be given to which method is most suitable, and a summary of the major reportable outcomes, advantages, and disadvantages of the techniques is provided in Table 1. While not necessarily critical for defense against *M. tuberculosis*, the role of vaccine-induced humoral immunity is of great importance for defenses against other bacterial pathogens. Therefore, we also detail assessment of antibody responses to vaccine antigens, using isotype-specific enzyme-linked immunosorbent assays (ELISA).

Finally, while assessment of the immune response generated by a vaccine is required to progress a vaccine toward clinical development, ultimately, it is critical to determine whether a vaccine can prevent or limit infection with the pathogen of interest in a suitable preclinical model. This is particularly the case if the immune correlates of protection for a particular pathogen are not well-defined, as is the case against *M. tuberculosis* [26]. While providing detailed methods for efficacy testing is beyond the scope of this chapter, it is important to identify immune responses correlating to protection in preclinical animal models by performing similar immunological assessment of the response generated by the vaccine prior to and postinfectious challenge.

2 Materials

Prepare all solutions using ultrapure water and store at room temperature (~22 °C), unless indicated otherwise. Standard cell culture consumables and plasticware, including 70 μm filters, should be sterile. Follow all institutional biosafety and chemical hazard regulations, including when disposing of waste.

2.1 Leukocyte Preparation

All reagents should be sterile. Saline buffer may be autoclaved, otherwise filter buffers or stock solutions through a nonpyrogenic 0.22 μm membrane filter.

1. 1× phosphate buffered saline (PBS), pH 7.4.
2. Fluorescence activated cell sorting (FACS) buffer (keep at 4 °C): PBS, 2% v/v fetal bovine serum (FBS) or fetal calf serum (FCS), 2 mM ethylenediaminetetraacetic acid (EDTA).
3. Complete cell culture media (keep at 4 °C): Roswell Park Memorial Institute (RPMI) 1640 Medium (containing L-glutamine), 10% FBS or FCS, 0.05 mM 2-Mercaptoethanol, 100 U/ml penicillin–streptomycin.
4. PBS/Heparin (keep at 4 °C): PBS, 20 U/ml Heparin.

Table 1
Comparison of key methods for assessing T-cell cytokine responses to peptide or protein vaccine antigens

Method	Major reportable outcome(s)	Key advantages	Disadvantages
Leukocyte restimulation, ICS and flow cytometry	Proportion of antigen-responsive cytokine positive T-cells, including single-cell level analysis of multifunctionality and phenotype	Flexibility—Simultaneous single-cell level detection of multiple different phenotypes High sensitivity—Large numbers of cells can be analyzed to allow detection of rare antigen-responsive populations	Cannot provide quantitation of the cytokines released Requires access to a multiparameter flow cytometer and analysis software Depending on experiment design, may be costly
ELISpot	Enumeration of antigen-specific cells capable of releasing the detected cytokine	High sensitivity Cost-effective High-throughput	Limited to detection of one cytokine only (unless variant methods such as dual-color [40] or FluoroSpot [41] are utilized) Cannot determine the phenotype of cells releasing the detected cytokine
Sustained leukocyte restimulation—Recall of memory antigen-specific T-cell responses	Quantitative detection of cytokine released Can measure multiple cytokines if multiple ELISAs are performed on cell supernatants, or if multiplex methods such as cytokine bead array are used Can also examine antigen-specific cell proliferation in the same assay, either simply by enumerating cells, or by radioisotope or fluorescent methods	High-sensitivity, particularly useful for detection of memory T-cell responses (not recently activated) Cost-effective High-throughput Cell supernatants can be stored frozen and cytokines quantitated when convenient Offers several ways to determine cellular proliferation, which can be selected based on the resources available	Takes longer Cannot determine the phenotype of cells releasing the detected cytokine Assessment of cellular proliferation by radiological or fluorescent methods may be costly

5. Ammonium-Chloride-Potassium (ACK) lysis buffer: 150 mM ammonium chloride (NH_4Cl), 1 mM potassium hydrogen carbonate ($KHCO_3$), 100 mM EDTA (pH 7.3).
6. Tissue digestive enzymes, collagenase IV and DNase I, in PBS.
7. Trypan blue 0.01% w/v in 1× PBS.

2.2 Antigen Restimulation of Leukocyte Suspensions and Flow Cytometry

1. Nonspecific T-cell stimulant, such as anti-CD3 and anti-CD28 monoclonal antibodies.
2. Purified peptide or protein vaccine antigens of interest.
3. Brefeldin A.
4. Fluorescently labeled monoclonal antibodies for cell surface or intracellular immunostaining.
5. Intracellular staining buffer kit, such as BD Cytofix/Cytoperm™ Fixation/Permeabilization Kit.

2.3 ELISpot

1. Nonspecific T-cell stimulant, such as Concanavalin A.
2. ELISpot plate (PVDF membrane preferable).
3. Cytokine-specific primary coating antibody.
4. Complementary cytokine-specific biotinylated secondary antibody.
5. PBS-Tween (PBST): 1× PBS, 0.1% v/v Tween 20.
6. PBS/bovine serum albumin (BSA): 1× PBS, 0.5% w/v BSA.
7. Avidin Alkaline Phosphatase (AAP) enzyme conjugate.
8. Alkaline Phosphatase Substrate.

2.4 ELISA

1. Purified peptide or protein vaccine antigens of interest.
2. Carbonate/bicarbonate coating buffer 50 mM, pH 9.6.
3. Blocking buffer (store 4 °C): PBS, 1% w/v BSA.
4. Wash buffer: PBS, 0.05% v/v Tween 20.
5. Detection antibody (e.g., anti-mouse IgA), horseradish peroxidase (HRP)-conjugated (optimize concentration for each detection antibody).
6. Tetramethylbenzidine (TMB) substrate.
7. Stop solution: 2 M HCl.

3 Methods

Carry out all procedures at room temperature (RT; ~22 °C), unless indicated otherwise.

3.1 Isolation of BAL and Lung Leukocytes from Vaccinated Mice

Leukocytes should be collected and processed using aseptic techniques, with sterile reagents and equipment. Keep tissue or cells at 4 °C or on wet ice as much as is practical, to slow down metabolic processes that alter cell phenotype and decrease the rate of cell death. Keep FACS buffer and complete cell culture media at 4 °C.

1. Euthanize the mouse as per the requirements of your institutional animal ethics committee. Avoid cervical dislocation, as

3.1.1 Bronchoalveolar Lavage Collection and Cell Isolation

this is likely to cause injury to the trachea, chest cavity, and blood vessels, resulting in difficulty performing the lavage.

2. Immediately after confirmation of euthanasia, spray 70% ethanol on the neck to sterilize and make an incision in the skin. Incise away the muscles to expose the trachea. Use two pairs of curved forceps to gently tease away the membrane covering the trachea.
3. Slide one pair of curved forceps under the trachea, to provide stability during intubation.
4. Very carefully, with the bevel facing upward, punch a hole in one side of the trachea with a 21 GA needle.
5. Insert a ~1.8 in. 20 GA flexible cannula, attached to a 1 ml syringe containing 1 ml sterile PBS, into the hole created by the needle to 0.5–1.5 cm depth. Do not exceed this, as it may damage the lungs. With a second pair of fine forceps, pinch the trachea to seal the cannula firmly in place. Remove the pair of forceps under the trachea.
6. Perform lavage: slowly inject the 1 ml PBS into the lungs—you should see the lungs inflate and the chest expand as the lungs fill with PBS. There should be no leakage of PBS from the trachea or lungs. If this occurs, it is likely the seal with the forceps on the trachea is not sufficiently firm, or, there has been a rupture in the trachea.
7. Pull back on the syringe to collect as much fluid from the lungs as possible, usually 700–900 μl. Dispense into an Eppendorf or Falcon tube. In a healthy naïve mouse, there should be no obvious evidence of blood in the bronchoalveolar lavage fluid (BALF); however, this may occur in mice with inflammation or infection in the lungs.
8. Centrifuge the BAL for 5 min (300–500 × *g*, 4 °C). Collect the supernatant—this is the BALF and can be used in assays to quantitate the presence of cytokines/chemokines or antibodies in the airways by ELISA. If not used immediately, this should be stored at −80 °C to −30 °C. It may be beneficial to add protease inhibitors to prevent degradation of protein analytes.
9. Resuspend the cells in ~200 μl FACS buffer or complete cell culture media by gently pipetting up and down. Count viable cells by trypan blue exclusion. The number of cells will be variable, but in a healthy mouse expect ~1×10^5 or less, and up to 1×10^6 following pulmonary immunization or if inflammation or infection is present.

3.1.2 Lung Leukocytes

1. Euthanize the mouse as per the requirements of your institutional animal ethics committee, avoiding cervical dislocation.

2. Immediately after confirmation of euthanasia, expose the lungs and heart, using blunt edged scissors to cut away the rib cage and surrounding tissue, being careful not to pierce the lungs.
3. If you wish to remove circulating blood from the lung tissue, immediately perfuse. Holding the heart with tweezers, inject into the apex or right atrium 10 ml PBS/Heparin. The lung should inflate and clear of blood until they are white. If lungs do not inflate, try pulling back on the needle a little to make sure you are not blocking the major blood vessels between the lung and heart. If you pierce the lungs while removing the rib cage, you will have difficulty getting the lungs to inflate.
4. Remove the lungs in pieces and place in 1.5 ml complete cell culture media in a 24-well tissue culture plate or 5 ml flat bottom tube.
5. Using a pair of fine pointed small scissors, chop the lungs into small pieces (1–2 mm in diameter). In between samples, dip the scissors in 70% ethanol, then sterile PBS, to prevent cross-contamination. Alternatively, use a scalpel and a Petri dish to slice-up the lung. The aim is to increase the surface area of the tissue for digestive enzymes.
6. Once the lung is chopped, place the lung pieces and associated media into a tube using a sterile transfer pipette. Use one pipette per experimental group. Add RPMI only to make up to a total volume of 5 ml.
7. Add enzymes from sterile stock solutions to achieve a final concentration of 50 U/ml collagenase and 13 μg/ml DNAse I.
8. Incubate for 30–45 min at 37 °C on a gentle shaker, rocker or rotating wheel. Alternatively use a 37 °C water bath and gently invert samples every 10–15 min to mix.
9. Pass media and tissue (using a 5 ml syringe plunger or similar) through a 70 μm mesh sieve to disperse clumps and rupture larger pieces of tissue, into a new V-bottom 10 or 15 ml tube. Use 1 sieve and syringe for every 2 to 3 samples, changing between groups. Wash through thoroughly with complete media.
10. Centrifuge for 10 min (500 × *g*, 4 °C). Discard supernatant. Whenever discarding supernatants, do not invert tube multiple times, as this will resuspend the cell pellet and result in inadvertent loss of cells.
11. If all the blood was removed by perfusion, the cell pellet will be white. If blood remains, lyse erythrocytes with ACK lysis buffer. Resuspend the pellet by flicking the tube, then add ~2 ml ACK lysis buffer for 1–2 min (this time will need to be optimized for each batch of ACK lysis buffer—do not incubate for longer than necessary to lyse red blood cells, as this will

increase cell death). Dilute with cold (4 °C) PBS to at least five times the volume to stop lysis.

12. Filter again through a 70 μm sieve—one sieve and syringe per experimental group. Spin cells down for 5 min (500 × *g*, 4 °C).
13. Resuspend pellet by flicking tube. Add 500 μl complete cell culture media and then resuspend cells by gently pipetting up and down.
14. Count viable cells by trypan blue exclusion. A normal lung will yield between 1 and 3 million cells; inflamed or infected lungs will yield between 30 and 40 million cells.

3.2 Assessment of Antigen-Specific T-cell Responses Via Ex vivo Antigen Restimulation, Intracellular Cytokine Immunostaining and Flow Cytometry

Flow cytometric methods provide enormous scope to determine the pattern of T-cell responses to the vaccine and the complex phenotypes of T-cells, at a single-cell level. For instance, cell surface immunostaining can be utilized to enumerate vaccine-specific α/β TcR $CD4^+$ and $CD8^+$ T-cells, using epitope-specific MHC Class II and I-tetramer complexes [27, 24]. Their location in the lung vasculature or parenchyma may be determined by prior intravascular labeling. This can be achieved by intravenous injection of 5 μg anti-CD45-APC/Cy7 (or another fluorophore), 3–5 min prior to euthanasia, to distinguish labeled vascular leukocytes from unlabeled lung parenchymal cells [28–31]. Lung resident memory T cells stimulated by pulmonary vaccines can also be identified within the lung parenchyma by cell surface markers, which may include CD69, CD103, CXCR3, and CXCR6 [31, 32]. Other T-cell populations may be stimulated by the vaccine adjuvant, such as donor-unrestricted T cells including γ/δ TcR T cells and MAIT cells [33], and these may also be enumerated with fluorochrome-labeled specific mAbs. Phenotyping of these various T-cell subtypes should be performed by immunostaining freshly collected leukocyte suspensions that have been kept at 4 °C, or on wet ice, from the time of collection. The prolonged incubation period at 37 °C required to assess antigen-specific cytokine responses, as in the following protocol, may change the surface expression of these memory or phenotypic markers.

To assess $CD4^+$ and $CD8^+$ T-cell cytokine responses specific to a vaccine peptide or protein antigen, single-cell suspensions of leukocytes isolated from mice (e.g., from the lungs, blood, spleen, lymph nodes) must first be restimulated in culture with the antigen. A reagent that blocks intracellular protein transport processes, such as Brefeldin A, is added so that cytokine from responding T-cells accumulates within the Golgi complex. Immunostaining of cell surface and intracellular markers is then performed with fluorophore-labeled antibodies, and data are acquired by flow cytometric analysis.

This method provides the most flexibility for detection of different types of T-cell cytokine responses, including multifunctional responses, but is limited to providing the proportion of cytokine positive responsive cells. That is, it cannot provide a quantitative report of the amount of cytokine released in response to antigen.

Subheading 3.2.1 of the protocol must be performed under aseptic conditions to reduce the risk of T-cells responding to contaminants; however, the steps from Subheading 3.2.2 onward need not be sterile.

3.2.1 Restimulation of Antigen-Specific T-cells in Ex vivo Culture

1. Add 2–4 × 10^6 leukocytes (*see* **Note 1**) from each sample into a well of a 96-well round- or V-bottom plate. Use one well for each antigen you wish to restimulate the sample with.
2. Centrifuge the plate to pellet cells (3–5 min, 300–500 × *g*) and discard the supernatant (*see* **Note 2**). Resuspend the cells, by gently pipetting up and down (*see* **Note 3**), in 200 μl per well complete culture media containing the peptide or protein vaccine antigen of interest (5–10 μg/ml is generally suitable). Also prepare negative and positive control samples (*see* **Note 4**).
3. Incubate the plate at 37 °C, 5% CO_2 to allow the antigen to be processed for presentation on MHCI/II to T-cells. This incubation time may require optimization. As a guide: for peptide restimulation 1–2 h; for protein restimulation, 3–4 h.
4. Add Brefeldin A to each well (10 μg/ml; this will block transport of proteins from the Golgi) and pipette gently to resuspend the cells. Incubate (humidified) at 37 °C, 5% CO_2 for 4–6 h to allow intracellular accumulation of the cytokines (*see* **Note 5**).

3.2.2 Immunostaining Cells for Detection of Intracellular Cytokine

The number and type of cell surface or intracellular markers selected for immunostaining will vary greatly depending on the experimental design. In addition, the monoclonal antibody panels will vary between laboratories, accounting for the configuration of the cytometer available for data acquisition, as well as more practical considerations such as antibody stocks already within the laboratory and the budget available. We recommend the following papers to assist in the critical steps of selecting antibody combinations for use in flow cytometry, optimization of antibody dilutions and tips for data acquisition on the cytometer [34, 35]. The antibody clones and fluorophores utilized to generate the sample data in Fig. 1 are included in **Note 6**, but this is by no means restrictive.

From this step onward, keep the cells on wet ice or at 4 °C to prevent any further metabolic changes or increasing cell death. As vaccine experiments often involve testing many samples, to improve efficiency and consistency, we strongly recommend performing cell staining in a 96-well plate, rather than staining cells in individual FACS tubes.

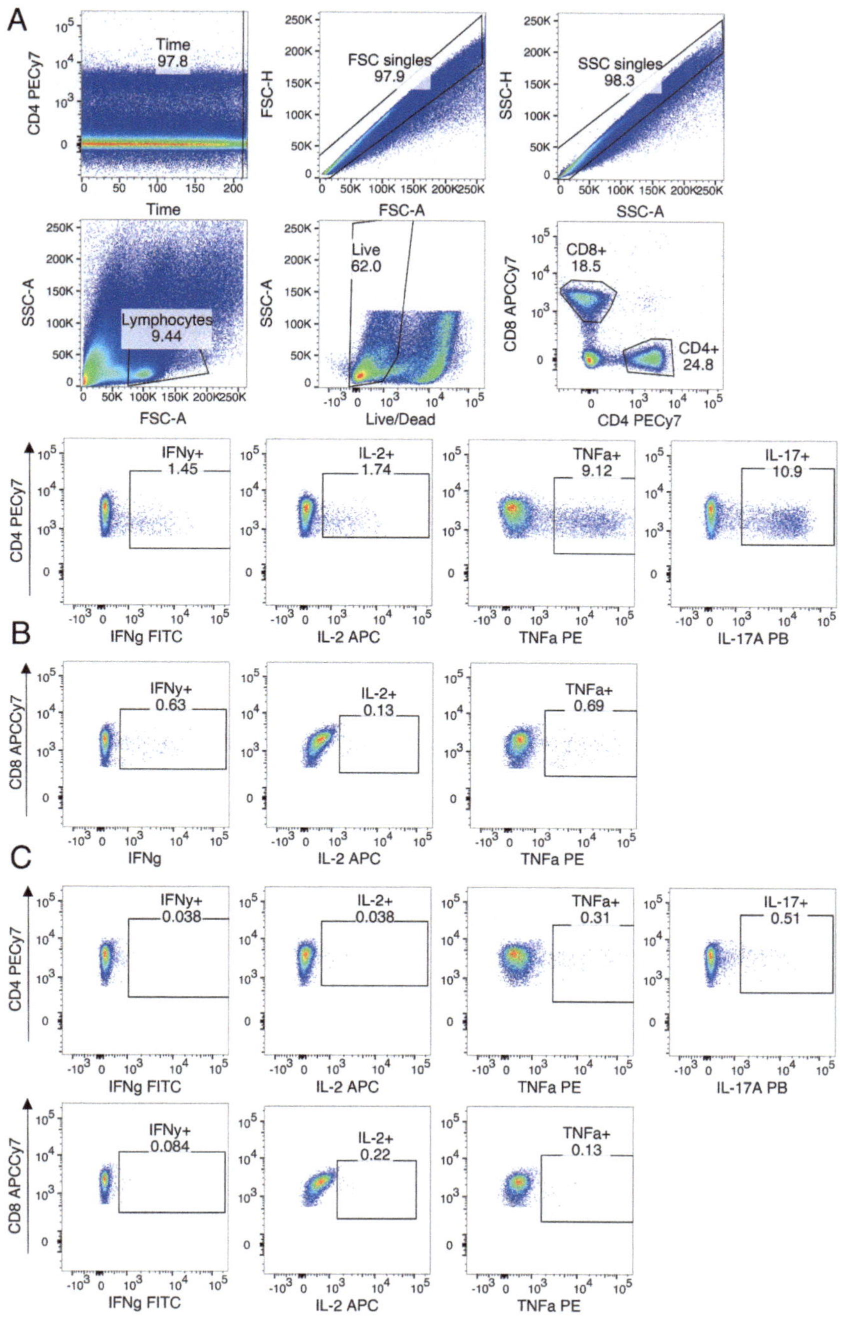

Fig. 1 Gating strategy for analysis of intracellular stained flow cytometry data, to quantitate the proportion of antigen-specific cytokine producing T-cells and determine polyfunctional T-lymphocyte responses to immunization. Debris and dead cells are excluded, $CD4^+$ or $CD8^+$ T-cells selected, cytokine positive cells gated, followed by Boolean gating to enumerate the frequency of polyfunctional populations. A representative immunostained lung sample from a Pam_2Cys-$ESAT6_{1-20}$-$TB10.4_{3-11}$ intranasally immunized mouse is shown after ex vivo stimulation in the presence of Brefeldin A (10μg/ml) with (**a**) $ESAT6_{1-20}$ peptide (10μg/ml) recall, (**b**) $TB10.4_{3-11}$ peptide (10μg/ml) recall and (**c**) no peptide recall, used as a negative control to set gates for cytokine producing cells

1. Centrifuge the cells to pellet (3–5 min, 300–500 × *g*), and discard supernatant as before. Add 200 μl cold (4 °C) PBS to each well and pipette gently to resuspend, washing the cells, and centrifuge as before.
2. To allow identification of live cells, stain dead cells with a fixable cell viability dye (e.g., LIVE/DEAD Fixable Blue Dead Cell Stain Kit, for UV excitation, available from Life Tech), diluted as per the manufacturer's instructions (*see* **Note** 7). Also include Fc blocking reagent (e.g., Mouse BD Fc Block—purified rat anti-mouse CD16/CD32, clone 2.4G2, at 1:100 dilution) to prevent nonspecific binding of antibodies used in the immunostaining steps to Fc receptors on leukocytes. Pipette gently to resuspend cells in 50 μl of a stock solution (i.e., for consistency, prepare enough staining stock to provide sufficient volume for every sample). Incubate on ice for 20–30 min, or with gentle agitation on a plate shaker, at 4 °C.
3. To halt staining with the viability dye, add 200 μl cold FACS wash directly into each well, then centrifuge the cells to pellet (3–5 min, 300–500 × *g*).
4. To perform immunostaining of cell surface markers, prepare a stock staining solution consisting of FACS buffer and your selection of appropriately diluted monoclonal antibodies and gently resuspend cells in 50 μl of the stain. Incubate on ice for 20–30 min, or with gentle agitation on a plate shaker at 4 °C, in the dark as the fluorophores are light sensitive (e.g., wrap plate in aluminum foil).
5. Remove excess stain by adding 200 μl cold FACS wash directly into each well, then centrifuge the cells to pellet (3–5 min, 300–500 × *g*). Repeat wash once more.
6. After discarding the wash buffer, gently vortex the plate to resuspend the cell pellets, and add 100 μl fixative/permeabilization agent (*see* **Note 8**). Pipette the cells gently to resuspend thoroughly then incubate for 20–30 min, at 4 °C, in the dark.
7. Centrifuge cells to pellet and discard supernatant appropriately. Wash cells twice as before, but in 200 μl BD Cytowash (dilute buffer stock as per the manufacturer's instructions).
8. Prepare a stock staining solution consisting of BD Cytowash buffer and your selection of appropriately diluted monoclonal antibodies to stain intracellular cytokines or markers, and gently resuspend the cells in 50 μl of the stain. Incubate on ice for 30 min, or with gentle agitation on a plate shaker at 4 °C, in the dark.
9. Wash cells twice with 200 μl BD Cytowash per well as before, before resuspending cells in FACS wash at desired volume. If required, resuspend instead with a fixative agent as per the manufacturer's instructions.

10. Samples may be stored at 4 °C in the dark for up to 48 h although we recommend performing data acquisition on the cytometer as soon as possible (*see* **Note 9**). Be sure to filter the sample through a 70 μm mesh into a FACS tube immediately prior to loading onto the cytometer, to ensure a single cell suspension and reduce the risk of blockages. Aim to acquire at least 2×10^6 events to allow accurate quantitation of small populations of cytokine producing T-cells.
11. For an example gating strategy to identify cytokine-producing antigen-specific T-cells, *see* Fig. 1. *See* [18, 19] for an example of the final analysis of cytokine producing T-cell populations, using additional Boolean gating analysis in FlowJo version 10 (BD). Report the data as the percentage (or proportion) of $CD4^+$ or $CD8^+$ T-cells producing a particular cytokine, or combination of cytokines (*see* **Note 10**).

3.3 IFNγ ELISpot

First described by Czerkinsky in 1983, ELISpot provides a quantitative measure of cells in the sample capable of releasing antibody or a particular cytokine, specifically in response to a recall antigen [36–38]. It is widely considered a sensitive and reproducible technique and is utilized in both research and clinical diagnostic settings [39]. Similar to plate-based sandwich ELISA, ELISpot utilizes complementary antibody pairs that recognize different epitopes of the desired analyte, in this case IFNγ, and combines this with membrane-based Western blotting techniques. The first of the antibody pairs is coated onto a polyvinylidene difluoride (PVDF)-backed microtiter plate, then leukocytes are cultured in the plate with the vaccine peptide or protein antigen of interest. During the incubation, if the sample contains antigen-specific cells capable of releasing IFNγ, this is released and captured in place on the membrane by the coating antibody. Leukocytes are then washed out of the plate and IFNγ is detected with a biotinylated or directly enzyme-conjugated complementary antibody. If needed, avidin or streptavidin conjugated enzyme is bound to the antibody, then substrate solution is added to form colored spots at the sites of cytokine secretion on the membrane. These can then be counted to quantitate the number of spot-forming units per well, and extrapolated to provide the number of responding cells per million leukocytes.

While the protocol presented here provides the materials and method for detection of murine IFNγ, this may be adjusted to detect other cytokines or antibodies of interest using complementary monoclonal antibody pairs at optimized concentrations. Modification of the cell culture conditions and incubation period may also be required.

Subheadings 3.3.1 and 3.3.2 should be performed under aseptic conditions. The subsequent steps need not be sterile.

3.3.1 Preparation and Coating of the ELISpot Plate

1. Wet the membrane at the base of the wells of the ELISpot plate with 40 μl of 35% ethanol (v/v prepared in milliQ sterile water) for less than 1 min (*see* **Note 11**). Wash the plate three times by pipette (*see* **Note 12**) with 200 μl sterile PBS per well, adding the first wash straight on top of the ethanol. Ensure the membranes do not dry out for the remainder of the protocol. Leave final PBS wash on until ready to commence the next step.
2. Coat each well of the ELISpot plate with the primary antibody, purified anti-mouse IFNγ antibody (clone AN18) by pipetting 100 μl antibody diluted in sterile PBS into each well. Incubate overnight at 4 °C. Alternatively incubate at room temperature for a minimum of 2 h.
3. Pipette out the primary antibody solution and wash with 200 μl sterile PBS three times as before.
4. To ensure proteins do not bind nonspecifically to the membrane, add 200 μl complete cell culture media into each well as a blocking agent. Incubate at 37 °C for 2 h, or at room temperature (22 °C) if for longer.

3.3.2 Plating of Leukocytes for Antigen Restimulation

1. To restimulate vaccine antigen-specific cells, prepare dilutions of desired peptide or protein antigen in complete cell culture media. Prepare these, as well as negative and positive control stimuli, at double the desired final concentration for the cell culture (*see* **Note 13**). If not already sterile, filter-sterilize the antigen stocks using a nonpyrogenic 0.22 μm syringe filter.
2. Remove the blocking solution from the ELISpot plate by gentle pipetting, then quickly (do not allow membranes to dry) pipette 100 μl per well of either negative control media, diluted antigen, or positive control stimulant.
3. Gently plate 100 μl of a 2×10^6 per ml leukocyte suspension prepared in complete cell culture media into each well (final density of 2×10^5 leukocytes per well; *see* **Note 14**).
4. Incubate (humidified) the plate at 37 °C, 5% CO_2 for 20 h (minimum 18 h).

3.3.3 Staining and Developing the ELISpot Plate

1. Decant cells and wash plate six times with PBST (0.1% Tween 20), 200 μl per well, with a pipette. Leave final wash on while you make up the secondary antibody solution.
2. Add 100 μl per well biotinylated anti-mouse IFNγ antibody (clone XMG.12, *see* **Note 15**) prepared at an appropriate dilution in 0.5% BSA (w/v)/PBS and incubate at room temperature for greater than 2 h, or alternatively leave overnight at 4 °C.
3. Decant secondary antibody solution and wash plate six times with PBST as before. Leave final wash on while you make up secondary antibody label solution.

4. Add 100 μl of secondary antibody label solution Avidin Alkaline Phosphatase (Sigma) at 1:1000 (v/v) in 0.5% BSA/PBS and incubate for 45 min at room temperature.
5. Decant solution and wash plate three times with PBST, then three times with PBS, as before. Before final PBS wash, remove the plastic backing of the plate and fill with PBS, rinsing the base of membranes. Leave final PBS wash on while you prepare the substrate solution.
6. Add 100 μl of alkaline phosphatase conjugate substrate solution to every well and allow color development by incubating at room temperature in the dark (wrap in foil) for ~11 min, or until spots are just visible. Stop reaction by decanting solution and immediately washing thoroughly with distilled water.
7. Allow plates to air-dry in the dark at 4 °C overnight to allow the spots to darken. Allow further drying in the dark at room temperature for a few hours before reading with an ELISpot plate reader (*see* **Note 16**).
8. To report the data, multiply the number of spots per well by 5, to provide the spot forming cells per 10^6 leukocytes for that sample.

3.4 Sustained Leukocyte Restimulation for Recall of Memory Antigen-Specific T-cell Responses

To optimize the sensitivity for detecting antigen-specific T-cell responses to the vaccine peptide or protein antigen, in particular memory responses, it can be beneficial to perform an extended cell culture incubation period, allowing cytokine released from antigen-specific T-cells to accumulate in the culture supernatant. This method, in contrast to ELISpot, allows a quantitative measure of the amount of cytokine released from the leukocytes in response to antigen restimulation. During this extended incubation time, it is expected that antigen-specific T-cells will have proliferated, and these may also be quantitated to give a measure of the magnitude of the immune response.

This assay should be performed under aseptic conditions.

1. Prepare single cell leukocyte suspensions in complete cell culture media at 2×10^6 cells/ml.
2. In a 96-well round bottom plate, plate in duplicate or triplicate, 100 μl per well of appropriately diluted recall antigen, or negative (media only) or positive (ConA 3–5 μg/ml) controls, as per Subheading 3.3.2, **step 1**. Then add 100 μl per well of the leukocyte suspension (2×10^5 cells/well), taking care not to cross-contaminate between wells. Incubate (humidified) at 37 °C, 5% CO_2, for 72 h (*see* **Note 17**).
3. Centrifuge the plate (300–500 × *g*, 3–5 min) to pellet the cells, and carefully remove the supernatant by pipette into a new 96-well plate, taking care not to disturb the cell pellet. The

supernatant can be used immediately, or frozen at −30 °C until required, for detection of cytokine release by ELISA or other method such as cytokine bead array.

4. Resuspend the cell pellets in complete culture media. Proliferation of the cells may be determined by quantitating the cells per well (*see* **Note 18**).

3.5 Quantitation of Vaccine-Induced Antigen-Specific Antibodies by ELISA

The majority of clinically utilized vaccines aim to induce humoral immunity against the pathogen of interest, with a level of vaccine-specific antibodies determined that correlates with protective immunity. Antibodies generated against the vaccine antigen can be readily detected in serum samples by straightforward ELISA methods. As mice may be bled routinely throughout vaccination experiments, serum may be collected sequentially from individual animals to monitor changes to the isotype and titer of antibodies over time. Take care to detect isotypes specific to the experimental mouse strain, for instance, C57BL/6 mice produce IgG2c instead of the IgG2a produced by BALB/c. However, vaccine-induced antibody may also be found in many peripheral tissues and samples taken at euthanasia, such as BALF as described in Subheading 3.1.1, nasal washes or peritoneal washes, may also be assessed for the presence of vaccine-induced antibodies. This may be particularly valuable to detect mucosal IgA if vaccination is delivered by inhalation [19].

Here we present a standard indirect ELISA-based method to quantitatively determine vaccine antigen-specific antibody titer in serum or BALF.

1. Coat high-binding 96-well ELISA plates with 100 μl per well of the vaccine peptide or protein antigen, at 1 μg/ml diluted in coating buffer. Seal plate and incubate for 2 h at 37 °C or overnight at 4 °C.
2. Decant coating solution, then thoroughly wash the plate by filling wells completely with wash buffer then decanting, 4–6 times. This may be performed using a squirt bottle or an automated plate washer, to forcefully wash the wells. Tap plate firmly on absorbent paper toweling to blot.
3. Add 200 μl blocking buffer per well, seal plate and incubate at 37 °C for 1 h.
4. While the ELISA plate is blocking, prepare serial dilutions of samples in blocking buffer, in a separate nonbinding 96-well plate. For serum, we recommend a starting dilution of 1:100, performing an 8 to 12-point curve and 2–5 fold dilution series. For BALF samples, start at 1:20 dilution. All samples should be run in technical duplicate (*see* **Note 19**).

5. Decant blocking buffer and wash plate as described in **step 2**. Add 100 μl sample per well, taking care to include the relevant biological and technical negative controls on every plate.
6. Seal the plate and incubate at 37 °C for 1 h. Decant sera/BALF and wash plate as before.
7. Detect bound antigen-specific antibodies via addition of 100 μl HRP-conjugated detection antibody, diluted in blocking buffer.
8. Seal the plate and incubate at 37 °C for 30 min. Decant antibody and wash plate as before.
9. Develop ELISA via addition of 100 μl TMB substrate and monitor for color development (*see* **Note 20**). Neutralize the reaction by addition of 100 μl stop solution (2 M HCl).
10. Read absorbance at 450 nm with an ELISA plate reader. If possible, also take a reference reading at 570 nm, and subtract this from the absorbance reading at 450 nm.
11. Analyze data: reporting antibody titer is more informative than providing raw absorbance values. Plot dilution curves for each sample, averaging the technical duplicates or triplicates for each dilution, and generate an equation to fit the curve. Solve to find the antigen-specific antibody titer, determined as the dilution at which the absorbance equals the mean absorbance (+1–3 SD) of 1:100 sera or 1:20 BALF from unvaccinated or adjuvant-only vaccinated control mice.

4 Notes

1. We recommend performing antigen restimulation and staining for every test sample separately, rather than pooling samples for an experimental group, to assess biological variability. The number of cells required for restimulation, staining, data collection and subsequent analysis, will vary depending on the experimental system, in particular the magnitude of the vaccine-induced response. This may require optimization; however, we would not recommend performing this protocol with less than 2×10^6 leukocytes per sample. If the investigator needs to restimulate more than 4×10^6 leukocytes per sample, it will be necessary to do this across multiple wells of a 96-well plate or utilize a larger plate size, as if the cells are more concentrated, the risk of cell death increases which is detrimental to the assay.
2. Supernatant may be discarded by decanting the supernatant from the plate (inside a biosafety hood) onto absorbent paper towel, within a biohazard bag or container. However, if the samples contain a pathogen or chemical that represents an

aerosol hazard, instead remove the supernatant by aspiration, for example, manually with a pipette. Take care not to cross-contaminate between wells; this can be aided by spacing samples one well apart in the plate.

3. Do not pipette cells vigorously, or introduce air bubbles to the cell suspension, as this can induce cell death.
4. Negative controls—to confirm that cytokine responses are specific to the peptide or protein antigen used for restimulation, and not due to nonspecific activation of the cells in culture, culture cells from the sample in the same manner but in the absence of any antigen. This may be performed for each sample individually; however, to conserve time and reagents, this may be performed using a pooled sample containing an equal number of cells from each mouse within the same experimental group. Positive controls—to ensure cells are viable and responsive, as well as to verify the intracellular cytokine staining steps, stimulate a sample for each organ type processed in the experiment with a non–antigen-specific T-cell stimulant. Purified monoclonal antibodies against murine CD3e (clone 145-2C11) and CD28 (clone 37.51), at 5 μg/ml, are effective at inducing T-cell activation and cytokine release. We also recommend including a few additional wells of spare cells, treated similarly, that may be used for single-stain control samples for acquisition on the flow cytometer.
5. Do not shorten the incubation time, as insufficient cytokine may accumulate to obtain clear immunostaining. For practical reasons, however, it may be necessary to extend the incubation time, for example allowing the cells to incubate with Brefeldin A overnight. This will still provide acceptable cell stimulation and cytokine accumulation; however, note that cell death will increase progressively. Where possible incubation time with Brefeldin A should be limited to 4–6 h. Other Golgi block reagents are also commercially available and should be used according to the manufacturer's instructions.
6. In the example provided in Fig. 1, surface stain antibodies were anti-mouse CD8-APC/Cy7 (53-6.7; BD Pharmingen), CD4-PE/Cy7 (RM4-5; BD Pharmingen) and CD3-PerCP/Cy5.5 (17A2; BioLegend). Intracellular stain antibodies were anti-mouse IFNγ-FITC (XMG1.2; BD Pharmingen), IL-17A-PB (TC11-18H10.1; BioLegend), TNFα-PE (MP6-XT22; BioLegend), and IL-2-APC (JES6-5H4; BioLegend).
7. It is expected that a proportion of the cells will have died either during the extraction from tissue, or during the culture restimulation period. Dead cells should be excluded from flow cytometry data analyses as they can nonspecifically take up antibody

providing a false positive signal. We strongly recommend optimizing the concentration of fixable viability dyes before use in an experiment. Test a dilution series of the fixable dead cell stain on the same number of cells to be used in each sample in the experiment to establish optimal concentration, for example, dilutions of 1:200, 1:500, 1:1000, 1:5000, 1:10000, on 2–4 $\times$ 10^6 cells. Prepare in the buffer recommended by the manufacturer; often this is PBS, not FACS buffer. Do not use the dye at a greater concentration than is necessary to clearly distinguish live and dead cell populations, and do not exceed the recommended staining time. Live cells may take up excess stain, increasing the fluorescent background signal.

8. We find the BD Cytoperm/Cytofix buffer system works well for intracellular cytokine staining protocols. It may be necessary to centrifuge cells at an increased speed of 500 $\times$ g, as components of the Cytoperm/Cytofix buffers result in a looser cell pellet.
9. As with all acquisitions on a flow cytometer, you will require single-color stained control samples for every fluorophore in your antibody panel, including the cell viability dye. We find the use of compensation beads to be preferable (with the exception of the cell viability stain) as in general, they will provide a stronger fluorescent signal than cells. Regardless of whether cells or compensation beads are utilized for single-color controls, ensure that they have been stained and fixed using exactly the same protocol as your experimental samples to provide the best controls for optimized compensation of fluorescence spillover signal. Different staining or fixation protocols will alter the fluorescent signatures of some fluorophores.
10. As the cells have been cultured, it is expected that there may have been proliferation or cell death of certain immune populations; therefore, it is not reliable to present the data as a quantitation of the number of T-cells producing a particular cytokine in the sample.
11. It is essential that the entire membrane of each well of the ELISpot plate is completely wet. You will notice a color change from a dry white to a gray-white tone as the membrane is saturated. However, do not exceed 1 min incubation with the 35% ethanol, as this may damage the membrane and reduce the capacity for binding to the antibodies in subsequent steps.
12. To reduce the risk of contamination, we recommend not decanting wash buffer at this stage, but rather use a multichannel pipette to aspirate washes. However, take great care not to scrape or puncture the membrane at the bottom of the well with the pipette tip. This may be avoided by angling the plate, and pipetting at a corner of the well, just above the membrane.

13. The concentration of antigen may need to be optimized, but in general a final concentration of 0.1, 1, 3, and 10 μg/ml will provide a suitable gradient of responses, with maximum response expected when restimulating with 3 or 10 μg/ml antigen. As a negative control, you will also need complete cell culture media with no antigen. As a positive control to ensure cell viability and responsiveness, we recommend the use of a compound to nonspecifically activate T-cells in the culture, such as ConA at a final concentration of 3–5 μg/ml in complete culture media. We recommend performing the assay with at least technical duplicates, preferably triplicates, for each condition for every experimental sample.
14. If a very strong vaccine response has been generated, it may be necessary to reduce this cell concentration, in order to provide sufficient resolution for counting of the spots of the ELISpot membranes. This may particularly be the case in nonlymphoid organs where a strong immune response has occurred, for example, the lungs of mice receiving a potent pulmonary vaccine, or following infection with a respiratory pathogen.
15. It is critical to use a different antibody clone to that used to coat the ELISpot plate in Subheading 3.3.1, that recognizes a different portion of the IFNγ protein, for example, XMG1.2. If the antibodies used at the two different steps recognize a similar epitope, or bind to a similar site on the protein, then binding of the secondary antibody will be prevented directly either from blocking of the binding site or by steric hindrance.
16. If necessary, spots can instead be counted manually under a microscope, or imaged and counted using computer software.
17. The 72 h incubation period is optimized for accumulation of murine IFNγ. If other cytokines are of primary interest, this timepoint may need to be altered to optimize accumulation.
18. Cell quantitation may be performed by a variety of methods. Most simply, perform a viable cell count by hemocytometer or automated cell counter. Alternatively, radioisotope methods such as tritiated thymidine incorporation may be performed. Prelabeling of the leukocyte suspensions prior to the culture with a cell proliferation dye, such as carboxyfluorescein succinimidyl ester (CFSE) or similar fluorescent dye, would at this stage allow acquisition of the cells on a flow cytometer to calculate the proportion of proliferated cells per well.
19. We recommend running every sample individually rather than pooling samples for an experimental group, to provide a measure of biological variability. On every ELISA plate, you will need to run sera/BALF from the experimental negative control group, that is mice receiving no vaccination or adjuvant-

only vaccination, at 1:100 for serum or 1:20 for BALF to provide a measure of the baseline background signal. You will also need to leave 2–3 wells spare on every ELISA plate as a technical negative control—do not add any samples to these wells, simply blocking buffer.

20. The time taken to develop the ELISA will vary—as a guide, when wells containing the most concentrated samples are sky blue, or after a maximum of 20 min, neutralize the reaction. The color should change to yellow. Take care to add substrate and stop solution to wells in the same order, and with similar pace, so that development time is consistent between wells.

Acknowledgments

We acknowledge support from the National Health and Medical Research Council (Project APP1044343, and Centres of Research Excellence in TB Control, APP1043225, APP1153493) to WJB and the Australian Research Council and its Centre of Excellence for Innovations in Peptide and Protein Science to RJP.

References

1. Krammer F (2020) SARS-CoV-2 vaccines in development. Nature 586(7830):516–527
2. WHO (2019) Global Tuberculosis Report. World Health Organization, Geneva
3. Andersen P, Doherty TM (2005) The success and failure of BCG—implications for a novel tuberculosis vaccine. Nat Rev Microbiol 3 (8):656–662
4. Sable SB, Posey JE, Scriba TJ (2019) Tuberculosis vaccine development: Progress in clinical evaluation. Clin Microbiol Rev 33(1): e00100–e00119
5. Zhu X, Ramos TV, Gras-Masse H, Kaplan BE, BenMohamed L (2004) Lipopeptide epitopes extended by an Nϵ-palmitoyl-lysine moiety increase uptake and maturation of dendritic cells through a toll-like receptor-2 pathway and trigger a Th1-dependent protective immunity. Eur J Immunol 34(11):3102–3114
6. McDonald DM, Hanna CC, Ashhurst AS, Corcilius L, Byrne SN, Payne RJ (2018) Synthesis of a self-adjuvanting MUC1 vaccine via diselenide-selenoester ligation-deselenization. ACS Chem Biol 13(12):3279–3285
7. Abdel-Aal A-BM, El-Naggar D, Zaman M, Batzloff M, Toth I (2012) Design of Fully Synthetic, self-Adjuvanting vaccine incorporating the tumor-associated carbohydrate Tn antigen and Lipoamino acid-based toll-like receptor 2 ligand. J Med Chem 55 (15):6968–6974
8. McDonald DM, Byrne SN, Payne RJ (2015) Synthetic self-adjuvanting glycopeptide cancer vaccines. Front Chem 3:60
9. Bettahi I, Dasgupta G, Renaudet O, Chentoufi AA, Zhang X, Carpenter D, Yoon S, Dumy P, BenMohamed L (2008) Antitumor activity of a self-adjuvanting glyco-lipopeptide vaccine bearing B cell, CD4+ and CD8+ T cell epitopes. Cancer Immunol Immunother 58 (2):187
10. McDonald DM, Wilkinson BL, Corcilius L, Thaysen-Andersen M, Byrne SN, Payne RJ (2014) Synthesis and immunological evaluation of self-adjuvanting MUC1-macrophage activating lipopeptide 2 conjugate vaccine candidates. Chem Commun 50 (71):10273–10276
11. Belnoue E, Mayol JF, Carboni S, Di Berardino BW, Dupuychaffray E, Nelde A, Stevanovic S, Santiago-Raber ML, Walker PR, Derouazi M (2019) Targeting self and neo-epitopes with a modular self-adjuvanting cancer vaccine. JCI Insight 5(11):e127305
12. Moyle PM, Toth I (2008) Self-adjuvanting lipopeptide vaccines. Curr Med Chem 15 (5):506–516

13. Anderson RJ, C-w T, Daniels NJ, Compton BJ, Hayman CM, Johnston KA, Knight DA, Gasser O, Poyntz HC, Ferguson PM, Larsen DS, Ronchese F, Painter GF, Hermans IF (2014) A self-adjuvanting vaccine induces cytotoxic T lymphocytes that suppress allergy. Nat Chem Biol 10:943
14. Azmi F, Ahmad Fuaad AA, Giddam AK, Batzloff MR, Good MF, Skwarczynski M, Toth I (2014) Self-adjuvanting vaccine against group a streptococcus: application of fibrillized peptide and immunostimulatory lipid as adjuvant. Bioorg Med Chem 22(22):6401–6408
15. Batzloff MR, Hartas J, Zeng W, Jackson DC, Good MF (2006) Intranasal vaccination with a Lipopeptide containing a Conformationally constrained conserved minimal peptide, a universal T cell epitope, and a self-Adjuvanting lipid protects mice from group a streptococcus challenge and reduces throat colonization. J Infect Dis 194(3):325–330
16. Rai PK, Chodisetti SB, Maurya SK, Nadeem S, Zeng W, Janmeja AK, Jackson DC, Agrewala JN (2018) A lipidated bi-epitope vaccine comprising of MHC-I and MHC-II binder peptides elicits protective CD4 T cell and CD8 T cell immunity against *Mycobacterium tuberculosis*. J Transl Med 16(1):279
17. Xu Z, Rivera-Hernandez T, Chatterjee O, Walker MJ, Moyle PM (2020) Semisynthetic, self-adjuvanting vaccine development: efficient, site-specific sortase A-mediated conjugation of toll-like receptor 2 ligand FSL-1 to recombinant protein antigens under native conditions and application to a model group A streptococcal vaccine. J Control Release 317:96–108
18. Ashhurst A, McDonald D, Hanna C, Stanojevic V, Britton WJ, Payne RJ (2019) Mucosal vaccination with a self-adjuvanted lipopeptide is immunogenic and protective against *Mycobacterium tuberculosis*. J Med Chem 62(17):8080–8089
19. Hanna CC, Ashhurst AS, Quan D, Maxwell JWC, Britton WJ, Payne RJ (2021) Synthetic protein conjugate vaccines provide protection against *Mycobacterium tuberculosis* in mice. Proc Natl Acad Sci U S A 118(4): e2013730118
20. Tyne AS, Chan JGY, Shanahan ER, Atmosukarto I, Chan H-K, Britton WJ, West NP (2013) TLR2-targeted secreted proteins from *Mycobacterium tuberculosis* are protective as powdered pulmonary vaccines. Vaccine 31 (40):4322–4329
21. Sabin AB (1983) Immunization against measles by aerosol. Rev Infect Dis 5(3):514–523
22. Satti I, Meyer J, Harris SA, Thomas Z-RM, Griffiths K, Antrobus RD, Rowland R, Ramon RL, Smith M, Sheehan S, Bettinson H, McShane H (2014) Safety and immunogenicity of a candidate tuberculosis vaccine MVA85A delivered by aerosol in BCG-vaccinated healthy adults: a phase 1, double-blind, randomised controlled trial. Lancet Infect Dis 14(10):939–946
23. WHO (2012) Measles aerosol vaccine project - report to SAGE. WHO, Geneva, Switzerland
24. Counoupas C, Ferrell KC, Ashhurst A, Bhattacharyya ND, Nagalingam G, Stewart EL, Feng CG, Petrovsky N, Britton WJ, Triccas JA (2020) Mucosal delivery of a multistage subunit vaccine promotes development of lung-resident memory T cells and affords interleukin-17-dependent protection against pulmonary tuberculosis. NPJ Vaccines 5(1):105
25. O'Garra A, Redford PS, McNab FW, Bloom CI, Wilkinson RJ, Berry MPR (2013) The immune response in tuberculosis. Annu Rev Immunol 31(1):475–527
26. Counoupas C, Triccas JA, Britton WJ (2019) Deciphering protective immunity against tuberculosis: implications for vaccine development. Expert Rev Vaccines 18(4):354–364
27. Flórido M, Pillay R, Gillis CM, Xia Y, Turner SJ, Triccas JA, Stambas J, Britton WJ (2015) Epitope-specific CD4+, but not CD8+, T-cell responses induced by recombinant influenza a viruses protect against *Mycobacterium tuberculosis* infection. Eur J Immunol 45(3):780–793
28. Barletta KE, Cagnina RE, Wallace KL, Ramos SI, Mehrad B, Linden J (2012) Leukocyte compartments in the mouse lung: distinguishing between marginated, interstitial, and alveolar cells in response to injury. J Immunol Methods 375(1–2):100–110
29. Sakai S, Kauffman KD, Schenkel JM, McBerry CC, Mayer-Barber KD, Masopust D, Barber DL (2014) Cutting edge: control of *Mycobacterium tuberculosis* infection by a subset of lung parenchyma–homing CD4 T cells. J Immunol 192(7):2965–2969
30. Flórido M, Muflihah H, Lin LCW, Xia Y, Sierro F, Palendira M, Feng CG, Bertolino P, Stambas J, Triccas JA, Britton WJ (2018) Pulmonary immunization with a recombinant influenza a virus vaccine induces lung-resident CD4+ memory T cells that are associated with protection against tuberculosis. Mucosal Immunol 11:1743–1752
31. Ashhurst AS, Flórido M, Lin LCW, Quan D, Armitage E, Stifter SA, Stambas J, Britton WJ (2019) CXCR6-deficiency improves the control of pulmonary mycobacterium tuberculosis

and influenza infection independent of T-lymphocyte recruitment to the lungs. Front Immunol 10:339

32. Mackay LK, Kallies A (2017) Transcriptional regulation of tissue-resident lymphocytes. Trends Immunol 38(2):94–103
33. Joosten SA, Ottenhoff THM, Lewinsohn DM, Hoft DF, Moody DB, Seshadri C (2019) Harnessing donor unrestricted T-cells for new vaccines against tuberculosis. Vaccine 37 (23):3022–3030
34. Ashhurst TM, Smith AL, King NJC (2017) High-dimensional fluorescence cytometry. Curr Protoc Immunol 119:5.8.1–5.8.38
35. Maciorowski Z, Chattopadhyay PK, Jain P (2017) Basic multicolor flow cytometry. Curr Protoc Immunol 117(1):5.4.1–5.4.38
36. Saletti G, Çuburu N, Yang JS, Dey A, Czerkinsky C (2013) Enzyme-linked immunospot assays for direct ex vivo measurement of vaccine-induced human humoral immune responses in blood. Nat Protoc 8 (6):1073–1087
37. Czerkinsky CC, Nilsson L-Å, Nygren H, Ouchterlony Ö, Tarkowski A (1983) A solid-phase enzyme-linked immunospot (ELISPOT) assay for enumeration of specific antibody-secreting cells. J Immunol Methods 65 (1):109–121
38. Czerkinsky C, Andersson G, Ekre HP, Nilsson LA, Klareskog L, Ouchterlony O (1988) Reverse ELISPOT assay for clonal analysis of cytokine production. I. Enumeration of gamma-interferon-secreting cells. J Immunol Methods 110(1):29–36
39. Zhang W, Caspell R, Karulin AY, Ahmad M, Haicheur N, Abdelsalam A, Johannesen K, Vignard V, Dudzik P, Georgakopoulou K, Mihaylova A, Silina K, Aptsiauri N, Adams V, Lehmann PV, McArdle S (2009) ELISPOT assays provide reproducible results among different laboratories for T-cell immune monitoring--even in hands of ELISPOT-inexperienced investigators. J Immunotoxicol 6(4):227–234
40. Boulet S, Ndongala ML, Bernard NF (2010) Dual-color ELISPOT assay for the simultaneous detection of IL-2 and/or IFN-γ secreting T cells. Cold Spring Harbor Protocols 2010(1):pdb.prot5369
41. Gazagne A, Claret E, Wijdenes J, Yssel H, Bousquet F, Levy E, Vielh P, Scotte F, Goupil TL, Fridman WH, Tartour E (2003) A Fluorospot assay to detect single T lymphocytes simultaneously producing multiple cytokines. J Immunol Methods 283(1–2):91–98

Chapter 17

Determination of Maternal and Infant Immune Responses to Pertussis Vaccination in Pregnancy

Thomas Rice and Beth Holder

Abstract

The *Bordetella pertussis* bacterium is the causative agent of whooping cough (pertussis disease). Following recent outbreaks of pertussis, disproportionately affecting young infants, several countries have introduced maternal pertussis vaccination strategies, aimed at boosting transplacental transfer of protective antibodies during pregnancy. Given historical associations between high maternal antibody and blunted infant responses to vaccination, subsequent research studies have investigated the impact of maternal pertussis vaccine on infant humoral responses. However, far less is known about the potential impact of the vaccine on innate immunity. Here, we describe methods to detect in vitro cellular responses to *B. pertussis* in mothers and their infants using a *B. pertussis* stimulation assay and multiplex cytokine assays to address this research question.

Key words Maternal vaccination, Pregnancy, Bacterial infection, Innate immunity, Cytokines, Pediatrics

1 Introduction

Epidemics of pertussis (whooping cough) have been observed in many countries including the USA [1], Australia [2], and the UK [3], resulting in the deaths of infants too young to be protected by immunization. As a result, maternal pertussis vaccination during pregnancy has become the foremost tool in protecting vulnerable young infants from pertussis in the first weeks of life, through the transplacental transfer of maternal antibody to the fetus during pregnancy [4, 5].

The bacterial agent of pertussis, *Bordetella pertussis*, elicits an innate immune response upon infecting the host [6, 7], characterized by the release of cytokines from innate immune cells such as natural killer cells [6, 8, 9]. These innate cytokines help direct the adaptive immune response as the infection progresses, promoting cytokine release from T cells.

Fadil Bidmos et al. (eds.), *Bacterial Vaccines: Methods and Protocols*,
Methods in Molecular Biology, vol. 2414, https://doi.org/10.1007/978-1-0716-1900-1_17,

In this chapter we describe how whole blood collected from mothers and infants can be utilized to investigate the impact of tetanus-diphtheria-acellular pertussis (Tdap) vaccination on innate immune cell responses. Maternal, cord and infant blood collected from unvaccinated pregnancies, and from pregnancies where mothers received Tdap vaccination are stimulated in vitro with heat-killed *B. pertussis.* Cytokine responses are detected by multiplex assay or ELISA performed on supernatants. In addition, we describe how this assay can be adapted to investigate the contribution of the plasma compartment in cord blood cytokine responses to *B. pertussis* in a plasma removal assay.

2 Materials

2.1 Patient Samples

Samples are required from mother-infant pairs, with equal numbers of Tdap-vaccinated and unvaccinated pregnancies. The gestational age at time of vaccination of the mother, and the brand of vaccine administered should be recorded.

1. Maternal sample(s): at minimum, one sample is required from the mother; usually at term. For practical reasons, this sample is often taken near the time of birth, when mothers attend a healthcare setting. The timing of the sample (before/after birth) and the mode of delivery should be considered, as labor is an inflammatory process. If samples are taken after birth, it is preferable to use samples from caesarean sections. Likewise, if patients are delivering vaginally, it is preferable to collect blood prior to commencement of labor (*see* **Note 1**). Ideally, gestation-matched controls from unvaccinated pregnancies should also be obtained.
2. Infant samples: at minimum, cord blood taken at birth can serve as a proxy for neonatal blood (*see* **Note 2**). If the aim is to investigate impact on infant vaccine responses, infant blood is additionally required at 7 weeks of age (1 week prior to primary pertussis immunization at 8 weeks) and at 5 months of age (1 month after completion of the primary pertussis immunization at 4 months; *see* **Note 3**).

2.2 Collection and Processing of Whole Blood

1. Blood collection tubes: commercially available sodium or lithium heparin tubes. Tube type should be kept consistent.
2. Needles: 21-gauge for maternal and cord blood samples, 23-gauge for infant samples.
3. 20 mL syringes.
4. Gauze.

2.3 Heat-Killed B. pertussis

1. *B. pertussis* bacteria.
2. 500 mL shake flasks.
3. THIJS medium: 994.169 mL ultrapure H_2O, 3.32 g NaCl, 0.11 g NH_4Cl, 0.5 g KH_2PO_4, 0.5 g KCl, 0.1 g $MgCl_2{\cdot}6H_2O$, 1.53 g Tris–HCl, 1.87 g Na glutamate, 40% w/v L-lactate 3.76 mL, 2.071 mL 5 M NaOH. Heat-sterilize at 110 °C for 20 min and store at 4 °C (*see* **Note 4**).
4. 100× THIJS supplement: 0.2 g L-cystine, 0.13 g $CaCl_2{\cdot}2H_2O$, 0.5 g L-glutathione reduced, 0.05 g $FeSO_4{\cdot}7H_2O$, 0.02 g nicotinic acid, 0.1 g L-ascorbic acid. Add 44 mL of ultrapure water and 6 mL of 1 M HCl. Filter-sterilize with a 0.22-μm filter and store as 1 mL aliquots at −20 °C for up to 1 year.
5. Water bath.
6. 2.0 mL cryovials.

2.4 In Vitro Whole Blood Stimulation Assay

1. 96-well sterile round-bottom tissue culture plate with lid.
2. Tissue culture hood.
3. Humidified cell culture incubator set to 5% CO_2 and 37 °C.
4. S-RPMI: Commercially available Roswell Park Memorial Institute (RPMI 1640) medium supplemented with 10% decomplemented fetal calf serum and 1% penicillin–streptomycin.
5. Heat-killed *B. pertussis* bacteria.
6. 5 mg commercially available bacterial Lipopolysaccharide (LPS).
7. 3.5 mL bijou.
8. Wet ice.

2.5 Detection of Cytokines by Multiplex and Singleplex Assays

1. Meso Scale Discovery Proinflammatory Panel 1 Human kit (*see* **Notes 5** and **6**).
2. MESO QuickPlex SQ 120 plate reader and Discovery Workbench 4.0 software (free to download).
3. Commercial IL-8 ELISA.
4. High-binding 96-well ELISA plates.
5. Non–high-binding 96-well plate.
6. Reagent reservoirs.
7. Multichannel pipette—ranges 10 μL–200 μL.
8. Adhesive plate sealers.
9. Phosphate buffered saline (PBS): 137 mM NaCl, 10 mM phosphate, 2.7 mM KCl, in H_2O. Adjust to a final pH of 7.4.
10. ELISA wash buffer: 1× PBS, 0.05% Tween 20.
11. Wet ice.

12. Vortexer.
13. 1.5 mL microcentrifuge tubes.
14. Plate shaker.
15. Rocking platform or tube rotator (optional).
16. 2 N H_2SO_4.
17. Absorbance microplate reader capable of reading 450 nm and accompanying analysis software.

2.6 Contribution of the Plasma Compartment in Cord Blood Cytokine Responses to B. pertussis

This assay is an adaption of the in vitro whole blood *B. pertussis* stimulation assay. As such, reagents and equipment as described in Subheadings 2.2–2.4 are required, in addition to those listed below.

1. Serum or plasma blood collection tubes.
2. Pools of cord blood serum or plasma from unvaccinated and Tdap-vaccinated pregnancies—40 μL required per assay (*see* **Note 7**).
3. Centrifuge.
4. Microcentrifuge tubes.

3 Methods

3.1 Preparation of Heat-Killed B. pertussis and LPS Stocks

1. Grow liquid cultures of *B. pertussis* overnight in 200 mL THIJS medium at 37 °C and 200 rpm (*see* **Note 8**).
2. Harvest *B. pertussis* at mid-log growth phase (OD_{620} 0.5–0.6) by centrifugation at 3200 × *g* for 10 min at room temperature.
3. Wash the resulting pellet in PBS, centrifuging at 3200 × *g* for 10 min at room temperature.
4. Resuspend in 10 mL THIJS medium and determine the concentration of the culture, either using a spectrophotometer or the plate count method (*see* **Note 9**).
5. Adjust the concentration of the *B. pertussis* culture to 1×10^8 CFU/mL and heat-inactivate 1 mL aliquots of *B. pertussis* for 30 min at 56 °C in a water bath.
6. Flash freeze the 1 mL aliquots of heat-inactivated *B. pertussis*, with addition of 10% v/v glycerol, and store at −80 °C.
7. LPS is provided as a lyophilized powder. Reconstitute to 5 mg/mL by the addition of 1 mL of sterile endotoxin-free water, leave to dissolve for at least 5 min and vortex thoroughly.
8. Dilute to intermediate stock concentration of 2 μg/mL in S-RPMI and store at −80 °C in 800 μL aliquots for future use.

3.2 Bulk Preparation of *B. pertussis* Stimulation Plates

Plates are prepared in bulk and frozen to reduce variability and to simplify the logistics of working with fresh blood samples. Treatments are prepared at 2× the final required concentration, as they will be diluted by the addition of an equal volume of prepared blood sample in the stimulation assay (*see* Subheading 3.4, **step 3**).

1. To prepare ten plates (each with one set of treatments), thaw one aliquot of frozen heat-killed *B. pertussis* stock and four aliquots of frozen LPS stock on ice (*see* **Note 10**).
2. Label ten 96-well round-bottom tissue culture plates with "*B. pertussis* stimulation" and the date of preparation.
3. Working in a tissue culture hood, prepare 3.5 mL of bacteria at 1×10^5 CFU/mL. From a stock of 1×10^8 CFU/mL (*see* **Note 11**), transfer 35 μL of bacteria to 3465 μL of S-RPMI. Gently vortex to mix.
4. Pour carefully into a reagent reservoir and, using a multichannel pipette, add 100 μL per well in triplicate (in three adjacent wells in a row) in each of the ten 96-well plates (*see* **Note 12**).
5. For the LPS positive control, pool four aliquots of LPS stock in a sterile 3.5 mL bijou. Vortex gently to mix (*see* **Note 13**).
6. Pour into a reagent reservoir and, using a multichannel pipette, add 100 μL per well in triplicate, adjacent to the wells containing the heat-killed *B. pertussis*, in each of the ten 96-well plates.
7. For the no-treatment (i.e., negative control) wells, add 100 μL per well of S-RPMI to the plates in triplicate, adjacent to the wells containing the LPS positive control aliquots.
8. Seal plates with the accompanying lid and freeze at −80 °C for future use (*see* **Note 14**).

3.3 Processing of Maternal, Cord, and Infant Whole Blood

1. Peripheral venous blood samples (minimum 1 mL) from mums and babies are collected into blood collection tubes (*see* **Note 15**).
2. Immediately invert the blood sample tubes end to end at least twice to ensure blood is mixed with the anticoagulant.
3. Keep tubes at room temperature and process within 2 h (*see* Subheading 3.4, **step 2**).

3.4 In Vitro *B. pertussis* Stimulation Assay

1. Thaw preprepared stimulation plates on ice.
2. In a tissue culture hood, for each sample dilute blood 1:5 by adding 200 μL of whole blood to 800 μL S-RPMI in a sterile bijou.
3. Pour into a reagent reservoir and, using a multichannel pipette, add 100 μL diluted blood to each of the nine wells containing a treatment (i.e., *B. pertussis*, LPS control and negative control) per sample. Mix by gently pipetting up and down twice and

change tips with each addition. This results in a final blood dilution of 1:10 and a *B. pertussis* concentration of 5 × 10^5/mL.

4. Place the plate in the incubator at 37 °C, 5% CO_2 for 24 h.
5. After 24 h, carefully remove the supernatant into two new 96-well cell culture plates (*see* **Note 16**).
6. Place the tissue culture plate lids onto the plates and freeze at −80 °C for use in future assays (*see* Subheadings 3.5 and 3.6).

3.5 Meso Scale Discovery (MSD) Multiplex Cytokine Assay

MSD plates can only be used once. Therefore, it is important to accumulate samples before analyzing cytokine levels in supernatants using this assay.

1. Prepare a plate plan for your experiment (*see* **Note 17**).
2. Thaw eight sets of supernatant samples (from Subheading 3.4) on ice and allow all assay reagents to reach room temperature before being used (*see* **Note 18**).
3. Reconstitute the cytokine calibrator provided in the kit by adding 1 mL of Diluent 2 to lyophilized MSD Calibrator Blend and vortexing (*see* **Note 19**). Allow the reconstituted calibrator to sit for a minimum of 5 min before using.
4. In this time, add 300 μL of Diluent 2 to each of eight 1.5 mL microfuge tubes.
5. Prepare a dilution series of the calibrator by transferring 100 μL to the first of the tubes containing Diluent 2, prepared above, to create a fourfold dilution. Mix well by vortexing.
6. Transfer 100 μL to the subsequent tube and repeat this process to serially dilute 6 more times, leaving the eighth tube with Diluent 2 alone. This final tube will serve as the calibration blank.
7. Using a multichannel pipette, aliquot 90 μL Diluent 2 per well into ten columns of wells in a 96-well plate (not supplied in the MSD kit; use any non–high-binding plate), according to the MSD plate layout (*see* **Note 17**). This plate will be used for the dilution of supernatant samples only.
8. Dilute thawed supernatant samples (from **step 2** above) 1:10 by transferring 10 μL of each sample, using a multichannel pipette, into appropriate wells (according to prepared plate layout) containing 90 μL Diluent 2 and mix by pipetting up and down at least five times. Change tips between sets of samples. All samples are tested at a single dilution of 1:10.
9. Add 50 μL of standard and samples to the MSD plate according to the plate layout (*see* **Note 20**).
10. Cover plates with an adhesive plate sealer and incubate for 2 h at room temperature on a plate shaker at ≥300 rpm.

11. Discard samples and standards by flicking the plate into a laboratory sink.
12. Using a multichannel pipette, wash each well with 150 μL PBS–0.05% Tween, allowing the well to soak for at least 30 s. IMPORTANT: Do not use an automatic plate washer. Afterward, blot-dry by tapping plate well-side down on tissue paper.
13. Prepare a cocktail of detection antibodies for all cytokines (apart from IL-8) at 1× concentration in Diluent 3. The total volume required for a full plate is 3 mL, which equates to 60 μL per antibody in 2.46 mL diluent 3 (*see* **Note 21**).
14. Add 25 μL detection antibody cocktail per well, ensuring the whole surface is covered and there are no bubbles. Incubate on plate shaker at ≥300 rpm for 2 h at room temperature.
15. Wash the plate three times, as described in **step 12** above, with PBS–0.05% Tween and blot-dry.
16. Add 150 μL 2× Read Buffer to each well (*see* **Note 22**).
17. Read plates within 30 min on a MESO QuickPlex SQ 120 and analyze data using Discovery Workbench 4.0.

3.6 IL-8 ELISA

1. Prepare a plate plan for your experiment (*see* **Note 23**).
2. Prepare 1× IL-8 capture antibody in 1× coating buffer (*see* **Note 24**), by adding 40 μL antibody to 10 mL coating buffer per 96-well plate (*see* **Note 25**).
3. Coat ELISA plate with 100 μL per well of IL-8 capture antibody (*see* **Note 26**).
4. Seal the plate with an adhesive plate sealer and incubate overnight at 4 °C.
5. Aspirate and wash wells three times with 200 μL ELISA wash buffer to remove unbound antibody (*see* **Note 27**). Allow wells to soak in wash buffer for at least 30 s. Afterward, blot on tissue paper to remove and excess wash buffer from wells.
6. To prevent nonspecific binding of antigens and antibodies, block wells with 200 μL 1× diluent (*see* **Note 28**).
7. Seal and incubate the plate at room temperature for a minimum of 1 h.
8. Whilst the plate is being blocked, thaw supernatants on ice (*see* **Note 29**).
9. Using standard nonsterile 96-well plates (not high-binding), dilute assay supernatants to an intermediate dilution of 1:50 in PBS according to the plate plan designed in **step 1** (*see* **Note 30**). For example, in one plate, dilute 10 μL of sample in 90 μL of PBS for a 1:10 dilution. In a separate plate, take 20 μL of the diluted sample from plate one, and add to 80 μL PBS (making an overall intermediate dilution of 1:50; *see* **Note 31**). Ensure

during dilution steps that sample and PBS are well mixed by pipetting up and down at least five times. When used in **step 15** below, the final dilution for samples will be 1:500 (*see* **Note 32**).

10. Reconstitute the IL-8 standard provided in the kit using the amount of deionized water stated on the vial, and place on a rocking platform or tube rotator for at least 15 min to solubilize (*see* **Note 33**).
11. Whilst the standard is being reconstituted, wash the blocked ELISA plate twice as described in **step 5**, and blot-dry. It is important not to leave the wells dry for long periods of time between steps.
12. To columns 1 and 2 of the plate, apart from wells A1 and B1 which will contain the reconstituted standard, add 100 μL of 1× diluent. In addition, 100 μL of 1× diluent should be added to the two designated blank wells G12 and H12.
13. To all wells that will receive samples, add 90 μL of 1× diluent (*see* **Note 34**).
14. Add 200 μL of the reconstituted IL-8 standard to wells A1 and B1. Using a multichannel pipette, take 100 μL from A1 and B1 into A2 and B2, pipetting up and down to mix with the diluent present in the well. Change tips, and repeat the process a further six times, discarding 100 μL from A8 and B8. This will create a serially diluted eight-point standard curve.
15. Add 10 μL of prediluted 1:50 sample from **step 9** above to designated sample wells containing 90 μL of 1× diluent, making a final sample dilution of 1:500.
16. Once all samples and standards are added, seal the plate with an adhesive plate sealer and incubate at room temperature for 2 h.
17. Prepare biotin-conjugated IL-8 detection antibody by adding 40 μL antibody to 10 mL 1× diluent per 96-well plate, or the same volumes as for capture antibody if running partial plates (*see* **Note 25**).
18. Wash and aspirate plates four times with wash buffer as described in **step 5**.
19. Add 100 μL diluted IL-8 detection antibody per well, seal plate and incubate at room temperature for 1 h (*see* **Note 35**).
20. Prepare avidin-conjugated horseradish peroxidase (HRP) enzyme by adding 40 μL HRP conjugate to 10 mL 1× diluent for a 96-well plate, or same volumes as for capture antibody if running partial plates (*see* **Note 25**).
21. Wash and aspirate plates four times with wash buffer as described in **step 5**.

22. Add 100 μL prepared HRP per well, seal plate and incubate at room temperature for 30 min.
23. Wash and aspirate plates six times with wash buffer as described in **step 5**.
24. Add 100 μL of 3,3′,5,5′-tetramethylbenzidine (TMB) per well (*see* **Note 36**).
25. Monitor the colour change of sample and standard wells from clear to blue to ensure samples and standards do not become saturated. When the standard curve is colored blue, with a visible difference between the middle standards (*see* **Note 37**), stop the reaction by the addition of 100 μL 2 N H_2SO_4 (*see* **Note 38**).
26. Immediately read the plates at 450 nm using an absorbance microplate reader. If wavelength subtraction is available, subtract the values of readings at 570 nm from those at 450 nm and analyze data (*see* **Note 39**).

3.7 Contribution of the Plasma Compartment in Cord Blood Cytokine Responses to B. pertussis

1. Collect blood in blood collection tubes. The assay requires 400 μL of cord whole blood.
2. Dilute cord blood 1:5 in S-RPMI. After dilution, remove 700 μL of this into a microcentrifuge tube for use in control wells.
3. Centrifuge the rest of the diluted blood at 500 × *g* for 5 min.
4. Remove the top clear plasma layer, without disturbing the cells beneath and record the volume taken off. Replace this volume with S-RPMI (*see* **Note 40**).
5. Prepare 96-well cell culture stimulation plate (*see* Fig. 1 for layout). To eight wells (1A-1D and 2A-2D), add 80 μL S-RPMI. To eight wells (3A-3D and 4A-4D), add *B. pertussis* prepared in 80 μL S-RPMI. To 2 wells (5A and 6A), add 100 μL of 2 μg/mL LPS.

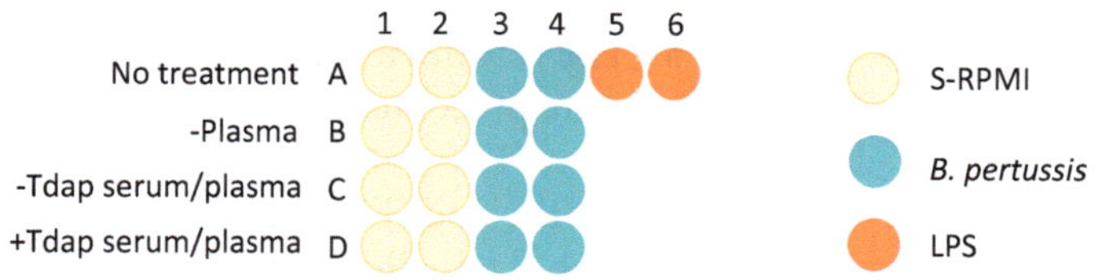

Fig. 1 Plate layout showing the stimulation conditions for the plasma removal assay. Cord blood is stimulated with S-RPMI, *B. pertussis* and LPS. Stimulations are done with plasma removed from cord blood and replaced with S-RPMI (−plasma), with plasma removed and replaced with cord plasma/serum from unvaccinated pregnancies (−Tdap serum/plasma) or with plasma removed and replaced with cord plasma/serum from Tdap-vaccinated pregnancies (+Tdap serum/plasma)

6. Add a further 20 μL S-RPMI to wells (except 5A and 6A) in rows A and B.
7. Add 20 μL of unvaccinated (to wells in row C) and Tdap-vaccinated (to wells in row D) cord blood serum or plasma, as shown in Fig. 1.
8. Add 100 μL of blood with plasma to all wells in row 1. These are the no-treatment control wells for the assay. Mix up and down using a multichannel pipette.
9. To the rest of the wells on the plate, add 100 μL of cord blood without plasma. Mix up and down using a multichannel pipette.
10. Incubate at 37 °C and 5% CO_2 for 24 h.
11. Remove the supernatant as described in Subheading 3.4, **step 5** (*see* **Note 16**), and store at −80 °C.
12. Analyse cytokine levels in supernatants by multiplex assay or ELISA as described in Subheadings 3.5 and 3.6, respectively.

4 Notes

1. If desired, additional samples can be taken. These could comprise samples taken before vaccination (which can be the day of vaccination) and 2 weeks post vaccination. This will enable comparison of innate cell responses before and after vaccination in the same patient, as well as the measurement of the antibody response, if desired.
2. The neonatal immune system changes quite rapidly post-birth, and this should be recognized when using cord blood [10].
3. Pediatric vaccine schedules differ between countries, so it is important to check the national vaccine schedule for infants in your study and record the dates and the doses of vaccines infants receive.
4. For background on THIJS medium, please *see* Ref. 11.
5. The Proinflammatory Panel 1 Human kit contains an MSD plate precoated with capture antibodies for each target cytokine (i.e., IFN-γ, IL-1β, IL-2, IL-4, IL-6, IL-8, IL-10, IL-12p70, IL-13, and TNF-α) that are located on independent spots in each well, allowing individual samples to be tested in a single well for this set of cytokines using a cocktail of the detection antibodies provided. The kit also contains the necessary standards, diluents, and buffers for the assay.
6. The chemokine IL-8 is found at high levels and is regularly out of range of detection in the MSD assay. Levels of IL-8 can be measured separately by ELISA.

7. To prepare serum pools, thaw ten serum samples on ice. Pool and mix, aliquoting and freezing at −80 °C for use in future experiments.
8. All laboratory work with *B. pertussis* should be carried out in Biosafety level 2 laboratories.
9. It is helpful to freeze *B. pertussis* at a working stock concentration of 10×10^8 colony forming units (CFU)/mL. The plate count method consists of serially diluting a sample with sterile saline or phosphate buffer diluent until the bacteria are dilute enough to count accurately. CFU/mL can then be calculated as (no. of colonies × dilution factor)/volume of culture plate.
10. This protocol describes preparing ten plates to be thawed and used for one blood sample each. To save on plastic laboratory consumables, and if laboratories expect to receive more than one blood sample in a day, plates can be prepared to include more than one sample per plate.
11. *B. pertussis* stimulation will be performed in triplicate at a final concentration of 5×10^5 CFU/mL. For ten plates, with one sample per plate, a total of 3 mL of bacteria is required. The bacteria will be diluted in the plate by the addition of an equal volume of blood.
12. Any set of nine contiguous wells on the plate can be used for the triplicate aliquots of the three treatments, as this will allow use of a multichannel pipette for downstream processing. Surround the stimulant wells with water or PBS to help prevent evaporation in these wells once blood samples are added and the plate is placed in an incubator. This can be done prior to freezing, or once the plate has been thawed.
13. LPS is a component of gram-negative bacteria that elicits an innate immune response through toll-like receptor 4. This serves as a positive control in cytokine assays. The final concentration used in our assay is 1 μg/mL.
14. Plates can be prepared in batches and frozen at −80 °C to limit variability between plates. This also fits in with the logistics of pregnancy studies, where the timing of sampling can be unpredictable.
15. For the collection of cord blood from the placenta, fetal veins are usually larger, whereas arteries are smaller and more superficial. Arteries usually cross over veins at some point in their chorionic plate arrangement. Venous cord blood can be taken from the umbilical cord of the placenta, or the fetal veins on the surface of the placenta. Ensure the placenta has been wiped with gauze to prevent contamination with any maternal blood that may be present of the surface of the placenta.

16. Be careful not to disturb the cell pellet sitting at the bottom of the well. Approximately 160 μL of supernatant can be removed using a multichannel pipette without disturbing the cell pellet. This can be done in two 80 μL draws into one plate, mixing by pipetting up and down, and then removing 80 μL into the second plate.
17. MSD plates are precoated with capture antibodies for each target, located on independent spots within each of the 96 wells on the plate. MSD plates are expensive and can only be used once; therefore, it is good to make maximal use of the available wells. For the experiment described in this chapter, each blood sample requires nine wells in order to test the effects of three treatments, each tested in triplicate. A suggested MSD plate plan that can accommodate eight sets of samples is shown in Fig. 2. As columns 1 and 2 will contain duplicates of standards, the eight sets of samples must be arranged in in consecutive rows (A to H) of nine columns, which would leave one column empty. In order not to waste these wells, other samples, unrelated to the experiment described in this chapter may be tested in this column. A different layout will be required for the samples from Subheading 3.7.
18. Upon the first thaw of MSD assay diluents, aliquot into smaller volumes for future use to avoid repetitive freeze–thaw cycles. 8 mL aliquots of Diluent 2 and 3 mL aliquots of Diluent 3 can be made. Aliquots can be kept at −20 °C.
19. Each kit contains a vial of lyophilized calibrator per MSD plate. The reconstituted calibrator is stable for 1 day at 2–8 °C.

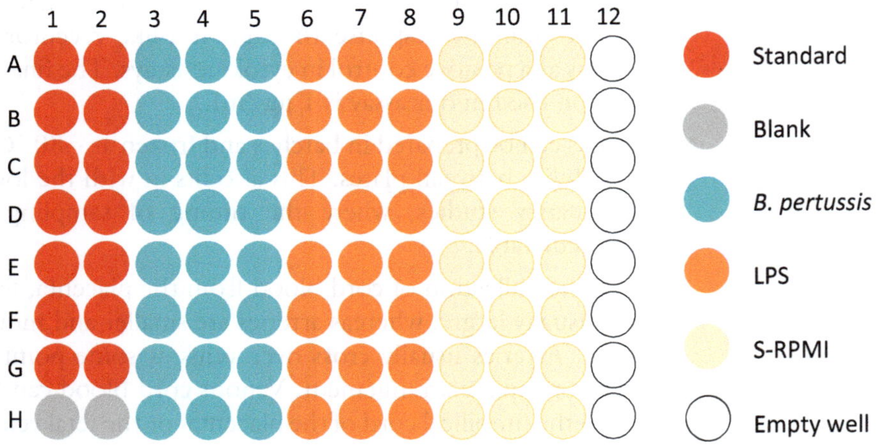

Fig. 2 MSD plate layout. Shown is the suggested layout for testing of eight sets of blood samples, each stimulated with the three treatments (S-RPMI, *B. pertussis*, and LPS) run in triplicate. The assay standards and blanks are applied to wells in columns 1 and 2, and the eight sets of samples are applied, one per row, in wells of columns 3 to 11, as shown

20. It is important that the full surface of the well is covered with sample. Reverse pipetting can help to prevent bubbles in the wells. To reverse pipette, set the pipette to the desired volume. When pressing the pipette plunger down to take up a solution, go past the first stop. Immerse the tip in the liquid, and slowly release the plunger to full extension. The takes up more volume of solution than set on the pipette. When dispensing, press the plunger down to the first stop. A small volume of liquid will remain in the tip, which helps to prevent bubbles.
21. For the detection antibody cocktail, use 60 μL per antibody, or $60 \times 9 = 540$ μL. For 3 mL total volume, add 2.46 mL Diluent 3. NB: the IL-8 antibody is not included in the cocktail, as sample levels are regularly out of range in the MSD assay, so are tested separately by ELISA.
22. Read Buffer is provided at 4× concentration, dilute to 2× in deionized water. For a 96-well plate, add 7.5 mL of Read Buffer to 7.5 mL of deionized water.
23. A suggested ELISA plate plan that can accommodate eight sets of samples, tested at a single dilution, is shown in Fig. 3. Columns 1 and 2 will contain duplicates of standards, and the eight sets of samples are tested in consecutive rows (A to H) of columns 3 to 11. The final column contains two wells used for blanks and six empty wells. A different layout will be required for the samples from Subheading 3.7.
24. Coating buffer is provided as a 10× stock, dilute to 1× in deionized water. For one full plate, dilute 1 mL coating buffer in 9 mL deionized water.

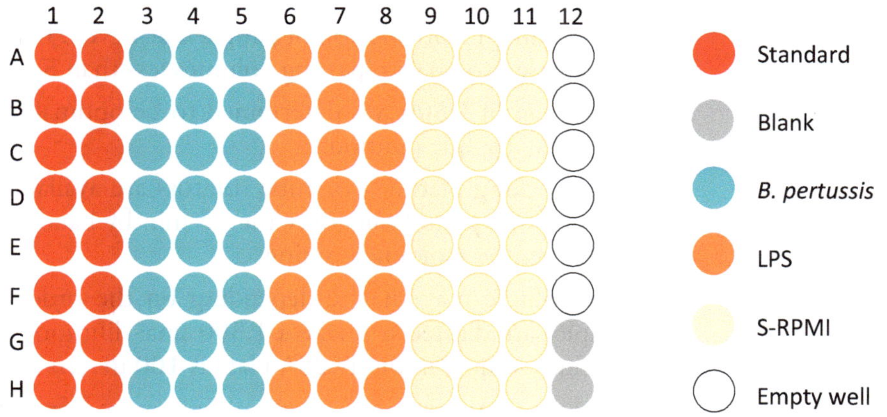

Fig. 3 IL-8 ELISA plate layout. Shown is the suggested layout for testing of eight sets of blood samples, each stimulated with the three treatments (S-RPMI, *B. pertussis*, and LPS) run in triplicate. The assay standards are applied to wells in columns 1 and 2, and the eight sets of samples are applied, one per row, in wells of columns 3 to 11, as shown. Two blank wells are included in column 12

Table 1
Volumes of reagents for partial ELISA plates

Number of ELISA plate columns	12	11	10	9	8	7	6	5	4	3
Volume of antibody (μL)	40.0	36.7	33.3	30.0	26.7	23.3	20.0	16.7	13.3	10.0
Volume of diluent (mL)	10.0	9.2	8.3	7.5	6.7	5.8	5	4.2	3.3	2.5

25. Partial ELISA plates can be run using the reagent volumes shown in Table 1.
26. Pipette into wells by placing the tips of the pipettes on the rim of one side of the well. This will help prevent contamination between wells in later steps.
27. Wells can be washed and aspirated using an automatic plate washer, or the contents can be tipped out into a laboratory sink and washed with a multichannel pipette as described earlier in the protocol (*see* Subheading 3.5, **step 12**).
28. Diluent is provided as a 5× stock to be diluted to 1× in deionized water. Add 10 mL diluent to 40 mL deionized water. The excess 1× diluent can be saved for later steps in the ELISA experiment.
29. The IL-8 ELISA and MSD assays can be performed separately, as stimulation supernatants are frozen in two separate plates, to minimize freeze-thawing.
30. Standards are prepared separately, so there is no need to add PBS to columns 1 and 2 for the sample predilution steps.
31. A 1:50 dilution can be done in 96-well plates in two steps (1:10 followed by 1:5) as described, with a final (1:10) dilution being done in the ELISA plate, giving an overall sample dilution of 1:500. However, these volumes can be changed, as long as the desired final 1:500 dilution of sample is achieved. For accuracy, it is advised to avoid pipetting volumes less than 5 μL.
32. A dilution of 1:500 was determined to be optimal for sample testing at a single dilution.
33. If no rocking platform or tube rotators are available, mixing can be done periodically over the 15 min by hand. Invert the bottle gently, avoiding foaming.
34. This volume may change depending on the order in which samples are diluted to make a 1:500 final dilution, as long as the final volume in the ELISA plate is 100 μL (*see* **Note 31**).
35. After the sample incubation, all subsequent reagents should be added at a consistent speed and order across the plate to make the incubation time in each well as consistent as possible.
36. TMB is a chromogenic substrate that reacts with HRP, resulting in a colour change when the assay target is present in each

sample. When pipetting this into the wells, lean the tips on the opposite side of the well to previous steps to avoid cross-contamination.

37. It is important to ensure the reaction is not stopped too early, before wells with low-level concentrations have developed, or too late, when the colour change becomes saturated leading to inaccurate standard curve and/or plateauing of samples. It can be helpful to use standard rows C and D as a guide; as long as there is still a difference in colour between these wells, the reaction can continue. If samples wells turn brightly colored immediately and before the standard curve can develop, stop the reaction as the sample concentrations will be too high and out of range. If needed, repeat experiment with a new dilution.
38. Sulfuric acid is a highly corrosive chemical that is potentially explosive in concentrated form. It can cause severe skin burns, can irritate the nose and throat and cause difficulties breathing if inhaled. Laboratory personal protective equipment should be worn when using, including gloves and a lab coat.
39. The accompanying analysis software to the microplate reader will produce a concentration of IL-8 in each sample by extrapolating from the standard curve values, taking into account the sample dilution factor.
40. Be careful not to disturb the blood layer below the plasma. It can help to remove the final volume of plasma with a smaller volume pipette, such as P200 micropipette.

References

1. Sawyer M, Liang JL, Messonnier N, Clark TA (2012) Updated recommendations for use of tetanus toxoid, reduced diphtheria toxoid, and acellular pertussis vaccine (Tdap) in pregnant women--advisory committee on immunization practices (ACIP). Morb Mortal Wkly Rep 62 (7):131–135
2. Wiley KE, Massey PD, Cooper SC et al (2013) Pregnant women's intention to take up a post-partum pertussis vaccine, and their willingness to take up the vaccine while pregnant: a cross sectional survey. Vaccine 31(37):3972–3978
3. Donegan K, King B, Bryan P (2014) Safety of pertussis vaccination in pregnant women in UK: observational study. BMJ 349(9526): g4219
4. Gall SA, Myers J, Pichichero M (2011) Maternal immunization with tetanus–diphtheria–pertussis vaccine: effect on maternal and neonatal serum antibody levels. Am J Obstet Gynecol 204(4):334. e1
5. Palmeira P, Quinello C, Silveira-Lessa AL et al (2012) IgG placental transfer in healthy and pathological pregnancies. Clin Dev Immunol 2012:985646
6. Byrne P, McGuirk P, Todryk S, Mills KHG (2004) Depletion of NK cells results in disseminating lethal infection with *Bordetella pertussis* associated with a reduction of antigen-specific Th1 and enhancement of Th2, but not Tr1 cells. Eur J Immunol 34 (9):2579–2588
7. Higgins SC, Lavelle EC, McCann C et al (2003) Toll-like receptor 4-mediated innate IL-10 activates antigen-specific regulatory T cells and confers resistance to *Bordetella pertussis* by inhibiting inflammatory pathology. J Immunol 171(6):3119–3127
8. Mahon BP, Sheahan BJ, Griffin F et al (1997) Atypical disease after *Bordetella pertussis* respiratory infection of mice with targeted disruptions of interferon-γ receptor or

immunoglobulin μ chain genes. J Exp Med 186 (11):1843–1851

9. Kroes MM, Mariman R, Hijdra D et al (2019) Activation of human NK cells by *Bordetella pertussis* requires inflammasome activation in macrophages. Front Immunol 10:2030
10. Lee AH, Shannon CP, Amenyogbe N et al (2019) Dynamic molecular changes during the first week of human life follow a robust developmental trajectory. Nat Commun 10 (1):1–14
11. Thalen M, Van Den Ijssel J, Jiskoot W et al (1999) Rational medium design for *Bordetella pertussis*: basic metabolism. J Biotechnol 75 (2–3):147–159

Chapter 18

Generation of a Universal Human Complement Source by Large-Scale Depletion of IgG and IgM from Pooled Human Plasma

Frances Alexander, Emily Brunt, Holly Humphries, Breeze Cavell, Stephanie Leung, Lauren Allen, Rachel Halkerston, Elodie Lesne, Elizabeth Penn, Stephen Thomas, Andrew Gorringe, and Stephen Taylor

Abstract

Complement is a key component of functional immunological assays used to evaluate vaccine-mediated immunity to a range of bacterial and viral pathogens. However, standardization of these assays is complicated due to the availability of a human complement source that lacks existing antibodies acquired either through vaccination or natural circulation of the pathogen of interest. We have developed a method for depleting both IgG and IgM in 200 mL batches from pooled hirudin-derived human plasma by sequential affinity chromatography using a Protein G Sepharose column followed by POROS™ CaptureSelect™ IgM Affinity resin. The production of large IgG- and IgM-depleted batches of human plasma that retains total hemolytic and alternative pathway activities allows for improved assay standardization and comparison of immune responses in large clinical trials.

Key words Complement system, IgG depletion, IgM depletion, Human plasma, FPLC

1 Introduction

Complement-mediated immunity is an important mechanism in the host defense against invading pathogens and its measurement is a valued tool in evaluating vaccine-derived immunity. Serum bactericidal activity (SBA) has long been established as a correlate of protection against meningococcal disease [1, 2] and many immunoassays have been developed to understand the role of complement in protection against a variety of bacterial and viral pathogens [3–13]. One of the challenges to such functional immunoassays is the availability of a standardized complement source that lacks intrinsic bactericidal and/or opsonophagocytic antibodies against the pathogen of interest which could interfere with the

Fadil Bidmos et al. (eds.), *Bacterial Vaccines: Methods and Protocols*,
Methods in Molecular Biology, vol. 2414, https://doi.org/10.1007/978-1-0716-1900-1_18,

assay. Vaccination or natural circulation of the bacteria or virus within the human population means that many researchers use an animal source of complement such as baby rabbit or guinea pig serum that is claimed to lack this intrinsic activity [14, 15]. However, in the case of meningococcal SBA, the high specificity of meningococcal factor H binding protein for human factor H is known to result in elevated bactericidal titers when rabbit complement is used as the complement source [16]. Poor correlation in bactericidal titers obtained using rabbit or human complement with meningococcal serogroup A, W, and Y strains [17] also suggests other species-specific interactions may play a role. Depending on the pathogen of interest it may be possible to screen individuals to identify a source of human serum with low cross-reactive antibodies to the test strain. Alternatively, small batches (1–2 mL) of human serum could be IgG-depleted with a Protein G column immediately prior to addition to immunoassays [7, 18, 19]. While more biologically relevant, these methods make it difficult to standardize assays and compare across different strains. Different donors may be required for each target strain and day-to-day variation in small batches of IgG-depleted serum can easily be introduced, if they are not fully characterized and adjusted accordingly [20]. An alternative option to avoid the impact of pre-existing antibody may be the reconstitution of a functioning complement system using purified proteins [21].

We have previously described a method to deplete IgG from 300 mL batches of pooled lepirudin-derived human plasma that retains key complement cascade components and total hemolytic and alternative pathway activities [22]. This method allows for greater availability of large batches of human complement for improved assay standardization and comparison of immune responses. Here, we present an updated method whereby pooled hirudin-derived human plasma is depleted of both IgG and IgM in 200 mL batches by sequential affinity chromatography using Protein G Sepharose column followed by POROS™ CaptureSelect™ IgM Affinity resin.

Due to the delicate interplay between the complement and coagulation cascades [23, 24], plasma rather than serum was chosen as the complement source. Commonly used anticoagulants can adversely affect the complement cascade [25] and in our initial study lepirudin (Refludan™) derived plasma was chosen over heparinized plasma as it showed the greatest total hemolytic activity [22]. Unfortunately, due to lack of clinical demand for irreversible non-heparin anticoagulant, the manufacturer ceased production of lepirudin. The use of bivalirudin (Desirudin™) was trialed but the reversible nature of the anticoagulant meant the complement source was not suitable for assays with incubation times that extended beyond the binding half-life unless excess anticoagulant was added to the assay. In this updated method recombinant hirudin, from which lepirudin is derived, is used as the anticoagulant.

Protein G Sepharose was chosen as it was able to completely remove IgG below detectable levels from the plasma compared to either Protein A or Protein L Sepharose [22]. We also found that there was a significant reduction in IgM following IgG depletion but no change in IgA levels [22]. While IgA is a poor activator of complement [26], and so should not have intrinsic bactericidal/opsonophagocytic activity, IgM is able to interact with C1q to initiate complement activation [27]. The importance of also depleting IgM varies depending on the bacterial pathogen of interest. Our initial studies showed the suitability of IgG-depleted human plasma as a complement source in complement binding (measuring deposition of complement proteins C3b/iC3b and C5b) and SBA assays with serogroup B meningococci. However, unencapsulated human pathogens such as *Bordetella pertussis* and nontypeable *Haemophilus influenzae* are easily killed in vitro using human complement due to activation by IgG [10, 28]. With the additional removal of IgM, we have since shown complement-mediated immunity to be important in protection against *B. pertussis* and nontypeable *H. influenzae* [10, 11, 28, 29]. Removal of IgM also allows assessment of functional immunity to commonly carried encapsulated Gram-positive bacteria such as Group B *Streptococcus* [9, 30] for which donors have high levels of both IgG and IgM.

The Poros CaptureSelect™ IgM resin has a monoclonal anti-IgM antibody covalently linked to the resin. Screening of other resins gave poor recoveries of IgM from plasma whereas Poros CaptureSelect™ IgM resin gave consistent and reproducible results, removing IgM below detectable levels. The resin is also sanitizable with 70% ethanol (minimum of 12 h contact time) and has a linear flow rate (not just volumetric flow rate) that is compatible with the Protein G resin, allowing the plasma to run through both columns in a single run and so reducing the time needed for the double depletion.

As described in Brookes et al. [22], C1q and C5 are also removed following IgG-depletion with Protein G Sepharose and a NaCl gradient elution step is required to remove these proteins from the affinity matrix [31]. While the loss of C5 during this step was unexpected, the removal of C1q is likely due to its activation to IgG bound to Protein G Sepharose [32, 33]. Both C1q and C5 were identified in the same eluted fractions, concentrated to pre-depletion levels and added back into the final IgG- and IgM-depleted plasma. No further loss of complement components was identified after passing the IgG-depleted plasma through the POROS™ CaptureSelect™ IgM Affinity column.

By using large pools of plasma from >20 volunteers it is possible to achieve consistency between batches of IgG- and IgM-depleted human plasma, with similar concentrations of all tested complement proteins; as well as comparable total hemolytic and alternative pathway complement activities as assessed by functional analysis of serum bactericidal activity and antibody-mediated deposition of complement proteins C3b/iC3b and C5b-9. However, it should be noted that not all the complement components have been exhaustively tested and it is possible that there are some that are depleted during processing.

2 Materials

2.1 Plasma and Blood Plasma Preparation

1. Freshly drawn blood from volunteer donors (*see* **Note 1**), using butterfly cannulas and 50 mL syringe.
2. Recombinant hirudin mature variant from *Pichia pastoris* freeze dried by request, >1 × 10^4 antithrombin units (ATU)/mg (Creative BioMart). Reconstitute hirudin to 20 mg/mL using sterile water for irrigation (*see* **Note 2**). Add sterile hirudin to 50 mL polypropylene tubes to give 2 mg per 50 mL whole blood.
3. Benchtop centrifuge: speeds up to 3000 × *g*.

2.2 Sanitization of Chromatography System and Columns (See Note 3)

1. Water: HPLC grade, filtered at 0.4 μm.
2. Phosphate buffered saline (PBS): 0.1369 M sodium chloride, 0.0081 M disodium hydrogen orthophosphate, 0.0015 M potassium dihydrogen phosphate, 0.0027 M potassium chloride pH 7.4 ± 0.2. Prepared in-house.
3. 0.1 M sodium hydroxide, filtered.
4. 20% ethanol in HPLC grade water, filtered.
5. 70% ethanol in HPLC grade water, filtered.
6. 70% ethanol in HPLC grade water, filtered in spray bottle.

2.3 Chromatography Buffers (See Note 3)

1. Water: HPLC grade filtered at 0.4 μm.
2. Phosphate buffered saline (PBS): 0.1369 M sodium chloride, 0.0081 M disodium hydrogen orthophosphate, 0.0015 M potassium dihydrogen phosphate, 0.0027 M potassium chloride pH 7.4 ± 0.2. Prepared in-house.
3. PBS supplemented with 1.5 M sodium chloride.
4. Elution buffer: 0.2 M glycine-HCl. Adjust pH to 2.7.

2.4 Chromatography Columns

1. IgG depletion resin: 360 mL Protein G Sepharose . Linear flow rate = 150–250 cm/h.
2. IgG depletion column: 400 × 50 mm h × internal diameter.
3. IgM depletion Resin: 70 mL POROS™ CaptureSelect™ IgM Affinity resin. Linear flow rate = 150 cm/h.

4. IgM depletion column: 400 × 26 mm height × internal diameter.
5. Manual switches ×8, SRV-1 (GE/Cytiva).

2.5 Chromatography Systems

1. AKTA FPLC, fraction collector placed within a Class II microbiological safety cabinet (MSC II), at 22 °C with UV monitor of 280 nm, conductivity cell and pH monitor.
2. AKTA Purifier, in cold room (or 4 °C) with UV monitor of 280, 254, 215 nm, conductivity cell, pH, and temperature monitor.

2.6 Concentration of Plasma, Complement, and C1q

1. Sterilized dialysis tubing: Snakeskin 3.5 K molecular weight cut off (MWCO) tubing, autoclaved in HPLC water.
2. Polyethylene glycol 20,000 Da.

2.7 Ancillary Items

1. Collection vessels: 100 mL to 5 L borosilicate glass bottle, dry-heat sterilized.
2. Collection vessel lids: Bottle lids individually autoclaved.
3. Aluminum foil: 200 × 200 mm folded and autoclaved.
4. Blunt ended forceps and scissors: dry-heat sterilized in first instance, and autoclaved thereafter.
5. Sterile lint-free paper: Autoclaved.
6. Tubes: Polypropylene cryotubes with internal thread.
7. Cryogenic storage box with hollow prong dividers.
8. Dry ice and ethanol bath for snap freezing (*see* **Note 4**).

2.8 QC Testing—Bioburden

1. Columbia agar with 5% horse blood media plates.
2. Chocolate agar with PolyViteX media plates.
3. Trypticase soy agar media plates.

2.9 QC Testing—Presence of Ig and C1q

1. Standard SDS PAGE equipment.
2. Standard Western blot equipment.
3. Anti-human IgG, goat alkaline phosphatase.
4. Anti-human IgM, goat alkaline phosphatase.
5. C1q protein.

2.10 QC Testing—Complement Component Activity

1. Optilite IE700, Optimised Protein System.
2. Optilite C1 inactivator kit.
3. Optilite C3c kit.
4. Optilite C4 kit.
5. Optilite CH50 kit.
6. Optilite prealbumin kit.

7. Optilite IgG kit.
8. Optilite IgM kit.
9. Radial Immunodiffusion Assays (RID).
   (a) Total hemolytic complement.
   (b) Alternative pathway hemolytic.
   (c) C1 inactivator.
   (d) C1q.
   (e) C2.
   (f) C3.
   (g) C4.
   (h) C4 binding protein.
   (i) C5.
   (j) C6.
   (k) C7.
   (l) C8.
   (m) C9.
   (n) Factor B.
   (o) Factor H (β 1H).
   (p) Factor I.
10. RID Reader, or microscope with micrometer and imager.

3 Method

All steps should be carried out using aseptic technique and where possible within an MSC II to maintain sterility, the exterior of all equipment should be sprayed with 70% ethanol (*see* **Note 5**). To prevent activation of complement during the processing care must be taken to avoid certain plastics (*see* **Note 6**).

3.1 Plasma Preparation

1. Collect venous human blood from multiple donors using butterfly cannulas and 50 mL syringes and immediately decant into polypropylene centrifuge tubes containing hirudin (final concentration 0.04 mg/mL) on ice. Mix the blood gently with the hirudin by inversion of the tube, without frothing.
2. Centrifuge at 3000 × *g* for 10 min to separate red blood cells. All particulate matter is removed, and the native plasma is snap frozen in small batches, *see* **Notes** 7 and **8**.
3. Defrost the small batches under running water and pool before snap freezing in 45 mL aliquots.
4. On day of depletion, defrost the plasma in approximately 200 mL batches and centrifuge again at 3000 × *g* for 10 min to remove any particulates or clots, *see* **Note 9**. Take 5 × 1 mL

aliquots of native plasma for QC, transfer remaining plasma to 4 °C and use immediately, *see* **Note 9**.

3.2 Column Setup and Sanitization

3.2.1 Column Preparation

1. Place the columns in a clamp within a laminar flow cabinet or MSC II and dismantle and check the O rings and adaptors (top and base) for integrity.
2. Wash the columns with sterile HPLC grade water (also sanitize with 0.1 M sodium hydroxide if previously used) and immerse all wet areas in sterile 70% ethanol for 12 h minimum. Immerse the top adaptor separately (not within the column).
3. Discard the 70% ethanol and replace with a sterile 20% ethanol wash before packing the columns with the appropriate resin, *see* **Note 10**.
4. Place the top adaptor on a sterile surface (foil or paper) and insert and seal when the column has been packed.

3.2.2 Sanitization of Chromatography Systems with Sodium Hydroxide (See **Notes 11** and **12**)

1. Assemble the chromatography systems with blank lines in place of columns, with dual end connector switches.
2. Set the pressure limit for system to 0.5 Pa.
3. Purge the system with water or 20% ethanol followed by water to neutralize the pH and remove all traces of salt from the system. Monitor the pH readings as the valves and lines change; a plateaued and neutral trace is required and may take 3–6 system volumes (sv) to achieve. A program can be used to prime pumps and then purge all lines, valves, ports and fraction collector in sequence.
4. Prime the system with 0.1 M sodium hydroxide and run a program to sanitize all pumps, valves, ports, lines fraction collector, and dead space for 1 h contact time.
5. Using a program (*see* **Note 13**), prime and flush with HPLC grade water to remove sodium hydroxide from all pumps, valves, lines, and dead space settings until the absorbance, conductivity, and pH traces indicate a neutral pH or the trace has stabilized. Ensure the sample line is included.
6. The cold room setting may require use of 20% ethanol to clear Bernoulli effects and enhance clearance of edge effect of the sodium hydroxide from the tubing and valves—this should be run through as previously stated. Gather the collection lines in a single sterile collection vessel and allow these lines to void 20% ethanol.
7. The system is now sterile and ready to be equilibrated. The columns may now be connected into separate column ports if they are already in 20% ethanol.

3.2.3 Sanitization of Columns at 4 °C with 70% Ethanol (See **Note 14**)

1. Ensure columns are primed with 20% ethanol prior to transfer to 4 °C.
2. Set pressure limit for 4 °C system to 0.35 Pa.
3. Insert the columns into separate column ports in the system, with 20% ethanol running at a low flow rate to avoid air in lines. Reverse flow may be applied to the columns. Spray connection points externally with 70% ethanol when connecting columns to maintain sterility.
4. Run columns separately for 3–6 column volumes (cv) in 20% ethanol or until conductivity has plateaued.
5. Prime the pump and flush bypass lines with 70% ethanol.
6. For the Protein G column: run 70% ethanol through the column for 1 h contact time, and then flush with 20% ethanol. Close the column top and bottom manual switch valves.
7. For the POROS™ CaptureSelect™ IgM column: equilibrate the column for 12 h contact time then flush with 20% ethanol. Close the column top and bottom manual switch valves.
8. Ensure both columns are equilibrated in 20% ethanol. This will take a minimum of 6× cv.
9. When both columns have been sanitized and equilibrated back to 20% ethanol, they are ready for use or storage.

3.3 IgG and IgM Depletion of Plasma

3.3.1 Equilibration of the Chromatography Systems and Columns at 4 °C

1. Ensure the system and columns have been sanitized and are ready for use.
2. Set the pressure limits for the system and column to 0.35 Pa for the Protein G column and 0.5 Pa for POROS™ CaptureSelect™ column.
3. Open the manual switches and run 20% ethanol slowly through both column ports.
4. Each column should be connected separately to the chromatography system, *see* **Note 15**. Flush columns and all parts of the system individually with water, including the sample pump. Continue to flush each column with water until both the UV and conductivity traces have stabilized.
5. Repeat **step 4** using PBS for a minimum of 6× cv and ensure pH trace has reached 7.4.

3.3.2 Loading the Plasma onto the Columns

1. Ensure columns and system have been sanitized and equilibrated in PBS at 4 °C and plasma is prepared for use (as described in Subheading 3.2).
2. Set the pressure limits to 0.35 Pa for the system and columns.
3. Connect the columns together with the Protein G (for IgG depletion) in front of the POROS™ CaptureSelect™ (for IgM depletion) (*see* Fig. 1). Allow the pressure and monitor traces to

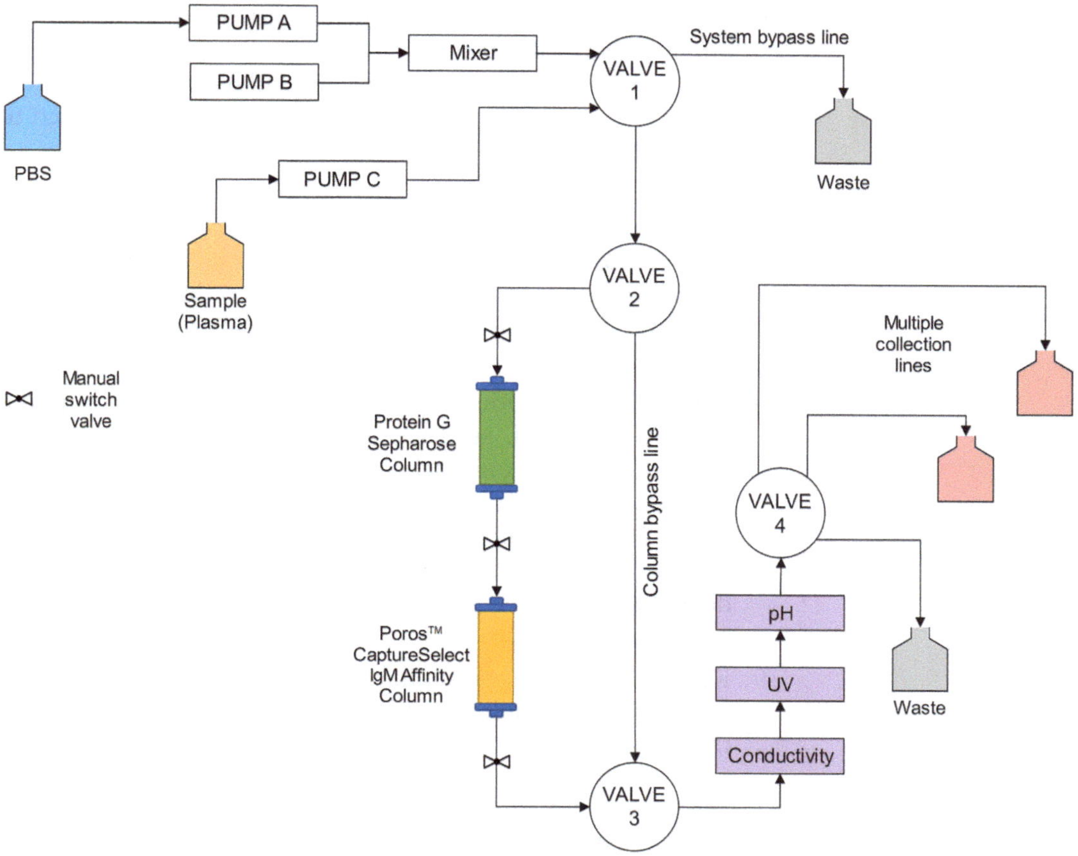

Fig. 1 Schematic diagram showing the setup of the chromatography system and columns for IgG and IgM depletion at 4 °C. The system is equilibrated and primed with PBS using pump A or B. Plasma is loaded onto the columns via pump C and passes first over the Protein G Sepharose column, then the POROS™ CaptureSelect™ IgM Affinity column. UV, conductivity, and pH monitors are used to assess the composition of column breakthrough, which is then directed to waste or collected accordingly

adjust and stabilize by flushing the columns with PBS for 3×–6× cv (use cv of both columns combined) and autozero the UV detector once this is achieved. Use manual switch valves and spray externally with 70% ethanol to ensure sterility when connecting the columns.

4. Aseptically place the collection lines into separate sterile collection vessels.

5. Set pump C running with the sample feed directed to the system bypass waste and use pump A or B on a slower flow rate, to pump PBS over the columns to waste. Once pumps are running, apply the plasma to the columns using pump C. Gradually raise the flow rate of pump C to approximately 5 mL/min.

6. When all plasma has been loaded onto the system, transfer pump C into PBS, taking care not to allow air into the lines and clear plasma from the sample line and pump. Once cleared, pump C can be switched off and pump A or B can be used to direct PBS over the columns.
7. Monitor the UV 280 nm trace and collect the breakthrough from 20 mAU until the traces indicate the breakthrough has cleared and the monitors have plateaued again. Collect the breakthrough until the 280 nm trace has dropped to 20 mAU. This indicates that large proteins have passed over the columns and been collected.
8. Once the breakthrough is collected, flush the sample pump with PBS to void, stop all pumps and disconnect columns from the system and each other and seal with the manual switches. Transfer columns to the second chromatography system at 22 °C and allow to acclimatize—this may take up to 3 h. Sanitize the 4 °C system as described in Subheading 3.2 (this does not have to be carried out immediately after removal of columns).
9. Keeping the breakthrough at 4 °C, pool fractions before removing QC samples (*see* Subheading 3.5) and concentrating the depleted plasma (*see* Subheading 3.4.3).

3.4 C1q Elution, Concentration of Components and Reconstitution

3.4.1 Elution of C1q at 22 °C with Salt Gradient (See **Note 16**)

1. Set up the fraction collector within an MSC II (to maintain sterility) and sanitize both the system and fraction collector. Aseptically load 15 mL tubes (minus lids) into the fraction collector carousel and only insert this once the system has been sanitized (avoid sodium hydroxide splashing into tubes).
2. Ensure columns are acclimatized to 22 °C and equilibrated in PBS.
3. With the system acclimatized to 22 °C, set the pressure limits to 0.35 Pa for the system and columns (*see* **Note 17**), and autozero the UV. Autozero with PBS flowing through the bypass line.
4. Insert the Protein G column between column selection valves (*see* Fig. 2) and start pump A (PBS) at a flow rate of 1 mL/min.
5. Apply a linear salt gradient of 0–1.5 M NaCl, at a flow rate of 1.6 mL/min for 1100 min (total 1760 mL). Collect 12 mL fractions in an MSC II to maintain sterility.
6. Identify fractions containing C1q using the UV monitor trace and SDS PAGE (*see* **Note 18**). Pool all fractions with C1q and concentrate (*see* Subheading 3.4.3).

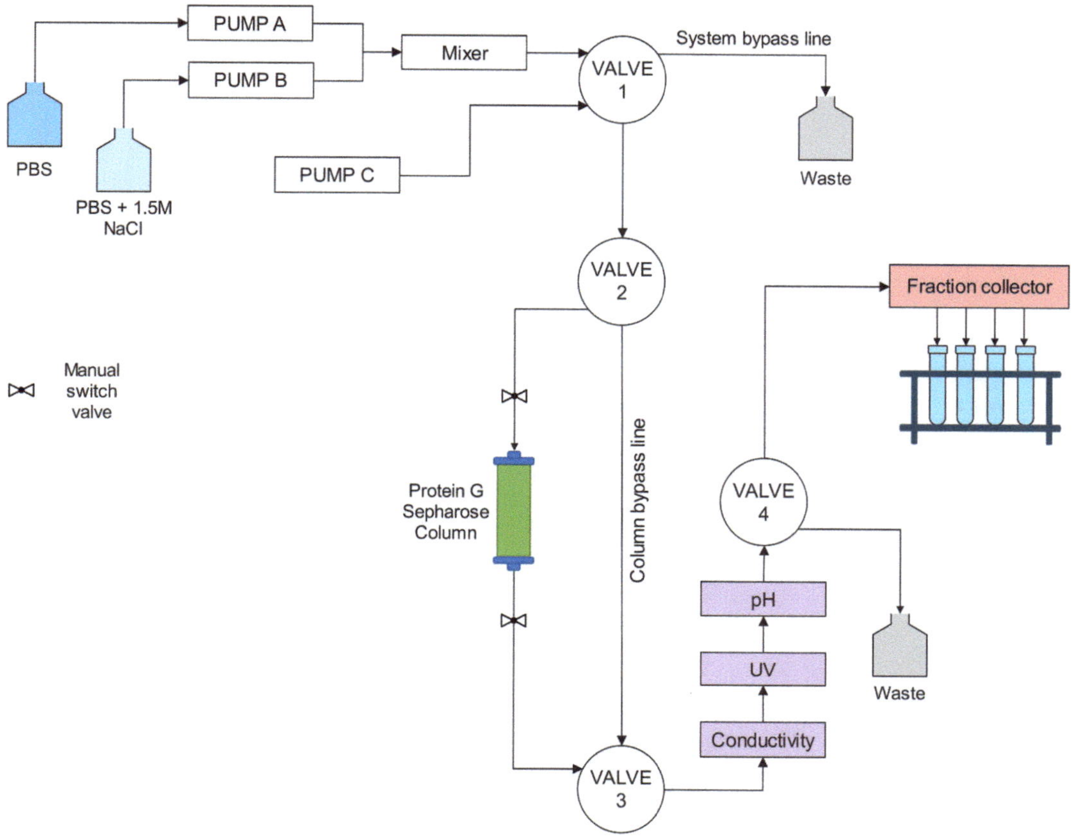

Fig. 2 Schematic diagram showing the setup of the chromatography system and columns for the elution of C1q at room temperature. Pump A is primed with PBS, and pump B is primed with PBS supplemented with 1.5 M NaCl. C1q is then eluted from the Protein G Sepharose using a salt gradient and collected in 12 mL fractions. The UV monitor trace can be used in identifying which fractions contain eluted C1q

3.4.2 Elution of IgG and IgM for Reuse of Column (See ***Note 19****)*

1. Ensure the salt gradient has been completed for the Protein G column and C1q has been collected (*see* Subheading 3.4.1). Ensure the POROS™ CaptureSelect™ column is acclimatized to room temperature.
2. Connect columns to the system in parallel (*see* Fig. 3).
3. Equilibrate the Protein G column separately back into PBS.
4. Apply the elution buffer, using pump C, to the Protein G column for a maximum of 1 h or until a peak has eluted.
5. Apply PBS to the column until it is equilibrated before flushing with water and 20% ethanol.
6. Collect the peaks manually and assess the protein profile by SDS PAGE and western blotting, *see* Subheading 3.5.1, to ensure antibodies have been eluted.
7. Repeat **steps 3–6** for the Poros™ CaptureSelect™ column to elute the IgM separately.

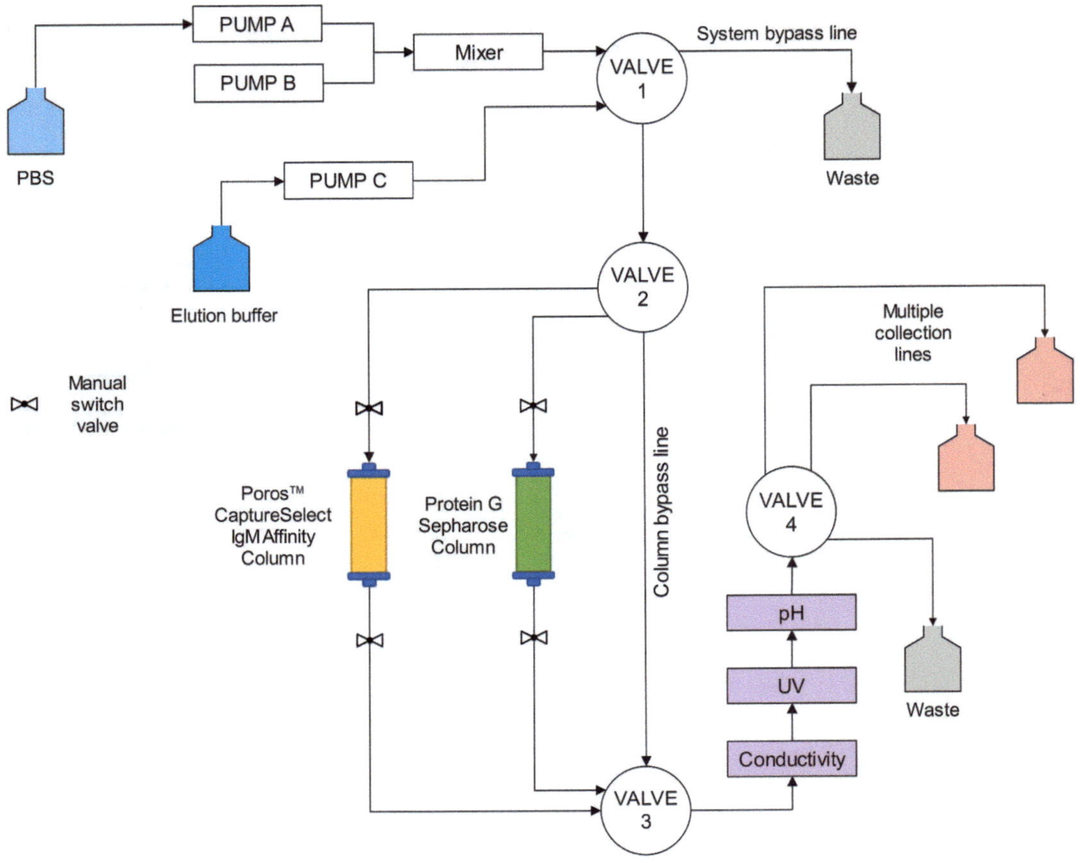

Fig. 3 Schematic diagram showing the setup of the chromatography system and columns for the IgG and IgM elution at room temperature. Pump C is primed with elution buffer and antibodies are eluted from the respective columns into multiple collection lines. The UV monitor trace can be used in identifying which fractions contain eluted antibody

8. Sanitize columns and chromatography system as described in Subheading 3.2. Columns should be transferred back into 20% ethanol promptly to prevent growth of contaminants.

3.4.3 Concentration of Plasma and C1q and Recovery from Dialysis Tubing

1. Set up sterile dialysis tubing, forceps, scissors, and lint-free paper in an MSC II.
2. Work aseptically in an MSC II but **do not** spray the outside of tubing containing depleted plasma or C1q with 70% ethanol.
3. Fill the dialysis tubing with the depleted plasma breakthrough or C1q fractions, void any residual air and knot the ends, leaving 5 cm lengths of tubing post knots. Dry down the outside of the tubing. Weigh the tubing once filled, for use as a guide to the rate of concentration.

4. Lay the sealed dialysis tubing onto a bed of chilled polyethylene glycol (PEG) 20,000 Da and cover the tubing with PEG. Leave at 4 °C until desired volume is reached.
5. The depleted plasma breakthrough is concentrated to 90% of the original volume (i.e., 180 mL for a 200 mL batch), while C1q is concentrated to 10% of original plasma volume (i.e., 20 mL for a 200 mL batch of native plasma). C1q may be concentrated overnight at 4 °C by placing in a bag/vessel and enveloping the tubing in a thin film of concentrated PEG.
6. Wash excess PEG off the dialysis tubing with water and recover contents aseptically within a MSC II, *see* **Note 20**. Snap freeze components in a sterile container after removing 1 mL aliquots for QC.

3.4.4 Reconstitution of Complement with C1q and Aliquoting for Storage

1. Pre-chill the labeled internally threaded cryovials and racks.
2. Defrost the required depleted plasma and the paired C1q.
3. Centrifuge the depleted plasma at 3000 × *g* for 10 min to remove all particulates or clots.
4. Within an MSC II, pool all depleted plasma and C1q into a sterile glass Duran bottle on ice, mix gently and the aliquot into 50 mL falcon tubes (for ease of aliquoting) and store at 4 °C for immediate aliquoting.
5. Aseptically pipette into 0.5 mL or 1 mL aliquots as required. Snap-freeze the vials (*see* **Note 4**) and store at −80 °C.

3.5 Batch Quality Testing

3.5.1 SDS-PAGE and Western Blotting

1. Samples are loaded onto 4–12% SDS-PAGE under reduced conditions; native plasma and reconstituted complement are loaded at a 1:20 dilution, while C1q and all other samples are loaded neat.
2. One gel should be stained for protein and the subsequent gels should be transferred to polyvinylidene difluoride (PVDF) membrane by western blotting and probed with anti-IgG and anti-IgM primary antibodies to show complete removal of the immunoglobulins from the final depleted plasma.

3.5.2 Bioburden Check

1. Apply 20 μL of each sample to each of the three types of agar plate.
2. Incubate at 37 °C for 5 days at 5% CO_2.
3. Observe contaminant growth.
4. Complete sterility is required for use of the depleted plasma in functional assays therefore discard any batch where any microbial contamination is seen.

3.5.3 Confirmation of Complement Activity

1. Measurement of individual complement protein levels and CH50 activity (*see* Fig. 4). All assays on the Optilite system (The Binding Site) are performed according to the manufacturer's instructions. All radial immundiffusion (RID) assays are performed to the manufacturer's instructions using either a calibrated microscope or the available Digital RID Reader (The Binding Site), which greatly improves the clarity of the zones. Critical assays are albumin, IgG, IgM, CH50, AH50, C1q, C3, C5, and FH, all others as listed in materials are optional. Batches are acceptable if critical assays show ±30% of native plasma.
2. Functional immunoassays (*see* Table 1). Functional activity against bacteria is measured using a serum bactericidal assay

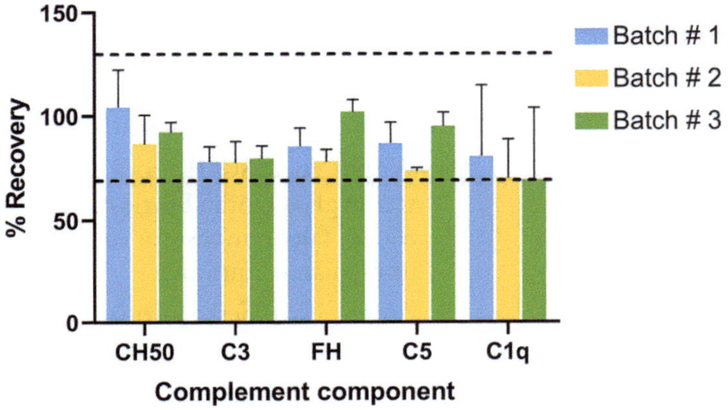

Fig. 4 Comparison of percentage recovery compared to native plasma for three batches of IgG- and IgM-depleted plasma for CH50 activity, C1q, C3, C5, and FH. Bars represent the standard deviation of a minimum of three measurements, dotted lines show 30% above and below full recovery, which represent acceptance cut-off levels

Table 1
Serum bactericidal activity reciprocal titers against *Neisseria meningitidis* strain NZ98/254 and *B. pertussis* strain B1917 using three different batches of IgG- and IgM-depleted plasma. Reciprocal bactericidal titers are accepted within 1 doubling dilution of the mean assigned titer calculated through repeated testing; NT, not tested

Depleted plasma batch	*N. meningitidis* NZ98/254 (Serum from 1 year post-Bexsero vaccination)	*B. pertussis* B1917 (International pertussis standard NIBSC 06/140)
Batch #1	32	128
Batch #2	32	NT
Batch #3	NT	128
Batch #4	32	128

(SBA). Batches are acceptable if the killing titer of a control test sera is within 1 dilution of the known mean titer. The SBA against *N. meningitidis* strain NZ98/254 is performed as described by Brookes et al. [22] while the SBA against *B. pertussis* strain B1917 is performed as described by Lesne et al. [11].

4 Notes

1. The volunteer blood and plasma have not been screened for infectious agents and should always be handled wearing gloves, safety glasses, and lab coat by staff who have been vaccinated against Hepatitis B.
2. Add sterile water to the hirudin vial and leave for 30 min before aliquoting the resuspended hirudin to required polypropylene tubes. The tubes may be frozen if not used immediately. **Safety note**: This is a concentrated and irreversible anticoagulant; care must be taken when handling and the use of sharps should be avoided.
3. All buffers must be filtered, prepared from the highest quality grade components possible, and using HPLC grade water. When preparing the buffers, a sterile lid is aseptically placed on the filtered buffer bottle and covered with a sterile foil cover. All buffers and solutions are prepared and filtered ahead of the purification, the only one prepared fresh on the day is glycine buffer for the antibody elution. Buffers are left to equilibrate to required temperature for 3 h prior to use.
4. Snap freezing is performed using an ethanol bath containing dry ice (solid carbon dioxide pellets). The solution may be as cold as −70 °C so care must be taken to not freeze fingers or splash ethanol over the body, face, or clothes. The cryotubes used should be selected to avoid both activation of the complement and plasma coagulation; the "O" ring and lid composition should be considered, and in our hands we found internally threaded cryovials gave the better results. Tubes are filled to 80% maximum volume, tightly sealed, and immersed to the level of the liquid to ensure even and immediate supercooling of the contents. After reconstitution of the 200 mL complement, the solution is subdivided into 4 or 5 prechilled polypropylene tubes or glass vessels and remains chilled until dispensed. The complement is dispensed as 0.2–1 mL aliquots into pre-chilled and pre-labeled cryovials ready for use. The tubes are tightly sealed and placed in racks sitting in the ethanol bath sufficient to cover the contents but not the lid or the seal. The use of cryogenic storage boxes with fixed hollow pronged dividers allows rapid permeation of the freezing mixture

between the tubes, and supercooling from beneath and within the prong. As the depleted plasma freezes it changes color from orange, becoming more yellow in colour.

5. To maintain the sterility of the final product: aseptic technique should be practiced at all times to reduce the risk of contamination, which could lead to activation of the complement components and the interference with assays in which the final product is used. 70% ethanol spray is used in preference to 70% isopropanol (IPA) as the ethanol evaporates faster. The 70% ethanol spray is used directly on open chromatography lines and valves when connecting the columns and lines to the manual switch valves. Collection lines should be placed to the base of the fraction vessel (glass borosilicate bottle) to lessen the complement–air interface formed and so prevent complement activation.
6. Materials to avoid: Complement activation within the native or IgG- and IgM-depleted plasma from different surfaces should be avoided. Polyethylene, polyurethane and polystyrene should be avoided wherever possible, or the contact time reduced to a minimum. The use of polypropylene tubes and pipettes/pastettes should be used wherever possible, with borosilicate glassware for larger volumes. Dialysis tubing made from regenerated cellulose with a small MWCO (3.5 kDa) should be used; the use of dialysis cassettes has resulted in the coagulation and/or activation of the complement.
7. For plasma manipulations, use sterile polypropylene pastettes, which leads to better product. It is important that the plasma/complement is not frothed when manipulated; when pipetting (by pastette), when transferring to vessels for loading onto column, or when preparing dialysis tubing to recover contents, etc.
8. Avoid hemolysis of the red blood cells and do not transfer any red blood cells with the plasma. After centrifugation the plasma should appear as a translucent/transparent solution layer above the cell layer, with a particulate interface and top surface. Only the upper plasma layer should be harvested, and the top surface/skim should first be removed using polypropylene pastettes before accessing the plasma layer below, to avoid contamination. All particulate matter must be gently removed using a polypropylene pastette prior to freezing. If not, on defrosting, the particulates will cause the plasma to clot/coalesce. Blood should be processed and snap frozen with 1 h of donation.
9. When defrosting the plasma, do so under running water and centrifuge as soon as possible at 4 °C to keep chilled. The centrifugation step is required to clarify the plasma and remove

particulates again, as a pellicle may form on the surface and base of the tube and this must be removed.

10. The choice of chromatography system, columns and resin was determined by the biocompatibility of all wettened surfaces with the plasma and complement coagulation and activation pathways. The scalability and reliability of the systems, columns, and resins were an important consideration for downstream processing with regard to sanitization and chromatography running conditions. The two chromatography systems, independently equilibrated at 4 °C and 22 °C, facilitated both systems being synchronized to optimize both the C1q recovery at 22 °C, and the column and system sanitization. The transfer of chromatography systems to different temperature-controlled environments risks electrical or mechanical failure, while the independent cooling of columns does not chill the system lines and surfaces, should the ambient temperature not reflect the required temperature. Although not recommended, a 1 cm buffer head space may be left above the resin bed to prevent air entering the resin during the run. Should this occur, the direction of flow can be reversed and the air voided. Later chromatography systems have an air sensor that will shut down the system, preventing air entering the resin and damaging the resin.

11. System sanitization with sodium hydroxide should **never** be performed with the columns in place. If this is not possible, a manual inline switch should be present to protect the columns and the columns removed from the inner surface of the switches. A period of 1 h contact time should be allowed for sanitization of all lines, valves, ports, and pumps in the system. All fractionating lines should be sanitized as well as the fraction collector to allow versatility if problems occur. Each column port should have a blanking line connected by a manual switch that is sanitized with the system in situ (i.e., port–blanking line–switch–blanking line–port). When the 4 °C system is sanitized and placed in 20% ethanol, the columns (already equilibrated in 20% ethanol) are inserted in tandem for the base prep elution.

12. Precipitation of crystals during sanitization of columns and systems may occur if insufficient water is used between the use of PBS/sodium chloride and ethanol (and vice versa). Sanitization of the system with sodium hydroxide requires not only the use of water but also a 20% ethanol flush within the water wash to prevent crystal formation in the lines especially when sanitizing the cold room system. Complete flushing of the system is required for each step when the columns are in place as the precipitate will result in the chromatography run being void due to the blockage requiring aseptic clearing.

Precipitates are dispersed by sonicating the offending line or valve in warm water while applying gentle pressure using a hand-held syringe.

13. Programs to sanitize the system and columns may be used to run the columns overnight/weekend to allow fast turnaround of columns and systems while the complement base prep and C1q are being concentrated, recombined and product snap-frozen. Both columns may be sanitized independently by using different flow rates. The pressure setting should reflect the lower column criteria.

    Manual switches should be placed either end of each column, and these, together with the blanking lines, must be sanitized with sodium hydroxide, not just 70% ethanol. This is done by rotation of the manual switch valves as the column lines are inserted or removed.

    Chromatography system programs are written specifically for the configuration of the system used. Many systems have a default pump wash and prime program, and this may be incorporated into a bespoke program that reflects the configuration of the system valves, ports, manual valves, bypass lines, sample pump lines, and fractionating lines; together with the tubing length and internal diameter. The program reflects the time to equilibration by assessing UV monitor at 280, 254, and 215 nm, conductivity, and pH; all of which are allowed to plateau and stabilize. These are checked after being stationary for at least 5 min; when residual buffers may leach from the tubing or connections. The program should be written to cover lines sequentially, as any spike seen after a stationary sequence results in lengthening of the time of wash, until the spike is no longer seen. The only program that is not lengthened is the sodium hydroxide wash which must be completed within the hour to allow the water wash program to flow after 1 h contact time.

14. Sanitization of the columns with ethanol is best performed at the end of a chromatography run. Dependent upon the time left, the system may be sanitized with 0.1 M sodium hydroxide ahead of the columns. However, the columns should **never** be inline when sodium hydroxide is used. Columns are probably best sanitized separately after transfer to the cold room system after a depletion has been performed. This prevents the accidental sanitization of the column with sodium hydroxide and the degradation of the ligand. If required, the columns may be run in reverse flow mode during sanitization—this allows the slow buildup of residual protein to be cleared, which maintains the flow rate and, in turn, lessens the back pressure thereby prolonging the life of the column.

15. Columns should only be added to the systems when the system contains a sterile and compatible buffer/solution. The flow rate should be set to a slow flow rate (e.g., 0.5–1.0 mL/min). The upper blanking line switch should be opened, and the blanking line uncoupled downstream of the valve so the liquid flows through; these are held in one hand and sprayed with 70% ethanol. The column top adaptor line is uncoupled downside of the column top switch while the base switch remains closed and the top line sprayed immediately with 70% ethanol. The top adaptor line is inserted into the downstream of the flowing blanking line. The base adaptor switch is uncoupled to prevent back pressure. This process is repeated with the base switches. The column switches should be periodically rotated into the system to ensure they become regularly sanitized with sodium hydroxide.
16. C1q is in an equilibrium within the plasma and bound to circulating antibody [34, 35]. When the base plasma preparation is collected in the column breakthrough, a proportion of the C1q remains attached to the antibody which is now bound to the column. Antibody binds to the column by the Fc region, and so to avoid possible leaching of the antibody by the sudden addition of high salt to the column, a slow and gentle salt gradient is applied, thereby resulting in the C1q ionically dissociating from the antibody, and consequently leaving the antibody attached to the Protein G matrix. In our hands, transferring the Protein G column from 4 °C and equilibrating to 22 °C gives clear separation of the C1q from other proteins along the gradient. The C1q peak is very minor (~40 mAU) and may be easily missed (M Pangburn personal communication and Ref. 31) without analysis by SDS PAGE. Pure C1q is observed in the second peak of the elution profile corresponding to ~40% pump B elution. Alternatively, an isocratic elution may be made to clear all proteins by stepping directly to 40% high salt buffer concentration, and then 100% to clear all. Elution of the C1q may be scaled up or down using the ratio of the gradient over time for different column sizes.
17. Use of newer AKTA systems (or other chromatography systems) allows the selection of the column delta pressure and so finely tunes the pressure exerted across the column without being tripped by the internal pressures of the pump. This is noticeable when eluting the C1q from the Protein G column at low flow rates when the piston pump reverse flow creates a pressure in excess of that set for the pressure limit.
18. Sampling of the fractions is made from every sixth tube and the fractions collected for all six either side of the last C1q sample identified by SDS PAGE. The C1q remains in the MSC II until the SDS PAGE gel profile has been confirmed (using neat

samples). Once the C1q profile is confirmed and fractions pooled (samples taken), the concentration step occurs and may take a few hours. Performing the C1q concentration in narrow bore tubing increases the surface area–volume ratio and, thereby, the rate of concentration. Agitation of the tubing ensures a gradient across the dialysis membrane and encourages the rate of concentrations. Analysis of the C1q fractions should start as soon as possible in the morning as this fraction should elute overnight and be ready for pooling once the fractions can be identified. The peak profile is distinctive for each column used.

19. IgG and IgM elution: The IgG is released from the column by use of low pH elution releasing the Fc from the Protein G and the IgM is recovered from the column by low pH isocratic elution from the resin ligand monoclonal. Both columns are cleared of bound antibody using a low pH glycine buffer and pre-equilibrated back to PBS within 1 h contact time to preserve the life of the column and ligand.

20. Recovery from dialysis tubing: when the concentrated components are ready, the dialysis tubing is removed from the PEG and gently washed with HPLC grade water to remove all residual PEG. Wash the main length including the knots. Avoid frothing the contents. Lay out the sterile lint-free paper on the MSC II floor and place the tubing flat upon it. Gently massage with a gloved hand to void the PEG film from the dialysis tubing, working finger tips along the length of the tubing and massaging the sides to ensure all contents are released—especially the C1q. Pick up the tubing and allow the contents of the tubing to fall to one end. Twist the tubing over the contents allowing the empty tubing to fall away on the other side of your hand. Spray the empty end with 70% ethanol and, using a pair of sterile scissors, make an angled cut (not right angled) and cut away the empty end above the fresh twist. Pour the contents into a fresh container in one smooth maneuver.

References

1. Goldschneider I, Gotschlich EC, Artenstein MS (1969) Human immunity to the meningococcus. I. The role of humoral antibodies. J Exp Med 129:1307–1326. https://doi.org/10.1084/jem.129.6.1307
2. Goldschneider I, Gotschlich EC, Artenstein MS (1969) Human immunity to the meningococcus. II. Development of natural immunity. J Exp Med 129:1327–1348. https://doi.org/10.1084/jem.129.6.1327
3. Romero-Steiner S, Libutti D, Pais LB, Dykes J, Anderson P, Whitin JC, Keyserling HL, Carlone GM (1997) Standardization of an opsonophagocytic assay for the measurement of functional antibody activity against *Streptococcus pneumoniae* using differentiated HL-60 cells. Clin Diagn Lab Immunol 4:415–422. https://doi.org/10.1128/cdli.4.4.415-422.1997
4. Borrow R, Balmer P, Miller E (2005) Meningococcal surrogates of protection - serum

bactericidal antibody activity. In: Vaccine. Elsevier BV, Amsterdam, pp 2222–2227

5. Terajima M, Cruz J, Co MDT, Lee J-H, Kaur K, Wrammert J, Wilson PC, Ennis FA (2011) Complement-dependent lysis of influenza a virus-infected cells by broadly cross-reactive human monoclonal antibodies. J Virol 85:13463–13467. https://doi.org/10.1128/JVI.05193-11
6. Townsend K, Ladhani SN, Findlow H, Borrow R (2014) Evaluation and validation of a serum bactericidal antibody assay for *Haemophilus influenzae* type b and the threshold of protection. Vaccine 32:5650–5656. https://doi.org/10.1016/j.vaccine.2014.08.010
7. Aase A, Herstad TK, Merino S, Brandsdal KT, Berdal BP, Aleksandersen EM, Aaberge IS (2007) Opsonophagocytic activity and other serological indications of *Bordetella pertussis* infection in military recruits in Norway. Clin Vaccine Immunol 14:855–862. https://doi.org/10.1128/CVI.00081-07
8. Guttormsen H, Liu Y, Paoletti L (2008) Functional activity of antisera to group B streptococcal conjugate vaccines measured with an opsonophagocytosis assay and HL-60 effector cells. Hum Vaccin Vaccin 4:370–374
9. Herbert J, Thomas S, Brookes C, Turner C, Turner P, Nosten F, LeDoare K, Hudson M, Heath PT, Gorringe A, Taylor S (2015) Antibody-mediated complement C3b/iC3b binding to group B streptococcus in paired mother and baby serum samples in a refugee population on the Thailand-Myanmar border. Clin Vaccine Immunol 22:319–326. https://doi.org/10.1128/CVI.00803-14
10. Thomas SR, Leung S, Knox K, Wilkinson TMA, Staples KJ, Lestrate P, Wauters D, Gorringe A, Taylor SC (2018) Development of flow cytometric opsonophagocytosis and antibody-mediated complement deposition assays for non-typeable *Haemophilus influenzae*. BMC Microbiol 18:167. https://doi.org/10.1186/s12866-018-1314-5
11. Lesne E, Cavell BE, Freire-Martin I, Persaud R, Alexander F, Taylor S, Matheson M, van Els CACM, Gorringe A (2020) Acellular pertussis vaccines induce anti-pertactin bactericidal antibodies which drives the emergence of Pertactin-negative strains. Front Microbiol 11:2108. https://doi.org/10.3389/fmicb.2020.02108
12. Mehlhop E, Whitby K, Oliphant T, Marri A, Engle M, Diamond MS (2005) Complement activation is required for induction of a protective antibody response against West Nile virus infection. J Virol 79:7466–7477. https://doi.org/10.1128/JVI.79.12.7466-7477.2005
13. Ji X, Olinger GG, Aris S, Chen Y, Gewurz H, Spear GT (2005) Mannose-binding lectin binds to Ebola and Marburg envelope glycoproteins, resulting in blocking of virus interaction with DC-SIGN and complement-mediated virus neutralization. J Gen Virol 86:2535–2542. https://doi.org/10.1099/vir.0.81199-0
14. Maslanka SE, Gheesling LL, Libutti DE, Donaldson KBJ, Harakeh HS, Dykes JK, Arhin FF, Devi SJN, Frasch CE, Huang JC, Kriz-Kuzemenska P, Lemmon RD, Lorange M, Peeters CCAM, Quataert S, Tai JY, Carlone GM, Mills EL, Ashton FE, Diepevaen J, Bielecki J, Bybel M, Cloutier M, Poolman JT (1997) Standardization and a multilaboratory comparison of *Neisseria meningitidis* serogroup a and C serum bactericidal assays. Clin Diagn Lab Immunol 4:156–167. https://doi.org/10.1128/cdli.4.2.156-167.1997
15. Yu J, Tostanosk LH, Peter L, Mercad NB, McMahan K, Mahrokhia SH, Nkolol JP, Liu J, Li Z, Chandrashekar A, Martine DR, Loos C, Atyeo C, Fischinger S, Burk JS, Slei MD, Chen Y, Zuiani A, Lelis FJN, Travers M, Habibi S, Pessaint L, Van Ry A, Blade K, Brown R, Cook A, Finneyfrock B, Dodson A, Teow E, Velasco J, Zahn R, Wegmann F, Bondzi EA, Dagotto G, Gebr MS, He X, Jacob-Dolan C, Kirilova M, Kordana N, Lin Z, Maxfiel LF, Nampanya F, Nityanandam R, Ventur JD, Wan H, Cai Y, Chen B, Schmid AG, Weseman DR, Bari RS, Alter G, Andersen H, Lewi MG, Barou DH (2020) DNA vaccine protection against SARS-CoV-2 in rhesus macaques. Science 369:806–811. https://doi.org/10.1126/science.abc6284
16. Granoff DM, Welsch JA, Ram S (2009) Binding of complement factor H (fH) to *Neisseria meningitidis* is specific for human fH and inhibits complement activation by rat and rabbit sera. Infect Immun 77:764–769
17. Gill CJ, Ram S, Welsch JA, De Tora L, Anemona A (2011) Correlation between serum bactericidal activity against *Neisseria meningitidis* serogroups a, C, W-135 and Y measured using human versus rabbit serum as the complement source. Vaccine 30:29–34. https://doi.org/10.1016/j.vaccine.2011.10.068
18. Ray TD, Lewis LA, Gulati S, Rice PA, Ram S (2011) Novel blocking human IgG directed against the pentapeptide repeat motifs of *Neisseria meningitidis* lip/H.8 and Laz lipoproteins. J Immunol 186:4881–4894. https://doi.org/10.4049/jimmunol.1003623

19. Giuntini S, Reason DC, Granoff DM (2012) Combined roles of human IgG subclass, alternative complement pathway activation, and epitope density in the bactericidal activity of antibodies to meningococcal factor H binding protein. Infect Immun 80:187–194. https://doi.org/10.1128/IAI.05956-11
20. Beernink PT, Shaughnessy J, Braga EM, Liu Q, Rice PA, Ram S, Granoff DM (2011) A meningococcal factor H binding protein mutant that eliminates factor H binding enhances protective antibody responses to vaccination. J Immunol 186:3606–3614. https://doi.org/10.4049/jimmunol.1003470
21. Mutti M, Ramoni K, Nagy G, Nagy E, Szijártó V (2018) A new tool for complement research: in vitro reconstituted human classical complement pathway. Front Immunol 9:2770. https://doi.org/10.3389/fimmu.2018.02770
22. Brookes C, Kuisma E, Alexander F, Allen L, Tipton T, Ram S, Gorringe A, Taylor S (2013) Development of a large scale human complement source for use in bacterial immunoassays. J Immunol Methods 391:39–49. https://doi.org/10.1016/j.jim.2013.02.007
23. Berends ETM, Kuipers A, Ravesloot MM, Urbanus RT, Rooijakkers SHM (2014) Bacteria under stress by complement and coagulation. FEMS Microbiol Rev 38:1146–1171. https://doi.org/10.1111/1574-6976.12080
24. Amara U, Rittirsch D, Flierl M, Bruckner U, Klos A, Gebhard F, Lambris JD, Huber-Lang M (2008) Interaction between the coagulation and complement system. Adv Exp Med Biol 632:71–79
25. Mollnes TE, Brekke O-L, Fung M, Fure H, Christiansen D, Bergseth G, Videm V, Lappegård KT, Köhl J, Lambris JD (2002) Essential role of the C5a receptor in *E coli*-induced oxidative burst and phagocytosis revealed by a novel lepirudin-based human whole blood model of inflammation. Blood 100:1869–1877
26. Vidarsson G, van der Pol W-L, van den Elsen JMH, Vilé H, Jansen M, Duijs J, Morton HC, Boel E, Daha MR, Corthésy B, van de Winkel JGJ (2001) Activity of human IgG and IgA subclasses in immune defense against *Neisseria meningitidis* serogroup B. J Immunol 166:6250–6256. https://doi.org/10.4049/jimmunol.166.10.6250
27. Kojouharova M, Reid K, Gadjeva M (2010) New insights into the molecular mechanisms of classical complement activation. Mol Immunol 47:2154–2160
28. Brookes C, Freire-Martin I, Cavell B, Alexander F, Taylor S, Persaud R, Fry N, Preston A, Diavatopoulos D, Gorringe A (2018) *Bordetella pertussis* isolates vary in their interactions with human complement components. Emerg Microbes Infect 7:81. https://doi.org/10.1038/s41426-018-0084-3
29. Staples KJ, Taylor S, Thomas S, Leung S, Cox K, Pascal TG, Ostridge K, Welch L, Tuck AC, Clarke SC, Gorringe A, Wilkinson TMA (2016) Relationships between mucosal antibodies, non-Typeable *Haemophilus influenzae* (NTHi) infection and airway inflammation in COPD. PLoS One 11:e0167250. https://doi.org/10.1371/journal.pone.0167250
30. Le Doare K, Faal A, Jaiteh M, Sarfo F, Taylor S, Warburton F, Humphries H, Birt J, Jarju S, Darboe S, Clarke E, Antonio M, Foster-Nyarko E, Heath PT, Gorringe A, Kampmann B (2017) Association between functional antibody against Group B Streptococcus and maternal and infant colonization in a Gambian cohort. Vaccine 35:2970. https://doi.org/10.1016/J.VACCINE.2017.04.013
31. Kolb WP, Kolb LM, Podack ER (1979) C1q: isolation from human serum in high yield by affinity chromatography and development of a highly sensitive hemolytic assay. J Immunol 122:2103–21011
32. Björck L, Kronvall G (1984) Purification and some properties of streptococcal protein G, a novel IgG-binding reagent. J Immunol 133:969–974
33. Tao MH, Canfield SM, Morrison SL (1991) The differential ability of human IgG1 and IgG4 to activate complement is determined by the COOH-terminal sequence of the CH2 domain. J Exp Med 173:1025–1028. https://doi.org/10.1084/jem.173.4.1025
34. Wang G, de Jong RN, van den Bremer ETJ, Beurskens FJ, Labrijn AF, Ugurlar D, Gros P, Schuurman J, Parren PWHI, Heck AJR (2016) Molecular basis of assembly and activation of complement component C1 in complex with immunoglobulin G1 and antigen. Mol Cell 63:135–145. https://doi.org/10.1016/j.molcel.2016.05.016
35. Feinstein A, Richardson N, Taussig MI (1986) Immunoglobulin flexibility in complement activation. Immunol Today 7:169–174

Chapter 19

Assessment of Serum Bactericidal and Opsonophagocytic Activity of Antibodies to Gonococcal Vaccine Targets

Evgeny A. Semchenko, Freda E.-C. Jen, Michael P. Jennings, and Kate L. Seib

Abstract

There is no vaccine available to prevent *Neisseria gonorrhoeae* infection, however there is currently a high level of interest in developing gonococcal vaccines due to the increasing number of cases and continuing emergence of antimicrobial resistance worldwide. A key aspect of vaccine development is the investigation of the functional immune response raised to the vaccine targets under investigation. Here, we describe two assays used to assess the functional immune response raised against gonococcal vaccine targets: the serum bactericidal assay (SBA) and the opsonophagocytic assay (OPA).

Key words *N. gonorrhoeae*, Vaccine, Antigen, Antibody, Complement, Polymorphonuclear leukocyte (PMN), Neutrophil, Immune response, Serum bactericidal assay (SBA), Opsonophagocytic assay (OPA)

1 Introduction

Neisseria gonorrhoeae, the causative agent of the sexually transmitted infection gonorrhoea, is a major public health problem worldwide for which a vaccine is urgently needed [1]. There is an estimated incidence of 106 million cases of *gonorrhoea* worldwide each year [2], infection rates are rising in many parts of the world (e.g., 67% increase in cases in the USA [3], 80% increase in Australia [4] over the past 5 years) and infection is increasingly hard to treat due to emerging antimicrobial resistance [5]. There are various challenges to developing a gonococcal vaccine, including the high level of phase and antigenic variation of *N. gonorrhoeae* surface structures, and the fact that there is no protective immunity following infection, which means there are no established correlates of protection to guide preclinical vaccine studies (reviewed in [6, 7]). Key aspects of the general immune response to bacterial pathogens include generation of antibodies which kill via complement-

Fadil Bidmos et al. (eds.), *Bacterial Vaccines: Methods and Protocols*,
Methods in Molecular Biology, vol. 2414, https://doi.org/10.1007/978-1-0716-1900-1_19,

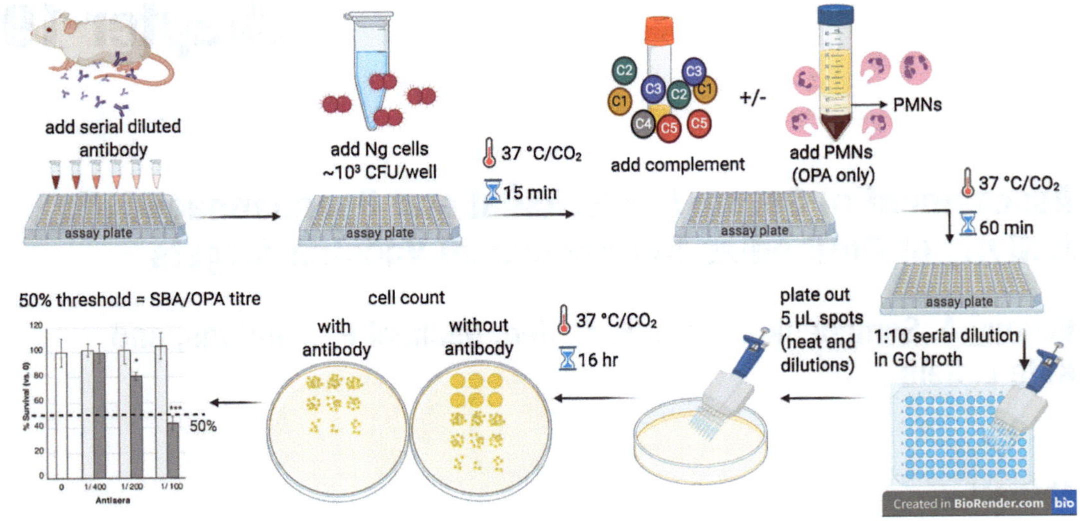

Fig. 1 Schematic overview of the serum bactericidal assay (SBA) and the opsonophagocytic assay (OPA). For both SBA and OPA, serial dilutions of antibody are prepared, then *Neisseria gonorrhoeae* (Ng; ~10^3 colony forming units (CFU)) are added and incubated at 37 °C/5% CO_2 for 15 min. Then complement is added (and polymorphonuclear leukocyte (PMNs) are added for OPA only) and incubated for 1 h. Samples (5 μL neat, and dilutions) are then plated and incubated for 16 h. Finally, CFU counts and the SBA or OPA titer are determined

mediated lysis and opsonophagocytic killing. Assays such as serum bactericidal assay (SBA) and opsonophagocytic assay (OPA) are commonly used as a correlate or surrogate vaccine-induced immunity [8]. The functional immune response raised in animals to potential gonococcal vaccine antigens has been evaluated in several studies using different variations of SBA and OPA [9–13] (*see* Fig. 1).

The SBA measures antibody-mediated, complement-dependent killing of a target bacteria. SBAs are performed by incubating an appropriate bacterial target strain with serial dilutions of serum (animal or human serum containing the antigen-specific antibody) and a complement source [14]. Antibody binds to the bacterial surface and activates the classical pathway of complement, resulting in lysis death of the bacteria. The SBA titre for each serum is the reciprocal of the serum dilution that results in ≥50% killing of the bacteria, relative to the number of target cells present before incubation with serum and complement. Although the role that bactericidal antibodies play in protecting against *N. gonorrhoeae* is unknown, SBA titres correlate with immunity to the closely related bacteria *Neisseria meningitidis* and have been used extensively in development and licensing of meningococcal vaccines [15, 16].

The OPA assay measures antibody-mediated, complement-dependent uptake and killing by phagocytic cells [17]. Similar to the SBA, OPA are performed by incubating an appropriate bacterial target strain with serial dilutions of serum and a complement source, with the addition of phagocytic cells such as neutrophils. The OPA titre for each serum is the reciprocal of the serum dilution that results in ≥50% killing of the bacteria. The OPA has been used to support pneumococcal vaccine licensure [18].

2 Materials

1. Benchtop block heater.
2. Biological Safety Cabinet Class II.
3. Calibrated plastic inoculation loops (10 μL).
4. CO_2 incubator (set to 37 °C and 5% CO_2).
5. Microcentrifuge.
6. Temperature controlled centrifuge with swinging bucket rotor.
7. Dry ice–ethanol bath.
8. Light microscope.
9. Hemocytometer.
10. Sterile, round bottom 96-well plates with lids.
11. Polypropylene 1.5, 15, and 50 mL tubes.
12. EDTA blood collection tubes.
13. Serum clot activator tubes with gel separator.
14. Serological pipette.
15. Spectrometer and spectrometer cuvettes (0.5 mL).
16. GC agar plates supplemented with 1% (v/v) IsoVitaleX.
17. GC broth (GCB) supplemented with 1% (v/v) IsoVitaleX.
18. PolymorphPrep™.
19. Hanks's Balanced Salt Solution (HBSS).
20. Human serum albumin (HSA).
21. Roswell Park Memorial Institute (RPMI) media.
22. Trypan Blue solution (0.4% w/v).
23. 2.5% (w/v) formaldehyde.
24. Sterile Phosphate Buffered Saline (PBS).
25. Sterile solution of 1.7% NaCl (w/v).
26. Sterile solution of 1 M $CaCl_2$.
27. Sterile solution of 1 M $MgCl_2$.
28. Sterile solution of 10% (w/v) saponin in GCB.
29. Sterile milli-Q H_2O.

3 Methods

3.1 Preparation of Assay Media

1. Prepare and autoclave the following: PBS, 1.7% NaCl (w/v), 1 M $CaCl_2$, 1 M $MgCl_2$, milli-Q H_2O.
2. Prepare 10% (w/v) saponin in GCB and sterilize by passing through a 0.22 μm filter.
3. Supplement RPMI with 0.15 mM $CaCl_2$, 0.5 mM $MgCl_2$ and 0.5% (w/v) HSA. Sterilize media through a 0.22 μm filter and store in the fridge (*see* **Note 1**).
4. Bring media to room temperature (RT) before using in the assays.

3.2 Preparation of Bacterial Inoculum

1. *N. gonorrhoeae* is grown on GC agar plates overnight (16 h) at 37 °C/5% CO_2.
2. Harvest approximately 10 *N. gonorrhoeae* colonies from the plate and restreak onto a fresh GC plate. Grow the bacteria for 4 h (*see* **Note 2**).
3. Harvest bacteria from the plate with a sterile loop into 1 mL of HBSS in 1.5 mL tube and resuspend by pipetting up/down gently.
4. Wash cells by centrifuging the tube at 4000 × *g* for 5 min at RT to pellet bacteria. Discard the supernatant.
5. Resuspend cells in HBSS and measure optical density (OD) (*see* **Note 3**).
6. Prepare bacterial inoculum in assay media to $OD_{600} = 0.001$ (~1×10^5 colony forming units (CFU)/mL) (*see* **Note 4**).

3.3 Heat Inactivation of Immune Sera

1. If using immune serum (mouse, rabbit, human etc.) as a source of antibodies for the assay, inactivate heat labile complement proteins by incubating serum on a block heater (set to 56 °C) for 1 h, then briefly centrifuge (*see* **Note 5**). This step is not necessary if using purified antibodies.
2. Test bactericidal activity of heat-inactivated serum against *N. gonorrhoeae* using the same conditions as in the assay (i.e., heat-inactivated serum alone should not kill *N. gonorrhoeae* after 1 h at 37 °C).

3.4 Preparation of Serum for the Complement Source

1. Collect blood from healthy volunteers (e.g., using Vacuette® Z or equivalent Serum Clot Activator Tubes with Gel Separator) and clot as per the manufacturer's instructions (e.g., invert several times and incubate at room temperature for 10 min before centrifuging at 2000 × *g* for 10 min, room temperature). Transfer serum into tubes on ice before combining, aliquoting and storing (*see* **Note 6**).

2. Adsorb serum against formaldehyde-fixed bacteria to remove cross-reactive antibodies as described below (**steps 3–10**).
3. Grow the *N. gonorrhoeae* strain(s) being tested in the assay as in Subheading 3.2, **step 1**. Harvest bacteria from a GC agar plate with a sterile loop into 1 mL of PBS in 1.5 mL tube and resuspend by pipetting up/down gently.
4. Wash cells by centrifuging at 4000 × *g* for 5 min (4 °C) to pellet bacteria. Discard the supernatant. Resuspend cell pellet in 1 mL of PBS. Repeat twice, then after final wash discard the supernatant.
5. Resuspend the cell pellet in 1 mL of 2.5% (w/v) formaldehyde (diluted in PBS) and incubate at room temperature for 15 min. Wash cells (repeat **step 4** above).
6. Resuspend fixed bacteria in PBS and measure optical density (OD) as described in **Note 3**.
7. Prepare 1 mL of OD_{600} 3 fixed bacteria and pellet cells at 4000 × *g* for 10 min (discard supernatant).
8. Resuspend cell pellet in 5 mL of prepared pooled serum (*see* **Note 7**). Incubate on ice for 1 h, then pellet the bacteria (centrifuge precooled to 4 °C). Collect and pass the supernatant through a syringe filter (0.22 μM) to remove residual cells.
9. Aliquot prepared serum into 1.5 mL tubes and snap freeze in an ethanol–dry ice bath (*see* **Note 8**).
10. Test the bactericidal activity of the prepared serum using the SBA assay method in Subheading 4 below, with the absence of added antibody (*see* **Note 9**).

3.5 Preparation of Polymorphonuclear-Cells

1. Collect blood from healthy volunteers (e.g., using Vacuette® K_3EDTA blood collection tubes or equivalent). Invert tubes several times to ensure adequate mixing with the anticoagulant.
2. Dispense 6 mL of PolymorphPrep™ or equivalent into a 15 mL tube and carefully overlay with an equal volume of freshly collected blood (*see* **Note 10**).
3. Centrifuge the tube in a swinging-bucket rotor centrifuge at 500 × *g* for 30 min (RT, no brakes). After centrifugation carefully transfer the tube to the work bench.
4. Using a serological pipette, carefully collect the PMN buffy coat and transfer it into a 50 mL tube (*see* **Note 11**).
5. Gently add HBSS to the PMN buffy coat, bringing the final volume to 50 mL. Invert the tube several times to mix the contents.
6. Centrifuge the tube at 200 × *g* for 10 min.
7. Carefully remove the supernatant and resuspend the PMN pellet in 10 mL of sterile milli-Q H_2O.

8. Approximately 20 s later, add an equal volume of sterile 1.7% NaCl and invert the tube several times to mix the contents.
9. Centrifuge the tube at 200 × *g* for 10 min and remove the supernatant.
10. Wash PMNs three times by resuspending the cell pellet in HBSS and centrifuging at 200 × *g* for 10 min.
11. After the final wash resuspend PMNs in assay media and measure cell concentration with haemocytometer (*see* **Note 12**).
12. Prepare a suspension with ~3 × 10^5 PMNs/mL.

3.6 SBA Assay

1. In a round-bottom 96-well plate prepare two-fold serial dilutions of antibody (triplicate wells) (*see* **Note 13**).
2. To all appropriate test wells, add bacterial inoculum (~10^3 CFU). Place the lid on the plate and incubate at 37 °C/CO_2 for 15 min.
3. While incubating the assay plate, plate out the bacterial inoculum to determine the bacterial CFU (*see* **Note 14**). In a separate 96-well plate, prepare serial dilutions of the bacterial inoculum (e.g., add 5 μL of inoculum to 45 μL of GCB in the first set of triplicate wells). Mix the contents by repeated pipetting and transfer 5 μL into subsequent wells containing 45 μL of GCB (1/10 dilution). After performing several tenfold dilutions, plate 5 μL spots from all wells onto a GC agar plates. Allow the spots on the agar plate to dry and then incubate in 37 °C/CO_2 incubator overnight.
4. Remove the assay plate from the incubator and add required volume of serum as a complement source (*see* **Note 9**). Without delay, make up the final volume to 100 μL with assay media and return the plate to the 37 °C/CO_2 incubator. Incubate the plate for 1 h.
5. While the assay plate is incubating, prepare a 96-well dilution plate by adding 45 μL of GC broth into appropriate wells for serial dilutions and plating of bacteria. Also, prepare two GC agar plates for each set of triplicate wells (i.e., neat/undiluted and 1/10 dilution).
6. Remove the assay plate from the incubator and mix the contents of wells by gently, repeated pipetting. Transfer 5 μL from all the wells in the assay plate into prepared 96-well dilution plate. Plate out 5 μL spots on prepared GC agar plates. Allow the spots on the agar plates to dry and then incubate in 37 °C/ CO_2 incubator overnight.
7. Next day, count CFU and determine the SBA titer, defined as the lowest concentration/dilution of antibody that caused ≥50% killing relative to no treatment control.

3.7 OPA Assay

1. In a round-bottom 96-well plate prepare serial dilutions of antibody (triplicate wells) (*see* **Notes 13** and **15**).
2. To all appropriate wells add bacterial inoculum (~10^3 CFU). Place the lid on the plate and incubate at 37 °C/CO_2 for 15 min.
3. While incubating the assay plate, plate out the bacterial inoculum (*see* **step 3** of SBA assay).
4. Remove the assay plate from the incubator and to all wells add appropriate volume of serum as a complement source (*see* **Note 9**).
5. Add PMNs (10^5 cells/well) and assay media (final volume 100 μL/well). Return the assay plate to the 37 °C/CO_2 incubator and incubate for 1 h.
6. While incubating, prepare a 96-well dilution plate (*see* **step 5** of Subheading 4, SBA assay).
7. Remove the assay plate from the incubator and add 10 μL of saponin (10% solution) to all wells to lyse the PMNs. Vigorously mix the contents of all wells by repeated pipetting up/down. Transfer 5 μL from all the wells in the assay plate into a prepared 96-well dilution plate and perform 1/10 dilutions. Plate out 5 μL spots on prepared GC agar plates. Allow the spots on the agar plates to dry and then incubate in 37 °C/ CO_2 incubator overnight.
8. Next day, count CFU and determine the OPA titre, defined as the lowest concentration/dilution of antibody that caused ≥50% killing relative to no treatment control.

4 Notes

1. Complete assay media is stable for 1 week (stored in the fridge, 4 °C).
2. Replating ensures bacteria in the assay are in the early log phase (i.e., to minimize number of nonviable and dead bacteria in the assay). Only light growth will be apparent.
3. Prepare a spectrophotometer and a cuvette. To a cuvette add 0.9 mL of PBS and take a "blank" measurement at 600 nm. Add 0.1 mL of resuspended bacterial cells to the cuvette and apply paraffin tape to seal the top. Invert the cuvette several times to mix the contents. Take a reading of the cuvette and record the absorbance. Calculate the OD by adjusting for the dilution factor (1:10).
4. Bacteria will remain viable for several hours in assay media if kept at room temperature or 37 °C.

5. Heating the serum inactivates its innate bactericidal activity. Serum volume should be no less than 50 μL per 1.5 mL tube. Store heat-inactivated serum short term at 4 °C (<2 weeks) or long term at −80 °C. Alternatively, to avoid repeated freeze–thaw cycles that are damaging to antibodies add equal volume of sterile glycerol (final concentration of glycerol is 50% (v/v)), mix thoroughly and store at −20 °C (solution will remain liquid). Using serum with 50% (v/v) glycerol does not affect bacterial viability or performance of assays.
6. Combine serum from several donors (consider using equal volume and ratio of male/female donor). Serum can be aliquoted into appropriate tubes and stored in the −80 °C freezer.
7. The serum for the complement source can be scaled up if large quantities are required. Use the same ratio of bacteria to serum.
8. Store serum prepared for the complement source at −80 °C. Complement is stable at −80 °C for over 6 months and freeze–thaw cycles are not recommended. When conducting assay, thaw complement immediately prior to use in the assay and keep on ice.
9. Typically, *N. gonorrhoeae* with nonsialylated lipooligosaccharide (LOS) will tolerate 10–2.5% (v/v) complement (i.e., 100% survival after 1 h). In SBA/OPA use highest concentration of complement that does not induce killing of *N. gonorrhoeae* after 1 h.
10. Bring both the blood and the PolymorphPrep™ to room temperature before use.
11. After centrifugation of the blood and PolymorphPrep™, there will be several distinct layers in the tube. Starting from the top of the tube these are: blood plasma, mononuclear cell buffy coat, first PolymorphPrep™ layer, PMN buffy coat, second PolymorphPrep™ layer and erythrocyte layer.
12. In a 1.5 mL tube mix 20 μL of PMNs with equal volume of Trypan Blue (0.4%) before applying to hemocytometer. Determine the viability and cell concentration. Typically, cell viability will be >95%.
13. For example, pipet 40 μL of prepared serum into the top wells and transfer 20 μL into subsequent wells containing 20 μL of assay media. Include one set of wells that contain media only for the no treatment control (i.e., bacteria and complement only, no antibody). Additional control wells can be also added (e.g., media only and serum only).
14. Label GC agar plates and leave with lids open at the back of the Biological Safety Cabinet to dry (5–10 min) before plating spots.

15. The OPA assay plate set up is identical to SBA, however this assay may require additional control wells (i.e., PMN and serum only, PMN only). The presence of complete complement in the OPA assay means that some killing may also be due to bacterial lysis via SBA. As such it is recommended that SBA and OPA assays are performed simultaneously if direct comparison is required.

References

1. Gottlieb SL, Ndowa F, Hook EW 3rd, Deal C, Bachmann L, Abu-Raddad L et al (2020) Gonococcal vaccines: public health value and preferred product characteristics; report of a WHO global stakeholder consultation, January 2019. Vaccine 38:4362–4373. https://doi.org/10.1016/j.vaccine.2020.02.073
2. WHO (2012) Global incidence and prevalence of selected curable sexually transmitted infections - 2008. World Health Organization, Geneva
3. Braxton J, Davis D, Emerson B, Flagg E, Grey J, La Zetta et al (2018) Sexually transmitted disease surveillance 2017. Centers for Disease Control and Prevention, Atlanta. https://www.cdc.gov/std/stats17/2017-STD-Surveillance-Report_CDC-clearance-9.10.18.pdf. Accessed 18 Jan 2021
4. Kirby Institute (2018) HIV, viral hepatitis and sexually transmissible infections in Australia: annual surveillance report 2018. The Kirby Institute, UNSW Sydney
5. Unemo M (2020) Optimizations to keep gonorrhoea treatable and reduce antimicrobial resistance selection. Nat Rev Urol 17:609–610. https://doi.org/10.1038/s41585-020-00378-5
6. Edwards JL, Jennings MP, Apicella MA, Seib KL (2016) Is gonococcal disease preventable? The importance of understanding immunity and pathogenesis in vaccine development. Crit Rev Microbiol 42:928–941. https://doi.org/10.3109/1040841X.2015.1105782
7. Rice PA, Shafer WM, Ram S, Jerse AE (2017) *Neisseria gonorrhoeae*: drug resistance, mouse models, and vaccine development. Annu Rev Microbiol 8(71):665–686. https://doi.org/10.1146/annurev-micro-090816-093530
8. Plotkin SA (2010) Correlates of protection induced by vaccination. Clin Vaccine Immunol 17:1055–1065. https://doi.org/10.1128/CVI.00131-10
9. Semchenko EA, Day CJ, Seib KL (2017) MetQ of *Neisseria gonorrhoeae* is a surface-expressed antigen that elicits bactericidal and functional blocking antibodies. Infect Immun 85: e00898–e00816. https://doi.org/10.1128/IAI.00898-16
10. Semchenko EA, Day CJ, Seib KL (2020) The *Neisseria gonorrhoeae* vaccine candidate NHBA elicits antibodies that are bactericidal, opsonophagocytic and that reduce gonococcal adherence to epithelial cells. Vaccines (Basel) 8:219. https://doi.org/10.3390/vaccines8020219
11. Jen FE, Semchenko EA, Day CJ, Seib KL, Jennings MP (2019) The *Neisseria gonorrhoeae* methionine sulfoxide reductase (MsrA/B) is a surface exposed, immunogenic, vaccine candidate. Front Immunol 10:137. https://doi.org/10.3389/fimmu.2019.00137
12. Gulati S, Beurskens FJ, de Kreuk BJ, Roza M, Zheng B, DeOliveira RB et al (2019) Complement alone drives efficacy of a chimeric antigonococcal monoclonal antibody. PLoS Biol 17: e3000323. https://doi.org/10.1371/journal.pbio.3000323
13. McQuillen DP, Gulati S, Rice PA (1994) Complement-mediated bacterial killing assays. Methods Enzymol 236:137–147. https://doi.org/10.1016/0076-6879(94)36013-8
14. Findlow J, Balmer P, Borrow R (2019) A review of complement sources used in serum bactericidal assays for evaluating immune responses to meningococcal ACWY conjugate vaccines. Hum Vaccin Immunother 15:2491–2500. https://doi.org/10.1080/21645515.2019.1593082
15. Borrow R, Aaberge IS, Santos GF, Eudey TL, Oster P, Glennie A et al (2005) Interlaboratory standardization of the measurement of serum bactericidal activity by using human complement against meningococcal serogroup b, strain 44/76-SL, before and after vaccination with the Norwegian MenBvac outer membrane vesicle vaccine. Clin Diagn Lab Immunol 12:970–976. https://doi.org/10.1128/CDLI.12.8.970-976.2005
16. Lucidarme J, Louth J, Townsend-Payne K, Borrow R (2019) Meningococcal serogroup A, B, C, W, X, and Y serum

bactericidal antibody assays. Methods Mol Biol 1969:169–179. https://doi.org/10.1007/978-1-4939-9202-7_12

17. Dwyer M, Gadjeva M (2014) Opsonophagocytic assay. Methods Mol Biol 1100:373–379. https://doi.org/10.1007/978-1-62703-724-2_32

18. Jódar L, Butler J, Carlone G, Dagan R, Goldblatt D, Käyhty H et al (2003) Serological criteria for evaluation and licensure of new pneumococcal conjugate vaccine formulations for use in infants. Vaccine 21:3265–3272. https://doi.org/10.1016/s0264-410x(03)00230-5

Chapter 20

Opsonophagocytic Killing Assay to Measure Anti–Group A Streptococcus Antibody Functionality in Human Serum

Helen Wagstaffe, Scott Jones, Marina Johnson, and David Goldblatt

Abstract

The opsonophagocytic killing assay (OPKA) is designed to measure the functionality of strain-specific antibodies and, therefore, assess protective immunity or the immunogenicity of Group A Streptococcus (GAS) (type A *Streptococcus pyogenes*) vaccines. Opsonization of GAS for phagocytosis is an important mechanism by which antibodies protect against disease in vivo. The Opsonophagocytic Index or Opsonic Index (OI) is the estimated dilution of antisera that kills 50% of the target bacteria. Here, we describe the protocol of the standardized GAS OPKA developed by Jones et al., 2018.

Key words Antibody, Phagocytosis, Opsonophagocytic killing assay, *S. pyogenes*, HL-60 cells, Baby rabbit complement, In vitro, Functional assay

1 Introduction

Group A Streptococcus (GAS), a Gram-positive bacterium, can cause asymptomatic infections, mild and severe disease and is the leading causative agent of pharyngitis in children and adolescents worldwide [1, 2]. There is currently no licensed vaccine despite a large global burden of disease resulting in more than 0.5 million deaths per year [3]. A lack of standardized immunoassays to measure postvaccination GAS immunity has hindered vaccine development to date; the WHO GAS Vaccine Development Technology Roadmap highlighted this as a key priority activity for vaccine development in 2019 [3]. The Lancefield assay, widely used in the past, measures growth and survival of GAS in fresh human or animal blood (plus immune or nonimmune serum) [4]. Interdonor variation in neutrophil and complement activity disqualify this type of assay from full standardization and therefore use in large, multicenter vaccine studies. The opsonophagocytic killing assay (OPKA) described in this chapter, first developed by Jones et al., 2018 [5], is an important tool for the reliable detection and quantification of

Fadil Bidmos et al. (eds.), *Bacterial Vaccines: Methods and Protocols*,
Methods in Molecular Biology, vol. 2414, https://doi.org/10.1007/978-1-0716-1900-1_20,

functional anti-GAS antibodies, which is important for the continued development of GAS vaccines.

The GAS OPKA assay makes use of baby rabbit complement (BRC) as a standard source of complement and continuous human promyelocytic leukemia cell line (HL-60) as an exogenous source of phagocytic cells. This method removes the major sources of variation present in many current and past functional assays used to measure GAS vaccine immunogenicity. The assay was optimized for seven clinically relevant GAS strains of *emm*-type 1, 12, and 6, but other strains can be tested and optimized to be used in the assay.

2 Materials

Prepare all reagents at room temperature. Pyrogen-free or deionized water is required for buffer preparations (unless stated otherwise). Volumes can be adjusted accordingly.

1. HL-60 cell culture medium: 500 ml RPMI 1640 medium, 50 ml fetal calf serum (FCS), 5 ml L-glutamine (200 mM).
2. Freezing medium: 9 ml FCS, 1 ml dimethyl sulfoxide (DMSO).
3. Todd-Hewitt-yeast extract broth (THY broth): 6 g Todd-Hewitt broth, 1 g yeast extract, 200 ml water. Mix until all components are dissolved and filter-sterilize using a 0.22 μm bottle-top filter into a sterile 200 ml bottle.
4. Bacteria storage buffer: 3 g tryptone soya broth, 0.5 g glucose, 10 ml glycerol, 100 ml water. Mix until all components are dissolved and autoclave.
5. THY agar (THY plates): 48 g Todd-Hewitt broth, 8 g yeast extract, 24 g bacteriological agar, 1600 ml water. Autoclave and bring to 50 °C in a water bath. Pour 25 ml agar into 100 × 100 mm square agar plate. Leave on a flat surface to dry for ~20 min. Invert stacked plates and store at 4 °C for up to 1 month.
6. THY overlay agar: 48 g Todd-Hewitt broth, 8 g yeast extract, 12 g bacteriological agar, 1600 ml water. Make fresh on day of assay. Autoclave and store in a 50 °C water bath until required.
7. 1% gelatin solution: 4 g gelatin, 400 ml water. Dissolve gelatin in water and autoclave. Store at room temperature (RT) for ≤2 months.
8. 2,3,5-Tetraphenyltetrazolium chloride (TTC) stock: 1.25 g TTC, 50 ml water. Dissolve TTC in 40 ml water and make up to a final volume of 50 ml with the remaining water.

Sterile-filter with a 0.22 μm filter. Liquid should have a yellowish tinge. Store at 4 °C for ≤1 month.

9. Opsonization buffer (OPS buffer): 5 ml FCS, 40 ml 1× Hank's Buffered Salt Solution (HBSS, +Ca/Mg), 5 ml 1% gelatin solution. Prepare on day of assay and discard after use.
10. 96-well round-bottom microtiter plates.
11. GAS bacteria isolates.
12. HL-60 Cells, preferred source; American Type Culture Collection (ATCC) (*see* **Note 1**).
13. BRC, preferred source; Pel-Freez (*see* **Note 1**).

3 Methods

3.1 HL-60 Master Stock Propagation from Master Cell Bank Stock

HL-60 cells are promyelocytic leukemia cells which are differentiated to a neutrophil-like cell with 0.8% dimethylformamide (DMF) for use in the assay. All cell culture is to be undertaken in a culture hood under sterile conditions. Warm culture medium to 37 °C before use.

1. Add 10 ml culture medium to a 15 ml centrifuge tube. Thaw the master stock of HL-60 cells rapidly by swirling in a 37 °C water bath. Transfer the cells into the culture medium.
2. Centrifuge the tube at 350 × *g* for 5 min at room temperature (RT). Remove the supernatant.
3. Resuspend the cells in culture medium to a final concentration of 2×10^5 per ml and transfer to a culture flask. Incubate at 37 °C, 5% CO_2.
4. When the cell density reaches 5×10^5 cells per ml, add further fresh culture medium to readjust the concentration to 2×10^5 per ml (*see* **Note 2**). Use multiple flasks when the volume reaches the capacity of the flask. To avoid risk of contamination, the medium must not reach the cap when the flask is horizontal.
5. When the concentration reaches 5×10^5 cells per ml in 10 flasks, freeze the cells. Transfer the contents of the flasks into 50 ml centrifuge tubes and spin at 350 × *g* for 5 min.
6. Extract and discard the supernatant, being careful not to disturb the pellet.
7. Add 2.5 ml freezing medium to each 50 ml centrifuge tube and resuspend each pellet.
8. Combine the contents of all tubes in a single culture flask and aliquot 1 ml into cryovials (each vial should contain 1×10^7 cells).

9. Transfer cryovials into controlled-rate freezing containers and into a −80 °C freezer.
10. After a minimum of 2 h, transfer the cryovials into a liquid nitrogen tank (maximum 1 week at −80 °C).

3.2 HL-60 Working Stock Preparation from Master Stock

1. Take a 1 ml aliquot of HL-60 master stock prepared as in Subheading 3.1, from the liquid nitrogen tank and transfer into a controlled-rate freezing container.
2. Defrost cells quickly by swirling in a 37 °C water bath.
3. Transfer into a 50 ml centrifuge tube containing 50 ml of fresh culture medium.
4. Spin at 350 × *g* for 5 min at RT. Pour off supernatant and resuspend pellet in 10 ml fresh culture medium.
5. Transfer into a tissue culture flask and incubate at 37 °C, 5% CO_2, overnight.
6. Monitor the cell density using a hemacytometer and readjust to 2×10^5 per ml in warm culture medium. Repeat after 2–3 days.
7. Maintenance, every 3–4 days, harvest cells into a 50 ml centrifuge tube, spin at 350 × *g* for 5 min at RT, pour off supernatant and resuspend pellet in warm culture medium. Reseed the required number of flasks with 2×10^5 cells per ml. Cell density must remain $\leq 1.2 \times 10^6$ per ml.

3.3 Differentiation of HL-60 Cells

This section describes the differentiation method in 200 ml volumes. To differentiate other volumes, the concentration of cells and DMF remains constant.

1. Count the cells and calculate the volume required to resuspend at 4×10^5 cells per ml.
2. Cell viability must be ≥90% prior to differentiation.
3. Add 1.6 ml DMF to 175 ml fresh culture medium (this will give a final concentration of 0.8% in 200 ml).
4. Spin cells at 350 × *g* for 5 min at RT and pour off the supernatant.
5. Resuspend cells in 25 ml culture medium and add to the 175 ml of culture medium containing DMF.
6. Incubate the cells at 37 °C, 5% CO_2 for 5 or 6 days. Do not feed the cells during this time.
7. Harvest cells for use in the assay (*see* Subheading 3.8).

3.4 Bacterial Master Stock Preparation and Maintenance

All bacterial culture to be undertaken in a bacterial culture hood under sterile conditions.

Frozen vials of GAS strains are stored at −80 °C. To maximize the integrity of the bacterial master stocks, the vials should remain frozen at all times by keeping on dry ice. Vials can be stored for several years.

3.4.1 From Clinical Swabs

Swabs can be stored at 4 °C for up to 4 months depending on the strain or for longer time if stored at −80 °C.

1. Streak the swab along the edge of a horse blood agar plate.
2. Rotate the plate by 90°, streak using a sterile loop and repeat once more using the same loop (optional, use a new loop, pass through the final streak and streak on clean zones toward the center of the plate to ensure isolated colonies).
3. Incubate overnight at 37 °C, 5% CO_2 until colonies appear.
4. To create a master stock, harvest enough colonies on a loop and mix well in 1 ml of storage buffer before storing at −80 °C.

3.4.2 From Plated Bacteria

Plated bacteria can be stored at 4 °C for up to 4 weeks, depending on the storage condition. To create a master stock, harvest enough colonies on a loop and mix well in 1 ml of storage buffer before storing at −80 °C.

3.5 Bacterial Expansion

Day 1:

1. Remove master vial from freezer and streak a fleck of the frozen bacteria onto 2 blood agar plates. Immediately return master vial to cold storage.
2. Label the plates with strain name and date and incubate at 37 °C, 5% CO_2 overnight.

Day 2:

3. Add 30 ml THY broth to two 50 ml Falcon tubes, label them A and B plus the bacterial strain name.
4. Harvest a single colony from one blood agar plate using an inoculating loop.
5. Add harvested bacteria to tube A, repeat for tube B. Use 50 ml THY broth alone to act as a blank.
6. Measure OD600 of the broth before and after inoculation, and of the blank. The ODs should be comparable.
7. Incubate tubes at 37 °C, 5% CO_2. The caps must be kept loose to allow gas exchange. Incubate for 2–3 h.
8. Label the desired number of sterile 1.5 ml microcentrifuge tubes per serotype with the bacterial strain and date.
9. Check the OD of the tubes after 2 h and then every hour/half-hourly until the OD reaches between 0.5 and 0.6 (bacteria are in exponential growth phase) (*see* **Note 3**). OD of blank must be ≤ 0.02.

10. On reaching required OD, harvest the top 10 ml of broth (do not mix) and gently mix 1:1 with storage buffer.
11. Aliquot 0.5 ml into the labelled microcentrifuge tubes and transfer to −80 °C freezer, this is the working bacterial stocks.

3.6 Human Serum Sample Preparation

Human serum must be handled in accordance with local guidelines and stored at −80 °C. Prior to testing, the serum must be thawed at RT and heat-inactivated by incubating at 56 °C for a minimum of 30 min. Allow the samples to cool to RT before use in an assay. Samples can be stored at 4 °C for up to 1 month during the testing process to avoid repeated freeze–thaw cycles (if sample requires retesting).

3.7 Preparing Working Aliquots of Baby Rabbit Complement (BRC)

Caution must be exercised when handling BRC as its components are extremely heat-sensitive. When receiving BRC stock from the supplier, ensure that the contents are entirely frozen and transfer to −80 °C storage immediately. There is the potential for high degrees of variation in performance between BRC lots; therefore, prospective lots need screening before use in an assay.

1. Defrost each bottle of BRC in cold water with constant agitation, for example, inside an ice box containing a mixture of ice and cold water, placed on top of an orbital shaker.
2. Label tubes with the lot number and aliquot date. Place tubes on ice to cool. Place the bottle of BRC on ice as soon as it has thawed.
3. Quickly aliquot the BRC placing the aliquots in ice until finished (this can be done in a culture hood to ensure sterility).
4. Store all aliquots at −80 °C until required.

3.8 Preparation of Differentiated HL-60 Cells (for Use in the Assay)

This procedure describes the process for one 200 ml flask; volumes can be altered accordingly. The concentration of cells required for use in the assay is 1×10^7 per ml. Do not use cells $>1.5 \times 10^6$ per ml at initial count. *See* **Note 4** for more detail on HL-60 cell acceptance criteria.

1. Resuspend differentiated cells by shaking gently to ensure equal distribution throughout the flask. Count cells and check viability by trypan blue exclusion. Cell viability must be ≥80%.
2. Decant cells into four 50 ml centrifuge tubes under sterile conditions. Spin tubes for 5 min at 350 × *g* at RT.
3. Remove the supernatant and resuspend each 50 ml of cells in 50 ml HBSS (without Ca/Mg). Centrifuge for 5 min at 350 × *g* at RT.
4. Remove the supernatant and resuspend each 50 ml tube in 50 ml HBSS (+Ca/Mg).

5. Centrifuge for 5 min at 350 × *g* at RT.
6. Remove the supernatant and resuspend cells at 1×10^7 cells per ml in OPS buffer. Count the cells and assess the viability. Cell viability must be ≥80%.
7. Store at room temperature until required.

3.9 Preparation of Bacteria (for Use in the Assay)

1. Remove one vial of bacteria from the −80 °C freezer and defrost.
2. Centrifuge the tube for 2 min at 13,000 × *g* at RT.
3. Remove the supernatant using a pipette, being careful not to dislodge the pellet of cells.
4. Add 1 ml of OPS buffer to each tube and vortex. Centrifuge the tube at 13,000 × *g* for 2 min at RT.
5. Carefully remove the supernatant and resuspend the pellet in 0.5 ml OPS buffer.

3.10 Procedure for Determining Optimal Dilution of Bacteria

Optimal Dilution Experiment 1 (OD1) will determine a rough estimate for the dilution of bacteria required in the assay without the inclusion of serum and controls. Optimal Dilution Experiment 2 (OD2) includes human serum with associated control A and B to determine more accurately the optimal dilution of bacteria required. Procedure is carried out in 96-well round-bottom microtiter plates. OD assays are run in the presence of BRC and HL-60 cells as they can influence the bacterial growth.

3.10.1 Optimal Dilution Experiment 1

1. Prepare THY overlay agar.
2. Remove working aliquot of BRC (Subheading 3.7) from the −80 °C freezer and defrost on ice.
3. Dilute bacteria (working stock prepared as Subheading 3.9) twofold (1:1) by mixing 75 μl bacteria with 75 μl of OPS buffer in a single well (row A) of a microtiter plate (plate 1). Add 120 μl of OPS buffer to rows B to H. Two columns of the plate will be required per serotype *see* Fig. 1.
4. Prepare fivefold serial dilutions (1:9, 1:49, 1:249, 1:1249, 1:6249, 1:31,249, 1:156,249) by diluting 30 μl of diluted bacteria from row A with 120 μl OPS buffer in row B. Repeat to column H (Fig. 1).
5. In a second microtiter plate (plate 2), add 20 μl OPS buffer to two columns for each serotype of bacteria to be tested.
6. Transfer 10 μl of diluted bacteria prepared in the first plate to the appropriate wells in the second plate (in duplicate). Plate 1 can be discarded.
7. Incubate plate 2 at RT on a mini-orbital shaker for 30 min at 700 rpm.

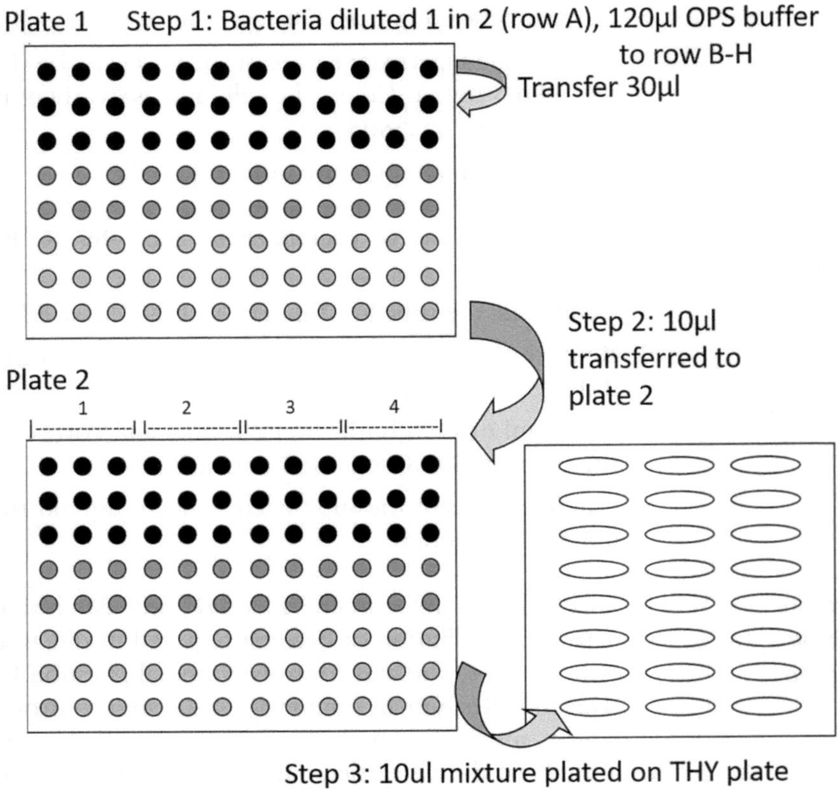

Fig. 1 Optimal dilution 1 experiment procedure. Bacteria is serially diluted in a plate 1 then transferred to plate 2 when BRC and HL-60 cells are added. After incubation, the mixture is spotted onto THY agar and incubated overnight

8. Following the incubation period, add 10 µl of active BRC to each column in use on plate 2. BRC may be prediluted in OPS buffer prior to addition to the plate to ascertain a final in-well predetermined concentration. The optimal BRC concentration was determined for seven different clinically relevant GAS strains (*see* **Note 5**).
9. Resuspend the HL-60 cells. Add 40 µl to each column in use on plate 2.
10. Incubate on a mini-orbital shaker at 37 °C, 5% CO_2 for 90 min at 700 rpm. Incubate multiple plates in a single layer to maintain equal CO_2 exposure.
11. During this time, remove THY agar plates from the fridge, remove lids, and lay on the bench to dry for 30 min to 1 h.
12. Place plate 2 on ice for ≥20 min to halt the phagocytic process.
13. Vortex the plate at a low speed. Using a manual eight-channel pipette, spot 10 µl from each column of plate 2 side by side on a THY plate. Tilt the plate to the left and right so the spots measure approximately 1–1.5 cm across, being careful not to

allow the spots to merge. If there are multiple serotypes to be assessed, use a separate agar plate for each serotype.

14. Repeat this procedure for the next column, spot next to the previous column on the agar plate. Repeat until there are three columns of spots on the agar plate.
15. Leave agar plates at RT for ~20 min to allow the spots to dry (*see* **Note 6**).
16. Remove the overlay agar from the water bath. Measure out the amounts of THY overlay agar and TTC (1 μl/ml) required. Mix well and add 20 ml to each THY agar plate and allow the agar to solidify. TTC must be added after the agar has cooled to below 50 °C as it turns red upon heating.
17. Invert the stacked plates and incubate at 37 °C, 5% CO_2 for 16–18 h. Bacteria will form red-colored colonies.
18. After the incubation period, count the colonies using an automated colony counter (plates can be stored at 4 °C for a maximum of 72 h before reading).
19. Determine the average colony-forming unit (CFU) count of the duplicate rows and identify the 2 dilutions between which the average CFU count yields 50–200 CFU.
20. Choose a number of dilutions (maximum of six) between the range identified in OD1. For example, if the OD1 result was between 1:249 and 1:1249, dilutions chosen could be 1:249, 1:499, 1:749, 1:999 1:1499 and 1:1999. Proceed to OD2.

3.10.2 Optimal Dilution Experiment 2

1. Make up each dilution of bacteria working stock determined in OD1 required for the OD2 experiment.
2. Assay a known positive quality control (QC) serum, Control A (heat inactivated BRC, no serum) and Control B (active BRC, no serum) as detailed in Subheading 3.11 (each dilution of bacteria will be added to four columns of one plate) (Fig. 2).
3. Select the dilution that yields ~100 CFU in Control B and the OI of the known positive QC serum within the predetermined accepted range.

3.11 OPKA Procedure

The following procedure details the quantities required for four 96-well plates. Unknown test sera can be used neat or prediluted in OPS buffer, for example, at 1:4 or 1:30 dilution.

1. Prepare 400 ml overlay agar. Store in a 50 °C water bath until required. Ensure a minimum of 16 THY agar plates are available for use.
2. Remove the BRC from the −80 °C freezer and defrost on ice.
3. Label each round-bottomed 96-well microtiter plate with plate ID (A, B, C or D) and sample layout. The procedure described

	1	2	3 4	5	6	7 8	9	10	11 12
A	Heat inactivated complement	Active complement	Dilution 8	Heat inactivated complement	Active complement	Dilution 8	Heat inactivated complement	Active complement	Dilution 8
B			Dilution 7			Dilution 7			Dilution 7
C			Dilution 6			Dilution 6			Dilution 6
D			Dilution 5			Dilution 5			Dilution 5
E			Dilution 4			Dilution 4			Dilution 4
F			Dilution 3			Dilution 3			Dilution 3
G			Dilution 2			Dilution 2			Dilution 2
H			Dilution 1			Dilution 1			Dilution 1
	Control A	Control B	Positive serum	Control A	Control B	Positive serum	Control A	Control B	Positive serum
	Bacteria dilution 1			Bacteria dilution 2			Bacteria dilution 3		

Fig. 2 Optimal dilution 2 experiment plate layout. Each test dilution of bacteria is assayed with a dilution series of known positive serum, Control A and Control B

	1	2	3 4	5 6	7 8	9 10	11 12
A	Heat inactivated complement	Active complement	Dilution 8	Dilution 8	Dilution 8	Dilution 8	Dilution 8
B			Dilution 7	Dilution 7	Dilution 7	Dilution 7	Dilution 7
C			Dilution 6	Dilution 6	Dilution 6	Dilution 6	Dilution 6
D			Dilution 5	Dilution 5	Dilution 5	Dilution 5	Dilution 5
E			Dilution 4	Dilution 4	Dilution 4	Dilution 4	Dilution 4
F			Dilution 3	Dilution 3	Dilution 3	Dilution 3	Dilution 3
G			Dilution 2	Dilution 2	Dilution 2	Dilution 2	Dilution 2
H			Dilution 1	Dilution 1	Dilution 1	Dilution 1	Dilution 1
	Control A	Control B	Test serum 1	Test serum 2	Test serum 3	Test serum 4	QC sample

Fig. 3 OPKA plate layout

below utilizes 8 dilutions of serum serially diluted threefold, in duplicate.

4. Add 20 μl of OPS buffer to columns 1 and 2 of each plate, rows A–H inclusive.
5. Add 20 μl OPS buffer to rows A–G, columns 3–12 inclusive.
6. Add 30 μl of serum sample 1 into row H, columns 3 and 4 of plate A. Add 30 μl of serum sample 2 into row H, columns 5 and 6. Continue adding a maximum of 5 samples per plate (4 test samples and a QC sample).
7. Perform threefold serial dilutions in columns 3–12 by transferring 10 μl from row H to row G making sure no bubbles form while mixing. Then transfer 10 μl from row G to row F and continue up the plate. Once 10 μl is transferred from row B to row A and mixed, remove 10 μl from row A and discard (*see* Fig. 3).
8. Add 10 μl of prediluted bacteria working stock (prepared as Subheading 3.9 and prediluted to optimal dilution determined in Subheading 3.10) to each well including all control wells. When adding bacteria, pipet 10 μl by nonreverse pipetting directly into the liquid at the bottom of the well, release all the liquid by pressing the pipette to the second stop.

9. Incubate the 96-well plates for 30 min at RT on a mini-orbital shaker at 700 rpm.
10. Incubate an aliquot of BRC in a 56 °C degree water bath to heat inactivate for a minimum of 30 min.
11. Following the incubation period, add 10 μl of heat inactivated BRC (prediluted in OPS buffer if indicated) to column 1 of all plates. Add 10 μl of active BRC (also prediluted) to all other wells (columns 2–12). When adding BRC, reverse-pipet onto the side of the well, then tap the plate to mix in the well.
12. Resuspend the HL-60 cells and add 40 μl to all wells.
13. Incubate on a mini-orbital shaker at 700 rpm at 37 °C, 5% CO_2 for 90 min. Incubate the plates in a single layer to maintain equal CO_2 exposure.
14. During this time remove THY agar plates from the fridge, remove lids and lay on the bench to dry for 30 min to 1 h.
15. Place plates on ice for ≥20 min to halt the phagocytosis process.
16. Replace the covers on the dried agar plates and stack in piles of 4 (1 stack per 96-well plate). Label the side of each plate with the 96-well plate ID, the section of 96-well plate (1, 2, 3, or 4), technician initials and the date.
17. Vortex the plate at a low speed. Using a multichannel pipette, remove 10 μl from a single column and spot 10 μl on agar plate by reverse pipetting to minimize bubbles. Tilt the plate to the left and right so the spots measure approximately 1–1.5 cm across, being careful not to allow the spots to merge.
18. Repeat this procedure for the next column, spot next to the previous column on the agar plate. Repeat until there are three columns of spots on the agar plate.
19. Leave all plates at RT for ~20 min to allow the reaction mixture to soak into the agar (or until completely dry).
20. Remove the overlay agar from the water bath. Measure out amount of overlay and TTC (1 μl/ml) required. Mix well and add 20 ml to each THY agar plate, allow to solidify.
21. Invert the stacked plates and incubate at 37 °C, 5% CO_2 for 16–18 h. Bacteria will form red-colored colonies.
22. After the incubation period, count the colonies using an automated colony counter (plates can be stored at 4 °C for a maximum of 72 h before reading).

3.12 Data Analysis

1. The OI is the estimated dilution of serum that kills 50% of the target bacteria.

2. The OI is calculated using the Opsotiter software, which is a Microsoft Excel-based program developed to analyze data from OPKAs.
3. Raw colony counts generated from the automated colony counter are copied and pasted into the program. The program tabulates the opsonic indices for the individual samples as well as determining the percentage of nonspecific killing (NSK).
4. Each bacterial strain must be analyzed separately giving one file for each 96-well plate.

3.12.1 Limits of Detection

The following applies when samples are run neat.

1. The upper limit of detection (ULD) is 8748. This is calculated by taking the fold dilution of serum [3] to the power of 7 (number of fold dilutions), and multiplying by 4 (total volume in well (80 μl) divided by volume of serum in well (20 μl)).
2. However, Opsotiter automatically multiplies all 8748 results by 2 for statistical purposes. The "OI final" on the results sheet is therefore displayed as 17,496.
3. Samples that have an opsonic index higher than the upper limit of detection may be repeated at a predilution of 1 in 30.
4. The lower limit of detection (LLD) is 4, as this is the initial dilution of serum once all components have been added to the plate. However Opsotiter automatically divides all results of 4 by 2 for statistical purposes. The "OI final" on the results sheet is therefore displayed as 2.
5. The lower limit of quantification (LLOQ) was determined for seven different clinically relevant GAS strains (*see* **Note 7**). Samples below LLOQ can be reported as half LLOQ for statistical purposes.

3.12.2 Tentative Plate Acceptance Criteria

1. The level of NSK is determined as the killing induced by BRC and HL-60 cells alone, in the absence of serum. This is determined as the difference between Control A and Control B: 1—(average Control B CFU/average Control A CFU) × 100. This should be kept at ≤35%; *see* **Note 5**.
2. The CFU obtained in Control B should be kept between 50 and 150 to ensure accurate and reproducible counting on the automated colony counter.
3. A QC sample should be included on all plates run, the OI result should remain between a predetermined acceptability range.

3.12.3 Tentative Sample Acceptance Criteria

1. Sample replicates must cross the 50% killing limit within one 3-fold dilution. If they do not cross the 50% killing limit, then the sample will be retested.

2. Good killing curves are those that show a sigmoid shape. Curves may be irregular, and "N" or "U" shaped curves may be identified. Irregular curves may be treated individually as they may not conform to acceptance criteria.
3. For a sample to be considered positive, the maximum killing must be greater than 70%. If the sample falls between 40% and 70% maximum killing, the sample must be repeated.

4 Notes

1. HL-60 cells from ATCC are well characterized and qualified, this is the preferred source of HL-60 cells for this assay. Similarly, BRC from Pel-Freez is of good quality and is the preferred supplier, however BRC can be purchased from any supplier. Significant lot to lot variation exists in BRC reagents, new batches will need to be tested for suitability in the assay first.
2. HL-60 cell concentrations must not exceed 5×10^5 per ml at the master stock propagation stage. During working stock maintenance, cell density must remain $\leq 1.2 \times 10^6$ per ml. Cell viability must remain $\geq$80%. HL-60 cell viability tends to decline over 25–26 passages when a decrease in viability occurs and irregular cell shapes are seen under the microscope, a new vial of master stock should be thawed every 3–4 months. A small volume of culture should checked for *mycoplasma* and other microbial contamination using standard tissue culture mycoplasma screening techniques.
3. Different strains will grow at different rates. Inoculation from a single isolated colony will ensure that the broth culture is pure. To avoid contamination from other bacteria that may be present in the swab and increase purity of the strain of interest, it is recommended to inoculate single colonies from the blood agar plates and streak them onto a new blood agar plate and incubate overnight at 37 °C, 5% CO_2 until colonies appear.
4. HL-60 cells must be mycoplasma-free. On the day of the assay, cell viability must be $\geq$80%. Phenotype of HL-60 cells is determined by flow cytometry using mouse anti-human CD35 FITC conjugated antibody and mouse anti-human CD71 PE-conjugated antibody. Differentiated cells are accepted for use in the assay if the up-regulation of CD35 was $\geq$55% of the cell population and CD71 expressing was down-regulated by $\leq$12% when compared to the working stock preparation.
5. The concentration of BRC can be optimized to reduce NSK to below an acceptable level. NSK at $\leq$35% is commonly used as acceptance criteria in an optimized assay; however, levels up to 60% can also be accepted. The optimal concentration

determined for strains emm1 (43, 02-12, and GAS05134), emm12 (611,020, 611,025, and GAS 09437) was 2.1%, for emm6 (GASOPA6_02) was 3.1% [5]. Other factors that can be optimized to ensure maximal killing within acceptable range of NSK are plate shaking speed and the incubation time.

6. For the spots to dry, plates can be left to air-dry on a bench top. If excessive contamination occurs, plates can be dried in a laminar flow hood with a shortened length of time.
7. The LLOQ was determined for strains emm1 43; 9, 02-12; 15, GAS05134; 8, emm12 611,020; 7, 611,025; 5, GAS 09437; 7 and emm6 GASOPA6_02; 6 [5]. The average LLOQ across all strains was 8. The LLOQ can be determined by spiking low concentrations of known positive serum into heat-inactivated nonimmune (antibody-depleted) serum to produce a sample of OI 4–12. Generate each sample individually and run four times in duplicate.

References

1. Barth DD, Moloi A, Mayosi BM, Engel ME (2020) Prevalence of group a streptococcal infection in Africa to inform GAS vaccines for rheumatic heart disease: a systematic review and meta-analysis. Int J Cardiol 307:200–208
2. Sims Sanyahumbi A, Colquhoun S, Wyber R, Carapetis JR (2016) Global disease burden of group a streptococcus. In: *Streptococcus pyogenes*: basic biology to clinical manifestations. University of Oklahoma Health Sciences Center, Oklahoma City (OK)
3. Vekemans J, Gouvea-Reis F, Kim JH, Excler JL, Smeesters PR, O'Brien KL et al (2019) The path to group a streptococcus vaccines: World Health Organization Research and Development technology roadmap and preferred product characteristics. Clin Infect Dis 69(5):877–883
4. Lancefield RC (1957) Differentiation of group a streptococci with a common R antigen into three serological types, with special reference to the bactericidal test. J Exp Med 106 (4):525–544
5. Jones S, Moreland NJ, Zancolli M, Raynes J, Loh JMS, Smeesters PR et al (2018) Development of an opsonophagocytic killing assay for group a streptococcus. Vaccine 36 (26):3756–3763

Chapter 21

Neisseria lactamica Controlled Human Infection Model

Adam P. Dale, Diane F. Gbesemete, Robert C. Read, and Jay R. Laver

Abstract

Neisseria lactamica is a nonpathogenic commensal of the human upper respiratory tract that has been associated with protection against *N. meningitidis* colonization and disease. We have previously utilized the *N. lactamica* controlled human infection model to investigate the protective effect of *N. lactamica* colonization on *N. meningitidis* colonization, the nature of cross-reactive immune responses mounted toward *N. meningitidis* following *N. lactamica* colonization, and the microevolution of *N. lactamica* over a 5-month colonization period. More recently, we have assessed the possibility of utilizing genetically modified strains of *N. lactamica* to enable use of the commensal as a vehicle for prolonged exposure of the nasopharynx of humans to antigens of interest, expressed in carried organisms. A controlled infection with *N. lactamica* expressing the meningococcal antigen NadA has been executed and the results demonstrate that this strategy is effective at generating immune responses to the target antigen. Throughout this chapter, we outline in a step-by-step manner the methodologies utilized when performing controlled human infection with *N. lactamica* including procedures relating to: (1) the dilution of *N. lactamica* stock vials to derive intranasal inocula, (2) the delivery of intranasal inocula to human volunteers, (3) the determination of *N. lactamica* colonization status following intranasal inoculation using oropharyngeal swabbing and nasal wash sampling, (4) the microbiological procedures utilized to identify *N. lactamica* colonization among study volunteers, and (5) the identification of *N. lactamica* colonies as strain Y92-1009 using polymerase chain reaction.

Key words *Neisseria lactamica*, *Neisseria meningitidis*, Controlled human infection model, CHIME, Human challenge

1 Introduction

Oropharyngeal carriage of *Neisseria meningitidis* is a prerequisite for meningococcal disease, and there is a complex biological interaction between this organism and its host [1]. In most continents other than Africa, the highest carriage rates of *N. meningitidis* occur in young adults, particularly in universities and colleges [2]. Although glycoconjugate meningococcal vaccines such as the quadrivalent "ACWY" vaccine have had dramatic effects on disease incidence, this is mostly due to herd protection conferred by reduced carriage and transmission [3]. However, even prior to the

Fadil Bidmos et al. (eds.), *Bacterial Vaccines: Methods and Protocols*,
Methods in Molecular Biology, vol. 2414, https://doi.org/10.1007/978-1-0716-1900-1_21,

advent of meningococcal vaccines, meningococcal disease was relatively rare other than during epidemics despite the fact that meningococcal carriage was prevalent in a significant minority of asymptomatic individuals. It is widely considered that natural suppression of widespread dissemination of carriage of disease-associated *N. meningitidis* occurs in nature by carriage of non-pathogenic Neisseriaceae. *N. lactamica*, like *N. meningitidis*, is a member of the *Neisseriaceae* but in contrast, *N. lactamica* is non-capsulate and lacks pathogenic potential. It is a common commensal of the human nasopharynx, particularly in young children. Age-specific rates of *N. meningitidis* carriage and disease have been shown to be inversely associated with carriage of *N. lactamica* [4, 5] and mathematical modeling suggests a period of 4–5 years of protection from meningococcal carriage following carriage of *N. lactamica* [6].

To try to determine the mechanism of this relationship, we did a controlled intranasal infection study of young adults with *N. lactamica* Y92-1009 (sequence type 3493, clonal complex 613). Stocks of this strain had been manufactured at the Good Manufacturing Practices (GMP) pharmaceutical manufacturing facilities at Public Health England (PHE), Porton Down, United Kingdom, to generate the seed banks for an outer membrane vesicle vaccine. Theoretically, strain Y92-1009 is capable of generating immunity to a broad spectrum of *N. meningitidis* serogroups including serogroup B, which expresses the least immunogenic meningococcal polysaccharide. This intramuscular vaccine had been tested in a Phase 1 study, and while well tolerated and capable of generating antibody to *N. lactamica*, it did not produce cross-reacting bactericidal antibodies against a bank of pathogenic *N. meningitidis* serogroup B strains [7]. Using a *N. lactamica* controlled human infection model experiment (CHIME), we showed that intranasal inoculation with 10^4 colony forming units (CFU) of strain Y92-1009 was well tolerated and safe and caused carriage in 65% of young adults. This carriage of Y92-1009 persisted in most carriers for 6 months. Note that this original group included active smokers. Carriage was followed by development of humoral immunity to *N. lactamica* in those who carried the inoculated strain, but this did not induce significant cross-reactive bactericidal antibodies against *N. meningitidis* [8].

Subsequently, we sought direct evidence for a carriage prevention effect by conducting controlled infection with *N. lactamica* in healthy university students, the group that has the highest rates of acquisition of meningococci. To do this, 310 nonsmoking university students were inoculated with *N. lactamica* or were sham-inoculated, and carriage was monitored for 26 weeks. At baseline, natural *N. meningitidis* carriage in the control group of students was 22.4%, which increased to 33.6% by week 26. Two weeks after inoculation of *N. lactamica*, 33.6% of the challenge group became colonized with *N. lactamica*. In this group, meningococcal carriage reduced significantly from 24.2% at inoculation to 14.7% 2 weeks after inoculation. We found that the inhibition of meningococcal carriage was only observed in volunteers who were actively colonized with *N. lactamica*, was due both to displacement of existing meningococci and to inhibition of new acquisition, and persisted over at least 16 weeks. At the end of the 26-week period, we did crossover inoculation of controls with *N. lactamica* and this replicated the result in the original control group. The impact that we observed on carriage reduction of *N. meningitidis* was just as potent as that observed after glycoconjugate polysaccharide vaccination. Furthermore, genome sequencing showed that the inhibition affected multiple meningococcal sequence types [9].

A closed genome sequence of strain Y92-1009 was then defined [10] which enabled us to study microevolutionary changes of *N. lactamica* over the course of 5 months of carriage using the human cohort undergoing controlled infection. We found that most mutations are transient indels within repetitive tracts of putative phase-variable loci associated with host–microbe interactions (*pgl* and *lgt*) and iron acquisition (*fetA* promoter and *hpuA*). Recurrent polymorphisms occurred in genes associated with energy metabolism (*nuoN*, *rssA*) and the CRISPR-associated *cas1*. In volunteers who were naturally cocolonized with meningococci, recombination altered allelic identity in *N. lactamica* to resemble meningococcal alleles, including loci associated with metabolism, outer membrane proteins and immune response activators, but there was little evidence of recombination in the opposite direction [11].

Recently, we have delineated B cell responses to *N. lactamica* colonization and their cross-reactivity with *N. meningitidis* [12] and have investigated the possibility of genetic modification of *N. lactamica* to enable use of the commensal as a vehicle for prolonged exposure of the nasopharynx of humans to antigens of interest, expressed in carried organisms. A controlled infection with *N. lactamica* expressing the meningococcal antigen NadA has been planned and executed [13] and the results demonstrate that this strategy is effective at generating immune responses to the target antigen [14].

Throughout this chapter, we outline in a step-by-step manner the methodologies utilized when performing *N. lactamica* CHIMEs including procedures relating to (1) the dilution of *N. lactamica* stock vials to derive intranasal inocula, (2) the delivery of intranasal inocula to human volunteers, (3) the determination of *N. lactamica* colonization status following intranasal inoculation using oropharyngeal swab and nasal wash sampling, (4) the microbiological procedures utilized to identify *N. lactamica* colonization amongst study volunteers, and (5) the identification of cultured *N. lactamica* colonies as the inoculum strain (Y92-1009) using polymerase chain reaction (PCR).

2 Materials

2.1 Dilution of N. lactamica Stock Vials to 105 CFU/ml for Intranasal Inoculation

1. Stock vial containing 1 × 10^8 CFU/ml *N. lactamica* Y92-1009 (ST-3493, clonal complex 613, suspended in 1 ml Franz medium) produced in the GMP-accredited facilities at PHE's Porton Down facility, Salisbury, UK. Following transfer from PHE, stock vials of *N. lactamica* are stored in a dedicated, secure and remotely monitored −80 °C freezer.
2. Decontaminated class II microbiological safety cabinet (MSC), dedicated for *N. lactamica* inoculum preparation.
3. Decontaminated vortex mixer, dedicated for *N. lactamica* inoculum preparation.
4. Decontaminated set of calibrated Gilson-style pipettes, dedicated for *N. lactamica* inoculum preparation.
5. Incubator, set to 37 °C, 5% CO_2.
6. Sterile filter pipette tips.
7. 10 μl sterile, disposable plastic microbiological loops.
8. 70% alcohol solution (v/v).
9. Single use waste box.
10. Disposable gloves.
11. Sterile phosphate buffered saline (PBS) (1×: autoclaved and passed through a sterile 0.2 μm filter).
12. Columbia blood agar (CBA) plates: Columbia agar base, 5% defibrinated horse blood.
13. Gonococcal (GC) agar plates with 5-bromo-4-chloro-3-indolyl-B-D galactopyranoside (X-gal) (GC-X-gal): 36 g/l GC agar base, 100 ml/l lysed horse blood, 20 ml/l Vitox supplement, 4 g/l glucose, 40 mg/l X-gal, 2 mg/l vancomycin, 7.5 mg/l colistin, 3 mg/l trimethoprim, and 1 mg/l amphotericin B.
14. Appropriate laboratory source document (to log all stages of *N. lactamica* inoculum preparation in real time).
15. Sterile universal containers.

2.2 Intranasal Inoculation with 105 CFU N. lactamica

1. Intranasal inoculum containing 10^5 CFU *N. lactamica* suspended in 1 ml PBS (*see* Subheading 3.2).
2. Dedicated room within a clinical environment, for example, a clinical research facility (CRF), complete with an examination couch.
3. Disposable aprons and gloves.
4. Decontaminated and dedicated 1000 μl Gilson-style pipette.
5. Sterile pipette tips.

6. Clinical waste bin.
7. Appropriate laboratory source document (to log all stages of *N. lactamica* inoculation in real time).

2.3 Taking and Processing an Oropharyngeal Throat Swab to Determine Volunteer N. lactamica Colonization Status

1. Sterile throat swab for microbiological culture with an appropriate transport medium (e.g., TS/5-17 Probact Amies Clear medium in a tube with a polystyrene viscose tip swab).
2. GC-X-gal plate.
3. Dedicated room within a clinical environment, for example, a CRF, complete with an examination couch.
4. Class II MSC housed within an Advisory Committee on Dangerous Pathogens (ACDP) containment level 2 (CL2) laboratory.
5. Appropriate laboratory waste container.
6. Incubator set at 37 °C, 5% CO_2.

2.4 Taking and Processing a Nasal Wash Sample to Determine Volunteer N. lactamica Colonization Status

1. Sterile saline (0.9%) for irrigation or injection (20 ml) at room temperature.
2. 2× 10 ml sterile syringes.
3. Sterile plastic Petri dish.
4. 2× GC-X-gal plates.
5. Sterile PBS.
6. Sterile 50 ml centrifuge tube.
7. Set of calibrated Gilson-style pipettes and sterile tips.
8. Sterile microbiology plate spreaders.
9. Disposable gloves and aprons.
10. Tissues.
11. Transport box with ice.
12. Dedicated room within a clinical environment, for example, a CRF, complete with an examination couch.
13. Class II MSC housed within an ACDP CL2 laboratory.
14. Centrifuge.
15. Appropriate laboratory waste container.
16. Incubator set at 37 °C, 5% CO_2.

2.5 Identification of N. lactamica Colonies

1. Class II MSC.
2. GC-X-gal agar plate.
3. 10 μl sterile, disposable plastic microbiological loops.
4. Sterile Bacterial Storage Medium: 50:50 (v/v) mixture of Tryptone Soya Broth supplemented with 0.2% (w/v) yeast extract, 60:40 (v/v) mixture of glycerol and PBS. Stored in 0.5 ml aliquots in cryogenic storage vials.

5. Oxidase reagent-impregnated strips.
6. Glass microscope slides.
7. 0.5% crystal violet solution.
8. Gram's (Lugol's) iodine solution.
9. Carbolfuchsin or safranin solution.
10. Gram's acetone solution.
11. Distilled water.
12. Heat block.
13. Blotting paper.
14. Mineral oil.
15. Microscope with 100× lens (oil immersion).
16. API® NH kit (Biomerieux).
17. Genomic DNA extraction kit.
18. Set of calibrated Gilson-style pipettes.

2.6 Identification of *N. lactamica* Colonies as Strain Y92-1009 Using PCR

2.6.1 Preparation of PCR Master Mix

1. Dedicated, decontaminated PCR preparation area or laminar flow cabinet.
2. Set of calibrated Gilson-style pipettes and sterile filter tips.
3. Disposable gloves and dedicated, pre-PCR laboratory coat.
4. Tissues.
5. 10% (v/v) bleach solution in spray bottle.
6. Distilled water in spray bottle.
7. 70% (v/v) alcohol solution.
8. Q5 Hot Start High Fidelity 2× Master Mix.
9. Primer stocks (10 μM each). For list of primers and sequences, *see* Table 1.
10. DNase/RNase-free, molecular biology grade water.
11. Thin-walled, 0.2 ml PCR tubes and 0.2 ml tube racks.
12. Benchtop centrifuge.
13. Appropriate laboratory waste container.

2.6.2 Performance of Y92-1009-Specific PCR

1. Dedicated, decontaminated PCR preparation area or laminar flow cabinet.
2. Set of calibrated Gilson-style pipettes and sterile filter tips.
3. Disposable gloves and dedicated, pre-PCR laboratory coat.
4. Tissues.
5. 10% (v/v) bleach solution in spray bottle.
6. Distilled water in spray bottle.
7. Number of 24 μl aliquots of [1× Y92-1009-specific PCR Master Mix] equal to number of isolates for identification.

Table 1
Nucleotide sequences of primers for Y92-1009-specific, multiplex PCR. (Forward—FOR; Reverse—REV)

Name	Nucleotide sequence (5′ to 3′)	Product size (bp)
SeqA_FOR SeqA_REV	gtgctgaatttatagacgggc aagctaggtctacttggtttag	388
SeqB_FOR SeqB_REV	caattgtttaccgccctgcc gtttcctgccaatcaatccc	273
SeqC_FOR SeqC_REV	agggaccgacatctttcatac ttgcaggctctttccaaac	594
*lacZ*_FOR *lacZ*_REV	cgggcaaacttgcgcgg gcaaaccgaaacggggcagg	745

8. Genomic DNA extracted from colonies positively identified as *N. lactamica* (*see* Subheading 3.5).
9. Genomic DNA extracted from *N. lactamica* strain Y92-1009 (@50 ng/μl).
10. DNase/RNase-free, molecular biology grade water.
11. Appropriate laboratory waste container.
12. Laboratory marker pen.
13. Laboratory coat for conducting PCR.
14. Thermal cycler.

2.6.3 Analysis of PCR Products Using Agarose Gel Electrophoresis

1. Laboratory coat and disposable gloves.
2. Set of calibrated Gilson-style pipettes and sterile tips.
3. Certified PCR agarose.
4. Tris–acetate–EDTA (TAE) buffer (1×).
5. Nontoxic DNA intercalating agent.
6. DNA molecular weight ladder (200–1000 bp).
7. Digital gel imaging system.
8. Appropriate laboratory source document.

3 Methods

3.1 Dilution of N. lactamica Stock Vials to 105 CFU/ml for Intranasal Inoculation

N. lactamica inoculum preparation is performed in tandem by two trained members of the laboratory technical team (*see* **Note 1**).

1. Don a pair of disposable gloves and disinfect the class II MSC with 70% alcohol.

2. Label a series of universal containers in preparation for serial dilution of the *N. lactamica* stock inoculum vial contents, as follows: 10^{-1}, 10^{-2}, 10^{-3}, 10^{-4}, 10^{-5}, and 10^{-6} dilution.
3. Pipet 900 μl PBS into the universal containers labelled 10^{-1}, 10^{-2}, 10^{-4}, 10^{-5}, and 10^{-6} and 4.5 ml of PBS into the universal container labelled 10^{-3}.
4. Label the CBA and GC-X-gal plates and ensure they are at room temperature prior to use.
5. Remove a stock vial containing 1×10^8 CFU/ml *N. lactamica* Y92-1009 from the −80 °C freezer and transport this vial immediately on ice to the class II MSC. Take the stock vial off ice and place within the class II MSC and allow it to thaw at room temperature.
6. Immediately upon thawing, vortex the stock vial for 30 s and then pipet 100 μl of the stock vial contents into the universal container labelled 10^{-1} before vortexing for 30 s.
7. Having applied a new filter tip, pipet 100 μl of the 10^{-1} dilution into the universal tube labelled 10^{-2} before vortexing for 30 s.
8. Having applied a new filter tip, pipet 500 μl of the 10^{-2} dilution into universal container labelled 10^{-3} before vortexing for 30 s.
9. Having applied a new filter tip, pipet 1 ml volumes of the 10^{-3} dilution (containing 10^5 CFU *N. lactamica*/ml) into universal containers and transport to the CRF for use as intranasal inocula.
10. To verify the purity of the prepared batch of *N. lactamica* inocula, streak 10 μl volumes of the 10^{-3} dilution onto CBA and GC-X-gal agar plates using a microbiological loop and incubate for 48 h at 37 °C, 5% CO_2. Following incubation, assess for the pure growth of colonies morphologically in keeping with *N. lactamica* that are gray on CBA and blue on GC-X-gal plates.
11. To enable *N. lactamica* CFU/ml of the prepared inocula to be calculated formally, utilize the 10^{-3} dilution to immediately perform three further serial dilutions in the universal containers labeled 10^{-4}, 10^{-5}, and 10^{-6}, ensuring that a new pipette tip is utilized between dilution steps and that each dilution is vortexed for 30 s.
12. Pipet 3× 10 μl drops of the 10^{-4}, 10^{-5}, and 10^{-6} dilutions onto a GC-X-gal plate and allow to dry for 5 min before incubating for 48 h at 37 °C, 5% CO_2. Following incubation, identify the dilution with an easily countable number of colonies (10–50 colonies within each 10 μl drop). Use the Mean to calculate *N. lactamica* CFU/ml in the prepared inoculum

taking into account the dilution factor (CFU/ml = Mean colony count × 100 × dilution factor).

3.2 Intranasal Inoculation with 10^5 CFU N. lactamica

Prior to enrolment onto a *N. lactamica* CHIME study, volunteers undergo extensive screening to ensure they are medically fit and that it is safe for them to participate. While the full volunteer screening protocol is beyond the scope of this chapter, key inclusion and exclusion criteria used across the *N. lactamica* CHIME studies during the screening process are outlined in **Note 2**. Once a volunteer has passed the screening process and is enrolled onto the study, the methodology outlined below is followed to ensure a standardized approach to intranasal inoculation. It must be noted that all intranasal inoculation procedures must be performed in an appropriate clinical environment, for example a CRF, by a member of the clinical research team.

1. Ask the volunteer to drink only water in the 1-h period prior to their intranasal inoculation appointment.
2. Explain the procedure to the volunteer and obtain informed consent.
3. Ask the volunteer to don a disposable apron and lie supine on the clinical examination couch with their head tilted back (*see* Fig. 1).
4. Transport the prepared inoculum (10^5 CFU *N. lactamica* suspended in 1 ml PBS) from the laboratory to the CRF. The inoculum must be utilized within 30 min of preparation (*see* **Note 3**).
5. Wash hands and don a pair of gloves and a disposable apron.
6. Instill 0.5 ml of the prepared inoculum slowly in a dropwise fashion into each nostril, one at a time, using a Gilson-style pipette. Following intranasal inoculation, the volunteer is asked to remain supine for 5 min.
7. Dispose of the used pipette tip, apron, and gloves in the clinical waste.
8. Monitor the volunteer following intranasal inoculation for 30 min. Take a set of clinical observations including pulse, blood pressure, respiratory rate, and temperature. Assuming the volunteer is well and has normal observations following intranasal inoculation, they can then be discharged from the CRF.

1. Explain the procedure to the volunteer and obtain informed consent.
2. Wash hands and apply a clean pair of disposable gloves and an apron.

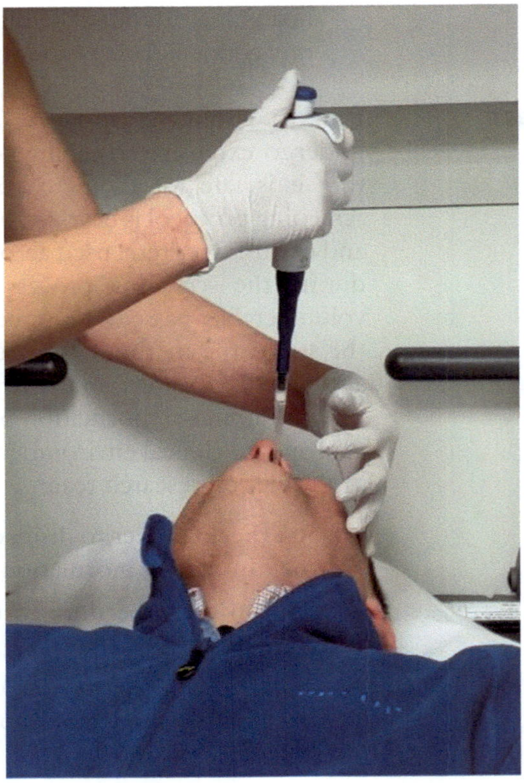

Fig. 1 Intranasal inoculation procedure. Prior to intranasal inoculation, the volunteer is asked to lie supine with their head tilted back

3.3 Taking and Processing Oropharyngeal Throat Swabs to Determine Volunteer N. lactamica *Colonization Status*

3. While sitting upright, ask the volunteer to tilt their head back and open their mouth wide.
4. When the head of the swab has made contact with the pharyngeal wall, behind the uvula, stroke it across the pharyngeal wall five times.
5. Remove the swab from the oral cavity, place it into the Amies transport medium, seal the swab and transfer it immediately to the class II MSC for further processing.
6. Working in the class II MSC, apply the swab contents directly to the GC-X-gal plate by streaking the swab vertically in a continuous fashion while rotating the plate and swab tip. Ensure the swab makes contact with all areas of the agar.
7. Dispose of the swab in the appropriate waste container and incubate the GC-X-gal plate immediately at 37 °C, 5% CO_2 for 48 h.
8. Following 48 h incubation, assess the GC-X-gal plate for growth of putative colonies of *N. lactamica* in line with the methodology outlined in Subheading 3.5.

3.4 Taking and Processing Nasal Wash Samples to Determine Volunteer N. lactamica Colonization Density

1. Explain the procedure to the volunteer and obtain informed consent.
2. Ask the patient to don a disposable apron to protect their clothing.
3. Position the patient supine with a pillow placed under their shoulders to enable extension of the neck backward.
4. Don an apron and a pair of gloves.
5. Draw up 10 ml of sterile 0.9% saline into the syringe and expel any air.
6. Ask the volunteer to hold the Petri dish on their chest to enable collection of the nasal wash fluid (leave the lid on the Petri dish at this point).
7. Ask the volunteer to open their mouth, extend their head backward, and position their tongue to avoid swallowing the saline. Remind them to breathe through their mouth and not to swallow.
8. Once in position and the volunteer has signaled they are ready, place the tip of the syringe in their nostril and gently instill 10 ml of saline. Ask them to hold the saline in their nasopharynx for 1 min.
9. After 1 min, ask the volunteer to open the lid of the Petri dish and to lean forward as fast as possible to allow the saline to exit the nose passively and be caught in the Petri dish.
10. Withdraw the nasal wash specimen from the Petri dish using the syringe and transfer it to the centrifuge tube.
11. Repeat **steps 5–10** using the alternate nostril.
12. Transport the nasal wash specimen on ice immediately to the microbiology laboratory for processing.
13. Measure and record the total volume of nasal wash fluid retrieved from the volunteer.
14. Centrifuge the nasal wash fluid at 5000 × *g* for 10 min to pellet the bacteria.
15. Carefully decant the supernatant and resuspend the pellet in a total volume of 300 μl PBS before vortexing for 30 s to loosen the mucous.
16. Inoculate 2× GC-X-gal plates with 250 μl and 25 μl volumes of the resuspended pellet, respectively, and lawn over the plates using individual microbiology spreaders.
17. Incubate the GC-X-gal plates immediately at 37 °C, 5% CO_2, for 48 h.
18. Following 48 h incubation, assess the GC-X-gal plates for growth of putative colonies of *N. lactamica* in line with the methodology outlined in Subheading 3.5.

19. Calculate the *N. lactamica* colonization density within the nasal wash where necessary as *N. lactamica* CFU/ml.

3.5 Identification of N. lactamica Colonies from Throat Swab and Nasal Wash Cultures

1. Working in a class II MSC, visually inspect the GC-X-gal agar plates inoculated with throat swab or nasal wash contents for evidence of putative *N. lactamica* colonies, that is, blue colonies (beta-galactosidase positive) with morphology consistent with *Neisseria* spp.
2. Pick a well-defined, accessible blue colony and subculture it by streaking it onto GC-X-gal agar with a 10 μl bacteriological loop. Incubate at 37 °C, 5% CO_2, for 24 h.
3. Assuming a pure subculture is obtained, produce a stock of the isolate in an aliquot of bacterial storage medium. Macerate colonies against the side of the cryogenic storage vial to produce a suspension, then incubate for 10 min and freeze at −80 °C.
4. Check the oxidase status of the organism by spreading a small amount of a single blue colony onto an oxidase strip using a 10 μl bacteriological loop. A positive oxidase test (strip turns blue/black where contact with colony is made) would be in keeping with potential *N. lactamica* and should prompt a Gram stain to be performed. If the bacteria are oxidase negative, then the isolate is not *N. lactamica* and the prepared stock (**step 3**, above) can be discarded.
5. To perform a Gram stain, the following steps should be followed:
   (a) Place a drop of distilled water onto a clean glass microscope slide and emulsify a small amount of a single blue colony into it using a 10 μl bacteriological loop.
   (b) Dry and fix the slide by placing on a heat block at 65 °C.
   (c) Remove the slide from the class II MSC and flood the slide with 0.5% crystal violet solution and leave for 30 s.
   (d) Wash off the crystal violet solution with tap water and then flood the slide with Gram's acetone and wash off with tap water rapidly.
   (e) Flood the slide with carbolfuchsin or safranin solution and leave for 30 s.
   (f) Wash off the carbolfuchsin or safranin with tap water and dry with clean blotting paper.
   (g) View under oil immersion at 100× magnification.
6. A Gram stain that reveals Gram-negative diplococci is consistent with *N. lactamica* and should prompt speciation using a biochemical test strip, for example, the API® NH (Biomerieux) (*see* **Note 4**).

7. Blue colonies on GC-X-gal agar that are oxidase positive, Gram negative diplococci with an API NH code of 5041 are *N. lactamica*. To confirm the cultured *N. lactamica* as strain Y92-1009, that is, the inoculum strain, PCR is used (Subheading 3.6).
8. To provide template material for amplification by the Y92-1009-specific PCR: return to the frozen stock of the isolate, culture the bacteria and then isolate genomic DNA using a genomic DNA extraction kit according to the manufacturer's instructions.

3.6 Identification of N. lactamica Colonies as Strain Y92-1009 Using PCR

3.6.1 Preparation of PCR Master Mix

1. Don pre-PCR laboratory coat and gloves.
2. If not already active, ensure the laminar flow cabinet is empty, and then turn on the laminar flow cabinet. Expose inside of cabinet to UV light for 15 min (if available).
3. Decontaminate the working area by liberal application of 10% (v/v) bleach solution.
4. Wait for 10 min for the bleach to inactivate environmental amplicons and other potential DNA contaminants. Change gloves.
5. Meanwhile: gather together the Q5 Hot-Start High Fidelity 2× Master Mix (in 500 μl aliquots), the primer stocks, the DNase/RNase-free molecular grade water and the 0.2 ml, thin-walled PCR tubes and racks. Note that you will need 20 × 0.2 ml PCR tubes per 500 μl aliquot of Q5 Hot-Start High Fidelity 2× Master Mix.
6. Spray the working area with distilled water and then mop up the resulting moisture with paper toweling. Carefully dispose of the paper toweling in the waste bin. Change gloves.
7. If not already deactivated, deactivate the UV light and then open the laminar flow cabinet.
8. Inside the laminar flow cabinet, add the following volumes of the appropriate primers into the 500 μl aliquot of Q5 Hot-Start High Fidelity 2× Master Mix:
   (a) 25 μl: SeqA FOR
   (b) 25 μl: SeqA REV
   (c) 25 μl: SeqB FOR
   (d) 25 μl: SeqB REV
   (e) 25 μl: SeqC FOR
   (f) 25 μl: SeqC REV
   (g) 25 μl: *lacZ*_FOR
   (h) 25 μl: *lacZ*_REV.

9. Add 30 μl of DNAse/RNAse-free molecular biology grade water to the 500 μl aliquot of Q5 Hot-Start High Fidelity 2× Master Mix supplemented with primers. This is the Y92-1009-specific 1× Master Mix.
10. Transfer 24 μl aliquots of the Y92-1009-specific 1× Master Mix into individual 0.2 ml, thin walled PCR tubes and stand in 0.2 ml tube racks. Make sure the tube rack is labelled as containing aliquots of the Y92-1009-specific Master Mix.
11. Place the aliquots of Y92-1009-specific 1× Master Mix into the fridge (4 °C) for use later that day, or proceed directly to Subheading 3.6.2.
12. If work in the laminar flow cabinet is completed, seal the laboratory waste container, remove all items from inside the laminar flow cabinet, spray the inside of the laminar flow cabinet liberally with 70% (v/v) alcohol solution and then close the sash on the laminar flow cabinet.

3.6.2 Performance of Y92-1009-Specific PCR

1. If using boiled bacterial lysates, either fresh or frozen, remove bacterial debris from each suspension by centrifugation of the lysates at 17,000 × *g* for 10 min. Note that the Y92-1009-specific PCR will also amplify target sequences from purified genomic DNA. Solutions of extracted genomic DNA are most easy to use at a concentration of 50 μg/ml.
2. Don pre-PCR laboratory coat, safety spectacles and gloves.
3. Decontaminate working area by liberal application of 10% (v/v) bleach solution.
4. Wait for 10 min for the bleach to inactivate environmental amplicons and other potential DNA contaminants. Change gloves.
5. Meanwhile: gather together the appropriate number of aliquots of the 'Y92-1009-specific 1× Master Mix', such that there is one aliquot of 1× Master Mix for each sample to be analysed. Gather additional aliquots of 1× Master Mix sufficient to include one positive and one negative control reaction for every two rows of wells that will be filled during agarose gel electrophoresis. Note that each PCR control tube will provide sufficient material to load 2 wells of control material. One well containing positive control material and one well containing negative control material must be loaded in each row of wells, to allow for: (1) visual comparison of sample amplicons to those generated from wild type genomic DNA extracted from *N. lactamica* strain Y92-1009, and (2) to show that no contaminating DNA was present in the 1× Master Mix, respectively.

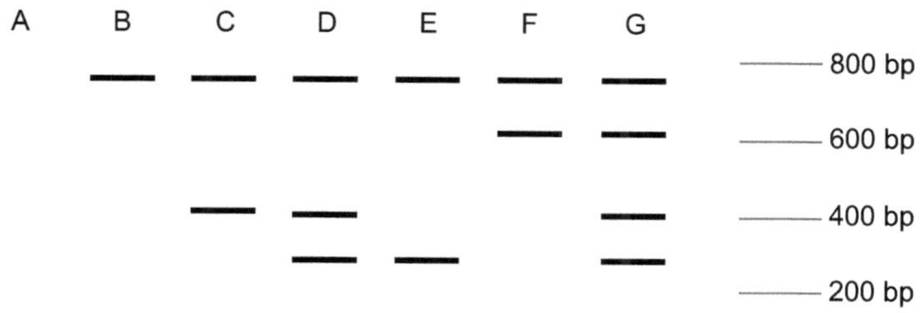

Fig. 2 Interpreting the banding pattern of reaction products from Y92-1009-specific PCR. Possible banding patterns generated in the Y92-1009-specific PCR: (**a**) PCR failure, empty lane, negative control or DNA present from species *other than N. lactamica*. (**b–f**) DNA present from strains of *N. lactamica other than* Y92-1009. (**g**) *N. lactamica* strain Y92-1009. Note that preliminary identification of *N. lactamica* colonies (i.e., oxidase test, Gram stain and API NH—*see* Subheading 3.5) will in most cases prevent PCR from being performed on species other than *N. lactamica*. Note that, in addition to amplification of the bands shown in this figure, strains of *N. lactamica* other than Y92-1009 may also produce additional bands of unknown and unpredictable size. Strains can only be identified as Y92-1009 if the banding pattern shown in (**g**) is present

6. Spray working area with distilled water and then mop up the resulting moisture with paper toweling. Carefully dispose of the paper toweling in the waste bin. Change gloves.
7. For each sample, carefully transfer 1 μl of the lysate or purified, genomic DNA (50 ng) into a single 24 μl aliquot of "Y92-1009-specific 1× Master Mix." Appropriately label each PCR tube with the laboratory marker pen.
8. Into one of the remaining 1× Master Mix aliquots, add 1 μl of *N. lactamica* gDNA derived from wild type strain Y92-1009 (50 ng), and into another add 1 μl of DNase/RNase-free water. Appropriately label each tube with the laboratory marker pen.
9. Retain all 1.5 ml microcentrifuge tubes containing the boiled lysates until a given lysate has been definitively identified. Boiled lysates can be refrozen at −20 °C.
10. Place all PCR tubes containing complete reaction mixtures (including controls) into the Thermal Cycler. Set the Thermal Cycler to run according to the parameters shown in **Note 5**.
11. Analyse PCR amplicons on an agarose gel and analyze the banding pattern present in each lane of the gel, with reference to Fig. 2.

4 Notes

1. Preparation of *N. lactamica* inocula from stock vials is performed by two trained members of the research team who have completed Good Clinical Laboratory Practice training.

Each step of the process is verified by both technicians and logged in real time using the appropriate study-specific laboratory source document. This document then forms part of the volunteer case report form.

2. Key volunteer inclusion and exclusion criteria utilized in *N. lactamica* CHIMEs in the United Kingdom are as follows:

   Inclusion criteria:

   (a) Healthy adults aged 18–45 years inclusive on the day of enrolment.

   (b) Fully conversant in the English language.

   (c) Able and willing (in the investigator's opinion) to comply with all study requirements.

   (d) Written informed consent to participate in the study.

   (e) For females only, willingness to practice continuous effective contraception during the study and a negative pregnancy test at the screening visit.

   Exclusion criteria:

   (a) Active smokers.

   (b) Individuals who have a current infection at the time of inoculation.

   (c) Individuals who have been involved in other clinical studies/trials involving receipt of an investigational product over the last 12 weeks or if there is planned use of an investigational product during the study period.

   (d) Any confirmed or suspected immunosuppressive or immunocompromised state, including HIV infection, asplenia, history of recurrent severe infections or use (more than 14 days) of immunosuppressant medication within the past 6 months (topical/inhaled steroids are allowed).

   (e) Allergy to yeast extract.

   (f) Any other significant disease, disorder, or finding which may significantly increase the risk to the volunteer because of participation in the study, affect the ability of the volunteer to participate in the study, or impair interpretation of the study data, for example recent surgery to the nasopharynx.

   (g) Occupational, household or intimate contact with immunosuppressed persons.

   (h) Pregnancy or lactation.

3. The viability of *N. lactamica* reduces following thawing and dilution in PBS. The dilution strategy referred to within Subheading 3.1 will reliably produce a 1 ml volume of inoculum

containing 10^5 CFU, with the viability being maintained up to 30 min following preparation. By 1 h following inoculum preparation, the viability reduces to 5×10^4 CFU/ml. Thus, to ensure intranasal inoculation with 10^5 CFU, the inoculum should be utilized immediately, that is, within 30 min of preparation.

4. The API® NH testing kit comes ready to use with all required reagents and clear step-by-step instructions. *N. lactamica* can be identified with the API® NH test code of 5041 with the whole process taking approximately 3 h [15].
5. Thermal cycling parameters for amplification of Y92-1009-specific target sequences are as follows:

STAGE A: 1×	5 min at 98 °C
STAGE B: 35×	20 s at 98 °C (melting)
	15 s at 63 °C (annealing)
	26 s at 72 °C (extension)
STAGE C: 1×	5 min at 72 °C
(*Optional*):	Infinite at 10 °C

References

1. Laver JR, Hughes SE, Read RC (2015) Neisserial molecular adaptations to the nasopharyngeal niche. Adv Microb Physiol 66:323–355
2. Christensen H, May M, Bowen L et al (2010) Meningococcal carriage by age: a systematic review and meta-analysis. Lancet Infect Dis 10:853–861
3. Maiden MCJ, Ibarz-Pavón AB, Urwin R et al (2008) Impact of meningococcal serogroup C conjugate vaccines on carriage and herd immunity. J Infect Dis 197:737–743
4. Gold R, Goldschneider I, Lepow ML et al (1978) Carriage of *Neisseria meningitidis* and *Neisseria lactamica* in infants and children. J Infect Dis 137:112–121
5. Cartwright KA, Stuart JM, Jones DM et al (1987) The stonehouse survey: nasopharyngeal carriage of meningococci and *Neisseria lactamica*. Epidemiol Infect 99:591–601
6. Coen PG, Cartwright K, Stuart J (2000) Mathematical modeling of infection and disease due to *Neisseria meningitidis* and *Neisseria lactamica*. Int J Epidemiol 29:180–188
7. Gorringe AR, Taylor S, Brookes C et al (2009) Phase I safety and immunogenicity study of a candidate meningococcal disease vaccine based on *Neisseria lactamica* outer membrane vesicles. Clin Vaccine Immunol 16:1113–1120
8. Evans CM, Pratt CB, Matheson M et al (2011) Nasopharyngeal colonization by *Neisseria lactamica* and induction of protective immunity against *Neisseria meningitidis*. Clin Infect Dis 52:70–77
9. Deasy AM, Guccione E, Dale AP et al (2015) Nasal inoculation of the commensal *Neisseria lactamica* inhibits carriage of *Neisseria meningitidis* by young adults: a controlled human infection study. Clin Infect Dis 60:1512–1520
10. Pandey AK, Cleary DW, Laver JR et al (2017) *Neisseria lactamica* Y92-1009 complete genome sequence. Stand Genomic Sci 12:41
11. Pandey AK, Cleary DW, Laver JR et al (2018) Microevolution of *Neisseria lactamica* during nasopharyngeal colonisation induced by controlled human infection. Nat Commun 9:4753
12. Dale AP, Theodosiou AA, Laver JR et al (2020) Controlled human infection with *Neisseria lactamica* induces B-cell responses that are cross-reactive with *Neisseria meningitidis*. J Immunol 204:231.21
13. Gbesemete D, Laver JR, de Graaf H et al (2019) Protocol for a controlled human

infection with genetically modified *Neisseria lactamica* expressing the meningococcal vaccine antigen NadA: a potent new technique for experimental medicine. BMJ Open 9: e026544

14. Laver JR, Gbesemete D, Dale AP et al (2021) A recombinant commensal bacteria elicits heterologous antigen-specific immune responses during pharyngeal carriage. Sci Transl Med 13:eabe8573
15. APIWEB™, Biomerieux. https://apiweb.biomerieux.com/login. Accessed 27 Sep 2020

Chapter 22

Analyzing Macrophage Infection at the Organ Level

Ryan G. Hames, Zydrune Jasiunaite, Joseph J. Wanford, David Carreno, Wen Y. Chung, Ashley R. Dennison, and Marco R. Oggioni

Abstract

Classical in vivo infection models are oftentimes associated with speculation due to the many physiological factors that are unseen or not accounted for when analyzing experimental outputs, especially when solely utilizing the classic approach of tissue-derived colony-forming unit (CFU) enumeration. To better understand the steps and natural progression of bacterial infection, the pathophysiology of individual organs with which the bacteria interact in their natural course of infection must be considered. In this case, it is not only important to isolate organs as much as possible from additional physiological processes, but to also consider the dynamics of the bacteria at the cellular level within these organs of interest. Here, we describe in detail two models, ex vivo porcine liver and spleen coperfusion and a murine infection model, and the numerous associated experimental outputs produced by these models that can be taken and used together to investigate the pathogen–host interactions within tissues in depth.

Key words Ex vivo perfusion, Murine infection model, Immunohistochemistry, Confocal microscopy, Fiji, InForm, Image analysis, Correlates of protection

1 Introduction

Correlates of protection are a key issue in preclinical vaccine research but are highly specific for different pathogens and may include protective humoral, innate and cellular responses [1]. Examples of correlates of protection include the titer of bactericidal antibodies in *Neisseria meningitidis* vaccine research [2, 3] or opsonophagocytic titers in pneumococcal vaccine research [4]. We have recently found that a subpopulation of splenic tissue-resident macrophages are not only key players in the clearance of encapsulated bacteria but also represent a location for pathogen replication and subsequent initiation of invasive disease [5, 6]. The present provides the protocols for an integrated evaluation of the role of splenic tissue macrophages in current in vivo and ex vivo models with the scope of allowing the testing of this novel step in cellular immunity for evaluation in vaccine research.

Fadil Bidmos et al. (eds.), *Bacterial Vaccines: Methods and Protocols*,
Methods in Molecular Biology, vol. 2414, https://doi.org/10.1007/978-1-0716-1900-1_22,

This work describes the methodology of the murine intravenous (IV) infection model and porcine ex vivo liver and spleen coperfusion model. Mouse infection models, in particular for *Streptococcus pneumoniae* research, are the preferred choice for their high reliability and excellent reproducibility in research on the pathogenesis of infection [7, 8]. The IV infection model has the advantage of inducing an almost instantaneous delivery of bacteria to blood-filtering organs such as the spleen and liver which both have prominent roles in the control of bacteremia [5, 9–16]. This allows for the study of the development of invasive pneumococcal infection while minimizing the effects of confounding factors such as the lung innate and cellular immunity following intranasal infection. Therefore, the murine IV infection model is particularly useful for the study of the early stages of blood borne pathogenesis [5]. Despite being widely available, practical and having much comparability to humans, the mouse model has several drawbacks. The housing of animals in a sterile animal unit environment is costly and while mice have a high genetic similarity to humans [17], their bodies have broad physiological and morphological differences as far down as the cellular population and structure of the organs [18, 19]. Therefore, while excellent at providing fundamental data, a more comparable model is required to validate and build upon these results, enhance their clinical importance for the human population, and introduce a closer biological match. These aims can be addressed with the ex vivo perfusion model of the porcine spleen and liver [6, 20]. The organs are sourced from a local abattoir, from animals which are to be used within the food industry, therefore reducing the use of live animals in laboratory settings. In addition, this approach does not require specialized animal housing and importantly the utilization of organs of animals slaughtered for food production means that these ex vivo organ perfusion experiments are excluded from the rules and regulations associated with the utilization of live animals for experimental research. Furthermore, the model can be adapted for use with various organs [21], most of which show increased microanatomic comparability to humans [22]. The expertise gained by running ex vivo perfusions of porcine organs is also instrumental for the translation to experimental ex vivo perfusion of human organs. Human ex vivo spleen perfusion to study the early steps of infection, in particular by *S. pneumoniae* and SARS-CoV-2 infection, has been recently granted (ClinicalTrials.gov identifier NCT04620824).

The samples collected from these infection models are prepared through a common approach of flash freezing and storage at −80 °C for future sectioning which is done using a cryostat—a cryogenic temperature maintaining cutting device often used in hematoxylin and eosin (H&E) and tissue sample preparation. The sectioned samples are then prepared for visual analysis using immunohistochemistry (IHC), a technique based on specific target binding of primary antibodies [23], which is visualized using fluorochrome conjugated secondary antibodies to detect the bound primary IHC antibodies. Entire immunofluorescence-stained samples are then visualized using an automated quantitative imaging system to acquire an image of the whole slice section as opposed to the limited field-of-view of classic confocal microscopes.

The scanned images are subsequently analyzed via inForm, an automated image analysis software, or through a more manual approach using Fiji, an open-source image analysis tool. This allows for an unbiased distribution assessment of bacteria within the tissue, leaving less room for unconscious bias by the researcher as is often encountered using classical confocal microscopy. By using these image analysis techniques, the colocalization of the pathogen to the various host tissue cells can be quantified, and the structural distribution of the immune cells within the tissue can be determined. When used alongside traditional CFU enumeration, these results provide a stable ground for more robust conclusions by way of increased experimental outputs, therefore allowing greater insight into the host–pathogen interactions within tissue on a cellular level. This method has allowed our group to assess bacterial distribution in various host cells in multiple experiments, allowing for a clearer overview of the steps in systemic infection throughout an infection time course.

2 Materials

All reagents and solutions should be stored as per the manufacturer's instructions, unless otherwise stated.

2.1 Bacterial Strains and Culture Conditions

1. Bacteria laboratory stock.
2. Broth medium.
3. Petri dishes.
4. Broth medium with agar.
5. Distilled water.
6. Autoclave.
7. Defibrinated horse blood.
8. Universal tube with tight cap.
9. Spectrophotometer.
10. 50% v/v glycerol in phosphate-buffered saline (PBS).
11. Cryovial.
12. 96-well plates.

2.2 Murine Infection Model

1. Broth medium.
2. Broth agar base (BAB) plates.
3. Bacterial infection stock (*see* Subheading 3.1).
4. Defibrinated horse blood.
5. PBS.
6. 1.5 mL centrifuge tubes.
7. 96-well plates.
8. HEPA-filtered cages for mice.
9. 29G insulin syringes.
10. 70% v/v ethanol or equivalent disinfectant (such as industrial methylated spirit (IMS)).
11. Heating chamber for mice.
12. Mechanical restraining device for mice.
13. Dissection board.
14. Pins.
15. Sterile forceps.
16. Sterile scissors.
17. 40 μm cell strainers.
18. Cell strainer pestle or 5 mL syringe.

2.3 Ex Vivo Porcine Spleen Perfusion

1. Perfusion circuit (*see* **Note 1**).
2. Major reservoir for sourcing of blood.
3. Centrifugal pump and corresponding control console (*see* **Note 2**).
4. Oxygenator.
5. Heat exchange unit.
6. Venous reservoir to stimulate portal vein flow.
7. Infusion set (one per vessel to be perfused).
8. Large sterile container for collection of blood.
9. 25,000 IU/mL heparin solution.
10. Scissors.
11. Forceps.
12. Scalpels.
13. Large durable sterile bag to hold porcine organs (*see* **Note 3**).
14. Porcine organs (*see* **Note 4**).
15. Urine collection catheters (*see* **Note 5**).
16. Soltran preservative solution (one 500 mL bag per organ).
17. Large ice container and ice.

18. Saline infusion bag (one 250 mL bag per organ).
19. Sterile 1 L measuring cylinder.
20. 8.4% w/v sodium bicarbonate solution.
21. 0.5 mg epoprostenol sodium solution.
22. 5 mL or 10 mL Syringes.
23. Suture kit.
24. Surgical glue.
25. 2 mL lithium heparin calcium-balanced tubes.
26. Blood gas analyzer.
27. Bacterial growth medium and agar plates (*see* Subheading 3.1).
28. 96-well microtiter plates.
29. 40 μm cell strainers.
30. Cell strainer pestle (optional).
31. Tissue embedding molds.

2.4 Sample Freezing and Sectioning

1. Metal freezing container.
2. Ice container with dry ice.
3. 2-methylbutane.
4. Embedding molds.
5. OCT embedding matrix.
6. Forceps.
7. Cryostat (with anti-roll plate and compatible disposable microtome blades).
8. Polylysine adhesion microscopy slides.
9. Relevant primary and secondary antibodies.
10. Blocking solution.
11. Hydrophobic pen or staining rack.
12. Fixation solution.
13. 0.1% v/v Triton™ X-100.
14. PBS.
15. Distilled water.
16. Antifade slide mountant.
17. Slide coverslips.
18. Clear nail polish or coverslip adhesive.

2.5 Whole Slide Scanning

1. Vectra® Polaris™ Automated Quantitative Pathology Imaging System.
2. 70% v/v IMS.

2.6 Image Analysis

1. Phenochart whole slide viewer.
2. inForm automated image analysis software.
3. Fiji image processing package.
4. Bio-Formats Fiji Plugin.

3 Methods

3.1 Bacterial Strains and Culture Conditions

Bacteria are cultured to the mid-logarithmic phase and aliquots are stored at −80 °C with a cryopreservative such as glycerol. The media selected will depend on the requirements of your bacterial strain. For example, *S. pneumoniae* grows well in Brain Heart Infusion (BHI) broth and BAB culture plates with 3% v/v defibrinated horse blood.

3.1.1 Broth

1. Prepare the relevant bacterial growth medium (e.g., BHI; 14.8 g in 400 mL distilled water) and autoclave at 121 °C 15 psi (103 kPa) for 20 min.
2. Ensure color and media pH is within the parameters recommended by the manufacturer.
3. Store media at room temperature (RT).

3.1.2 Blood Agar Base (BAB) Culture Plates + 3% v/v Horse Blood

1. Prepare BAB (16 g in 400 mL of distilled water) and autoclave at 121 °C 15 psi (103 kPa) for 20 min.
2. Once autoclaved, allow to cool to around 56 °C, then add 3% v/v sterile defibrinated horse blood (12 mL blood in 400 mL molten agar) and mix (*see* **Note 6**).
3. Pour the agar mix into Petri dishes. 400 mL of medium should make roughly twenty 100 mm × 15 mm plates (around 20 mL per plate).
4. Leave to set in a sterile environment; once set the plates can be inverted. For long-term storage, store inverted at 4 °C.

3.1.3 Bacterial Infection Stocks

1. From laboratory stocks, streak to single colonies on blood agar plates. Record strain identifier.
2. Confirm the strain, in case it carries an antibiotic resistant marker, by placing an antibiotic disc onto the plate at the site of the initial streak with flamed forceps.
3. Incubate the plates inverted at the ideal temperature and time conditions for the strain selected.
4. Inoculate a sweep of colonies into 10 mL broth in a universal tube with a tight cap.
5. After 3–3.5 h, check the OD_{600} nm of the culture. The OD_{600} nm of the culture should be between 0.4 and 0.6; this

corresponds to the mid-logarithmic phase of bacterial growth. The CFU can be predicted accurately during the logarithmic phase.

6. Add 50% v/v glycerol in PBS to the bacterial culture, such that the final concentration of glycerol is 10%.
7. Divide into 1 mL aliquots in sterile cryovials and freeze at −80 °C.
8. After aliquots have been stored at −80 °C for over 24 h, determine the viable CFU/mL by way of 1:10 serial dilutions and spot each dilution on a BAB plate to calculate the volume required for the infection dose (*see* **Note 7**).
9. Check for contamination by streaking to single colonies on BAB plates.

3.2 Murine Infection Model

All procedures performed herein were done in the UK in accordance with the UK Home Office license P7B01C07A, and were approved by the University of Leicester Ethics Committee. Procedures documented here are for example purposes only; the actual procedures utilized in experimental projects should be in line with the relevant legislation and regulations applicable to the country where experiments are undertaken.

3.2.1 Infection Dose

1. Thaw an aliquot of bacteria at RT and add the corresponding volume of broth or PBS, based on the viability counts of the stocks, to reach the desirable infection dose (e.g., 1×10^7 CFU/mL of *S. pneumoniae* is required for a dose of 1×10^6 CFU per mouse with an 100 μL inoculation volume).
2. Make a serial dilution of the infection dose in a BAB plate, and incubate the plate at the optimal time and temperature for the strain used, to retrospectively determine the viable CFU in the inoculum (*see* **Note 7**).
3. The prepared infection dose should be used within 30 min and kept at RT.
4. Prepare the individual doses of 100 μL in insulin syringes equipped with a 29G needle immediately prior to infection.

3.2.2 Intravenous Route of Infection

1. Acclimate 6–8-week-old, sex-matched mice in the animal facility environment for at least 1 week prior to infection. Animals should be acclimated under the standard lighting and temperature conditions in individually HEPA-filtered cages with sterile bedding and nesting and free access to food and water provided ad libitum.
2. Ear tag or tattoo the mice for identification.

3. On the day of infection, carefully remove the mouse from the cage using a plastic tube to reduce potential stress for the animal.
4. Place the animal in a heating chamber at 37 °C for between 5 and 10 min (*see* **Note 8**).
5. Restrain the animal using the mechanical restraint device of your choice with the tail protruding (*see* **Note 9**).
6. Wipe the tail with antiseptic solution such as 70% v/v ethanol.
7. Immobilize the tail and rotate 90° to access the lateral tail vein.
8. Align the needle parallel to the tail with the beveled edge of the needle facing up.
9. Insert needle into vein starting at the tip of the tail (distally) at about a 30° angle (*see* **Note 10**).
10. Administer the inoculum slowly, removing the needle after completing the injection.
11. Apply gentle pressure with gauze until bleeding has stopped. Ensure the mouse is appropriately marked and return it to the cage.
12. Monitor animal for 5–10 min to ensure homeostasis and normal behavior.

3.2.3 Monitor Signs of Disease and Euthanasia

1. Following IV inoculation, monitor the mice every 6–8 h for determination of signs of disease (*see* **Note 11**).
2. At predetermined time points after infection or upon reaching the severity limit outlined in the project license, euthanize the mice using an approved humane killing method such as cervical dislocation. Confirm the mouse is dead by cutting the femoral artery (*see* **Note 12**).

3.2.4 Organ Recovery and Bacterial Enumeration

1. Before organ collection, prepare and label two sterile tubes for each organ to be collected. Add a small volume of PBS or broth into each tube.
2. Disinfect the work surfaces using 70% v/v ethanol. On a sterile surgical field, prepare a cleaned dissection board with all sterile dissection instruments.
3. Pin the mouse carcass to the dissection board with the abdomen facing up. Spray the fur of the chest and abdomen with disinfectant such as 70% v/v ethanol to reduce contamination by hair.
4. Using sterilized forceps, grasp the skin above the urethral opening and cut along the ventral midline from the groin to the sternum using fine point sharp scissors. Peel the skin back from the peritoneal wall underneath using forceps.

5. With another pair of sterilized scissors and forceps, carefully cut the peritoneal wall and open up the abdomen. Collect the spleen, located behind the stomach, and the whole liver.
6. Divide the organs into two portions and place in the prelabeled tubes. One portion will be placed in a PBS-containing tube for subsequent embedding in OCT matrix for later microscopy analysis (*see* Subheading 3.4). The second portion will be placed in a broth-containing tube for subsequent bacterial enumeration after the sample has been homogenized.
7. Dispose of infected carcasses according to the appropriate guidelines and regulations in place at your institution.
8. Weigh the tubes containing tissue and calculate the weight of the sample in each tube.
9. For the tube containing the organ suspended in broth, thoroughly homogenize the sample by mechanically mashing into a 40 μm cell strainer with a cell strainer pestle or plunger of a 5 mL syringe before washing through with 1 mL bacterial growth media.
10. Determine the number of viable bacteria in the tissue by performing serial 1:10 dilutions of the homogenates in sterile broth and plating on a BAB plate (*see* **Note 7**).
11. After incubation, count the viable number of colonies and calculate the bacterial load in each organ as CFU per gram of tissue, to normalize for different sample weights.

3.3 Ex Vivo Porcine Liver and Spleen Coperfusion Model

3.3.1 Set up of the Ex Vivo Perfusion Circuit

To prevent extensive ischemia resulting from delays in perfusion, the circuit should be set up prior to organ retrieval. The circuit described herein consists of a reservoir which supplies blood to the portal circulation and one which supplies the systemic circulation. A diagrammatic representation of the liver–spleen perfusion circuit can be found in Fig. 1.

1. Remove the perfusion circuit from its packaging under aseptic conditions.
2. Connect the drainage lines which will originate from the inferior vena cava and splenic vein to the systemic reservoir.
3. Connect the systemic reservoir to the centrifugal pump followed by the oxygenator.
4. Following the oxygenator, the line is split into three outlets. These should be connected via a cannula to the hepatic artery, splenic artery, and the reservoir which supplies blood to the hepatic portal vein once the organ is retrieved.

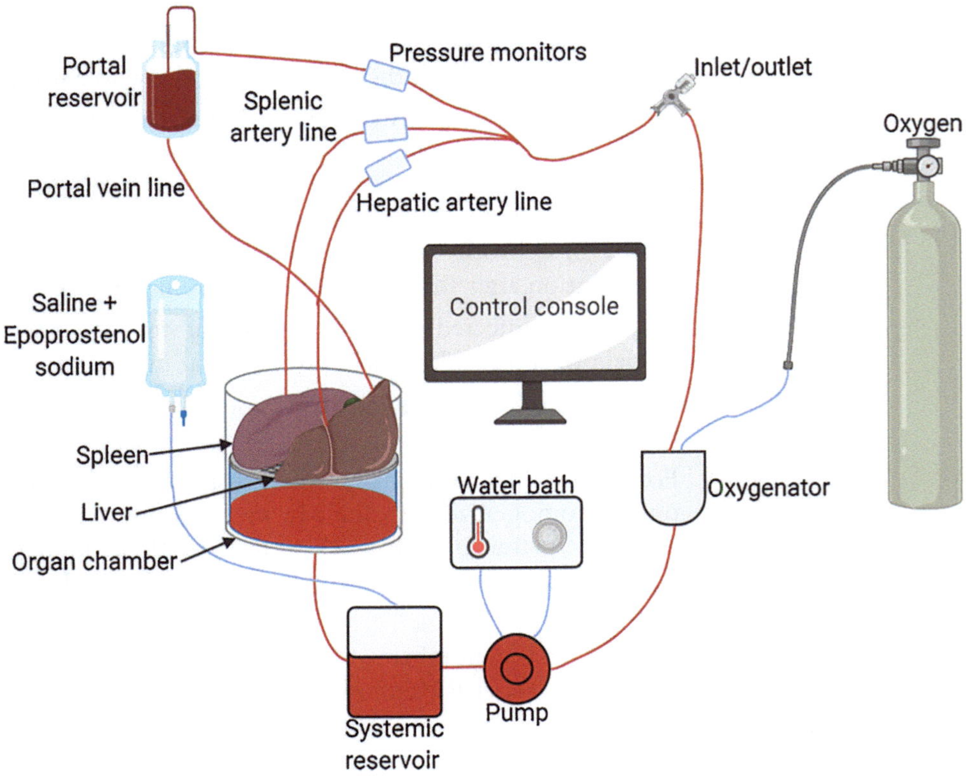

Fig. 1 The ex vivo normothermic porcine liver and spleen coperfusion. A schematic of the complete perfusion circuit. The circuit described here consists of a sterile organ chamber that houses the porcine spleen and liver, cannulated via the splenic artery, hepatic artery, and portal vein respectively. Autologous porcine blood supplemented with heparin drains passively into a blood reservoir which is fed by saline infusion bags supplemented with epoprostenol sodium. Blood subsequently flows through the pump and water bath which ensures physiological temperature and pressure. Following oxygenation, the blood line splits into three, one for each vessel: splenic artery, hepatic artery and portal vein (via a portal reservoir). The pressure of each vessel is relayed to, and can be controlled by, the control console. Once the perfusion is deemed stable, the circuit can be infected via syringe, and blood samples taken from, the inlet/outlet—commonly a three-way stopcock. (Figure was generated using BioRender)

3.3.2 Organ Retrieval

1. Allow the local qualified team at the abattoir to euthanize pigs by exsanguination from the jugular vein following stunning (UK common practice) and collect around 3 L of autologous blood in a container supplemented with 25,000 units of heparin.
2. For spleen retrieval: perform a laparotomy and sternotomy incision and access the celiac trunk to isolate the main splenic artery and vein.
3. Cannulate the main splenic artery in situ with a catheter of appropriate size and secure by tying in place with suture thread. Carefully ligate and divide other associated vessels.

4. Remove the spleen and allow a member of the team not involved in surgery to immediately infuse with 500 mL Soltran preservative solution by gentle squeezing of the infusion bag.
5. For liver retrieval: divide the diaphragm and pleurae, before ligating the suprahepatic inferior vena cava, thoracic aorta, and esophagus.
6. Cannulate the portal vein and hepatic artery in situ as above, before clamping and dividing additional vasculature and conjoining connective tissue.
7. Remove the liver and gall bladder en bloc, before infusing each vessel with 500 mL of Soltran preservative solution as above.
8. Put the organs into a sterile bag containing Soltran preservative solution, and place on ice for transportation back to lab. The ischemic time should be noted for consideration in analysis of results.

3.3.3 Ex Vivo Perfusion and Bacterial Infection

1. Flush organs with 250 mL saline by gentle squeezing of the infusion bag to remove excess preservative solution prior to plugging into the perfusion circuit.
2. Using a measuring cylinder, pour 3 L of autologous pig blood into the systemic reservoir, and apply a gentle flow rate to the circuit such that each organ input line is gently releasing blood. Start the saline infusion lines.
3. Plug the cannulated vessels into the circuit ensuring that no air bubbles are introduced.
4. Ensure normothermia by setting the heat exchange unit to 37 °C, and set the oxygenator to a physiological level of 2 L/min.
5. Once plugged in, set the perfusion pressure to 80 Hg/mm for the splenic artery and hepatic artery, and 10 Hg/mm for the portal vein for the length of the experiment. Monitor the flow rate through each vessel at hourly intervals.
6. Twenty to thirty milliliters sodium bicarbonate can now be added to the circuit perfusate to minimize the effect of tissue acidosis, in addition to epoprostenol sodium which should be added to the saline infusion bag to a final concentration of 2.5 μg/mL to facilitate vasodilation and thus effective oxygenation of the tissues (*see* **Note 13**).
7. Once a stable perfusion flow rate and blood gas parameters have been achieved, infect the circuit with bacteria via syringe into the circuit input, commonly a three-way stopcock (*see* **Note 14**).
8. At each predetermined time point, use a scalpel to excise a 1 cm^2 tissue biopsy from the organ, which should then be halved, and take 10 mL of blood via one of the circuit inputs (*see* **Note 15**).

9. Suture or surgically glue the subsequent wound to prevent leakage of blood from the biopsy site and a drop in flow pressure.
10. Transfer the blood sample to a lithium heparin calcium-balanced tube and analyze using a blood gas analyzer to determine parameters such as pH, hemoglobin oxygenation, lactate production and levels of physiological electrolytes to confirm maintenance of physiological blood gas parameters.
11. Serially dilute the blood sample into relevant bacterial agar growth media plates using a 96-well plate and plate 5 μL of each dilution on appropriate agar for blood CFU enumeration. Incubate under conditions appropriate for your bacterium (*see* **Note 7**).
12. To determine bacterial burden within the organ, weigh one half of the biopsy and thoroughly homogenize by mechanically mashing into a 40 μm cell strainer with a cell strainer pestle or plunger of a 5 mL syringe before washing through with 1 mL bacterial growth media. Serially dilute and plate the homogenate as above. Biopsy CFU should be calculated as CFU per gram to normalize for different biopsy weights.
13. For later microscopy analysis, flash-freeze the other biopsy half in OCT within a mold as outlined in Subheading 3.4.1.

3.4 Sample Preparation

3.4.1 Sample Freezing

1. Place a metal freezing container on dry ice and fill it with 2-methylbutane up to 1 cm in depth (*see* **Note 16**).
2. Fill approximately one third of an embedding mold with OCT.
3. Using forceps, place the sample in the middle of the mold. Ensure the biopsy is submerged in OCT by adding more if necessary.
4. Place the mold into the 2-methylbutane, being careful not to get any liquid onto the sample, and let it freeze.
5. Take the frozen sample out of the 2-methylbutane and store it at −80 °C until use.

3.4.2 Sample Sectioning

1. Switch on and set up your cryostat as per manufacturer's instructions, setting the internal temperature to −20 °C. Insert the blade and the glass anti-roll plate (*see* **Note 17**).
2. Once the temperature is reached, remove the OCT block from its mold. Add a small amount of OCT onto the specimen disk. Once the OCT starts to freeze, mount the sample onto the specimen disk. Once the OCT fully freezes (~5 min) and the sample is securely attached to the specimen disk, mount the specimen disk onto the stage.
3. Set the section thickness (*see* **Note 18**).
4. Move the specimen as close as possible to the blade by using the stage forward button (if applicable).

5. Begin sectioning. Handwheel rotation must be smooth and controlled in order to keep sections even (*see* **Note 19**).
6. Place the section onto a polylysine adhesion slide by suspending the slide just above the sample.
7. Once the section is dry, place the slide into a slide rack on dry ice for the remainder of the sectioning session. When all the samples have been sectioned, either thaw and begin staining or transfer the slides to −80 °C storage until use.

3.4.3 Indirect Immunohistochemistry Staining

It is important to keep the antibodies, dyes and section out of light as much as possible throughout the staining process to reduce the amount of photobleaching that occurs and thus increasing the quality and fluorescence intensity of the sample.

1. Select the relevant primary and secondary antibodies to label individual cell types and bacteria, as well as fluorescent dyes to label nuclei (commonly DAPI) and cell macromolecules (*see* **Note 20**).
2. Dilute the primary and secondary antibodies and dyes in blocking solution, and temporarily store away from light at 4 °C (*see* **Note 21**).
3. On the slide, circle around the section with a hydrophobic pen and let dry. Alternatively, place the slides into a specialized staining rack.
4. Add the fixation buffer (*see* **Note 22**).
5. Wash thoroughly with PBS then permeabilize the sample by adding 0.1% v/v Triton X-100 for 10 min at RT.
6. Block the sample with blocking solution for 30 min to 1 h.
7. Add the primary antibodies and incubate for 1 h at RT (*see* **Note 23**).
8. Wash thoroughly with PBS.
9. Add the secondary antibodies with any dyes and incubate for 45 min at RT (*see* **Note 24**).
10. Wash thoroughly with PBS and add DAPI for 10 min.
11. Wash thoroughly with PBS and one final time with distilled water.
12. Add a drop of mountant onto the sample followed by a coverslip (*see* **Note 25**).
13. Wipe away any excess mountant from the periphery of the coverslip and seal the around the edges of the coverslip with cover slip adhesive or clear nail polish. Let the adhesive dry.
14. The slide is ready to use or can be stored at either 4 °C or −20 °C until use (*see* **Note 26**).

3.5 Sample Microscopy and Image Analysis

3.5.1 Fluorescence Whole-Slide Scanning

1. Wipe the slides to be scanned with 70% v/v IMS.
2. With a fine permanent marker, draw a complete box on the underside of the slides around the sample.
3. Load the slides into the slide carriers (*see* **Note 27**).
4. Load the Vectra Polaris software, and load the carriers into the slide scanner (*see* **Note 28**).
5. Select "Edit Protocol"→"New..." to create a new protocol. A new protocol should be created for each combination of antibodies used on the slides to be scanned. Enter a suitable protocol name, ensure fluorescence imaging mode is selected, select "multispectral slide scan" (Opal 4 color) and select (or "create") a study in which to save the protocols and resulting scan images (*see* **Note 29**).
6. Under the slide scan settings, change the pixel resolution to "0.25 μm (40×)" and select "auto update" for the exposure times when prompted. Select "Select Scan Bands" and check the desired Opals depending on the fluorophores used in this particular protocol. The color of signal resulting from each Opal can also be changed here (*see* **Note 30**).
7. Select "Scan Exposures"→"Load Carrier" and select a rack containing a slide stained with the antibody panel corresponding to the protocol that is being created. With the DAPI channel automatically selected, locate the sample by clicking various locations on the appropriate slide depiction to the right of the live image, or by clicking on the sides of the live image itself. Alternatively, use "Take Overview" to scan the entirety of all slides in the carrier, hence showing exactly where the samples are located (*see* **Note 31**).
8. Move around the sample and use "Autofocus" and "Autofluorescence" until there are no more overexposed (red) areas in the image. Repeat this step for the remaining antibody channels, then select "Back" (*see* **Note 32**).
9. Save the protocol. Repeat **steps 5–9** to create a new protocol for each antibody combination used in the slides to be scanned. Once all protocols have been created, click "Back" (*see* **Note 33**).
10. Select "Scan Slides." Select the gray circle next to each slot to bring up the edit slides window. Change the task for all slides to be scanned to "Scan Slides," then enter the appropriate study, protocol and desired slide ID. Select "Scan."
11. Once scanning is complete, the scanned images of the sample (.qptiff file) and slide label photos (.tiff file) will appear in the study folder which was selected prior to scanning. The scans can then be opened with Phenochart, or analyzed using Fiji or inForm.

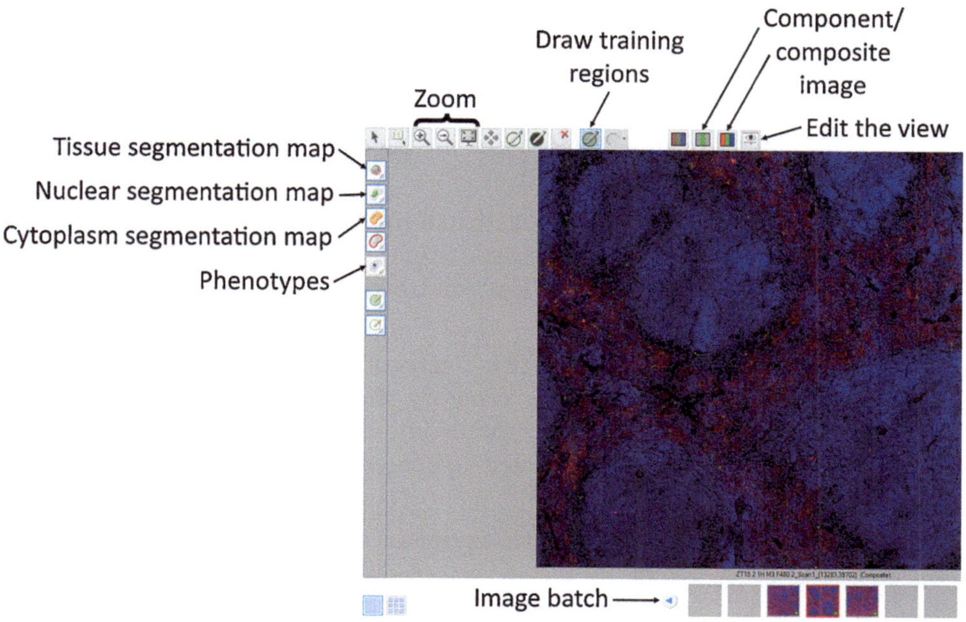

Fig. 2 inForm image analysis window. A screenshot of the inform image analysis window with tools utilized in this methodology labeled as appropriate. Options on the horizontal toolbar include zoom, draw training regions (used in the tissue segmentation step), component/composite view (used in the cell segmentation and phenotyping steps), and edit the view for adjusting channel brightness and contrast. Options on the vertical toolbar include buttons for the toggling on/off of the tissue segmentation, nuclear segmentation and cytoplasm segmentation maps along with the cell phenotypes. The image batch containing image stamps is shown at the bottom of the window

3.5.2 inForm Image Analysis

Whole slide scan images cannot be directly opened in the inForm software, therefore "stamp" sections of the images to be analyzed must be prepared beforehand in Phenochart. When analyzing the images in inForm, refer to Fig. 2 for a view of the processing window with labeled buttons that are used in the methodology described here.

1. Open the desired image(s) to be analyzed in the Phenochart software and login using the appropriate institutional password.
2. Select "Stamp" → "Select for: inForm Projects". Select an appropriate size in fields for the stamp and select areas on the image to be analyzed in inform (*see* **Note 34**).
3. Open the images to be analyzed in inForm through "File"→"Open"→"Image." The stamps of each image will be opened separately in the batch view.
4. Select "Configure…" to configure the analysis workflow. Select "Trainable Tissue Segmentation," "Adaptive Cell Segmentation," and "Phenotyping" while skipping the "Score" step.

5. Prepare the images by selecting which markers are to be unmixed under the "Spectra for Unmixing" tab, and use "Edit Markers and Colors..." to assign appropriate names and colors for each cell marker. Under the standard protocol, all markers will be included in unmixing. Select "Prepare All" (*see* **Note 35**).
6. Once the image batch has finished unmixing, alter the brightness and contrast of each component if needed by selecting the "Edit the view" button on the horizontal toolbar. Click "Advance."
7. Create new tissue categories under the "Tissue Categories" tab. A tissue category should be created to represent each marked cell type of interest, with an additional background category created to represent the blank spaces between cells (*see* **Note 36**).
8. Under the "Components for Training" tab, select which markers should be taken into consideration by the machine learning component when segmenting the tissue into its respective categories (*see* **Note 37**).
9. Select a pattern scale; in most instances a medium pattern scale will suffice (*see* **Note 38**).
10. Using the "Draw training regions" button above the image, and with a tissue category selected, draw around a region of that respective tissue category on an image. Repeat this step to create training regions for each tissue category, including a region with no cells (or an area of a blank slide) for the background category (*see* **Note 39**).
11. Select "Train Tissue Segmenter" and "Segment Image" once the training is complete. The tissue segmentation map will now overlay the image, displaying each tissue category in their respective color as shown under the "Tissue Categories" tab (*see* **Note 40**).
12. In many cases, the first tissue segmentation map will not be sufficiently accurate. In this scenario, additional training regions will have to be drawn and the tissue segmenter retrained until a segmentation map of sufficient accuracy is created. Once this is achieved, select "Segment All" and "Advance" onto the next step (*see* **Note 41**).
13. If a cytoplasmic marker is included in your staining panel, check the "Cytoplasm" option under the "Segment" tab alongside the preselected "Nuclei" option (*see* **Note 42**).
14. On the "Cell Segmentation Settings" tab, add a component under the "Components" section. The marker selected should be the nuclear marker (commonly DAPI). Selecting the ellipses adjacent to the marker will allow the alteration of typical

fluorescence intensity of the nuclei; this should be adjusted until the entirety of each nucleus within the image is highlighted (*see* **Note 43**).

15. Similarly, if a cytoplasmic stain or antibody is used, add an additional component, select the relevant antibody/stain from the dropdown menu and again select the neighboring ellipses to check the cytoplasm option.
16. Under the "Nuclear Component Splitting" tab, select the ellipses to select the representative quality of the nuclear staining within the image, and alter the splitting sensitivity such that adjacent nuclei can be separated while whole nuclei are not. Adjust the "Minimum Nuclear Size" to prevent partial selection of nuclei with a low fluorescence intensity (*see* **Note 44**).
17. For any cytoplasmic markers, alter the cytoplasm thickness under the "Cytoplasm Settings" tab such that the signal is overlaid sufficiently by the cytoplasm segmentation map (*see* **Note 45**).
18. Select "Segment Image." The nuclear and cytoplasmic segmentation map will be overlaid on the displayed image, indicating which areas the program has identified as nuclei in green, and cytoplasm (if segmented) in multicolor. If the selection and segmentation are accurate, select "Segment All." Reevaluate the segmentation on additional images and click "Advance" if appropriate. Alternatively, alter the cell segmentation settings and resegment the images until accurate.
19. Under the "Phenotypes" tab, add new cell phenotypes for bacteria positive and bacteria negative cells (*see* **Note 46**).
20. Select the "Edit a cell's phenotype" button on the horizontal toolbar above the image. Assign a phenotype to at least five cells which are bacteria positive and five cells which are bacteria negative. Train the classifier and select "Phenotype All" (*see* **Note 47**).
21. Under the "Export" tab, select an empty export directory folder in which to save the data and/or images. Select which images (if any) from the analysis workflow that you wish to save, along with the desired format, and select the cell segmentation data table. Select "Export for All."
22. A separate "cell_seg_data" and "cell_seg_summary_data" file will be created for each image stamp in the selected export directory. Opening the summary_data file will provide information on how many cells of each tissue category are bacteria positive or negative. In turn, this allows quick and simple enumeration of the percentage of each cell type which are infected, along with the percentage of total infected cells which belong to each cell type (*see* **Note 48**).

3.5.3 Fiji Image Analysis

1. Load the Fiji software and open the image to be analyzed by selecting "Plugins"→"Bio-Formats"→"Bio-Formats Importer."
2. In the "Bio-Formats Import Options," ensure that the color mode is set to composite (autoscale checked) and that channels are split into separate windows.
3. Select the series with highest resolution (Series 1) in the "Bio-Formats Series Options" screen. Click "OK."
4. On the nuclear marker (commonly DAPI) channel, use the "Polygon selections" tool in the toolbar to create a region of interest (ROI) around the entire sample. Add this ROI to the ROI manager by pressing Ctrl+T, and rename it as "whole slice ROI" and save.
5. On a macrophage marker channel, alter the threshold by selecting "Image"→"Adjust"→"Threshold." Ensure that the "Dark background" box is unchecked, and set the maximum fluorescence (bottom slider) to its highest level, and the minimum fluorescence (upper slider) to an appropriate level as to include the majority of fluorescent signal while negating background. Click "Apply" to apply the selected threshold levels (*see* **Note 49**).
6. The image should now be binary: black with white signal. If this is reversed, select "Process"→"Binary"→"Options," select "Black background" and reapply the threshold settings.
7. If there should be any artefacts in the image that have been included in the threshold selection, these can be omitted by drawing around them with the "Polygon selections" or "Freehand selections" tools and pressing the delete key.
8. Omit any signal outside of the sample by selecting the "whole slice ROI" from the ROI manager, inverting the selection by selecting "Edit"→"Selection"→"Make Inverse" and pressing the delete key (*see* **Note 50**).
9. Select "Edit"→"Selection"→"Create selection" to create ROIs around the macrophage marker signal. Ensure that the binary white signal and not the surroundings are selected and subsequently add the ROIs to the ROI manager using Ctrl+T. Rename this ROI as the appropriate macrophage marker ROI and save.
10. On the bacteria channel, apply a threshold to include all bacteria within the sample, as described in **step 5**. The image should have a black background with white signal; alter the Binary settings as outlined in **step 6** to reverse this (*see* **Note 51**).
11. On the bacteria channel, select the whole slice ROI. Select "Analyze"→"Analyze Particles…" and check the "Summarize" box. The size can be adjusted depending on the average

size of your bacteria present in the sample, although a degree of leeway should be incorporated (*see* **Note 52**).

12. Repeat **step 11** with the macrophage marker ROI selected.
13. The summary table gives two values of importance: Total Area (indicating the area of bacterial signal within the ROI), and % Area (indicating the area of the ROI that contains bacteria). This allows for the determination of bacterial location within the tissue at different time points by analyzing the colocalization of bacteria with different cell types. The % Area value can also be used to analyze the growth of bacterial foci within certain cell subtypes at different time points within a tissue, and can be used in conjunction with the infected cell percentage data from the inForm analysis to build a more detailed view of cell infection dynamics within samples over time (*see* **Note 53**).

4 Notes

1. The circuit described herein is similar to those used for extracorporeal bypass surgery, and was custom made by Medtronic (Netherlands). This circuit is optimized for coperfusion of the liver and spleen, but circuits can be adapted for different organ systems [6, 20, 24].
2. The pump facilitates the blood flow through the cannulated vessels while the control console provides pressure regulation and readings.
3. We use a 13 × 10 in sterile intestinal bag for this purpose.
4. The specific organs (liver and spleen) used in this protocol were sourced from Joseph Morris Abattoir (South Kilworth, Leicestershire, UK), which is in close proximity to the authors' laboratory. Organs were removed immediately postmortem following the UK guidelines for processing of animals for food production. Organs may be sourced from any abattoir provided sampling does not interfere with local best practice. This protocol is also amenable to use with experimental specific pathogen-free (SPF) reared animals in line with a relevant home office project license.
5. A catheter is required for each vessel and their size will depend on the vessel diameter. In this protocol the hepatic artery and splenic artery are cannulated with a 14F catheter, while the portal vein is cannulated with a 12F catheter.
6. The bottle should be warm but not hot to the touch. When mixing the agar, take care not to form bubbles which could set in the Petri dish. If the blood agar plates set brown, then the

bottle was too warm when the horse blood was added. If solid chunks of agar are poured, the agar was too cold when the horse blood was added.

7. Serial dilutions from neat to 1:100,000 can be achieved by diluting 20 μL into 180 μL media in a 96-well plate five times. A multichannel pipette can then be used to plate 5 μL of all dilutions onto the plate. To calculate CFU/mL, count the number of colonies in the highest dilution where bacteria can still be accurately counted, multiply by the dilution factor (e.g., 1×10^4 for the fourth dilution of the neat sample) and divide by 0.005.
8. This causes vasodilation and provides better vein visibility. Do not overheat the animal by leaving the animal in the chamber for more than 10 min. This will cause dehydration and make accessing the vein and administering the dose more difficult.
9. Sedation or anesthesia for intravenous inoculation is unnecessary for trained and experienced personnel. Placing food in the opposite end of the restraint tube can encourage the mouse to enter the tube themselves.
10. If the needle is correctly inserted, blood should flash into the syringe when the plunger is gently withdrawn and the inoculum will flow easily during administration. Do not force the inoculum—this means the dose is being improperly administered to tissue surrounding the vein and can cause distress to the animal. If the proper placement cannot be confirmed, attempt to place the needle in a more proximal position where the vein is wider in diameter.
11. The frequency of which the mice should be observed and/or scored will depend on the guidelines specific to your project.
12. The time points utilized in the study will depend on experimental objectives; however, a frequently used time course for the study of early stages of pathogenesis is up to 8 h postinfection.
13. The volume of sodium bicarbonate that should be added is dependent on the pH yielded by blood gas analysis and as such the actual volume required can vary. In our perfusion circuit, we utilize 200 mL infusion bags of saline, which requires the addition of 0.5 mg epoprostenol sodium.
14. In our experience, a stable flow rate is reached after around 30 min of perfusion. Our perfusion circuits are infected with around 1×10^7 CFU of *S. pneumoniae* or *Klebsiella pneumoniae*, although the infection bacteria and dosage should be specifically tailored to the experimental aims and objectives. Ensure blood is taken into the syringe and dispelled multiple times to ensure all the bacteria enter the circuit.

15. Sampling times during ex vivo perfusion will depend on the experimental set up, and the aims of the research project. In our experimental model, time points were every hour up to 6 h postinfection. Ensure a biopsy is taken from a well-perfused area of the organ by ensuring the area is dark red and is warm to the touch.
16. Metal containers work best as they are good conductors of heat and can withstand the cold temperatures. 2-methylbutane is used due to its high thermal conductivity and sample-freezing efficiency.
17. For spleen and liver samples, we commonly use −20 °C. The optimal sectioning temperature is dependent on the tissue and can be found in the literature, although −15 °C to −25 °C is a common range applicable to most organ types.
18. We commonly use 10 μm sections for liver and spleen samples, and this thickness is widely used in frozen IHC staining. Thinner sections tend to stain better and produce clearer images but tear easily. The thickness of the section is ultimately dependent on the tissue that is being sectioned, and literature should therefore be consulted for the optimal section thickness.
19. If the section bunches up around the blade, try adjusting the distance between the edge of the antiroll plate and the edge of the blade in small increments. In our experience, the antiroll plate should overhang the edge of the blade by ~2 mm. If the section rolls while underneath the antiroll plate, adjust the angle of the glass antiroll plate such that the antiroll plate becomes closer to the surface underneath. If the section immediately rolls once the antiroll plate is lifted, try briefly warming the sample by gently placing your sterile-gloved thumb pad on the specimen for 1–2 s, or raising the internal cryostat temperature slightly. A small paintbrush can also be used to carefully unroll the sections in these cases.
20. The antibodies and dyes will be dependent on the experimental aims. The primary antibodies should be specific for the antigen and species of interest (anti-pig for perfusion samples and anti-mouse for murine model samples) and all be raised in different species. Each secondary antibody should be specific for the species that the corresponding primary antibody was raised in. Secondary fluorophore-conjugated antibodies should be used in line with the excitation and emission profile of the confocal microscope to be used. It is also important that any dyes used do not have the same excitation wavelength as any of the fluorophore-conjugated secondary antibodies or other dyes used in the staining panel.

21. The blocking solution used here is 5% v/v goat serum in PBS; however, other type of sera used for blocking (i.e., bovine serum albumin) in IHC are also common. The antibodies and dyes should be diluted to a concentration outlined by the manufacturer, although 1:200 for primary antibodies and 1:500 for secondary antibodies are common dilutions.
22. 4% v/v formaldehyde is commonly used in immunohistochemistry; however, different fixation buffers such as methanol can be used in certain applications. The base can also vary depending on application and often yields different staining intensity and clarity. A common fixation buffer is PBS, as used here.
23. Alternatively, primary antibodies can often be incubated overnight at 4 °C or for 1 h at 37 °C. In some cases, manufacturers provide specific incubation times for antibodies which should be adhered to.
24. From this step onwards it is extremely important to keep the sample out of direct light as much as possible as the fluorochrome-conjugated secondary antibodies are light sensitive and can become photobleached when exposed to bright light for extended periods.
25. A coverslip with thickness that is compatible with the microscope should be used. When placing the coverslip onto the slide, lower down one end slowly to prevent the formation of air bubbles. If any bubbles form underneath the coverslip, a gentle rolling pressure to one end can expel the bubbles through the end of the coverslip.
26. Storage at −20 °C is recommended for long-term storage, to preserve the brightness of the fluorescent tags and dyes, although care should be taken to minimize the number of freeze–thaw cycles. 4 °C is suitable for short-term storage when planning to use the samples within the same week.
27. Take a photo of the slides in the carriers to refer back to during the scan exposures step, thus ensuring that the slides identity and the location of the sample on the slides can easily be found.
28. Document which slide carriers are placed at which position in the rack. You will refer back to these when imaging protocols are assigned to individual slides in the scanning step.
29. Oftentimes, similar sample types stained with the same panel of antibodies will yield similar fluorescence intensities, that is, mouse spleens stained with the same antibody combination in parallel will appear similar. In these cases, one protocol can be used for all slides harboring the same antibody panel. However, if the resulting scans show over/underexposed signal between slides, a protocol should be created for, and assigned to, each individual slide.

30. If secondary antibodies have not been conjugated to Opal fluorophores, the Opal corresponding to the fluorophore used can easily be found online.
31. Locate the sample on the slide by referring back to the photos taken of the slides in their carriers.
32. The focus can change when moving around the sample if the section is uneven on the slide. Ensure that autofocus is used before selecting autofluorescence, as in-focus images often display higher fluorescence signal intensity. If autofocus is unable to focus the image, manually use the stage height slider to bring the image into near-focus before using the autofocus function.
33. To load different slide carriers when scanning exposures, select "Unload Carrier," then "Load Carrier" with the relevant slide carrier.
34. To make the analysis of the stamps representative of the entire sample as a whole, we often utilize the largest 3 × 3 stamp field size wherever possible and create multiple random stamps per image. This will ensure that as many phenotypically different cells as possible can be included in the inForm analysis, while reducing any bias relating to which areas of the image are analyzed.
35. For our analysis workflow, bacteria are colored in green, macrophage marker in red, cell stains/additional macrophage markers in magenta, and autofluorescence in black.
36. For example, a sample stained with DAPI, an anti-bacteria antibody, and an anti-macrophage antibody will have tissue categories consisting of macrophage antibody positive, macrophage antibody negative, and background.
37. This usually consists of the cell markers and DAPI and should not include the bacteria or autofluorescence components as the presence or absence of these markers do not impact on determining a cell's phenotype.
38. The pattern scale relates to the size of the area of the image which will be selected to indicate the different tissue categories. If the area of one or more cell markers within the image is particularly small, the pattern scale can be set to small.
39. We recommend zooming in to the image and accurately drawing around a region of around 3–5 adjacent cells of the same marker. The program uses machine learning to determine the average fluorescence intensity of each of the markers selected under the "Components for Training" tab within the training region. Therefore, care should be taken to only include the markers of interest, as any area without the markers would reduce the average intensity of the markers of interest. When

selecting a training region to represent the background, we commonly use an area outside of the sample—this minimizes the chance of accidently including a nucleus with low fluorescence in the background training region.

40. We recommend only segmenting a single image after each training of the tissue segmenter as segmentation of the entire image batch can very lengthy, especially if multiple training rounds are required. We have found that the most efficient way of segmenting tissue is to carry out multiple training rounds (if required) on a single image until accurate, before training the rest of the image batch to confirm sufficient accuracy. Sometimes the trainer will not reach 100%, in this case stop the training when the accuracy ceases to increase.
41. The tissue segmentation map can be toggled on/off by selecting the appropriate button on the vertical toolbar beside the image. This will help in identifying areas of the tissue which have been miscategorized. If fluorescence intensity and/or staining quality of the images within the batch varies drastically, it is unlikely that tissue segmentation will be accurate for each image even after multiple training rounds. In this case, images with similar staining quality/fluorescence intensity should be grouped and subsequently analyzed as separate inForm projects.
42. Including a cytoplasmic marker into the analysis is especially important when analyzing tissues where nuclei may be dispersed—if only nuclei are used to segment cells, only cells with nuclei colocalizing with bacteria will be counted as positive in later phenotyping steps.
43. Select the nuclear marker channel option under the "Select a component Image" button on the horizontal toolbar to visualize the image as an H&E-style image. This makes altering the relative fluorescent intensity easier and more accurate than when viewing the color/composite image. The nuclear segmentation map can be toggled on/off with the appropriate button on the vertical toolbar—this can be used to ensure the nuclear selection is accurate.
44. Move the training box around different images including areas of differing fluorescence intensity to ensure that all nuclei throughout the entirety of each image is selected. This allows for more accurate nuclear selection.
45. Changing to a component image of the cytoplasmic marker or to a composite image, or by toggling the cytoplasm segmentation map, will aid when altering the cytoplasm thickness.
46. Do not use green as an indication of either phenotype as the nuclear segmentation map also appears green and can result in difficulty in visualizing cells of this phenotype against the nuclear segmentation map.

47. Toggle the segmentation maps on/off using the respective button on the horizontal toolbar to switch between locating bacteria and visualizing cells. Select the "Select a component image" button on the horizontal toolbar and select the bacteria channel to allow easier identification of bacteria. If bacteria cannot accurately be identified when looking only at the inForm image, use Phenochart to locate known bacteria within the stamp and subsequently locate the bacterium in the inForm image. We recommend selecting as many cells of each phenotype as possible across different images, thus allowing the machine learning classifier to be as accurate as possible.
48. More powerful programs such as "R" can also be used on the raw "cell_seg_data" files to carry out statistical analysis on additional parameters such as cell position to determine, for example, whether infected cells are grouped in close proximity or spread more randomly about the image. This additional analysis is beyond the scope of this chapter.
49. Changing the color of fluorescent signal from white to green ("Image"→"Lookup Tables"→"Green") can aid when setting the threshold.
50. To determine whether the inside or outside of an ROI is selected, the cursor will appear as a standard cursor arrow within selected areas, and a cross in the nonselected areas.
51. Extensively zoom in to the image for a more precise selection of bacteria—it is not uncommon for bacteria to have a faint "glow" around them and it is important that only the bacteria itself is used for setting the threshold. Utilizing a bacterium with particularly low fluorescence to set the minimum threshold will ensure that the majority, if not all, of the bacteria within the sample are included in the threshold selection. This can be done by finding a bacterium with weak signal in Phenochart (deselecting the other channels to leave only the bacteria channel can aid in locating bacteria with low signal), and subsequently locating and setting the minimum threshold to that bacterium in Fiji. Alternatively, repeating **steps 10–12** multiple times with a different bacterium each time can produce more accurate and representative mean values when the intensity of bacteria within the image vary significantly, particularly in thicker sample sections.
52. In our analysis, a lower size limit is set; however, the upper limit remains at infinity. This is especially important for intracellular bacteria that replicate to form foci as they may be omitted if the upper size limit is set too small.
53. The "Count" value should be used with caution—this parameter gives the number of particles in the given ROI; however, bacteria that are in close proximity and whose signal overlap would be counted as only a single bacterium.

Acknowledgements

We thank John Isherwood and Rohan Kumar for help with the perfusion of the porcine organs at explant, the staff of Joseph Morris Butchers, and Sarah Glenn and the staff of the Leicester Preclinical Research Facility for support with the mouse experiments. The grant was in part supported by a collaboration agreement with the University of Oxford and grants from the MRC MR/M003078/1 and BBSRC BB/S507052/1 to MRO. ZJ is funded by BBSRC BB/S507052/1.

References

1. Golding H, Khurana S, Zaitseva M (2018) What is the predictive value of animal models for vaccine efficacy in humans? The importance of bridging studies and species--independent correlates of protection. Cold Spring Harb Perspect Biol 10(4):a028902. https://doi.org/10.1101/cshperspect.a028902
2. Pizza M, Scarlato V, Masignani V et al (2000) Identification of vaccine candidates against serogroup B meningococcus by whole--genome sequencing. Science 287(5459):1816–1820. https://doi.org/10.1126/science.287.5459.1816
3. Borrow R, Carlone GM, Rosenstein N et al (2006) *Neisseria meningitidis* group B correlates of protection and assay standardization—international meeting report Emory University, Atlanta, Georgia, United States, 16–17 march 2005. Vaccine 24(24):5093–5107. https://doi.org/10.1016/j.vaccine.2006.03.091
4. Romero--Steiner S, Frasch CE, Carlone G et al (2006) Use of Opsonophagocytosis for serological evaluation of pneumococcal vaccines. Clin Vaccine Immunol 13(2):165–169. https://doi.org/10.1128/CVI.13.2.165-169.2006
5. Ercoli G, Fernandes VE, Chung WY et al (2018) Intracellular replication of *Streptococcus pneumoniae* inside splenic macrophages serves as a reservoir for septicaemia. Nat Microbiol 3(5):600–610. https://doi.org/10.1038/s41564-018-0147-1
6. Chung WY, Wanford JJ, Kumar R et al (2019) An ex vivo porcine spleen perfusion as a model of bacterial sepsis. ALTEX 36(1):29–38. https://doi.org/10.14573/altex.1805131
7. Kadioglu A, Weiser JN, Paton JC et al (2008) The role of *Streptococcus pneumoniae* virulence factors in host respiratory colonization and disease. Nat Rev Microbiol 6(4):288–301. https://doi.org/10.1038/nrmicro1871
8. Chiavolini D, Pozzi G, Ricci S (2008) Animal models of *Streptococcus pneumoniae* disease. Clin Microbiol Rev 21(4):666–685. https://doi.org/10.1128/CMR.00012-08
9. Kadioglu A, Cuppone AM, Trappetti C et al (2011) Sex--based differences in susceptibility to respiratory and systemic pneumococcal disease in mice. J Infect Dis 204(12):1971–1979. https://doi.org/10.1093/infdis/jir657
10. Gerlini A, Colomba L, Furi L et al (2014) The role of host and microbial factors in the pathogenesis of pneumococcal Bacteraemia arising from a single bacterial cell bottleneck. PLoS Pathog 10(3):e1004026. https://doi.org/10.1371/journal.ppat.1004026
11. Iannelli F, Chiavolini D, Ricci S et al (2004) Pneumococcal surface protein C contributes to sepsis caused by *Streptococcus pneumoniae* in mice. Infect Immun 72(5):3077–3080. https://doi.org/10.1128/iai.72.5.3077-3080.2004
12. Oggioni MR, Trappetti C, Kadioglu A et al (2006) Switch from planktonic to sessile life: a major event in pneumococcal pathogenesis. Mol Microbiol 61(5):1196–1210. https://doi.org/10.1111/j.1365-2958.2006.05310.x
13. Kerr AR, Paterson GK, McCluskey J et al (2006) The contribution of PspC to pneumococcal virulence varies between strains and is accomplished by both complement evasion and complement--independent mechanisms. Infect Immun 74(9):5319–5324. https://doi.org/10.1128/IAI.00543-06
14. Manso AS, Chai MH, Atack JM et al (2014) A random six--phase switch regulates pneumococcal virulence via global epigenetic changes. Nat Commun 5(1):5055. https://doi.org/10.1038/ncomms6055
15. Ricci S, Janulczyk R, Gerlini A et al (2011) The factor H--binding fragment of PspC as a vaccine antigen for the induction of protective

humoral immunity against experimental pneumococcal sepsis. Vaccine 29(46):8241–8249. https://doi.org/10.1016/j.vaccine.2011.08.119

16. Rukke HV, Kalluru RS, Repnik U et al (2014) Protective role of the capsule and impact of serotype 4 switching on *Streptococcus mitis*. Infect Immun 82(9):3790–3801. https://doi.org/10.1128/IAI.01840-14
17. Yue F, Cheng Y, Breschi A et al (2014) A comparative encyclopedia of DNA elements in the mouse genome. Nature (London) 515 (7527):355–364. https://doi.org/10.1038/nature13992
18. Steiniger BS (2015) Human spleen microanatomy: why mice do not suffice. Immunology 145(3):334–346. https://doi.org/10.1111/imm.12469
19. Mestas J, Hughes CCW (2004) Of mice and not men: differences between mouse and human immunology. J Immunol 172 (5):2731–2738. https://doi.org/10.4049/jimmunol.172.5.2731
20. Chung WY, Gravante G, Al-Leswas D et al (2013) The development of a multiorgan ex vivo perfused model: results with the porcine liver--kidney circuit over 24 h. Artif Organs 37(5):457–466. https://doi.org/10.1111/aor.12003
21. Daniel CR, Labens R, Argyle D et al (2018) Extracorporeal perfusion of isolated organs of large animals—bridging the gap between in vitro and in vivo studies. ALTEX 35 (1):77–98. https://doi.org/10.14573/altex.1611291
22. Meurens F, Summerfield A, Nauwynck H et al (2012) The pig: a model for human infectious diseases. Trends Microbiol 20(1):50–57. https://doi.org/10.1016/j.tim.2011.11.002
23. Ramos-Vara JA, Miller MA (2013) When tissue antigens and antibodies get along: revisiting the technical aspects of immunohistochemistry--the red, brown, and blue technique. Vet Pathol 51(1):42–87. https://doi.org/10.1177/0300985813505879
24. Butler AJ, Rees MA, Wight DGD et al (2002) Successful extracorporeal porcine liver perfusion for 72 h. Transplantation 73 (8):1212–1218

Chapter 23

Multicolor Flow Cytometry and High-Dimensional Data Analysis to Probe Complex Questions in Vaccinology

Megan E. Cole, Yanping Guo, Hannah M. Cheeseman, and Katrina M. Pollock

Abstract

Vaccines induce a highly complex immune reaction in secondary lymphoid organs to generate immunological memory against an antigen or antigens of interest. Measurement of post immunization immune responses generated by specialized lymphocyte subsets requires time-dependent sampling, usually of the blood. Several T and B cell subsets are involved in the reaction, including CD4 and CD8 T cells, T follicular helper cells (Tfh), and germinal center B cells alongside their circulating (c) counterparts; cTfh and antibody secreting cells. Multicolor flow cytometry of peripheral blood mononuclear cells (PBMC) coupled with high-dimensional analysis offers an opportunity to study these cells in detail. Here we demonstrate a method by which such data can be generated and analysed using software that renders multidimensional data on a two dimensional map to identify rare vaccine-induced T and B cell subsets.

Key words Flow cytometry, High-dimensional data analysis, t-SNE, FlowSOM, T cells, Vaccine, immunophenotyping, FlowJo™

1 Introduction

Novel techniques to investigate the cellular responses to immunization in humans, potentially leading to more profound mechanistic or kinetic insights into the immune responses, are of considerable interest, particularly when these techniques are coupled with the use of innovative vaccine technologies, such as virally vectored and RNA designs [1–3]. The response to vaccination is orchestrated by T and B cell cross talk. Among peripheral blood mononuclear cells (PBMC), activation and induction of several subsets, including memory T and B cells, plasmablasts (antibody secreting cells), and circulating Tfh (cTfh), have been observed after immunization [4–6].

The immune response to vaccination is studied by sampling specific timepoints in the days before and after immunization.

Fadil Bidmos et al. (eds.), *Bacterial Vaccines: Methods and Protocols*,
Methods in Molecular Biology, vol. 2414, https://doi.org/10.1007/978-1-0716-1900-1_23,

Design of the immunogenicity experiments is dependent on the research question(s) and is linked to the vaccine being studied, with immunogen, formulation, and prime- boost schedule all factors to be considered when selecting at which timepoints to take samples.

Phenotypic distinction of rare T and B cell subsets responding to immunization within PBMC can be achieved using multicolor flow cytometry. Detection of these cells is through staining with multiple fluorochrome-conjugated antibodies against the cellular proteins of interest and with a dead cell discriminator dye. Standard flow cytometry cell preparation techniques are used. The user is recommended to conduct this with limited exposure to light. The maximum number of fluorochromes that can be included in panel design will be dependent on the flow cytometer available. Oftentimes, multiple markers are needed to distinguish a particular subset, for example, cTfh are $CD3^{+}CD4^{+}CXCR5^{+}$ and variably express programmed cell death protein-1 (PD-1) and Inducible T-cell COStimulator (ICOS) in the peripheral circulation (Table 1).

Data generated from these analyses are multidimensional. To explore this data set, conventional two-dimensional gating can fall short for two reasons, one is the sequential loss of data and two is the rigidity of the a priori hypotheses driving the gating strategy. Rare subsets may be overlooked. An alternative is to use unsupervised machine learning to organize complex multidimensional data. Clusters of data are constructed through similarity of expressed proteins and these may correspond to cell subsets of interest. One such method is t-distributed stochastic neighbor embedding (t-SNE), a machine learning algorithm that can be used to visualize high dimensional flow cytometry data in two dimensions on a scatter plot [7]. This is based on stochastic neighbor embedding with a Student-t distribution to estimate the similarity between two data points. Rare subsets with previously unforeseen relationships can be highlighted. Cross-validation of results on the principal subsets of interest using alternative methods of data analysis including two-dimensional gating strategies, is advised. When combined, these approaches are powerful and innovative methods to explore the cellular immune response to immunization. Here we describe these methods, which were used to determine induction of circulating lymphocyte subsets in a study of influenza vaccine responses [8].

2 Materials

2.1 Equipment

1. Refrigerator 2–8 °C.
2. Freezer −20 °C.
3. Biological class II safety cabinet.
4. Timer.

Table 1
The fluorochrome for each marker has been selected based on the configuration of a Becton Dickinson (BD) LSRFortessa flow cytometer instrument, following the general principles of multicolor panel design BV is Brilliant Violet™. BUV dyes are BD Horizon Brilliant™ Ultraviolet. Fluorochrome-conjugated antibodies were obtained from BD Biosciences and BioLegend

Laser	Filter	Mirror	Fluorochrome	CD No.	Alternative
355 nM	379/28		BUV 395	CD8	
	530/30	505LP	BUV496	CD4	
405 nM	450/50		BV421	CD27	
	525/50	475LP	BV510	CD19	
	605/12	595LP	BV605	CD279	PD-1
	655/8	635LP	BV650	CD278	ICOS
	710/50	700LP	BV711	CD127	IL-7R alpha
	780/60	735LP	BV785	CD183	CXCR3
488 nM	530/30	505LP	FITC	CD45RA	
	710/50	685LP	PerCP-Cy5.5	CD38	
561 nM	582/15	570LP	PE	CD32	
	620/10	600LP	PE/Dazzle™ 594	CD185	CXCR5
	780/60	750LP	PE/Cy7	CD197	CCR7
633 nM	670/14		APC	CD28	
	720/40	710LP	AF700	CD3	
	780/60	750LP	Near-IR	NA	Dead cell stain

5. Bench top centrifuge, up to 24,000 × *g*.
6. Bench top microcentrifuge, up to 21,000 × *g*.
7. Bench top vortex mixer, speed approximately 2800 rpm.
8. Heat block capable of heating to 70 °C.
9. One channel air-displacement pipettes and tips in a range of sizes.
   (a) 0.2–2 ml.
   (b) 2–20 ml.
   (c) 20–200 ml.
   (d) 100–1000 ml.
10. Calibrated serological pipettes.
   (a) 5 ml.
   (b) 10 ml.
   (c) 25 ml.
   (d) 50 ml.

11. 96-well U bottomed plates.
12. Pipette controller.
13. 15 ml centrifuge tubes.
14. 50 ml centrifuge tubes.
15. Automated cell counter or light-field microscope with cell counting slides/hemocytometer and trypan blue dye.
16. Range of sterile lidded plastic tubes of different sizes from 1 to 5 ml.
17. Foil.
18. Foil-lined sealable plastic box.
19. FACS tubes.
20. Flow cytometer with multiparameter capability.

2.2 Software

1. BD FACSDiva™ and FlowJo (FlowJo™ Software, for Windows or for Mac, Version 10.5 or above).
2. Microsoft Excel or similar.

2.3 Reagents

1. Fetal Bovine Serum (heat inactivated) (HI-FBS).
2. Sterile phosphate-buffered saline (PBS); $MgCl_2/CaCl_2$ free.
3. Fluorochrome-conjugated antibodies (*see* Tables 1 and 2).
4. Fc block buffer (e.g., Human TruStain FcX™, BioLegend®).
5. Cell viability dye (e.g. LIVE/DEAD™ Fixable Dead Cell Stain Kits, ThermoFisher Scientific).
6. Compensation beads (e.g BD™ CompBeads, BD Biosciences).
7. Sodium azide (optional).
8. Commercial fixation buffer (e.g. BD Cytofix™ Fixation Buffer, BD Biosciences), or 4% paraformaldehyde.

3 Methods

Principles of Good Clinical Practice (GCP)

For investigations that involve samples from clinical trials, personnel must be appropriately qualified and have received the correct training.

Safety considerations

Local safety guidance for working with human samples should be followed. Consideration should be given to the possible presence of infectious material, particularly blood borne viruses. Recommendations for the use of personal protective equipment, for the safe handling of human samples, and for the safe disposal of waste material should be in place prior to embarking on sample handling. Use of liquid nitrogen should only be by those trained in

Table 2
Examples of antibody clones that could be used in a multicolor flow cytometry panel. Users should check with suppliers for the fluorescent antibody clone-conjugates available when designing the panel

Manufacturer	Target	Clone
BD Biosciences	CD4	SK3
BD Biosciences	CD8	RPA-T8
BD Biosciences	CD278 (ICOS)	DX29
BioLegend	CD19	HIB19
BioLegend	CD279 (PD-1)	EH12.2H7
BioLegend	CCR7	G043H7
BioLegend	CD38	HB-7
BioLegend	CD28	CD28.2
BioLegend	CD32	FUN-2
BioLegend	CD45RA	HI100
BioLegend	CD27	M-T271
BioLegend	CD127	A019D5
BioLegend	CXCR3	G025H7
BioLegend	CD3	OKT3
BioLegend	CXCR5	J252D4

its handling. In this protocol, live cells are fixed with 4% paraformaldehyde before acquiring on the flow cytometer.

Sterile working

Standard laboratory aseptic technique should be used throughout to avoid contamination of live samples. Surfaces, working area, and equipment are cleaned with 70% ethanol.

Before embarking on the wet-lab procedures, draw up a 96-well plate plan, including control wells and test wells. Use this as a reference throughout the experiment.

3.1 Preparation of Reagents

Prepare the following buffers (*see* **Note 1**):

1. FACS wash buffer (10% HI-FBS in PBS): 50 ml FBS plus 450 ml PBS kept at 2–8 °C or on ice during use.
2. FACS stain buffer (5% HI-FBS in PBS): 25 ml FBS plus 450 ml PBS kept at 2–8 °C or on ice during use.
3. Viability dye: According to the manufacturer's instructions, prepare one aliquot of the fluorescent reactive dye at room temperature (RT). Dissolve 1 μl of live–dead fixable dye in 1 ml PBS (1:1000 working solution).

4. Fc block buffer: Commercial FC block preparations are available (e.g., 5 μl Human TruStain FcX™ [BioLegend] is used per 100 μl of FACS stain buffer). Alternatively, to make Fc block buffer add 200 μl human serum to 9800 μl of FACS stain buffer.
5. Fixation buffer: Commercial preparations are available. The working solution should contain 3–4% paraformaldehyde (PFA).
6. Fluorescence-conjugated antibody cocktails: These must be prepared fresh on the day of use (*see* Tables 1 and 2 for examples).
   (a) Prepare the fluorescence-conjugated antibody full stain cocktail at 2× the optimized staining dilution in FACS stain buffer. Allow a sufficient overage for pipetting: 25 μl is required per well, with a final volume for staining of 50 μl. Store at 2–8 °C in the dark until needed (*see* **Notes 2** and **3**).
   (b) Prepare the fluorescence-conjugated antibody fluorescence minus one (FMO) control cocktails at 2× the optimized staining dilution in FACS stain buffer. Allow a sufficient overage for pipetting: 25 μl is required per well, with a final volume for staining of 50 μl. Store at 2–8 °C in the dark until needed (*see* **Note 4**).

3.2 Preparation of PBMC for Multidimensional Flow Cytometry Analysis

This method can be used on PBMC prepared from fresh whole blood or raised from viable cryopreservation.

1. Suspend the washed PBMC in FACS wash buffer and count the cells by trypan blue exclusion or using a cell counter and record viability. Adjust the concentration to 1×10^6 cells/100 μl.
2. Prepare the live–dead cell control sample. Program the heat block to 70 °C. Take 0.5×10^6 PBMC in FACS wash buffer and kill the cells by heating in a sealed plastic tube at 70 °C for 10 min. Combine the killed 0.5×10^6 PBMCs in FACS wash buffer with the same quantity of live cells and make up the volume to 200 μl. Transfer to a 96-well U bottomed plate.
3. Transfer 2×10^6 cells per well of live PBMC to the labeled 96-well U bottomed plate.
4. Centrifuge at 800 × *g* for 3 min and remove the supernatant.
5. Resuspend the cells in all wells in 100 μl diluted viability dye buffer, except for the unstained control well, where 100 μl PBS is used.
6. Incubate for 5 min at RT.
7. Stop the staining by adding 100 μl of FACS wash buffer.
8. Centrifuge at 800 × *g* for 3 min and remove the supernatant.

9. Resuspend cells in all wells in 25 μl of FC block buffer.
10. Incubate for 7 min at RT.
11. Add 25 μl of 2× concentrated antibody staining mix to full stained cells or 25 μl of 2× concentrated FMO antibody cocktail to FMO cells. Add 25 μl FACS buffer to unstained cells and live–dead cells control wells.
12. Incubate for 20 min in the dark at 2–8 °C.
13. Wash twice in 150 μl FACS buffer per well (final volume), and centrifuge at 800 × *g* 4 °C for 3 min and discard the supernatants.
14. Fix the cells by resuspending in 100 μl fixation buffer for a minimum of 20 min. Store the fixed cells at 2–8 °C in the dark for up to 24 h before acquiring the data on a flow cytometer. (Cells can either be washed before acquiring on the flow cytometer or run immediately in the fixation buffer).
15. When ready to acquire data on the flow cytometer, stain the compensation beads in the following steps.
16. Add one drop of each compensation bead set (negative and positive) into labelled FACS tubes, one for each antibody.
17. Add 1 μl of each antibody into the respective FACS tubes.
18. Incubate for 10 min at RT in the dark.

3.3 Acquiring Events on the Flow Cytometer

1. Open the instrument software (e.g., BD FACSDiva™ Software BD Biosciences) and set up the experiment layout on the instrument.
2. Run the compensation controls and adjust the voltages for each parameter for clear negative and positive signal separation with minimal compensation values (*see* **Note 5**).
3. Acquire and record at least one million events in the live cell lymphocyte gate for analysis of rare populations.

3.4 Data Analysis

In this section, we describe how to use the data analysis software, FlowJo™ (BD Life Sciences) for basic data analysis and plug-ins to visualize high-dimensional data on a two dimensional (2D) map, to enable the identification of cell clusters that share similar expression patterns but would be otherwise difficult to extract for analysis using conventional analysis methods (*see* **Note 6**).

3.4.1 Preliminary Analysis

1. Check data quality with FlowJo™ by choosing "Inspect" from the populations band of the workspace task in the workspace or by double-clicking on the circle badge to the left of a sample.
2. Cross-check the compensation matrix or, if necessary, recalculate a new compensation matrix with the compensation controls.

3. Build a suitable gating strategy to identify the key populations in the sample based on the expression of known markers ($CD4^+$ T cells in this study) using FMO controls to set the gates.
4. Export $CD4^+$ only T cells as new FSC files for further analysis.

3.4.2 Data Down Sampling and Concatenation

These steps define the number of events for analysis and combine them into one file.

1. Import all exported $CD4^+$ T cell FSC files into one new workspace.
2. Use the down sample plugin to define the number of events for analysis, this step will ensure the analysis is not biased due to an uneven number of events from a few samples or time points.
3. Concatenate the down-sampled populations into one new FCS file for further analysis, and gate the individual samples based on their sample ID with the concatenated file.

3.4.3 Running the t-SNE

1. Run the t-SNE plugin on FlowJo™ with the concatenated data sample to generate *x*- and *y*- axis parameters, which project similar populations from the high-dimensional space into a 2D plot. Select markers that are important and meaningful to the research question, to define the *x*- and *y*- axis. The running time for the algorithm depends on the computer RAM capacity, the number of events in the sample, and the number of markers selected for analysis. The plugin for t-SNE is fully integrated, and no R or extra package is needed.
2. The 2D plot created shows all events from the concatenated samples, or one can select the individual sample ID gate to show the individual sample on the plot. Choose the heatmap mode to show the expression level of each parameter.
3. To project the conventional populations onto the t-SNE plot, select the population with t-SNE *x*-axis and *y*-axis and copy the plot onto the experiment layout, and overlay other populations for comparison.
4. An example of simple data analysis using this method is shown below (Figs. 1, 2, and 3).

3.4.4 FlowSOM [9]

1. FlowSOM is an algorithm to visualize the cell subsets with Self-Organizing Maps (SOMs) and Minimal Spanning Trees, which can reveal how all markers are behaving on all cells and can detect subsets that might otherwise be missed. The FlowSOM Plugin needs R packages to run under the command of FlowJo™ software (*see* **Note 7 and 8**).
2. Run FlowSOM by clicking on the FlowSOM tab and select the parameters for clusters generation. The outcome of the algorithm run will include distinct populations that are used as

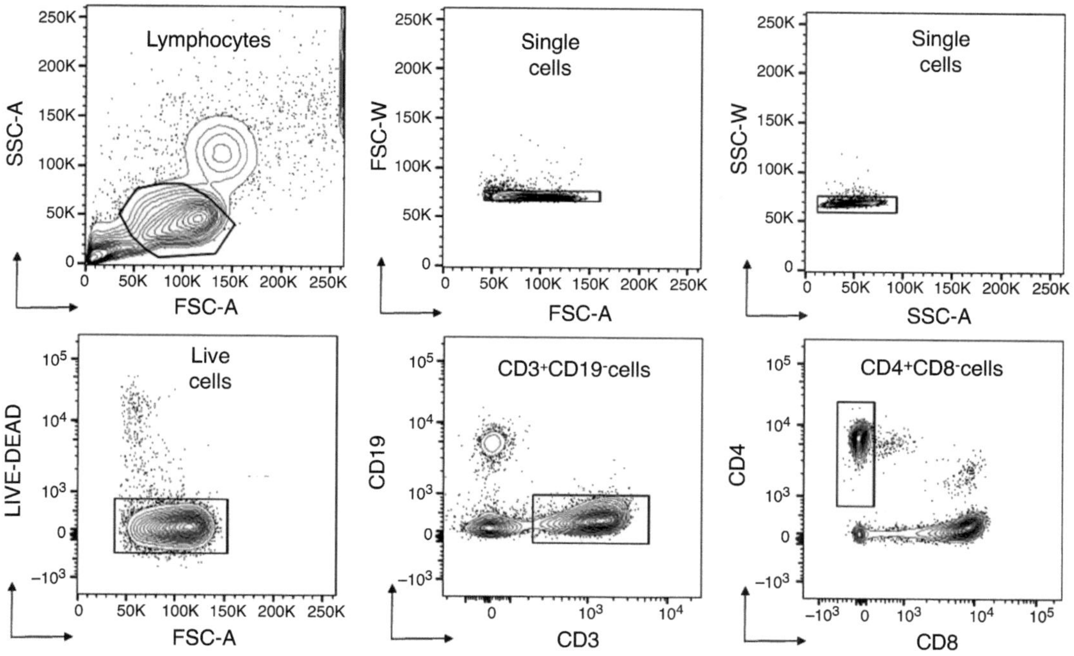

Fig. 1 Preliminary gating strategy for targeting populations. For each sample, check the sample quality and select the populations of interest using known markers. The gating strategy shown here is an example to enrich CD4$^+$ T cells for further analysis

gated population, Population Heatmap and clusters heatmap, clusters organized onto Minimal Spanning Trees and CSV files.

3. An example of simple data analysis using this method is shown below (Figs. 4, 5, and 6).

4 Notes

1. These buffers are optimal for cell viability and must be prepared fresh on the day. For storage up to 1 month at 2–8 °C, 0.25 g sodium azide (antibacterial agent) can be added.
2. *Panel design*: Successful flow cytometry experiments require robust panel design. This process is time consuming but, done correctly, reaps reward. The first stage is to decide how many target antigens will be in the panel. This will be dependent on the flow cytometer available so always check this first before designing the panel. The exact configuration and capability of the flow cytometer must be understood. Determine the possible fluorochromes in the panel using a matrix and include the excitation and emission spectra of these fluorochromes. A spectra analyzer or spectrum viewer can help with viewing the excitation and emission parameters. Spectral overlap can result in data that are difficult to interpret and should be

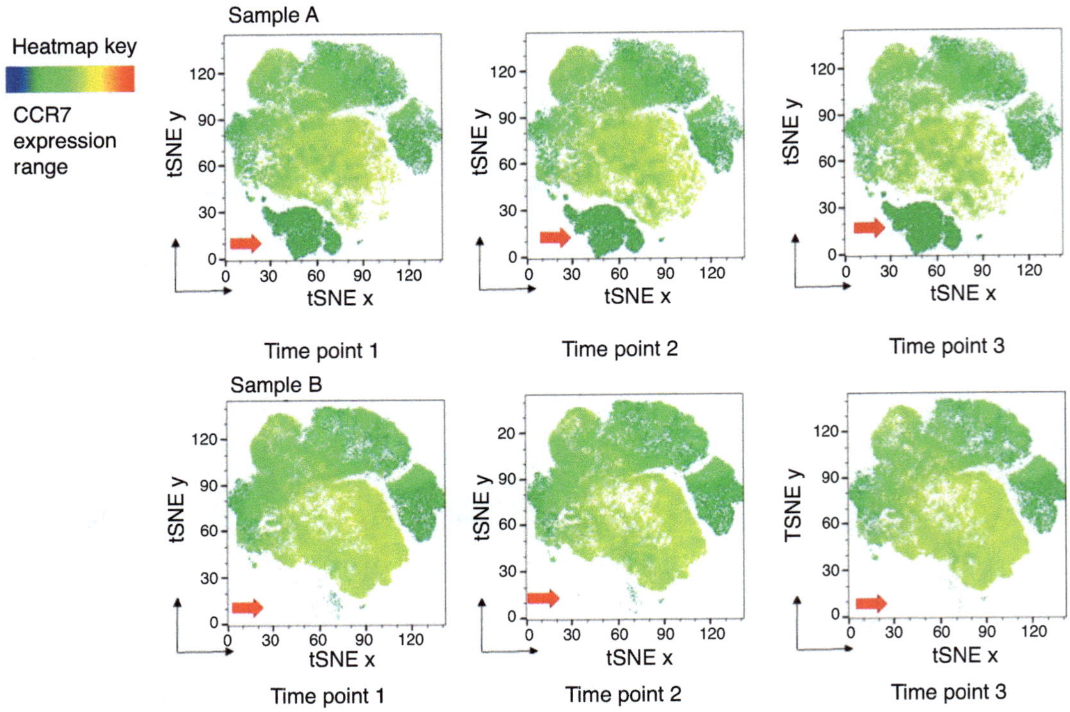

Fig. 2 t-SNE plots of individual samples to show the difference in cluster abundancy. The selected plots show a heatmap of CCR7 expression on the concatenated $CD4^+$ T cell FCS file. A cluster with relatively low expression of CCR7 (red arrow) is more abundant in sample A, compared with sample B at three different time points

avoided where possible. Some fluorochromes fluoresce more brightly than others. One approach to managing this is to put the lowest expressed target antigen on the brightest fluorochrome and the highest expressed on the weakest fluorochrome. Include a viability dye on a target fluorochrome that is not useful for other targets.

Spectrum viewer links are included below:

(a) https://www.bdbiosciences.com/en-us/applications/research-applications/multicolor-flow-cytometry/product-selection-tools/spectrum-viewer

(b) https://www.biolegend.com/en-us/spectra-analyzer

3. *Antibody titration*: To determine the optimal concentration for each fluorochrome-conjugated antibody selected for use in the panel, it is critical that each antibody is titrated. This should be performed on the cell/sample type to be used in the experiment. For example, if PBMC are used for the investigation, then the antibody needs to be titrated on PBMC. Stain as per the staining protocol using a serial dilution. To determine the best dilution, there are two calculations that can be used. First the titration data needs to be analysed by gating on the

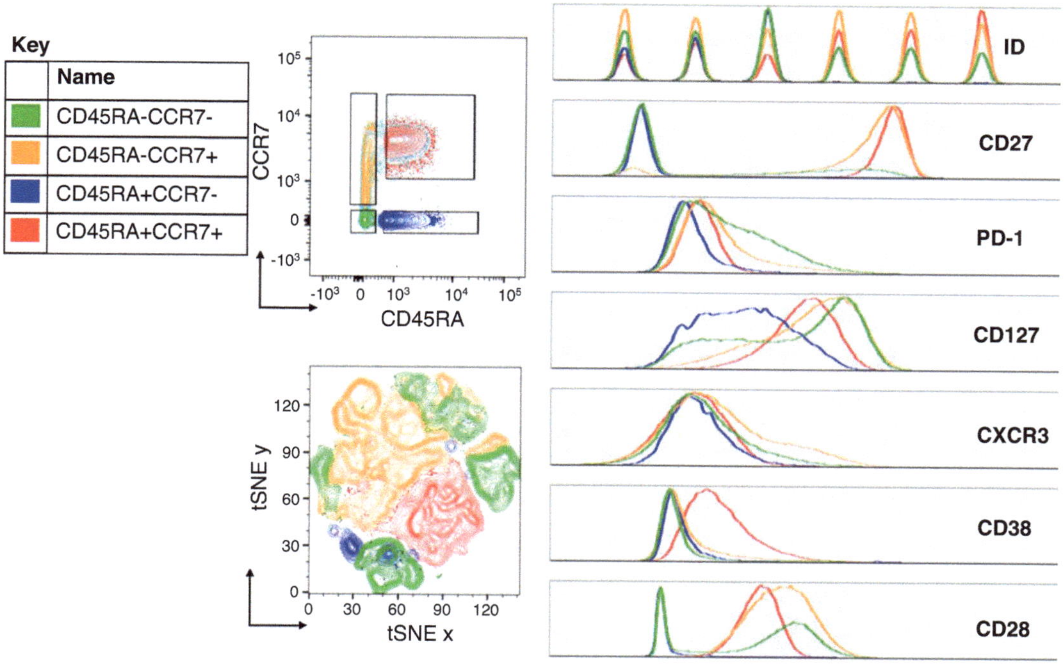

Fig. 3 Overlay of known cell subsets onto the t-SNE plot to reveal the complexity of the populations. In this example, the CD45RA$^-$CCR7$^+$ population (orange) and the CD45RA$^-$CCR7$^-$ population (green) distribute into different cluster areas on the t-SNE plot. The abundancy of each cell subset within individual samples is shown on the top of the histogram overlay

positively and negatively stained population. Calculate the mean fluorescence intensity (MFI) for each population and the standard deviation of the negative population. Export all the data into an Excel file and either calculate the stain to noise ratio or the staining index [10].

$$\text{Stain to noise ratio} = \text{MFI}_{\text{positive population}}/\text{MFI}_{\text{negative population}}$$

$$\text{Staining index} = \frac{\text{MFI}_{\text{positive population}} - \text{MFI}_{\text{negative population}}}{2 \times \text{standard deviation of MFI}_{\text{negative population}}}$$

Once the optimal concentrations are determined, these can be used to build the antibody cocktail.

4. *Controls in the experiment:* Several controls are recommended.
   (a) Dead cell control: This includes 50% heat killed PBMC and is single stained with the cell viability marker for setting compensation in this channel when acquiring on the flow cytometer.
   (b) Fluorescence minus one (FMO) controls: Optimization of the flow cytometry panel should include an experiment where these are run for every fluorochrome. These allow refinement of the gating strategy, particularly where a

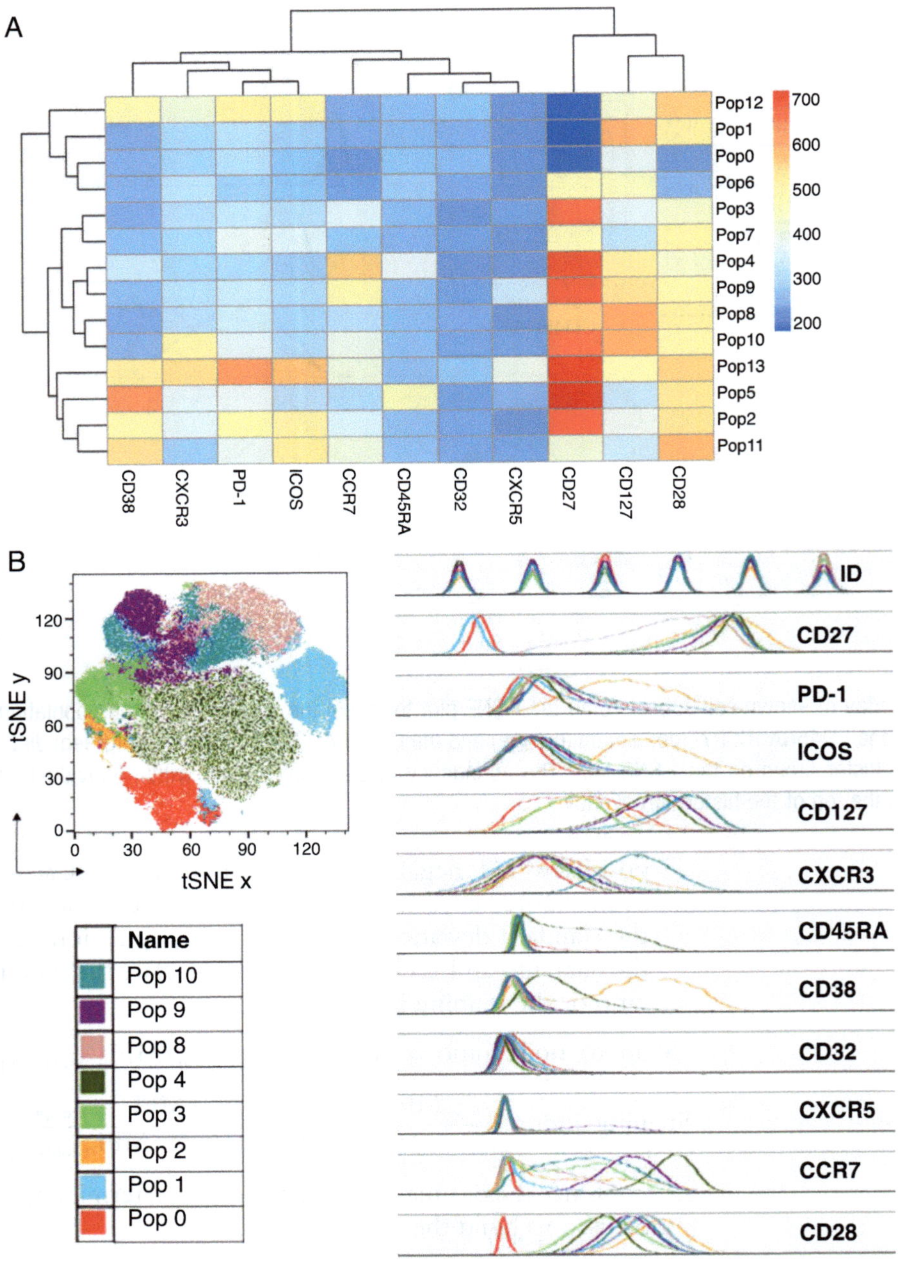

Fig. 4 Heatmap of metapopulations; diverse populations identified with the FlowSOM algorithm (**A**) and how those populations present on the t-SNE plot (**B**)

protein is expressed continuously as a spectrum across a particular cell subset of interest. Once the panel has been optimized, only selected FMOs may be required, depending on the data elicited for analysis.

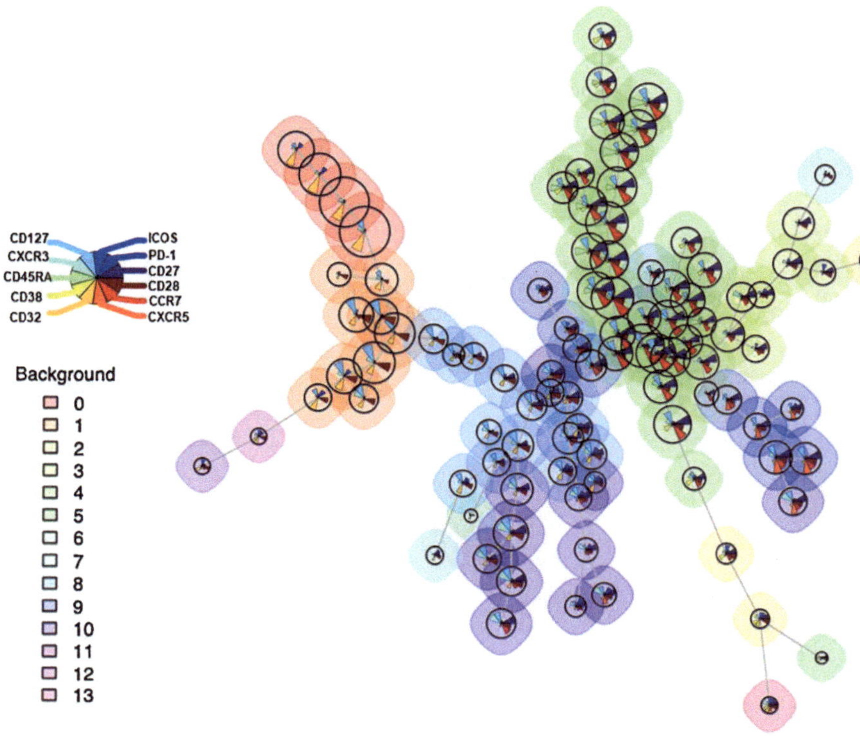

Fig. 5 The Minimum Spanning Tree generated by FlowSOM analysis. Each node is represented as a star chart, where the color-coded height of each segment shows the expression level of that marker. Every chart has a background colour indicating its meta-population assignment. On the tree, the distance of nodes clustersed together with each other within a branch shows their similarity to each other. The abundance of each node is represented by the size of its symbol

(c) Unstained controls: These are useful to gauge background fluorescence of the cells as treated during the preparation procedures.

(d) Compensation controls: These are needed to calibrate the flow cytometer during set-up. To preserve experimental samples and for ease of use, commercial compensation beads can be used. These should react with antibodies of the same species as the fluorochrome-conjugated antibody.

5. *Instrument calibration*: Instruments from different providers will have their own process for standardization and calibration. This process ensures that the performance of an instrument is consistent in alignments, signal accuracy and resolutions over a certain time-period. When the instruments are standardized, the data acquired on different dates are considered as standardized and can be objectively and quantitatively compared. The user may wish to run a few events from the unstained sample and the full stained sample briefly to check voltages initially

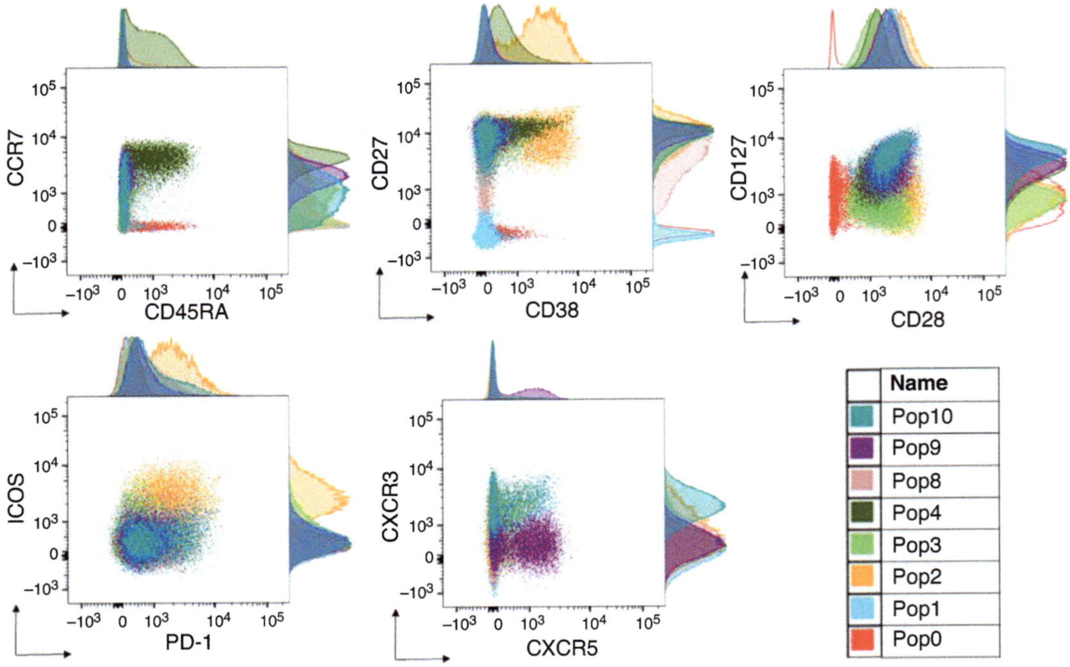

Fig. 6 Identification of FlowSOM populations with selected markers. Like the use of the heatmap, overlay of FlowSOM populations with selected markers can help to reveal the nature of the populations and possible gating strategy for future experiments. For example, population2 (orange) is high in CD27, CD38, PD-1, and ICOS expression

before running compensation controls. This will help the user to know in advance that the staining and signal detection are acceptable for further steps.

6. *Computer RAM requirement*: To run the plugins smoothly, the minimum requirement for the computer is 8G RAM, it is highly recommended to use a computer with more than 16G RAM to speed up the calculation time. The time needed for the calculation is based on the size of the data; a high number of parameters will increase the calculating time.
7. *FlowSOM plugin*: FlowSOM Plugin can be download from https://www.flowjo.com/exchange/#/. R and R packages should be installed to run FlowSOM. Please read the ReadMe document for the FlowSOM plugin to ensure the correct installation of R packages and R tools. Further information about FlowSOM can be found in Van Gassen, S. et al Cytometry A (2015) [9].
8. *YouTube videos—self-tutorials*: FlowJo have many tutorials online on their website with helpful directions and instructions on how to do T-SNE, FlowSOM and many other things, as well as a YouTube channel with videos under "FlowJo Media."

Acknowledgements

The FluAGE study was conducted with a grant from the National Institute for Health Research (NIHR) Imperial Biomedical Research Centre (BRC). The views expressed in this publication are those of the author(s) and not necessarily those of the NHS, the National Institute for Health Research or the Department of Health. The authors would like to thank the participants in the FluAGE study and the staff of the Imperial College Clinical Trials Centre, and the Department of Infectious Disease, Imperial College London." If a conflicts of interest statement is required then please write "K.P. reports grants from the National Institute for Health Research and the Medical Research Council UK Research and Innovation. K.P. is chief investigator for the Imperial College London COVID-19 vaccine development programme and principal investigator for the University of Oxford COVID-19 vaccine development trials. K.P. reports personal fees from Sanofi, and honoraria received from the British Society for Antimicrobial Therapy and ITV plc.

References

1. McKay PF, Hu K, Blakney AK et al (2020) Self-amplifying RNA SARS-CoV-2 lipid nanoparticle vaccine candidate induces high neutralizing antibody titers in mice. Nat Commun 11:3523. https://doi.org/10.1038/s41467-020-17409-9
2. Walsh EE, Frenck RW, Falsey AR et al (2020) Safety and immunogenicity of two RNA-based Covid-19 vaccine candidates. N Engl J Med 383:2439–2450. https://doi.org/10.1056/nejmoa2027906
3. Folegatti PM, Ewer KJ, Aley PK et al (2020) Safety and immunogenicity of the ChAdOx1 nCoV-19 vaccine against SARS-CoV-2: a preliminary report of a phase 1/2, single-blind, randomised controlled trial. Lancet 396:467–478. https://doi.org/10.1016/S0140-6736(20)31604-4
4. Ellebedy AH, Jackson KJL, Kissick HT et al (2016) Defining antigen-specific plasmablast and memory B cell subsets in human blood after viral infection or vaccination. Nat Immunol 17:1226–1234. https://doi.org/10.1038/ni.3533
5. Bentebibel SE, Khurana S, Schmitt N et al (2016) ICOS + PD-1 + CXCR3 + T follicular helper cells contribute to the generation of high-avidity antibodies following influenza vaccination. Sci Rep 6:26494. https://doi.org/10.1038/srep26494
6. Bentebibel SE, Lopez S, Obermoser G et al (2013) Induction of ICOS+CXCR3+CXCR5+ T H cells correlates with antibody responses to influenza vaccination. Sci Transl Med 5:176ra32. https://doi.org/10.1126/scitranslmed.3005191
7. Van Der Maaten L, Hinton G (2008) Visualizing data using t-SNE. J Mach Learn Res 9:2579–2625
8. Cole ME, Saeed Z, Shaw AT et al (2019) Responses to Quadrivalent influenza vaccine reveal distinct circulating CD4+CXCR5+ T cell subsets in men living with HIV. Sci Rep 9:1–16. https://doi.org/10.1038/s41598-019-51961-9
9. Van Gassen S, Callebaut B, Van Helden MJ et al (2015) FlowSOM: using self-organizing maps for visualization and interpretation of cytometry data. Cytometry A 87:636–645. https://doi.org/10.1002/cyto.a.22625
10. Swieboda D, Guo Y, Sagawe S et al (2019) OMIP-062: a 14-color, 16-antibody panel for Immunophenotyping human innate lymphoid, myeloid and T cells in small volumes of whole blood and pediatric airway samples. Cytometry A 95:1231–1235. https://doi.org/10.1002/cyto.a.23907

INDEX

Fadil Bidmos (eds.), *Bacterial Vaccines: Methods and Protocols*,
Methods in Molecular Biology, vol. 2414, https://doi.org/10.1007/978-1-0716-1900-1,

P

Q

R

S

MIX
Papier aus verantwortungsvollen Quellen
Paper from responsible sources
FSC® C105338
FSC
www.fsc.org

If you have any concerns about our products,
you can contact us on
ProductSafety@springernature.com

In case Publisher is established outside the EU,
the EU authorized representative is:
Springer Nature Customer Service Center GmbH
Europaplatz 3, 69115 Heidelberg, Germany

Printed by Libri Plureos GmbH
in Hamburg, Germany